25th Association for the Advancement of Animal Breeding and Genetics Conference 2023

Crawley, Washington, USA
26 - 28 July 2023

ISBN: 978-1-7138-7966-4

Printed from e-media with permission by:

Curran Associates, Inc.
57 Morehouse Lane
Red Hook, NY 12571

Some format issues inherent in the e-media version may also appear in this print version.

AAABG – Association for the Advancement of Animal Breeding and Genetics
University of New England
Armidale NSW 2351
Australia

Phone: 61 2 677 320 55
Fax: 61 2 677 332 66

AGBU@une.edu.au

Additional copies of this publication are available from:

Curran Associates, Inc.
57 Morehouse Lane
Red Hook, NY 12571 USA
Phone: 845-758-0400
Fax: 845-758-2633
Email: curran@proceedings.com
Web: www.proceedings.com

Association for the Advancement of Animal Breeding and Genetics

AAABG

Association for the Advancement of Animal Breeding and Genetics

Proceedings of the Twenty-fifth Conference

Animal Breeding at the Crossroads

Association for the Advancement of Animal Breeding and Genetics

The University Club of Western Australia
Hackett Drive, Crawley, WA 6009

26 – 28 July 2023

ISSN Number: 1328-3227

Produced by:
Association for the Advancement of Animal Breeding and Genetics
C/- AGBU
University of New England
Armidale NSW 2351

http://www.aaabg.org

PRESIDENT'S MESSAGE

On behalf of the organising committee, I am very pleased to welcome you to the 25th Conference of the Association for the Advancement of Animal Breeding and Genetics. We are meeting at the University of Western Australia which sits on sacred soil alongside the Derbal Yerrigan (Swan River) on Whadjuk Noongar Booja. We acknowledge that this has been a place of learning for tens of thousands of years and look forward to continuing that learning at AAABG.

The theme of our conference is Animal Breeding at the Crossroads. We will reflect on our achievements in animal breeding since the inaugural AAABG conference in 1979 and explore the role of genetics in a future with ever-increasing community concern for climate change, environmental impacts, animal welfare, and meat consumption.

A highlight of our conference will no doubt be the WA Livestock Research Council's Producer day where we aim to "link science with farm". We are taking the opportunity to provide a producer audience with the latest developments in animal breeding and genetics as well as showcasing emerging geneticists and launching the 2023 Farm a Friend Program. This program pairs emerging scientists with progressive farmers in a mentoring program designed to ensure science is grounded in practical farm reality and priorities. The Producer Day program will be meeting one of the major aims of AAABG, which is to "develop communication among all those interested in the application of genetics to animal production, particularly breeders and their organisations, consultants, extension workers, educators and geneticists".

Thank you to all our sponsors for your generous support of the conference and an enormous thank you to all members of our organising committee and the conference organising team from Conference Design. Finally, thank you to Dr Sue Hatcher, AAABG Editor and AAABG Executive Officer, for putting together these proceedings.

We hope you enjoy your time in Perth and take the opportunity to interact and network with students, researchers, and producers from all corners of the animal-breeding community.

Bronwyn Clarke
President

ASSOCIATION FOR THE ADVANCEMENT OF ANIMAL BREEDING AND GENETICS

2023

TWENTY FIFTH CONFERENCE COMMITTEE

CITATION OF PAPERS

Papers in this publication should be cited as appearing in the Proceedings of the Association for the Advancement of Animal Breeding and Genetics.

(Abbreviation: Proc. Assoc. Advmt. Anim. Breed. Genet.)

Example citation:
Bowley F.E., Amer P.R. and Meier S. (2013) New approaches to genetic analysis of fertility traits in New Zealand dairy cattle. *Proc. Assoc. Advmt. Anim. Breed. Genet.* 20: 37-40.

REVIEWERS

All papers, invited and contributed, were subject to peer review, by two referees. We acknowledge and thank those listed below for reviewing the AAABG papers contained in these proceedings and the AAABG Special issue of Animal Production Science.

Juliana Afonso
Johanna Aldersey
Pamela Alexandre
Hassan Aliloo
Jason Archer
Michelle Axford
Robert Banks
Doreen Becker
Amy Bell
Donagh Berry
Timothy Bilton
Ally Bird
Cynthia Bottema
Forbes Brien
Luiz Brito
Daniel Brown
Kim Bunter
Andrew Byrne
Amanda Chamberlain
Yizhou Chen
Evans Cheruiyot
Bronwyn Clarke
Schalk Cloete
Natalie Connors
James Copley
Nicholas Corbett
Roy Costilla
Brad Crook
Sara de las Heras Saldana
Florencia Di Rocco
Ken Dodds
Christian Duff
Rao Dukkipato
Madeline Facy
Mohammad Ferdosi
Peter Fitzgerald
Marina Fortes
Mehrnush Forutan
Daniel Garrick
Arthur Gilmour
Tom Granleese
Birgit Gredler-Grandl
Johan Greeff
Philip Gurman
Mekonnen Haile-Mariam
Sam Harburg
Bevin Harris
Sue Hatcher
Susanne Hermesch
Andrew Hess
Brad Hine
Gilbert Jeyaruban
Patricia Johnson
David Johnston
Nantapong Kamprasert
Jie Kang
Inan Kaya
Hamideh Keshavarzi
Kate Keogh
Majid Khansefid
Mehar Khatkar
Brian Kinghorn
Harry Lamb
Stephen Lee
Tosso Leeb
Li Li
Yutao Li
Nicolas Lopez-Villalobos
Zhi Loh
Iona MacLeod
Rudi McEwin
Aaron MacMillan
Karin Meyer
Flippo Miglior
Nasir Moghaddar
Suzanne Mortimer
John Morton
Kylie Munyard
Marina Naval Sanchez
Cornelius Nel
Jo Newton
Thuy Nguyen
Tuan Nguyen
Gert Nieuwhof
Brendon O'Rourke
Babatunde Olasege
Sameer Pant
Gertje Petersen
Wayne Pitchford
Laercio Porto-Neto
Jennie Pryce
Imtiaz Randhawa
Carlo Renieri
Toni Reverter
Elham Rezvannejad
Christina Rochus
Elizabeth Ross
Marie-Pierre Sanchez
David Scobie
Saleh Shahinfar
Erin Smith
Jen Smith
Warren Snelling
Bolormaa Sunduimijid
Imke Tammen
Andre Tan
Irene van den Berg
Phil Vercoe
Katharina Voss
Peter Wahinya
Sam Walkom
Bradley Walmsley
Christie Warburton
Franziska Weik
Matt Wolcott
Zehu Yuan

SPONSORS OF THE 25TH AAABG CONFERENCE 2023

The financial assistance of the following organisations is gratefully acknowledged.

Gold Sponsor

GOVERNMENT OF WESTERN AUSTRALIA
Department of Primary Industries and Regional Development

Silver Sponsors

awi Australian Wool Innovation Limited

mla MEAT & LIVESTOCK AUSTRALIA

THE UNIVERSITY OF WESTERN AUSTRALIA

Bronze Sponsors

merino Stud Merino Breeders' Association of WA

INTERNATIONAL BEEF RECORDING SCHEME BREEDPLAN

NEOGEN® NEOGEN.com

Murdoch University

WALRC WA Livestock Research Council

AAAS AUSTRALIAN ASSOCIATION OF ANIMAL SCIENCES WA BRANCH

Name Badge Sponsor

BUSINESS
EVENTS
PERTH

Keynote Speaker Sponsor

abacusbio.
BRIDGING SCIENCE & BUSINESS

THE ASSOCIATION FOR THE ADVANCEMENT OF ANIMAL BREEDING AND GENETICS[1] INCORPORATED[2]

OBJECTIVES

(i) to promote scientific research on the genetics of animals;
(ii) to foster the application of genetics in animal production;
(iii) to promote communication among all those interested in the application of genetics to animal production, particularly breeders and their organisations, consultants, extension workers, educators and geneticists.

To meet these objectives, the Association will:

(i) hold regular conferences to provide a forum for:
(a) presentation of papers and in-depth discussions of general and industry-specific topics concerning the application of genetics in commercial animal production;
(b) scientific discussions and presentation of papers on completed research and on proposed research projects;
(ii) publish the proceedings of each Regular Conference and circulate them to all financial members;
(iii) use any such other means as may from time to time be deemed appropriate.

MEMBERSHIP

Any person interested in the application of genetics to animal production may apply for membership of the Association and, at the discretion of the Committee, be admitted to membership as an Ordinary Member.

Any organisations interested in the application of genetics to animal production may apply for membership and, at the discretion of the Committee, be admitted to membership as a Corporate member. Each such Corporate Member shall have the privilege of being represented at any meeting of the Association by one delegate appointed by the Corporate Member.

Benefits to Individual Members.

- While it is not possible to produce specific recommendations or "recipes" for breeding plans that are applicable for all herd/flock sizes and management systems, principles for the development of breeding plans can be specified. Discussion of these principles, consideration of particular case studies, and demonstration of breeding programs that are in use will all be of benefit to breeders.
- Geneticists will benefit from the continuing contact with other research workers in refreshing and updating their knowledge.
- The opportunity for contact and discussions between breeders and geneticists in individual members' programs, and for geneticists in allowing for detailed discussion and appreciation of the practical management factors that often restrict application of optimum breeding programs.

Benefits to Member Organisations.

- Many of the benefits to individual breeders will also apply to breeding organisations. In addition, there are benefits to be gained through coordination and integration of their efforts.

---

[1] AAABG was formerly known as the Australian Association for Animal Breeding and Genetics. Following the 1995 OGM the name was changed when it became an organisation with a joint Australian and New Zealand membership.
[2] The Association for the Advancement of Animal Breeding and Genetics is incorporated in South Australia.

Recognition of this should follow from understanding of common problems, and would lead to increased effectiveness of action and initiatives.

- Corporate members can use the Association as a forum to float ideas aimed at improving and/or increasing service to their members.

General Benefits.

- Membership of the Association may be expected to provide a variety of benefits and, through the members, indirect benefits to all the animal industries.
- All members should benefit through increased recognition of problems, both at the level of research and of application, and increased understanding of current approaches to their solution.
- Well-documented communication of gains to be realised through effective breeding programs will stimulate breeders and breeding organisations, allowing increased effectiveness of application and, consequently, increased efficiency of operation.
- Increased recognition of practical problems and specific areas of major concern to individual industries should lead to increased relevance of applied research.
- All breeders will benefit indirectly because of improved services offered by the organisations which service them.
- The existence of the Association will increase appreciably the amount and use of factual information in public relations in the animal industries.
- Association members will comprise a pool of expertise – at both the applied and research levels – and, as such, individual members and the Association itself must have an impact on administrators at all levels of the animal industries and on Government organisations, leading to wiser decisions on all aspects of livestock improvement, and increased efficiency of animal production.

CONFERENCES

One of the main activities of the Association is the Conference. These Conferences will be structured to provide a forum for discussion of research problems and for breeders to discuss their problems with each other, with extension specialists and with geneticists.

ASSOCIATION FOR THE ADVANCEMENT OF ANIMAL BREEDING AND GENETICS

FELLOWS OF THE ASSOCIATION

"Persons who have rendered eminent service to animal breeding in Australia and/or New Zealand or elsewhere in the world, may be elected to Fellowship of the Association…"

Elected February 1990
R.B.M Dun
F.H.W. Morley
A.L. Rae
H.N Turner

Elected September 1992
K. Hammond

Elected July 1995
C.H.S. Dolling
J.R. Hawker
J. Litchfield

Elected February 1997
J.S.F. Barker
R.E. Freer

Elected June 1999
J. Gough
J.W. James

Elected July 2001
J.N. Clarke
A.R. Gilmour
L.R. Piper

Elected September 2005
B.M. Bindon
M.E. Goddard
H.-U. Graser
F.W. Nicholson

Elected September 2007
K.D. Atkins
R.G. Banks
G.H. Davis

Elected September 2009
N.M. Fogarty
A.R. Fyfe
J.C. McEwan
R. Mortimer
R.W. Ponzoni

Elected September 2011
B.P. Kinghorn
A. McDonald

Elected October 2013
H.M. Burrow
P.F. Fennessy
G. Nicol
P. Parnell

Elected October 2015
P.F. Arthur
D. Johnson
K. Meyer
B. Tier
R. Woolaston

Elected October 2019
S.A. Barwick
H.T. Blair
S.W.P. Cloete
I.W. Purvis

Elected November 2021
F.D. Brien
D. Garrick
J. Greeff
B. Hayes
J.E. Pryce
J.H.J van der Werf

Elected July 2023
K.G. Dodds
W.S. Pitchford
H.W. Raadsma
C.W. (Bill) Sandilands
A.A. Swan

HONORARY MEMBERS OF THE ASSOCIATION

"Members who have rendered eminent service to the Association may be elected to Honorary Membership…"

Elected September 2009
W.A. Pattie
J.R.W. Walkley

HELEN NEWTON TURNER MEDAL TRUST

The Helen Newton Turner Medal Trust was established in 1993 following an anonymous donation to the Animal Genetics and Breeding Unit to perpetuate the memory of Helen Newton Turner and to encourage and inspire those engaged in animal genetics.

Helen Newton Turner AO

THE HELEN NEWTON TURNER MEDAL

The Medal is named after Dr Helen Newton Turner whose career with CSIRO was dedicated to research into the genetic improvement of sheep for wool production.

The Helen Newton Turner Medal was first awarded in 1994 to Associate Professor John James and a list of all recipients to date is given below. The Helen Newton Turner Medallist is chosen by Trustees from the ranks of those persons who have made an outstanding contribution to genetic improvement of Australian livestock.

The recipient of the Medal is invited to deliver an Oration on a topical subject of their choice.

Medallists

1994	J.W. James	2003	F.W. Nicholas	2015	A.R. Gilmour
1995	L.R. Piper	2005	K. Hammond	2017	A. Collins
1997	J. Litchfield	2007	L. Corrigan	2019	K.D. Atkins
1998	J.S.F. Barker	2009	R. Hawker	2021	J.H.J van der Werf
1999	C.W. Sandilands	2011	R. Banks		
2001	G.A. Carnaby	2013	M. Goddard		

The Oration of the 2021 Medal recipient, Professor Julius van der Werf, is reproduced in the AAABG Special Issue of Animal Production Science that accompanies these proceedings.

THE HELEN NEWTON TURNER BRIGHT FUTURES AWARD

In 2021, the Trust established a new award, the Helen Newton Turner Bright Futures Award to recognise the achievements of an up-and-coming individual who is showing evidence of establishing a reputation for excellence in the field of animal genetics within Australia.

2021 S.A. Clark

TABLE OF CONTENTS

Preliminary pages

President's Message ii

25th Conference Committee and Citation of Papers iii

Reviewers iv

Sponsors v

AAABG Rules vii

AAABG Fellows ix

AAABG Honorary Members x

Helen Newton Turner Medal xi

Helen Newton Turner Bright Futures Award xii

Table of Contents xiii

John Vercoe Memorial Lecture

The role of animal genetic resources in the sustainable livestock transformation 1
P. Boettcher, R. Baumung, G. Leroy and B. Besbes

Plenary Papers

Animal breeding is part of the solution to environment, climate, and animal-welfare challenges facing animal production 5
M. Henryon T. Ostersen, P.E. Vercoe and A.C. Sørensen

Livestock breeding, where have we been and what lies ahead? 10
S.P. Miller

Breeding Plans A

An optimal contribution selection technique that utilises on-additive genetic components 23
Z. Loh, J.H.J. van der Werf and S. Clark

Ewe genotype effects in genetic evaluation of Merino fleece traits across ages 27
S.I. Mortimer, K.L. Egerton-Warburton, T.L. Bird-Gardiner and A.A. Swan

Evolution of sheep breeds within LAMBPLAN and the rise of the Composites 31
A.J. McMillan, S.F. Walkom and D.J. Brown

Genetic and economic effects of genomic selection amongst surplus dairy heifers in Australian dairy herds 35
J.M. Morton and G.J. Nieuwhof

Effects of selection and data truncation on estimates of genetic parameters obtained fitting a single-step model 39
K. Meyer

Genetic benchmarking of Maternal sheep flocks using genomic testing 43
D.J. Brown, P.M. Gurman and A.A. Swan

Preliminary evaluation of the impact of visual traits on lifetime ewe performance 47
P.K. Wahinya, D.J. Brown, S.F. Walkom, T. Bird-Gardiner, B.E. Clarke, J.L. Smith and A.A. Swan

Breeding for Reproductive Traits A

Comparison of udder and teat traits in Merino ewes recorded at lambing and weaning 51
E.G. Smith, G.A. Acton, A.M. Bell and J.L. Smith

Age at puberty, days to calving and first parity return to oestrus in Australian temperate beef breeds 55
K.A. Donoghue, R. Rippon, M. Wolcott, K.L. Moore, S.A. Clark and B.J. Walmsley New module for prediction of reproductive traits EBV in BREEDPLAN 59
M.G. Jeyaruban and D.J. Johnston
Fibre diameter variation as a measure of resilience in sheep 63
E.G. Smith, S.F. Walkom and S.A. Clark
Development of female fertility indicator traits for the Angus HeiferSELECT genomic tool 67
P.A. Alexandre, L.R. Porto-Neto, B.C. Hine, A.M. Samaraweera, A.I. Byrne, A.B. Ingham, C.J. Duff and A. Reverter

Multinomics
Library preparation method affects observed microbiome variation when using Oxford Nanopore sequencing 71
E.M. Ross, Z. Chen, L.T. Nguyen, S. Meale and C. T. Ong
Simultaneous investigation of genomic regions of interest - the use of adaptive sampling 75
R.M. Clarke, A. Hess, A, Caulton, R. Brauning, K.M. McRae, A. Chen and S.M. Clarke
Gene network prediction for bull fertility traits 79
W.L.A Tan, N. Hudson, L.R. Porto-Neto, A. Reverter, J. Afonso and M.R.S. Fortes
Methylome profiling in response to stress: mycotoxin (sporidesmin) exposure in sheep 83
A. Caulton, K.M. McRae, K.G. Dodds, R. Brauning, N.K. Pickering, P.L. Johnson and S. M. Clarke

Breeding Plans B
Resources assessment for determination of ancestral alleles in cattle 87
J. Dorji, L.R. Porto-Neto, A.J. Chamberlain, C.J. Vander-Jagt, J. Kijas and A. Reverter
Incorporating genomic information in evaluations for farmed deer in New Zealand 91
R. Costilla, R. Brauning, J. Ward, S. Newman, A. Hess, S. Hickey, T. van Stijn, J. McEwan, K. Dodds, S. Clarke and S. Rowe
Understanding trait preferences and views on genetic tools for the New Zealand beef industry 95
L. Kok, S. Harburg, B. Santos, P. Amer, J. Archer, A. Boyd and G. Jenkins
The economic value of gestation length 99
E. Ooi, N. Howes, F. Hely, M. Stephen, P. Amer and C. Quinton
Australian beef cattle breeding objectives 103
B.W. Gudex, P.J. Williams and B.J. Walmsley
Signatures of positive selection for scrotal circumference in three beef cattle breeds 107
Z. Manzari, D.J. Johnston, N.K. Connors and M.H. Ferdosi
BREEDPLAN single-step genomic evaluations deliver increased accuracies across all breeds and EBVs 111
D.J. Johnston, M.H. Ferdosi, N.K. Connors, J. Cook, C.J. Girard and A.A. Swan

Breeding for Reproductive Traits B
Genotype by trait-specific season interactions for farrowing rate and average piglet birthweight 115
A.M.G. Bunz, K.L. Bunter, J. Harper, R. S. Morrison, B.G. Luxford and S. Hermesch

Translating multibreed genomic prediction of bull fertility traits into on-farm selection of bulls 119
L.R. Porto-Neto, P. Alexandre, J. Dorji, M.R.S. Fortes and A. Reverter
The highs and lows of recording calf and calving traits 122
M. Axford, M. Khansefid E. Jewell, A. Chamberlain and J.E. Pryce
Using foetal age estimates to substitute birth date recording in beef cattle evaluations 126
J. Kang, F. Weik, N. Sanderson, D. Robertson and J.A. Archer
Prediction of gestation length and days to conception from foetal age scans for genetic improvement of maternal fertility in beef cow herds: a simulation study 130
F. Weik, J. Kang and J.A. Archer
Match birthweight ASBVs to flock fecundity for lamb survival 134
S. Hatcher, S. Robertson, D.J. Brown and K.L. Bunter
FecB carrier Deccani crossbred ewes in Maharashtra, India have moderately higher litter sizes than non-carrier ewes 138
C. Nimbkar, P.M. Ghalsasi, S.W. Walkden-Brown and J.H.J. van der Werf

Prediction/Genomic Prediction Beef
Using genomic and phenotypic data to characterise the genetic structure of Brahman cattle populations in Southern Africa 142
W.M.S.P. Weerasinghe and B.J. Crook
Application of an empirical approach for predicting accuracy for genomic evaluations 146
K.L. Moore, P.M. Gurman and D.J. Johnston
Genomic prediction using imputed whole-genome sequence in Australian Angus cattle 150
N. Kamprasert, H. Aliloo, J. van der Werf, C. Duff and S. Clark
Comparing genomic prediction accuracies for commercial cows' reproductive performance using GA2CAT and two machine-learning methods 154
Y. Li, S. Hu, L. Porto-Neto, R. McCulloch, S. McWilliam, J. McDonald, C. Smith, P. Alexandre, S. Lehnert and A. Reverter
Assessing the value of metafounders for genomic prediction in Australian Simmental beef cattle 158
D.J.A. Santos, N.K. Connors, P.M. Gurman, M.H. Ferdosi, S.P. Miller and A.A. Swan
Imputation accuracy for missing alleles in crossbred beef cattle 162
P.K. Wahinya and M.H. Ferdosi

Prediction/Genomic Prediction Dairy
Estimation of SNP effects in Lacaune dairy sheep depending on the reference population composition 166
M. Wicki, J. Raoul and A. Legarra
Genetic evaluation of longevity using productive life and survival scores in Australian Holstein cattle: prediction of early survival 170
M. Haile-Mariam, M. Khansefid, M. Axford, M.E. Goddard and J.E. Pryce

Novel Traits: Environment and Greenhouse Gas
Heritability and repeatability of methane emissions and feed intake in young dairy sires 174
R.C. Handcock, D.J. Garrick, D.P. Garrick, R.J. Spelman, P. van Elzakker, P. Beatson, G. Worth and L.R. McNaughton

Towards selecting for lower methane sheep 178
P.T Fitzgerald, E.H. Clayton, A.J. Donaldson, D.J. Brown, V.H. Oddy and J.H.J van der Werf
Using rumen microbial predictors for genomic prediction of feed efficiency 182
T.P. Bilton, P.L. Johnson, F. Booker, H.H. Henry, M.K. Hess, R. Jordan, S.M. Hickey, H. Baird, N. Amyes, J.C. McEwan and S.J. Rowe
Implementing genetic tools to reduce methane emissions in NZ sheep flocks 186
J.A. Archer, M.D. Aspin, D.E. Brier, S-J. H. Powdrell, D. Campbell, C.L. Winkworth-Lawrence, S.J. Rowe, J.C. McEwan and M. Lee

Genetic Diversity and Inbreeding A
A comparison between the use of pedigree or genomic relationships to control inbreeding in optimum-contribution selection 190
M. Sharif-Islam, M. Henryon, J.H.J. van der Werf, A.C. Sørensen, T.T. Chu, B. J. Wood and S. Hermesch
Detection of signatures of selection in Australian beef cattle 194
H. Aliloo, B.J. Walmsley, K.A. Donoghue and S.A. Clark
Demonstrating differences in survival and fertility of high and low genetic merit cows within and across herds 198
J.E. Newton, M. Haile-Mariam and J.E. Pryce
Population scale long-read sequence databaseses: are they useful for accurate SNP and indel discovery? 202
I.M. MacLeod, T.V. Nguyen, J. Wang, C.J. Vander Jagt and A.J. Chamberlain
Epigenetic repression of genes associated with ribeye area of Nelore cattle 206
J. Afonso, M.R.S. Fortes, W.J. Shim, T.F. Cardoso, J.J. Bruscadin, A.O. de Lima, W.J.S. Diniz, B. Silva-Vignato, W.L.A. Tan, A.S.M. Cesar, M. Boden, G.B. Mourão, A. Zerlotini, L.L. Coutinho and L.C. de Almeida Regitano

Meat Quality
Genomic prediction of consumer satisfaction traits of Australian beef 210
A.M. Lynn, P. McGilchrist, H. Aliloo, R. Polkinghorne, M. Forutan, B.J. Hayes and S.A. Clarke
Relationships of sire breeding values for Merino production traits with eating quality of lamb 214
S.I. Mortimer, B.W.B. Holman, S.M. Fowler, T.I.R.C. Alvarenga, D.L. Hopkins, K.L. Egerton-Warburton, J.L. Smith, B.C. Hine and A.A. Swan

Epigenetics, Structural Variants & Transcriptomes
Cost effective detection of deleterious structural variants in long-read sequence – How deep is enough? 218
T.V. Nguyen, J.Wang, A.J. Chamberlain and I.M. MacLeod
Heritability and repeatability of paternal haplotype recombination rate in beef cattle autosomes 222
M.H. Ferdosi, S. Masoodi and M. Khansefid

Genetic Diversity and Inbreeding B
Characterising heterozygosity of the X chromosome in the Australian Wagyu population 226
D.P. Garrick, R.C. Handcock and C. Teseling

Measured Goats in the Rangelands: An overview of a meat goat reference population 230
T. Granleese, S.I. Mortimer, T. Atkinson, G. Refshauge, T. Bird-Gardiner, F. Haynes, D.J. Brown, P. Alexandri and S.F. Walkom
Selecting for more methane efficient sheep 234
J.H.J. van der Werf

Novel Traits and Phenotyping Tools B
Investigating the genetic cause of Wry Face in the Australian Jersey cattle population 238
C.J. Vander Jagt, C.M. Reich, I.M. MacLeod, M.E. Goddard, B.J. Hayes, T.T.T. Nguyen, G.J. Nieuwhof and A.J. Chamberlain
Importance of incorporating feeding rate when developing predictions of feed intake 242
P.L. Johnson, K. Knowler, T.P. Biton and S.J. Rowe
Remote sensor collars measure age at puberty in tropical beef heifers in northern Australia 246
D.J. Johnston, M. Dayman, T.P. Grant, K. Hubbard, K. Goodwin, A.K. Doughty and J.D. Cook

Understanding and Utilising Genetic Diversity
Genetic diversity of domestic goats from Central Laos 250
S.V. Le, S. de las Heras-Saldana, P. Alexandri, L. Olmo, S.W. Walkden-Brown and J.H.J. van der Werf

Infectious Diseases/Disease Resistance
Immune competence and micro-environmental sensitivity 254
M.D. Madsen, J.H.J. van der Werf, A. Ingham, B. Hine, A. Reverter and S. Clark
An across-flock analysis on faecal worm egg counts in Merino sheep in South Africa 258
S.W.P. Cloete, Z. Mpetile and K. Dzama
Genetic variation in susceptibility to facial eczema in dairy cattle in New Zealand 262
A. Ismael, E. Donkersloot, F. Wallace, G. Worth, S. Davis, L. McNaughton and R.J. Spelman
Faecal microbiota of Angus cattle with divergent immune competence 266
B.N. Maslen, B.C. Hine, C. Duff, P.A. Alexandre, S.A. Clark, J.H.J van der Werf, J.D. White and S.D. Pant

GWAS
Multi-trait genome wide association meta-analysis of body weight, carcase composition and eating quality traits in Australian sheep 270
N. Moghaddar, A.A. Swan and J.H.J. van der Werf
Discovering the missing variation in the bovine genome; a long read sequencing pilot study into the structural variation in two dairy breeds 274
A. Chamberlain, T. Nguyen, J. Wang and I. Macleod
X and Y chromosome SNPs as indicators of sex in quality assurance checks for genomic analysis in Wagyu 278
C. Teseling and D.P. Garrick
Genome wide association study and heritability estimates for ram semen traits 282
M.J. Hodge, S. de las Heras-Saldana, S.J. Rindfleish, C.P. Stephen and S.D. Pant
Online Mendelian Inheritance in Animals (OMIA) – looking to the future 286
I. Tammen, M. Mather, D. Vanichkina, J. Nothman, Z. Li, S. Jufri and F.W. Nicholas

Novel Traits and Phenotyping Tools A

Genetic evaluation of coat type for Australian Angus 290
A.M. Samaraweera, H. Aliloo, A. Byrne, C.J. Duff and S.A. Clark

Genetic parameters for linear type traits in Hungarian Charolais beef cattle 294
B.J. Crook, W.M.S.P. Weerasinghe and M. Torok

Validation of breeding values for robustness in Australian Merinos 298
D.L. Waters, J.H.J. van der Werf, D.J. Brown, S.F. Walkom and S.A. Clark

Merits of using new intramuscular fat measurement technologies in genetic evaluation of Australian lamb 302
P. Alexandri, S.F. Walkom, S. Stewart, P. McGilchrist, C. Steel and D.J. Brown

Merits of using DEXA to measure lean meat yield for the genetic evaluation of Australian lamb 306
S.F. Walkom, P. Alexandri, S. Connaughton, G. Gardner, A. Williams and D.J. Brown

Progress of the Southern Multi-Breed Resource Population: Hard-to-measure phenotypes to drive genomic selection 310
B.J. Walmsley, K.L. Moore, S.F. Walkom, S.A. Clark, T. Granleese and K.A. Donoghue

Genetic Evaluation A

Estimate the genetic parameters and analysis of culling reasons in Iranian Holstein dairy cattle 314
H. Keshavarzi, E. Dehnavi and A. Small

Genetic parameters and lambda values for post-weaning production traits in Merino sheep 318
S. de las Heras-Saldana, P. Gurman, A.A. Swan and D.J. Brown

Longevity of reference populations in a trans-Tasman genetic evaluation: Review of the Angus Sire Benchmarking Program 322
S.F. Walkom, C.J. Duff, C. Girard and K. Moore

Angus BreedCHECK – Validation using industry data 326
C.J. Duff, A.M. Samaraweera, A.I .Byrne, A.B. Ingham, P.A. Alexandre, L.R. Porto-Neto and A. Reverter

Quantifying the linkage between genetics represented in the Southern Multi-Breed project and the wider Australian beef populations 330
K.L. Moore, S.F. Walkom, J.P. Siddell and B. Walmsley

Utility of pooled DNA samples for estimating a flock profile 334
P.M. Gurman, K. Gore and D.J. Brown

Assessing the influence of BayesR and GBLUP SNP effects in the correlation scan methodology 338
B.S. Olasege, I. van den Berg, E.J. Breen, M. Haile-Mariam, P.N. Ho, L.R. Porto-Neto, B.J. Hayes, M. Goddard, J.E. Pryce and M.R.S. Fortes

Health and Welfare A

The value of research and industry flocks for predicting breech strike resistance in Australian Merino sheep 342
E. Dehnavi, A.A. Swan, J.L. Smith, T.L. Bird-Gardiner, G. Burbidge and D.J. Brown

Development of a new BREEDPLAN objective body composition EBV to allow selection to improve cow survival 346
M.L. Wolcott, D.J. Johnston, M.G. Jeyaruban and C.J. Girard

Serum health biomarkers are significantly correlated with gene expression during the transition period of dairy cows 350
A. Kudriashova, A.J. Chamberlain, J.E. Newton, C.R. Bath, C.J. Vander Jagt, C.M. Reich, B.A. Mason, J.E. Hemsworth and M.E. Goddard

Preliminary genetic parameters for flystrike and its association with production traits in Australian Merino sheep 354
E. Dehnavi, A.A. Swan, A.M.M. Ramsay, G. Burbidge and D.J. Brown

Breeding for Reproductive Traits - Dairy

Milk production and fertility of spring-calved Holstein-Friesian, Jersey and crossbred cows milked once daily or twice daily in New Zealand 358
J.M.D.R. Jayawardana, N. Lopez-Villalobos, R.E. Hickson and L.R. McNaughton

Conception based fertility trait for genetic evaluation of New Zealand dairy cattle 362
K. Stachowicz, E.C. Ooi, B. Santos and P.R. Amer

Attributes of lactating cows ranked divergently for fertility using a milk mid-infrared spectroscopy based prediction model 366
A.R. Bird, J.E. Newton and P.N. Ho

Genetic Evaluation B

Appropriateness of combining carcass data from Angus sire benchmarking program and breeder herds in a single genetic evaluation 370
A.M. Samaraweera, A. Byrne and C.J. Duff

Infoherds, genomic selection for dairy cattle in New Zealand 374
B. Santos, N. Howes, S. Harburg, E. Ooi, S. Meier, A. Fear and P. Amer

Validation of calving ease EBVs examining the impact of genetic groups and single-step on predictive ability 378
P.M. Gurman, L. Li, M.G. Jeyaruban, D.J. Johnston, C.J. Girard, and A.A. Swan

Resource Allocation & Genetics of Intake and Efficiency

Remodelling the genetic evaluation of NFI in beef cattle - Part 1: Developing an equivalent genetic model 382
L.Vargovic, K.L. Moore, D.J. Johnston, G.M. Jeyaruban, C.J. Girard, J. Cook, J.A. Torres-Vázquez and S.P. Miller

Remodelling the genetic evaluation of NFI in beef cattle - Part 2: Shortening the length of the feed intake test 386
L.Vargovic, K.L. Moore, D.J. Johnston, G.M. Jeyaruban, C.J. Girard, J. Cook, J.A. Torres-Vázquez and S.P. Miller

Across-breeds systems biology analysis reveals key genes contributing to feed efficiency in beef cattle 390
K. Keogh, D.A. Kenny, P.A. Alexandre, M. McGee and A. Reverter

Producer Day

Identification of climate-resilient Merino sheep using satellite images 394
S. de las Heras-Saldana, L.A. Suarez, P.K. Wahinya, K.L. Bunter and D.J. Brown

Proc. Assoc. Advmt. Anim. Breed. Genet. 25: 1 - 4

THE ROLE OF ANIMAL GENETIC RESOURCES IN THE SUSTAINABLE LIVESTOCK TRANSFORMATION

P. Boettcher, R. Baumung, G. Leroy and B. Besbes

Food and Agriculture Organization of the United Nations, Rome, 00153 Italy

SUMMARY

The world's food systems must become more sustainable and equitable. The livestock sector must evolve and continue to deliver its benefits to humankind while competing for natural resources and maintaining resilience to changes in climate and other outside forces. The genetic diversity of livestock is a global public good that will underlie this transformation and demands collaborative stewardship. The FAO provides a forum for its member countries to discuss and agree upon priorities and actions for the proper management of animal genetic resources. It also facilitates the sharing of knowledge among countries and monitors the degree and impact of measures undertaken to safeguard and sustainably utilize livestock genetic diversity. Activities are guided by the Global Plan of Action for Animal Genetic Resources.

INTRODUCTION

Sufficient, nutritious and safe food is a need and right for everyone. Fortunately, due to advances in technology, improved agriculture and food policy, increases in income and greater international cooperation, among other factors, the proportion of undernourished people has continually declined across the recent decades. Proportions of incomes spent on food have also decreased steadily. Food systems are not perfect, however. Hundreds of millions of people still go hungry on a regular basis, while many others suffer from other dietary imbalances, including obesity. Many production systems have questionable sustainability from environmental, economic and/or social perspectives. The global COVID-19 pandemic revealed the fragility of many of the world's food systems. In 2021, the UN convened the Food Systems Summit as the climax of a comprehensive consultative process. Summit participants concluded that although current systems already produce billions of tons of food while considering the conservation of biodiversity and ecosystems, "business as usual" is not sufficient, and that a transformation of global food systems is needed. This process must consider "People, Planet and Prosperity" and align with the UN Sustainable Development Goals (SDG). FAO, a specialized agency of the UN, leads international efforts to defeat hunger and improve nutrition and food security, and will have a key role in this transition. Its current Strategic Framework aims to support its member countries in achieving the "Four Betters": better production, better nutrition, a better environment, and a better life, while leaving no one behind.

THE SUSTAINABLE LIVESTOCK TRANSFORMATION

Livestock production is an exemplary case. The sector makes a vital contribution to global food security and nutrition, livelihoods, and ecosystem services. It contributes to all of the SDGs (FAO 2018). At the same time, the sector utilizes vast amounts of natural resources, faces continual threats from epidemic, transboundary and zoonotic diseases, and both contributes to and is affected by climate change. Trade-offs abound, and disparities between regions and economies are commonplace. To provide just a few examples, animal source foods are a nutrient-dense source of protein, energy and many micronutrients, but are characterized by both under- and over-consumption. Ruminants can convert human-inedible plant matter into valuable foods, but they emit greenhouse gases in the process. Human-edible feeds improve efficiency of livestock diets, but compete with humans and require land for production. Wide differences exist in access to, and application of, technologies to help enhance productivity. In the future, livestock production will continue to play an important role in the diets and livelihoods of billions of people and to the economies of all countries. However, like food systems in general, evolution and transformation of the sector are required to ensure it achieves its goals in a sustainable manner, while addressing all of the many trade-offs.

LIVESTOCK GENETIC DIVERSITY

Within the livestock sector, wise management of animal genetic resources (AnGR) will be an essential part of this transformation. "Transformation" is just another way to say "change". The genetic diversity within and across breeds allows populations to adapt to changes in their production environment and breeders to improve the ability of their animals to achieve productivity objectives. This diversity will be critical in the future. The sustainable transformation will demand increased efficiency of resource utilization. This implies effective use of AnGR around the world and implementation of genetic improvement programmes. Climate challenges may lead to more movement of diverse genetic material across borders. This will depend upon increased knowledge on the characteristics of different AnGR, to properly match breeds with environments, both for productivity and welfare of animals and the sustainability of the natural biodiversity of the production environment. Greater equity in access to technology and the capacity to use it will also be critical. Conservation programmes must be strengthened to ensure diversity is maintained.

Although individual animals and breeds are private or "club" goods, the collection of genetic diversity is considered a global public good. Alas, recent assessments have determined that this diversity has been decreasing over time (FAO 2015). This suggests a need for global collaboration on its management, and as a UN agency, FAO has a natural role to play. FAO has a history of supporting countries on matters regarding AnGR since the 1960s. This work was formalized in the 1990s, when the Commission on Genetic Resources on Genetic Resources for Food and Agriculture (CGRFA) established its Intergovernmental Technical Working Group on AnGR (ITWG). The CGRFA and ITWG provide a forum for countries to discuss key issues regarding livestock genetic diversity and to advise FAO (and themselves) about steps to be taken to improve its use and conservation.

THE GLOBAL PLAN OF ACTION FOR ANIMAL GENETIC RESOURCES

Under the umbrella of the CGRFA and ITWG, FAO member countries developed the Global Plan of Action for Animal Genetic Resources (GPA; FAO 2007). The GPA is a policy document that includes 23 *Strategic Priorities* (SP), under four *Strategic Priority Areas* (SPA; described below) and provides the framework for sustainable management of AnGR. The SP address the most important actions to ensure sustainable use and conservation of AnGR that when implemented would ensure a substantial contribution of livestock genetic diversity. Table 1 provides an example of the actions under SP4 on breed development.

Although the GPA was adopted in 2007, it was prepared with a forward vision and remains fully valid today and highly relevant. In 2017, FAO members reaffirmed their commitment to its continued implementation. As sustainability was a key theme in its preparation, the priorities and actions of the GPA are highly appropriate for the evolution associated with the transformation of the livestock sector.

The GPA stipulates that the main responsibility for its implementation rests with national governments, but it also specifies that FAO has an essential role in supporting these country-driven efforts, as well as monitoring their progress and impact. In particular, FAO is key actor in coordinating international cooperation. FAO also promotes the importance of AnGR and their diversity and leads efforts in information sharing. With the contribution of international scientific experts, technical support to countries is provided by implementing and backstopping projects, organizing and participating in capacity building activities, developing international technical standards and protocols, and producing technical guidelines. The following paragraphs provide examples of current activities to support the integration of AnGR into the sustainable livestock transformation, organized according to the SPAs of the GPA.

Table 1. The actions associated with Strategic Priority 4 of the Global Plan of Action on Animal Genetic Resources

Strategic Priority	Actions
Establish national species and breed development strategies and programmes	Develop long-term planning and strategic breeding programmes
	Assess breed development programmes and revise, as appropriate, with the aim to meet foreseeable economic and social needs and market demands
	Establish and develop organizational structures of breeding programmes
	Incorporate consideration of the impacts of selection on genetic diversity
	Establish or strengthen recording schemes to monitor changes in non-production traits and adjust breeding goals accordingly
	Encourage the development of backup collections of frozen semen and embryos from current breeding schemes to ensure genetic variability
	Provide information to farmers and livestock keepers to assist in facilitating access to animal genetic resources

SPA1. Characterization, inventory and monitoring of trends and associated risks. Information about any entity is requisite for its proper management. A major role of FAO is curation and maintenance of the Domestic Animal Diversity Information System (DAD-IS), the web interface for the Global Database of Livestock Breeds. DAD-IS contains information on nearly 9,000 breeds from 37 livestock species plus managed bees. Among the data are inventories population sizes of breeds and material in gene banks, which are key indicators for risk of extinction. These data are used to indirectly monitor the impact of the GPA and inform official Indicators of the SDGs. Alas, these data are lacking for about half of the breeds, so FAO is currently developing low-cost methods to estimate population sizes. To complement census population size as a risk indicator, FAO is working with experts to facilitate the use of effective population size as an indicator of risk in DAD-IS. Developments in genomics have decreased costs substantially, making this possibility feasible. This effort corresponds to the recent release of guidelines on genomic characterization (Ajmone *et al.* 2023) and the recognition in the Kunming-Montreal Global Biodiversity Framework of the key importance of within-population genetic diversity. FAO has recently expanded DAD-IS to include data for populations of bees that are managed for food and agricultural purposes.

SPA2. Sustainable use and development. The profitable maintenance of breeds *in situ* is the optimal way to maintain livestock genetic diversity, as it ensures not only the survival of the breed, but the continued delivery of ecosystem services by the breed and its traditional production system. FAO is undertaking a study to collect best practices for establishing and operating breeding programmes on the community level and upscaling them to greater dimensions. Promotion of agroecosystems approaches in the management of AnGR is SP5 of the GPA. FAO has developed the Tool for Agroecology Performance Evaluation (TAPE) and is building capacity in its utilisation. TAPE can be used for self-diagnosis of existing production systems and for gathering of evidence on how agroecology can contribute to sustainability. The overwhelming majority of genetics-related technical cooperation projects of FAO and its joint centre with the International Atomic Agency address sustainable use. These projects support countries in the adoption of new technologies and practices to improve the management of livestock genetic diversity.

SPA3. Conservation. FAO has recently released guideline on cryoconservation of AnGR (Boes *et al.* 2023), which informs countries about recent innovations in animal gene banking. Major innovations include not only new methods for collection and cryopreservation of genetic material, but also in the application of genomics for management of collections. The guide also promotes a more active engagement with stakeholders in developing the gene banking strategy and in stimulating the wider use of gene banks in the active management populations *in situ*, rather than primarily as a safeguard against breed extinction. A series of webinars was implemented to build capacity on the content of the guidelines.

SPA4. Policies, institutions and capacity-building. Policy support is a key role of FAO. For livestock, FAO support to countries ranges from developing comprehensive Livestock Master Plans, to national strategy and action plans for all AnGR, to targeted conservation programmes for single breeds.

Previous global assessments on animal resources (e.g. FAO 2015) identified lack of technical capacity in developing countries as one of the factors hindering the sustainable use and conservation of AnGR. FAO both builds capacity directly and helps coordinate cooperation between countries. FAO has widely adopted the web-conferencing practices utilised as a necessity during the pandemic. More than a dozen webinars were presented in 2022 and more are planned for the future. Live or recorded presentations are also given in events of other organizations. However, many types of capacity building are difficult to do effectively online. In 2024 and 2025, FAO will benefit from support from the government of Germany to organize regional in-person training events on topics to be determined in direct consultation with the beneficiary regions.

In-person events also provide more visibility and opportunities for networking than can be offered by the virtual world. Therefore, from 25 to 27 September 2023, FAO will be hosting the first-ever Global Conference on the Sustainable Livestock Transformation. The conference will be held in Rome and will include scientific sessions on the contributions of livestock, including AnGR, to the Four Betters, as well as a high-level session for ministers of agriculture and livestock. Participants will be nominated by national governments, but the Conference will be webcasted to make sure that anyone who is interested can follow.

CONCLUSIONS

By improving the management of AnGR, countries will take an important step in the sustainable transformation of the livestock sector and contribute to better production, better nutrition, a better environment, and a better life for both human kind and their animals. FAO is looking forward to cooperating with all stakeholders in these efforts.

REFERENCES

Ajmone-Marsan, P., Boettcher, P.J., Colli, L., Ginja, C, Kantanen J. and Lenstra, J.A., eds. (2023) *Genomic characterization of animal genetic resources – Practical guide.* FAO, Rome, Italy.

Boes, J., Boettcher, P. and Honkatukia, M., eds. (2023) *Innovations in cryoconservation of animal genetic resources – Practical guide.* FAO, Rome, Italy.

FAO. (2007a) *The Global Plan of Action for Animal Genetic Resources and the Interlaken Declaration.* FAO, Rome, Italy.

FAO. (2015) *The Second Report on the State of the World's Animal Genetic Resources for Food and Agriculture.* Scherf B.D. and Pilling D. (eds) FAO, Rome, Italy.

FAO. (2018) *World Livestock: Transforming the livestock sector through the Sustainable Development Goals.* Rome, Italy.

Proc. Assoc. Advmt. Anim. Breed. Genet. 25: 5 - 9

ANIMAL BREEDING IS PART OF THE SOLUTION TO ENVIRONMENT, CLIMATE, AND ANIMAL-WELFARE CHALLENGES FACING ANIMAL PRODUCTION

M. Henryon[1,2], T. Ostersen[1], P.E. Vercoe[2], and A.C. Sørensen[1]

[1] Danish Pig Research Centre, Danish Agriculture and Food Council, Denmark
[2] School of Agriculture and Environment, University of Western Australia, Australia

SUMMARY

We argue that animal breeding is part of the solution to a major challenge facing animal production: community concerns for the environment, climate change, and animal welfare. Animal production will increasingly be expected to use fewer resources, reduce its impact on the environment and climate, and improve animal welfare. Animal breeders can provide animals with genetics that make them productive in future, reshaped, production systems by defining breeding objectives with traits that benefit the environment, climate, and animal welfare. Breeders are well-equipped to make gains in these breeding objectives because they can predict breeding values accurately. These accuracies will only increase as new genetic technologies become available, leading to even faster gains. However, faster gains also call for caution because they increase the risk of unintended side effects. To manage this increased risk, breeders should consider three safeguards: control of inbreeding, reliable selection criteria, and monitoring and surveillance of animals. Another safeguard is maintaining many populations of commercial breeds. It's an exciting time for animal production, and breeders must be there providing the genetics.

ANIMAL BREEDING IS PART OF THE SOLUTION

Animal breeders use selection to improve desirable traits in animal populations. The underlying principle is to rank animals for these traits and choose the best to be parents of the next generation while controlling rates of inbreeding at acceptable levels. This principle will not change in future. What is likely to change is the direction of this selection – the composition of traits in our breeding objectives – as animal production wrestles with community concerns for the environment, climate change, and animal welfare. We have little doubt that animal production has a future. Animals provide humans with high-quality protein, essential nutrients, and non-synthetic products; they convert biomass that is unsuitable for human consumption into food, manure, and ecosystem services; they utilise land that cannot be used to produce other types of food; and they are deeply embedded into the economies and cultures of societies around the world. However, like most other businesses, these benefits come at a cost. Animal production uses land, water, and energy, it degrades and pollutes terrestrial and aquatic ecosystems, it encourages deforestation, it emits greenhouse gasses, and it rears animals in captivity. Assuming communities are well fed and have their basic needs met, animal production will increasingly be expected to use fewer resources, reduce its impact on the environment and climate, and improve animal welfare. This is where animal breeding must play a key role by providing animals that are genetically suited to production systems of the future. Therefore, we argue that animal breeding is part of the solution to the challenges facing future animal production. Other solutions, which we do not address here, are to increase plant consumption, reduce animal consumption in wealthy countries, replace conventional meat with cultivated meat, and reduce food wastage. We see our paper as a summary of opportunity and a call to action. Our primary focus is on large, centralised breeding schemes as we believe that these schemes will provide most of the world's genetics in future.

ANIMAL PRODUCTION IS CHANGING

Modern animal production requires a "licence to produce". Animal products need to be produced and supplied in a way that eases the community's concern for the environment, climate change, and animal welfare. Governments, particularly in wealthy countries, are reacting to these concerns. They are introducing change in the form of legislation, incentives, and penalties to balance the economic benefits of animal production with its impact on the environment, climate, and animal welfare. For example, New Zealand's government will charge farmers for the greenhouse gases emitted by their livestock. The Dutch and Belgian Governments will halve nitrogen emissions by reducing livestock numbers. The German Government has already introduced strict requirements for animal welfare with a short phase-in period and did so without consulting any animal sectors. The European Commission will phase out cage production of farmed animals. In response to these changes, we must expand our definition of productivity to include economic incentives and penalties associated with the environment, climate, and animal welfare. They will almost certainly lead to new production systems with revamped management strategies and husbandry practices as producers cope with the new legislation, pursue the incentives, and avoid the penalties. The impact of these changes could be substantial. For example, producers that use cattle feedlots will need to improve animal welfare and reduce their impact on the environment and climate. Intensive pig, chicken, and fish enterprises may have reduced their impacts on the environment and climate, but they still cause animal-welfare concerns. Organic pig and chicken productions have improved animal welfare, but still have an impact on the environment and climate. We will probably also see new and efficient species introduced into production systems. A prime example is the growing interest in insects and microorganisms reared on waste products to generate food and animal feed. No matter what the production system, they all have one thing in common: they all require animals – including insects and microorganisms – with genetics that make them productive. So, animal production is changing because our definition of productivity is changing, requiring animals with genetics that make them productive in future, reshaped, production systems.

BREEDING OBJECTIVES FOR ENVIRONMENT, CLIMATE, ANIMAL WELFARE

Animal breeders can provide animals genetically capable of being productive in future production systems by defining new breeding objectives. Breeders define breeding objectives by identifying the traits they want to improve and deriving economic values that allocate an appropriate amount of selection pressure to each of these traits. New breeding objectives will almost certainly include most, if not all, of the traits in current breeding objectives, including growth rate, feed efficiency, meat and milk yields, fleece weight, litter size, and survival. Not only do these traits increase economic returns, they also benefit the environment, climate, and animal welfare by increasing production efficiency. So, new breeding objectives will reflect current breeding objectives, but there are likely to be two striking differences. First, these breeding objectives will also include new traits directed towards benefiting the environment, climate, or animal welfare. Possible examples include reduced emissions of nitrogen, phosphorus, and methane, lower production odours, and tail biting. Second, the economic values allocated to each trait will change to shift some selection pressure towards traits associated with the environment, climate, and animal welfare. Deriving some of these economic values could be particularly challenging for traits, such as survival and conformation disorders, that infer a "licence to produce". Economic values for these traits can have "non-market" values that are much larger than any profit margin when the phenotypic means of the traits fall below levels that are acceptable to the community. Non-acceptable standards can trigger government legislation, consumer boycotts and, in extreme cases, shut whole industries down. The problem for breeders is that they will be compelled to foresee "non-market" values for traits when the level of community acceptance in future is fraught with uncertainty. Therefore, defining new breeding objectives with traits that benefit the environment, climate, and animal welfare is certain to be challenging, but it is critical that we tackle these challenges because breeding objectives are the only lever breeders have to increase productivity in future production systems.

ANIMAL BREEDERS NEED STRONG SIGNALS

Animal breeders who practice good business management are unlikely to be "first movers" because they need certainty before they change their breeding objectives. Breeders make selection decisions based on projected market conditions but there is a time lag before genetic gains made from these decisions are realised and disseminated to producers. If these projected conditions are incorrect, breeders risk wasting selection pressure on improving traits that are not profitable. Governments and the community can assist all vested stakeholders in animal production by providing breeders with strong and early market signals. These signals are long-term legislation, incentives, and penalties directed at producers. They would enable breeders to define with confidence breeding objectives that provide a clear direction for selection, avoid selection for traits that can be improved by non-genetic methods, resist selection for traits that are merely indicators of productivity, and hasten the time before animals with improved productivity are disseminated to producers. So, we recommend that governments and the community provide breeders with strong and early market signals directed at producers. This is in the best interests of all vested stakeholders.

MAKING FAST GENETIC GAINS SAFELY

Modern animal breeders are well-equipped to make gains in their breeding objectives because they are good at ranking animals. They predict breeding values accurately by fitting sophisticated genetic-statistical models to phenotypes, pedigree relationships, and genomic information. This accuracy will increase further in future as breeders develop better genetic-statistical models, improve phenotyping strategies, and acquire new genetic technologies, such as intermediate phenotypes, genetic engineering, gene editing, and gene networking. This is good news for animal production because it implies faster genetic gains. However, faster gains also call for caution because animal breeding is, and will remain in the foreseeable future, a "black box" technique. Breeders make genetic gains without understanding the full genetic and physiological consequences of selection. Some of these consequences can be unintended behavioural, physiological, metabolic, reproductive, and immunological side effects caused by genetic correlations between these effects and the traits in breeding objectives. Faster genetic gains merely increase the risk of these side effects. Clearly, we need improved safeguards to manage the increased risk of unintended side effects with faster genetic gains. We suggest three safeguards that should be considered by animal breeders to address this problem.

1. Control of inbreeding. Controlling inbreeding within populations at acceptable rates is a safeguard against unintended side effects because it maintains genetic variation, reduces inbreeding depression, decreases the spread of deleterious recessives, and reduces variability in the rate of genetic gain. Unfortunately, control of inbreeding in selective breeding schemes is struggling to keep pace with the fast genetic gains being realised by highly accurate predictions using genomic information. We see three key issues that need to be resolved before we can control inbreeding effectively in these schemes. First, there is no consensus on the most appropriate definition of inbreeding following the advent of genomic information. Do we control identity-by-descent (IBD), loss of heterozygosity, or genetic drift? Second, we have not learnt to control inbreeding with genomic information in selective breeding schemes. Breeding schemes that use pedigree information to control inbreeding realise more genetic gain than genomic information at the same rate of IBD. This leads us to reason that pedigree control is the method-of-choice for inbreeding control in selective breeding programs. Genomic control is unlikely to realise more genetic gain than pedigree control until we understand which regions of the genome harbour quantitative trait loci and we can manage genetic variation along the genome. A notable caveat is that pedigree control tends

to underestimate rates of IBD when genomic information is used to predict breeding values. This implies that pedigree inbreeding should be controlled at rates lower than desired rates of IBD. Third, optimum-contribution selection (OCS) is the best method of selection because it maximises genetic gain for a given rate of inbreeding, but is not used in many breeding schemes because it can be difficult to implement in practice. With the promise of faster genetic gains, we urgently need to adapt OCS to conform to the practical aspects of animal breeding. Selection decisions made by OCS are not always optimal because reproductive biology and logistical constraints can be more complex than the input data we provide OCS software. There can be a mismatch between OCS decisions made centrally at discrete time points and the true optimum for any given day. So, there is clearly a lot of work to do before we have effective inbreeding control with fast genetic gains. Until then, we recommend pedigree control of inbreeding while correcting for the fact that pedigree underestimates rates of IBD.

2. Reliable selection criteria. Identifying reliable selection criteria for traits in the breeding objective provides a safeguard against unintended side effects by enabling breeders to allocate an appropriate amount of selection pressure to traits associated with the environment, climate, and animal welfare. The most reliable selection criteria are phenotypes that are easy to measure, express genetic variation, are genetically correlated with one or more traits in the breeding objective, and can be recorded for many selection candidates or their relatives. The challenge for breeders is that traits associated with the environment, climate, and animal welfare are often difficult to measure. Developing suitable and usable selection criteria for these traits must be a priority. Without them, we will forego potential gains in our breeding objectives by failing to allocate the correct amount of selection pressure to each trait. So, while it is key to include traits associated with the environment, climate, and animal welfare in breeding objectives, it is also important that we identify selection criteria that enable us to improve these traits by selection.

3. Monitoring, surveillance, and communication. Close monitoring and surveillance of animals is an important safeguard against unintended side effects by uncovering some of these effects before they spread through breeding populations. No monitoring or surveillance will uncover all unintended side effects, given that selection acts at the molecular level. However, we can increase the probability of uncovering them by routine evaluation using human assessment and surveillance technologies carried out by stakeholders with a vested interest in animal welfare. These stakeholders can be active at all levels of production and include animal breeders, producers, veterinarians, abattoir operators, retailers, and scientists. The side effects that they uncover are relayed back to the animal breeder who can then begin to rectify the effects. So, animal breeders can manage the increased risk of unintended side effects with faster genetic gains by communicating closely with vested stakeholders who routinely evaluate the animals generated by breeding.

MANY BREEDING POPULATIONS FOR BREED SECURITY

Like the safeguards against unintended side effects within animal populations, maintaining many populations of each commercial breed can provide a safeguard against production changes and market uncertainty. Maintaining many populations conserves genetic variation. It increases the probability that some populations will cope with change better than others. It also enables producers to choose animals from populations best suited to their production systems. However, maintaining many breeding populations is at odds with the business strategies and commercial goals of breeding companies for three reasons. First, like other businesses, breeding companies compete, go bankrupt, merge, exclude new entrants, and seek to monopolise global markets. For example, the world's genetics for broiler chickens is now supplied by only three companies and most of the pig genetics is supplied by just six companies. Second, breeding populations that do not make a return on investment are discontinued, and discontinued populations are seldom replaced. Third, breeding companies with the same commercial breed define similar breeding objectives for their populations so that these populations tend to converge genetically. The result is few breeding companies maintaining few breeding populations and these populations tend to resemble each other. This makes many commercial breeds vulnerable to market fluctuations and they risk being replaced by other breeds, species, and even alternative food sources. We need to balance the economic drive to concentrate breeding populations with the need to maintain populations. This balance could be achieved through government intervention to resist global monopolisation of genetic resources. So, we have a choice. We can leave breed security to the mercy of breeding companies and economic forces, or we can intervene to resist global monopolisation. We advocate for intervention.

ACKNOWLEDGEMENTS

We thank Søren Krogh Jensen, Anders Vernersen, Peer Berg, Christian Fink, David Lindsay, Vivi Aarestrup Moustsen, Bjarke Grove Poulsen, and Rob Banks for valuable discussions.

LIVESTOCK BREEDING, WHERE HAVE WE BEEN AND WHAT LIES AHEAD?

S.P. Miller

Animal Genetics Breeding Unit*, University of New England, Armidale, NSW, 2350, Australia

SUMMARY

Presented is an overview of recent advancements in livestock breeding, focussing post the implementation of genetic evaluation, which overlaps with the career of the single author. The rise of genomics is presented as a major turning point and the increased gains and future challenges with this technology is presented. Some history of the corporatisation of breeding programs is presented and parallels with the invention of an enabling technology, such as artificial insemination, is illustrated. It is suggested that further advancements in genetic engineering, such as surrogate sires, or the joining of embryonic stem cells to enable 'speed breeding', would be the next turning point. These technologies would create the environment for large corporate investment in sheep and beef cattle and could change the structure of those genetic industries forever.

EARLY FOUNDATIONS

Livestock breeding has changed considerably over time. A century ago selection was based primarily on phenotype and the 'eye' of the breeder played a major role. Compared to selection practices today, that are based heavily on quantitative data, these early selection programs can appear rudimentary. However, we should acknowledge that breeding in the early part of the 20th century was a great advancement from the century earlier and it is at this point in history when many of the livestock breeds were moved to 'colonies', which later became major food producing nations, one of which would be Australia. Indeed, the genetic improvement in the recent century has been made possible by the stock developed by the forebears of animal breeding.

Early developments included the establishment of breed societies in the 'new world' and an exportation of genetics from the old world. Initially, genetic improvement was focussed around bringing the best genetics into the new world and this was facilitated through the establishment of breed societies. The establishment of herd books within breed societies provided a way for buyers to verify the 'purity' of the stock they were purchasing, as the newly imported breed was an advancement over the local alternative, and it was this preservation of purity that was the main goal. Genetic advancement was achieved through a replacement of 'local' stock with 'improved' stock or the displacement of one breed with another.

Prior to the widespread use of artificial insemination, breeding was also a local affair. Since breeding was based in part on selection of a desired 'type', often set as a breed standard by the societies, the placement of stock in classes at exhibitions developed as an important ranking tool. Producers looking to advance their stock would seek out champions from an exhibition and the larger the exhibition (competition), the better the animal. However, the relative merits of producers in their ability to prepare animals for showing (eg feeding, grooming) may have been difficult to disentangle from genetic merit. Nevertheless, this culture of commerce supported a vibrant exhibition industry with local, state and national exhibitions. The remnants of these still exist today and some are still a marketplace for trade for some species, examples would be the Sydney Royal Easter show or the 'Ekka' Royal Queensland Show, to name a couple. These exhibitions were an important avenue for breeders to market their genetics. Without artificial insemination, commercial producers needed to purchase a natural service sire and a commercial producer's local genetics supplier was more likely to be a 'neighbour' in relative terms.

---

*A joint venture of NSW Department of Primary Industries and the University of New England.

THE RISE OF CORPORATISATION

The selection, marketing and trade of genetics today is very different to how it looked 50 to 100 years ago. The argument put forth in this paper is that this transition from a family enterprise, with a local selection and marketing program, to a more global corporate enterprise is based around four primary factors:

1. Reproduction rate
2. Production cost per breeder
3. Availability of reproductive technologies.
4. Ease of preservation and shipment of semen, embryos and stock

Progress in these four areas has created the corporatisation of breeding in some species more than others. These factors are contrasted across four species in Table 1 with indicative levels indicated for each. By comparing the corporatisation in these species and the contributing factors, we can make more informed predictions about how new technological developments may affect corporatisation in different species in the future.

Table 1. Indicative* levels of key factors leading to corporate investment in animal breeding programs across four major species

	Laying Hens	Pigs	Dairy Cattle	Merino Sheep
Reproductive rate	XXXXX	XXX	X	X
Low production cost per Breeder	XXXXX	XXX	X	XX
Availability of reproductive technologies.	XX	XX	XXX	X
Preservation and shipment of semen, embryos and stock	XXXXX	XX	XXX	XX
Hybridization, line crossing	XXXXX	XXX	X	X

* the more X's, the higher the level and contribution towards corporatisation of breeding programs

Let's first focus on the laying hen, where the reproductive rate is high, with each hen capable of laying 300 eggs per year, and a hen is low cost to maintain and support. A single corporate entity can finance the infrastructure to produce large quantities of commercial stock in a pyramid system. Also, although artificial insemination is somewhat limited to fresh semen, this provides little impediment to progress since considerable stock can be located in one facility. The fact that hatching eggs or newly hatched chicks can be shipped nationally and internationally at low cost also supports the corporatisation of breeding in that species. Finally, following the success seen in corn breeding, the development of inbred lines (which are crossed to form hybrid commercial stock) was a game-changer in the poultry industry. This could only really be achieved on a large scale with many lines of sufficient size, and this is where the corporate breeders pulled away from the smaller private enterprises in the last half of the 20th century. This same hybrid model was tried in other species such as pigs and beef cattle, to capitalize on the same 'hybrid vigour', but these attempts largely failed as the cost to maintain inbred lines was simply too high at the scale required. Also, these mammalian species lines that failed due to poor reproductive rates, as a result of the inbreeding depression, contributed to the downfall of these attempts. Similar to hybrid corn, these commercial birds were of limited value to keep as replacement stock by the commercial farmer, which means the commercial producer must keep coming back to the corporation for commercial chicks, which perpetuates the corporate model.

Within the scope of species considered for comparison in Table 1, pigs are next in line for the most corporate breeding model. Although the reproductive rate is much less than chickens, at 20-30 pigs per sow per year, they are far ahead of cattle and sheep that are naturally limited to one or a little over one progeny per year in natural mating situations. As a litter bearing species, the advantages of embryo transfer offer fewer gains than cattle and are less successful. The use of artificial insemination is widespread with the ability to ship fresh semen widely within a country. However, international shipment of stock and semen is limited by health status within some countries, with Australia restricting importation of new genetic material. Although the cost of maintaining a sow is considerably more than a hen, the intensive nature of swine housing in modern production practices enables large numbers of animals to be maintained with a moderate outlay of capital for land, which tends to be the limiting factor with more extensive species. This combination of factors has made the global swine breeding industry the next most corporate within the examples presented. Factors such as the cost and health restrictions to ship stock around the world has limited this corporatisation and as a result we find many more pig breeding companies, in more countries, compared to poultry. The cost of maintaining lines and perhaps challenges with inbreeding in a mammalian species has made the hybrid model that was successful in poultry infeasible in pigs. As a result, commercial pig rearing is dominated by dedicated lines that come together in a dedicated crossing program to produce commercial sows and feeder pigs.

Dairy cattle are the next most corporate of the species presented and this has been enabled through the widespread use of artificial insemination and the characteristic of essentially sex-limited breeding goals. Excellent conception rates from frozen semen and non-surgical techniques have allowed artificial breeding to become the standard in most developed dairy breeding industries world-wide. This has allowed global breeding businesses to be built around the sale and distribution of bull semen. The impact of artificial breeding is best realised when one answers the question "What would the dairy breeding industry look like if artificial insemination was never invented?". It is likely that without artificial insemination, that the dairy breeding industry would look a lot more like the beef or sheep industries, with many breeders and a structure that is much less 'corporate' by nature. Looking back at the dairy breeding industry in the 60's, before artificial breeding was widespread, the industry did indeed look more like the beef industry, with many more stud breeders selling bulls for natural service. Hindered by reproductive rate, crossbreeding of any kind has seen limited implementation, with the majority of cattle being milked commercially in the world's largest dairy producing nations being purebred, with New Zealand being a noted exception to this rule.

Finally, Merino sheep is the example of the species that is the least corporate, with many studs in operation and the primary market being the sale of rams for natural service matings. This is despite the ease with which semen can be stored; but perhaps reflecting the greater difficulty of AI (surgical) for ewes and the relative cost of AI compared to the value of the animal. Although Merino sheep was provided as this example, beef cattle breeding will share many similarities with sheep, but beef cattle has a greater degree of corporate influence. A notable difference with beef cattle is the more prominent availability of frozen semen and a viable export market. The export markets and channels in place to support the sale of dairy semen has been leveraged for beef semen sales globally. Also, unlike Merino wool production, where Australia dominates, beef production is a more global industry. This global aspect results in more corporate activity around semen purchase and sale.

Although crossbreeding is not common within a wool production system it is common in terminal and maternal sheep breeding (McMillan *et al.* 2023). The increased reproductive rate in sheep compared to beef, and the potentially reduced generation interval when ewes are lambed at a year of age, does promote greater implementation of cross and composite breeding systems in sheep than in beef cattle. This multi-breed nature of the breeding industry in sheepmeat productionhas enabled a national multi-breed genetic evaluation under Sheep Genetics in Australia (Brown *et al.* 2007), where beef cattle evaluations through BREEDPLAN in Australia have been dominated by within-breed evaluations as a result of the structure of the data coming primarily from breed societies.

CHANGING BREEDS

The change in breeds used over time has been dramatic in some instances. Take the Holstein-Friesian as an example, where it has dominated much of the developed world. The explosion of this breed began in the latter half of the 19th century at a time when farms were getting larger and milk supply and marketing moved to a more pooled system, with less scope for individual attributes. Changes to how milk was marketed favoured the Holstein and this displaced breeds such as Jersey, Guernsey and the Milking Shorthorn, that had an advantage for butterfat.

The change of breeds in the Australian beef industry has been even more dramatic in the past 30 years, as depicted in Figure 1. Presented is the population of registered cattle through the Australian Registered Cattle Breeders Association (ARCBA). Although this is not a perfect picture of the breeds in the commercial industry, it is logical to consider these numbers as a good indicator of change at a population level. The most remarkable change has been the move from an industry dominated by Hereford genetics to one dominated by Angus. The reason for this change cannot be proven but there are a number of theories. Considering the breed differences identified in America at the USDA Meat Animal Research Centre (Kuehn and Thallman 2022), the Angus breed is a clear leader for marbling, a product differentiator in many branded markets, one of which is Certified Angus Beef (CAB). CAB has grown into the world's largest beef brand, marketing over 1 billion pounds of beef annually (American Angus 2022). Although CAB is a brand that dominates in America, this same success paves the way for Angus brands operating in Australia as well. Secondly, the Angus breed leads for calving ease, making it a more solid choice for crossbreeding, especially when mating heifers. Some trait advantages, the rise in feedlot finishing in Australia and the associated access to key branded products are likely reasons for this rise in Angus over this time period.

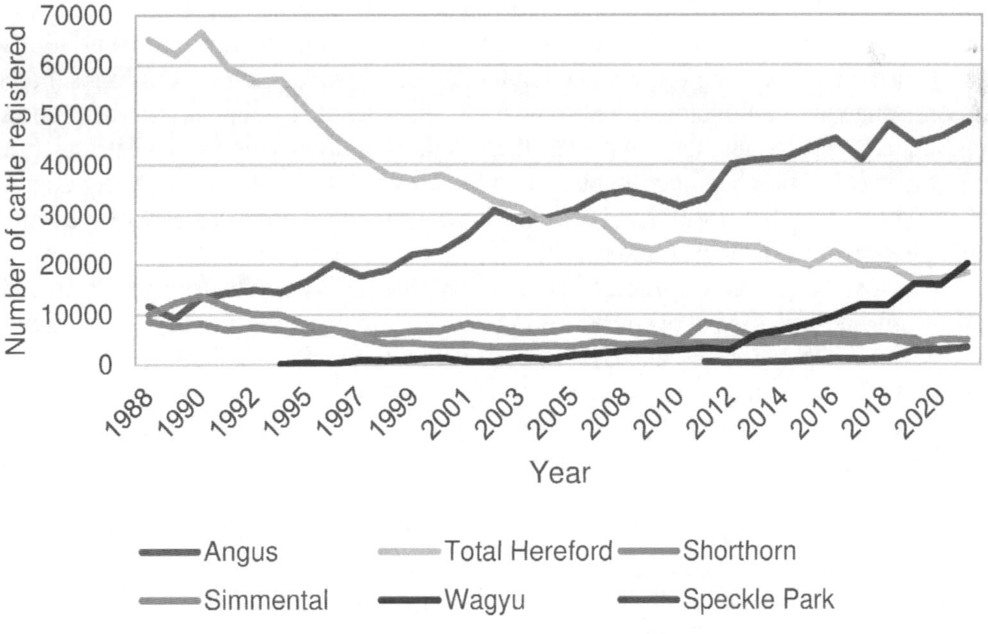

Figure 1. Changes in number of registrations of some beef breeds in Australia overtime
Source: Australian Registered Cattle Breeders Association (2023)

The total registered cattle population has fluctuated considerably overtime from a minimum of 127,000 in 2003 to 177,000 in 1990. Overall, the population is declining, where a linear regression estimates a decline of 513 animals per year in total registrations across all breeds. This gradual decline means that an increase in one breed is almost certainly gaining market share from another. Perhaps even more remarkable has been the rise of the Wagyu breed in Australia. This breed is also targeted at a premium market and lot feeding production system. The year 2000 was the first year to register over 1,000 Wagyu, and now Wagyu is the second largest breed for registrations, surpassing Herefords. It is remarkable to think that such dramatic changes in breed use is still taking place. The reason for this rise in Wagyu is likely to be similar to Angus, with a drive from the commercial market for a specific product, which in this case is one of very high marbling.

Changing the breed structure of an industry is perhaps the most dramatic example of genetic change. Changing breeds is certainly genetic change, but is it genetic progress? As animal breeders, much of the effort is focussed on within-breed selection in many instances, with very little input in the choice of breeds. This is despite breed choice having potentially the largest impact. The germplasm evaluation program at the USDA (Kuehn and Thallman 2022) and the recent Southern Multibreed project (Walmsley *et al.* 2021) in temperate Australia, along with Repronomics^{TM} in Northern Australia (Johnston *et al.* 2017) are meant to provide benchmarking for a limited set of current, more popular breeds.

In 1988 the Angus breed had a similar number of registrations to both Simmental and Shorthorn. Over time, Simmental and Shorthorn has retracted and Angus now registers 10-fold the numbers of either of these two breeds. The Speckle Park breed first registered animals in Australia in 2011 and has risen rapidly to now register a similar number to the Shorthorn breed. Clear objective information on the merit of the Speckle Park breed is not available. Although it might be seen as old fashioned, it would seem that objective comparisons of breeds is required for breeders to make informed choices on breed selection, as it continues to be in a state of change.

PERFORMANCE RECORDING AND THE BLUP ERA

It would be short sighted to suggest that performance recording started in the middle of the 20th century, as there has been recording of measurements for production and parentage going back well before that. However, it is during this period that more formal performance recording schemes were developed on a state and national level. Here in Australia, one such scheme was the National Beef Recording Scheme (NBRS) and there were similar schemes in other species in Australia and around the world. This was an era when phenotype truly was 'king' as it was the determining characteristic for selection. Then in the late 1980's and into the 1990's, schemes around the world transitioned to taking these performance databases and combining these with pedigree, that was typically recorded through a breed society, to create Estimated Breeding Values (EBVs). The technology to enable this was based on ground-breaking work (Henderson 1973) and Australia rapidly implemented these techniques (Graser 1982; Graser and Hammond 1985; Graser *et al.* 1987). At that time, the calculation of EBVs was brand new and required special skills in programming relatively large computational problems. Also, these problems required considerable computer power to run and such computer power was somewhat rare to access. These requirements resulted in the development work for EBV programs to be centred around Universities as they typically had the expensive computer hardware and the skilled staff required. The Animal Genetics and Breeding Unit (AGBU) at the University of New England, established in 1976 is one such entity. Around the world this race to implement EBV technology is what made the strong institutions in animal breeding that went on to make a significant impact in this area.

DAWN OF MOLECULAR GENETICS

During the mid-1990's, as EBVs were becoming entrenched, the next selection tool from the field of molecular genetics was also advancing. The future of animal breeding was in question. Was the calculation of breeding values using performance and pedigree information going to continue, or would this approach by surpassed with a purely molecular approach? Thoughts at the time were that once the genes controlling the traits were identified, selection could simply be to fix the desired variants. This was the beginning of somewhat of a divide in the field with two streams, Animal Breeding (quantitative) or molecular genetics. Those in the field of animal breeding did not simply bury their heads in the sand, but did what all good animal breeders do, and when faced with a lack of data, they 'simulated' what breeding would look like with molecular data and how the evaluation models would change to handle it. Some important papers resulting from this period related to the transition to molecular based breeding are Fernando and Grossman (1989) and Nejati-Javaremi *et al.* (1997).

Coming into the turn of the century it was becoming clear amongst the animal breeding community that molecular markers could have a significant role to play in practical breeding programs. 'Major genes" as they were commonly referred to at that time were starting to be identified. Some examples discovered included those affecting beef tenderness, including the related Calpain (Page *et al.* 2002; Casas *et al.* 2009) and Calpastatin genes (Schenkel *et al.* 2006) in this complex. The challenges facing the breeders was then how were we going to incorporate these new molecular tools into breeding programs? One early example of the integration of molecular information into breeding programs was the implementation of the Calpastatin genotypes in the Australia BREEDPLAN Brahman genetic evaluation for tenderness (Johnston *et al.* 2009). This was a challenging time for the animal breeding community as their funding sources were starting to fragment. Those wanting to fund genetic improvement in livestock were faced with a decision of funding the traditional programs that had been successful so far, or to start to direct money to this developing field of molecular genetics that just seemed more 'modern'. Unlike the field of quantitative genetics, that had been relatively low cost to date, research including genotyping and related laboratory costs was considerably more expensive. The result was animal breeders went through a phase where it was hard to secure research money unless their programs included something 'molecular'.

The first half decade into the twenty first century was one of very rapid advancement. Using beef cattle as an example, many molecular variants were being identified that were associated with economically important traits and companies were popping up that were now marketing these directly to farmers. Animal breeders found themselves sometimes in a position of validating these variants with independent data (Schenkel *et al.* 2005; Van Eenennaam *et al.* 2007; Johnston *et al.* 2010). In some cases single SNP tests were being sold for $80 USD. This quickly changed as more variants were discovered and genotyping companies realized that traits were influenced by multiple genes. The number of SNP in a test were quickly rising and were becoming limited by genotyping technology with plexes of 384 SNP being developed and sold. Meanwhile, alongside all this development of specific gene tests the animal breeding community was continuing to work on how this information would best be used and a landmark paper was released. This paper outlined the premise for what later became known as 'genomic selection' (Meuwissen *et al.* 2001). In 2005, while the search for specific variants continued to rage, a new genotyping array technology from Affymetrix (www.affymetrix.com) became available that enabled 10,000 SNP to be genotyped at reasonable cost, comparatively speaking. This technology was made possible by the Bovine Genome Project (Bovine Genome Sequencing Consortium 2009) along with contributing projects such as the Bovine HapMap project (Bovine HapMap Consortium 2009). With the ability to genotype large numbers of SNP effectively, the methods of genomic selection could then be applied. The 10K Affymetrix chip, ground breaking at the time, was soon replaced by the Illumina 50K (Matukumalli *et al.* 2009) with early access for research starting at the end of 2006, at an approximate cost of 300 dollars per sample, which was a dramatic reduction in cost per SNP, but is 10-fold the cost of genotyping today.

THE GENOMICS ERA

The 'standard' 50K chip changed the future of animal breeding and was a real turning point. During the next few years many changes occurred. The first was somewhat of a dissolving of the lines between quantitative animal breeding and molecular biology. Now instead of chasing individual genes, animal breeders could genotype their reference herds for 50K SNP and undertake meaningful Genome-Wide Association Studies (GWAS) analyses. This proliferation of large scale GWAS also discovered a number of new major variants. Genotyping companies were facing a cross roads, was the future bigger and better custom panels of hundreds of significant SNP, or were these mathematical approaches that use all 50K SNP, with no regard for which were significant, going to be better? Given the investment in specific panels (in beef cattle for example) to date, it seemed hard for the genotyping companies to believe the 50K shotgun approach could possibly be better. The length of time these debates raged was a blip on the overall timeline as the dairy industry soon proved beyond a doubt how powerful genomic selection could be (VanRaden *et al.* 2009). Where genomic selection was first applied I am sure is hotly debated, but I know it was applied in Canada in 2009 and although this may not have been the first implementation, it was not likely very far behind. The reason the dairy industry could apply genomic selection quickly was their extensive use of artificial insemination and long-standing progeny testing schemes provided them with a source of DNA (frozen semen) on thousands of bulls with highly accurate proofs, providing an instant genomic reference population.

The beginning of the second decade of the 21st century now saw other species looking at the success of genomic selection in the dairy industry and strategising how they could harness this same success. The key ingredient was clearly the reference populations and it was evident that these needed to be large and the bigger the better. The genotyping companies were also faced with the realization that their Intellectual Property (IP) in terms of specific marker panels could be displaced with this 50K product, something that was available to all. The implementation in dairy provided a stark example of a highly successful genomic product where genomic companies had no IP ownership. During this period the availability of low-density panels, first 3K, then 7K brought a new technology to the table in imputation (Sargolzaei *et al.* 2014; Browning and Browning 2016). Although the low-density panels did not last long as the cost of 50K genotypes came down, as global genotyping rates went up, the tool of imputation would prove important for the long-term.

The early implementations of genomic selection were predominantly multi-step approaches where predictions from the markers were combined with the traditional EBV, based only on phenotypes, in a blending approach. A popular blending approach used was the method of Harris and Johnson (2010) as applied to Australian beef cattle and sheep evaluations as described in Swan *et al.* (2011). In beef cattle, an approach applied in American Angus, as one example (Miller *et al.* 2018) was to bring the marker information into the genetic evaluation via a correlated trait with a heritability close to 1 and a correlation with the target trait in proportion to the prediction accuracy of the genomic trait (Kachman 2008). The multi-step approach allowed an expedited path for genomics to influence existing EBV procedures already in place bringing the technology to market with little delay. However, the multistep procedures were not optimal and relied on calibration steps that needed to be kept up to date (Johnston *et al.* 2010). The development of single-step procedures (Misztal *et al.* 2009; Aguilar et al. 2010) was a great advancement and allowed a simpler, more elegant approach, eliminating the need for separate calibration steps and enabled prediction with all the contributing information such as genomics, pedigree and performance information in a single procedure. In Australia, Sheep Genetics analyses went to Single Step in 2016 (Brown *et al.* 2018) and the first BREEDPLAN analyses transitioned to Single Step in 2017 (Johnston *et al.* 2018).

The implementation of genomic selection has been heralded as the greatest advancement in dairy cattle breeding since the widespread implementation of artificial insemination with frozen semen. Prior to genomic selection becoming fully implemented it was predicted that the rates of genetic progress would double in dairy cattle as a result of genomic selection (Schaeffer 2006). This has now been proven to be true (Scott *et al.* 2021; Fleming and Van Doormaal 2022). The early prediction of increased genetic progress by Schaeffer (2006) turned out to be an underestimate, possibly due to the fact it was based on early results with the Affymetrix 10K, whereas implementation has been with the Illumina 50K with more markers. Although higher density chips were also available at the time, these did not prove to increase the accuracy of genomic prediction.

Genomic selection has been a game-changer throughout many livestock industries. However, the basic implementation has not changed since it was implemented over a decade ago. Although there are some different variations being implemented, the basic model is via GBLUP, which is simply a better pedigree. In the past decade there has been much effort to increase genomic prediction accuracy through a better understanding of the genome. This era coincided with a great increase in whole genome sequencing resources being generated. The highly successful 1,000 bull genomes project (Hayes and Daetwyler 2019) is one example. Implementing sequence variants in the genetic evaluation has not increased selection accuracy considerably as demonstrated in dairy cattle by VanRaden *et al.* (2017) and in sheep by Li *et al.* (2021). The lack of papers purporting increases in selection accuracy following all the sequencing being done around the world at the recent (July 2022) World Congress on Genetics Applied to Livestock Production in Rotterdam (2022 WCGALP) was a testament to the disappointing progress in this area. The potential increase in prediction accuracy through models that more closely match the function of the genome still remains and should be pursued.

IMPACT OF GENOMIC SELECTION

Presented in Figure 2 are the genetic trends in Australian Angus, Friesian and Merino for a major respective economic index, each standardized to a genetic standard deviation. The increase in genetic progress in dairy is clearly evident in the graph, coinciding with the implementation of genomic selection in Australian Friesian in 2012 (Datagene 2022), and earlier for some other countries influential in dairy cattle genetics. In fact, a linear estimate of the trend 2001-2011 compared to 2012-2021 indicates the trend increase is over 4-fold in Friesian. The Merino trend is also increasing post-genomics with an increase of 1.57-fold pre and post genomics, which was first implemented in 2013 (blending Swan *et al.* 2011) with single step implemented in 2016 (Brown *et al.* 2018). The increase in trend in Angus is less dramatic with a 1.17-fold increase before and after the implementation of genomics in 2011 (blending) with single step implemented in 2017 (Johnston *et al.* 2018).

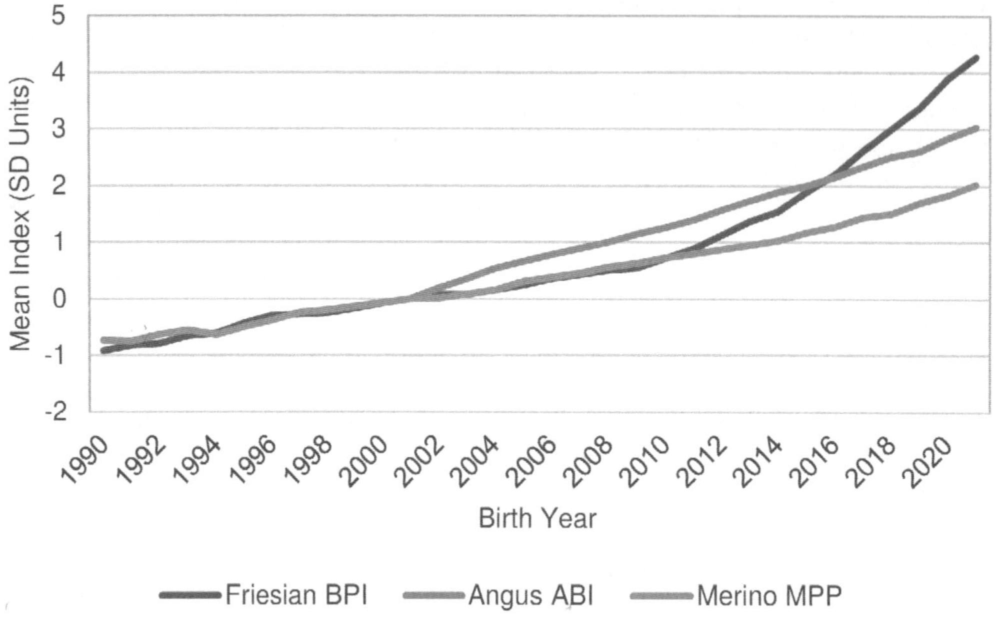

Figure 2. Standardized genetic trends for a prominent economic index in Australian Friesian, Angus and Merino populations
Source: Datagene, Angus Australia and Sheep Genetics (2023)

The dairy industry capitalized on genomics by greatly decreasing the generation interval on the male side, as sires no longer needed to be proven for lactation related traits through their daughters. The selection accuracy for females also greatly increased and was no longer plagued by preferential treatment of 'bull dams'. In contrast, the generation interval in Merinos and Angus could not be reduced to the same degree, as many of the economically important traits are measurable on the sire himself, promoting the use of young sires. In the case of Angus, carcass traits required progeny proving to some degree but ultrasound on the yearling bull was also available as a highly correlated predictor (Reverter and Johnston 2001), and similarly wool traits are measurable on young rams. Despite the already heavy use of young sires in Angus, Miller (2023) showed how the average age of sires is reducing in American Angus post the implementation of genomic selection.

Although the impact of genomic selection is starting to show in Merino and Angus, the results are far less dramatic than in dairy cattle. Despite biological differences such as generation interval and the levels of AI etc., it is reasonable to expect that there is much opportunity to further increase progress with the technology for both sheep and beef. One focus area could be the continual increase in selection accuracy that may be possible with further increases in the size of the reference population, which will accompany increases in genotyping, as long as breeders keep up the recording effort. One difference between dairy cattle breeding compared to sheep and beef cattle is that large breeding corporations have a much greater influence in dairy cattle breeding, compared to sheep and beef cattle. These large corporations run what is closer to a single-desk decision making process, compared to the thousands of decision makers in sheep and beef cattle. These same companies are able to hire specialised talent in the way of Ph.D. geneticists and implement the latest tools in selection. This rise in the corporate domination of breeding companies in dairy cattle globally was outlined by John Cole as part of the 2022 WCGALP plenary (no reference available). It was suggested that such companies will likely move to a more isolated model, with custom evaluations and reference populations and a potential withdrawal from industry wide evaluation schemes such as that provided by Datagene in Australia.

NEW TECHNOLOGIES AND A MORE CORPORATE FUTURE

Sheep and beef cattle breeders in Australia should be looking at the greater progress experienced in dairy cattle as an illustration of the potential threat to their business models. How would they compete with a large corporate breeder that is employing the tools available, such as genomic selection with custom reference populations and novel traits, and large-scale in-vitro embryo production programs with embryo genotyping? It is the suggestion of this paper that the reason why large companies have not entered this market is because of the lack of a technology, such as the deployment of artificial insemination in commercial farms, that is the major inhibitor. There is one technology on the horizon that has been in development for a number of decades and is described as a 'surrogate sire' in the review of reproductive technologies and their impact on genetic improvement by Mueller and Van Eenennaam (2022). This surrogate sire can be described as a walking artificial insemination delivery, where a natural service sire is breeding cows, but delivering the genetic material from an elite sire. This technology could provide the step change in technology needed for a significant entry of corporate investment into the largely untapped sheep and beef cattle genetic supply markets. Collectively this could be a very significant market for commercial genetics companies. There are about 10 times as many beef cattle as dairy cattle in Australia alone. The other advantage of this walking artificial insemination model would be the opportunity to disconnect the genetics of the walking bull from the genetics he is passing through his semen. This could be quite opportunistic for regions like northern Australia, where the sire will need to be adapted for the harsh tropical climate, but the resulting calves could be more suited to a feedlot system. A potential example could be a walking Brahman or tropical composite sire delivering elite Wagyu genetics.

Clearly the deployment of reproductive technologies can be transformational. A more recent technology that has had a large impact has been sexed semen as deployed in cattle breeding internationally. This has had recent significant ramifications for beef production in many countries, especially those with well-developed beef and dairy sectors. As outlined in Miller *et al.* (2021) sexed semen has created a significant increase in beef cross calves from the dairy herd, often referred to as 'beef on/from dairy'. Several factors have come together to facilitate this, among which is the availability of sexed semen, allowing dairy farmers to target dairy female replacements from the best cows in their herds and breed the remainder of the herd to beef sires to maximize their value. This trend was exacerbated by low global milk prices and a shrinking dairy herd, which decreased the demand for dairy replacements.

A step change in this 'beef on dairy market' could be possible with an improvement in embryo production. If a calf with half beef breed heritage is more profitable than a dairy calf, then a pure beef breed calf would be even more valuable. To accomplish this, bottom-end cows that are getting mated to beef semen could become pregnant with a pure beef embryo instead. At the moment, the cost of generating these embryo's and their decreased conception rate must make this proposition economically unattractive or it would have taken off. One pipeline that could be exploited would be the generation of IVF embryos from slaughter females. These could be culled beef cows, or slaughter heifers from feeding programs without drugs that prevent oestrus, which are purported to create difficulties in creating viable embryos. If commercial genotyping was widespread, the genetic merit of these females could be made available and linked to their mandatory national electronic ID in many countries. These best commercial females could be a source of the 'beef from dairy' animals, or even replacements in beef herds. The viability of replacements in beef herds would then depend on the specific herd genetic merit, the merit of the embryos and the increased costs per live replacement generated through embryos. The use of sexed semen could target females for replacements and males for dairy-beef as required. Another parallel application would be the production of more high value beef genetics from tropical environments. Similar to the surrogate sire scenario presented earlier, a well-designed beef embryo could lift the value of the calf generated from this system, with calves placed in easier finishing systems such as feedlots, while maintaining the indicus content required in the females.

Another step change in technology that could forever change animal breeding would be the realization of what is referred to as speed breeding. This was referred to as *In Vitro* Breeding (IVB) in the review by Mueller and Van Eenennaam (2022) and new breakthroughs to support this approach were recently reviewed by Goszczynski *et al.* (2019). The technique is called speed breeding as the generation cycle can be reduced to 3-4 months in cattle. Multiple embryos can be generated from elite parents and these embryos can be the start of multiple embryonic stem cell lines (ESC). The multiple ESC can be genotyped and through genomic selection, the best ESC can be selected. With viable gametes possible from ESC, the best ESC can be joined to create another generation of embryos, which will start another generation of selection. This technology, if implemented on a large scale, could create further opportunities for large corporate breeding companies, especially when combined with walking artificial insemination as previously described. Considering factors related to corporate investment in breeding as outlined in Table 1, walking artificial insemination allows genetics to be dispersed widely and speed breeding reduces the cost of maintaining the breeding female, as much of it will be done in the lab. Also, to undertake the breeding at a large scale will require investment in lab facilities. Currently in species such as sheep and beef the breeders with the land required to maintain the breeding herd dominate. Speed breeding could open this market to those with the capital to invest in lab facilities and is less tied to land ownership.

As genomics has shaped developments over the past two decades it is certain to continue to play a major role. It is making enabling technologies such as reproductive technologies more productive, which will also increase the corporatisation of breeding as outlined in Miller (2023). The cost of genomic sequencing continues to decline. Twenty years ago there was the push for the 1,000 dollar genome and this has been passed (NIH 2023) and the new horizon is a 100 dollar genome (Illumina 2023). With sequencing costs continuing to decline, genotyping by sequencing is poised to offer a low-cost genotyping alternative that is already being deployed (Snelling *et al.* 2020; McEwan *et al.* 2021). Such low-cost genotyping could also open up the market for widespread commercial genotyping. In beef cattle this could mean a genotype on every animal in key supply chains. A scoping study on the widespread use of genotyping in the Australian red meat industries for traceability purposes found that the biggest advantages to genotyping every animal would be the opportunities for supply chain efficiencies and better adoption of genetic improvement tools (Banks *et al.* 2022). Widespread commercial genotyping could change how reference populations are developed with a shift away from a seedstock focus to more dedicated commercial streams, that could be more private. This availability of private reference populations, with custom data collection streams, including novel traits, could also fuel a rise in more corporate breeding investment.

CONCLUSIONS

Animal breeding has gone through some transformative change in the past 50 years. From performance recording to BLUP and now genomic selection, advanced breeding programs today are making more progress than ever before. These improvements have not all come from the animal breeding community but in many cases development in other fields have been leveraged and successfully implemented. The development of large-scale performance recording schemes and genetic evaluation was made possible through the parallel developments in computing power. Advancements in reproductive technologies have played an important role in shaping industry structure including the rise of corporate ownership. Genomics was made possible through the development of low-cost, moderate density genotyping, following developments created for human genetic applications. Similarly, future opportunities are sure to leverage new technologies such as low-cost sequencing applications to reduce the cost of genotyping. Advancements in genetic engineering could make in-vitro breeding or the deployment of surrogate sires available on a commercial scale. These are technologies that could change the rate of genetic progress and also the structure of the industry, with a likely increase in corporate ownership. As new technologies continue to be deployed, new opportunities are created for more structured corporate ownership, which will continue to change the animal breeding industry as we know it.

ACKNOWLEDGEMENTS

The author is grateful to the AAABG conference organizers for the invitation to prepare this invited plenary talk and associated paper. Assistance from Alex MacDonald (ARCBA), Christian Duff and Malshani Samaraweera (Angus Australia), Peter Thurn (Datagene) and Andrew Swan (AGBU: Sheep Genetics Merino trends) is also appreciated.

REFERENCES

Aguilar I., Misztal I., Johnson D.L., Legarra A., Tsuruta S. and Lawlor T.J. (2010) *J. Dairy Sci.* 93: 743.

American Angus (2022). accessed 17 Feb 2023 https://www.angusjournal.com/salebooks/Annual_Report_2022/.

Banks R.B., Miller S.P., van der Werf JHJ., Ball A.J. and Woolaston R.R. (2022) accessed 17 Feb 2023 https://www.mla.com.au/contentassets/7ac2de474ff047eb9ff4a87e63fa057d/v.rda.2222-final-report-fa073.pdf.

Bovine genome sequencing and analysis consortium (2009) *Sci.* 324: 522.

Bovine HapMap consortium (2009) *Sci.* 324: 528.

Brown D.J., Huisman, A.E., Swan A.A., Graser H.U., Woolaston R.R., Ball A.J., Atkins K.D. and Banks R.G. (2007) *Proc. Assoc. Advmt. Anim. Breed. Genet.* 17: 187.

Brown D.J., Swan A.A., Boerner V., Li L., Gurman P.M., McMillan A.J., van der Werf J.H.J., Chandler H.R. and Tier B., Banks R.G. (2018) *Proc. World Congr. Genet. Appl. Livest. Prod., Auckland, New Zealand.* 11: 460.

Browning B.L. and Browning S.R. (2016) *Am J Hum Genet* 98: 116.

Casas E., Wheeler T.L., Shackelford S.D. *et al.* (2009). *J. Anim. Sci.* 87(E. Suppl.2): 533.

Datagene (2022) accessed 17 Feb 2023 https://datagene.com.au/sites/default/files/inline-files/Datagene%20Annual%20Update%202022%20Final.pdf

Fernando R.L. and Grossman M. (1989). *Genet. Sel. Evol.* 21: 467.

Fleming A. and Van Doormaal B. (2022) Available at https://lactanet.ca/en/impact-genomics/

Goszczynski D.E., Cheng H., Demyda-Peyrás S., Medrano J.F. and Wu J., Ross P.J. (2019) *Biol. of Reprod.* 100: 885.

Graser H.U. (1982) *Proc. Assoc. Advmt. Anim. Breed. Genet.* 3: 67.

Graser H.U. and Hammond K. (1985) *Aust. J. Agric. Res.* 36: 527.

Graser H.U., Tier B., Nicol D.C., Scarth R.D. and Hammond K. (1987) *Proc. Assoc. Advmt. Anim. Breed. Genet.* 6: 124.

Harris B.L. and Johnson D.L. (2010) *J. Dairy Sci.* 93: 1243.

Hayes B.J. and Daetwyler H.D. (2019) *Annu. Rev. Anim. Biosci.* 7: 89.

Hendersen C.R. (1973) *Proc. Animal Breeding and Genetics Symposium in Honour of Dr. Jay Lush. ASAS and ADSA, Champaign.* p.10.

Illumina (2023) accessed 17 Feb 2023. https://sapac.illumina.com/company/news-center/press-releases/2017/2236383.html

Li L., Gurman P.M., Swan A.A., Moghaddar N. and van der Werf J.H.J. (2021) *Proc. Assoc. Advmt. Anim. Breed. Genet*. 24: 200.
Johnston D.J., Tier B. and Graser H.U. (2009). *Proc Assoc. Advmt. Anim. Breed. Genet*. 18: 30.
Johnston D.J. and Graser H.U. (2010) *J. Anim. Sci*. 88: 1917.
Johnston D.J., Grant T.P., Schatz T.J., Burns B.M., Fordyce G. and Lyons R.E. (2017). *Proc Assoc. Advmt. Anim. Breed. Genet*. 22: 393.
Johnston D.J., Ferdosi M., Connors N., Boerner V., Cook J., Girard C., Swan A. and Tier B. (2018) *Proc. World Congr. Genet. Appl. Livest. Prod., Auckland, New Zealand*. 19: 269.
Kachman S.D. (2008). *Proc. 9th genetic prediction workshop of the beef improvement federation*
Kuehn, L. and Thallman T. (2022) accessed 17 Feb 2023 https://beefimprovement.org/wp-content/uploads/2022/02/21_ABEPD_pressreleaseandfactsheet.pdf
Legarra A., Christensen O.F., Aguilar I and Misztal I. (2014) *Livest. Sci*. 166: 54.
McMillan A.A., Walkom S.F. and Brown D.J. (2023). *Proc Assoc. Advmt. Anim. Breed. Genet*. 25: *These proceedings*.
Matukumalli L.K., Lawley C.T., Schnabel R.D., Taylor J.F., Allan M.F., Heaton M.P., O'Connell J, Moore S.S., Smith T.P.L., Sonstegard T.S. and Van Tassell C.P. (2009) *PloS One* 4: e5350.
McEwan J.C., Dodds K.G., Bilton T.P., Brauning R., Rowe S.J., Hess M.K., McRae K.M., McCulloch A.F., Anderson R.A. and S.M. Clarke (2021) *Proc. Assoc. Advmt. Anim. Breed. Genet*. 24: 14.
Meuwissen T.H.E., Hayes B.J. and Goddard M.E. (2001) *Genetics* 157: 1819.
Miller, S.P., Wang L., Retallick K.J. and Moser D.W. (2018) *Proc. World Congr. Genet. Appl. Livest. Prod., Auckland, New Zealand.*
Miller S.P., Archer J.A., Hely F., Quinton C., Retallick, K., Moser D. and Amer, P. (2021). *Proc. Assoc. Advmt. Anim. Breed. Genetic* 24: 234.
Miller (2022) *Reproduction, Fertility and Development* 35: 98.
Misztal I., Legarra A. and Aguilar I. (2009) *J. Dairy Sci*. 92: 4648.
Mueller M.L. and Van Eenennaam A.L. (2022) *CABI Agric. and Biosci*. 3: 13.
Nejati-Javaremi A., Smith C. and Gibson J.P. (1997) *J. Anim Sci*. 75: 1738.
NIH (2023) accessed 17 Feb 2023 https://www.genome.gov/about-genomics/fact-sheets/DNA-Sequencing-Costs-Data
Page B.T., Casas E., Heaton M.P. *et al.* (2002). *J. Anim Sci*. 80: 3077.
Reverter T. and Johnston D.J. (2001) *Proc. Assoc. Advmt. Anim. Breed. Genet*. 14: 155.
Sargolzaei M., Chesnais J.P., Schenkel F.S. (2014) *BMC Genomics* 15: 478.
Schenkel, F.S., Miller S.P., Jiang Z. *et al.* (2006). *J. Anim Sci*. 84: 291.
Scott B.A., Haile-Mariam M., Cocks B.G. and Pryce J.E. (2021) *J. Dairy Sci*. 104: 11832.
Schenkel F.S., Miller S.P., Ye X., Moore S.S., Nkrumah J.D., Li C., Yu J., Mandell I.B., Wilton J.W. and Williams J.L. (2005) *J. Anim. Sci*. 83: 2009.
Schaeffer L.R. (2006) *J. Anim. Breed. Genet*. 123: 218.
Snelling W.M., Hoff J.L., Li J.H., Kuehn L.A., Keel B.N., Lindholm-Perry A.K. and Pickrell J.K. (2020) *Genes* 11: 1312.
Swan, A.A., Johnston D.J., Brown D.J., Tier B. and Graser H.U. (2011) *Anim. Prod. Sci*. 52: 126.
Van Eenennaam A.L., Li J., Thallman R.M. *et al.* (2007). *J. Anim Sci*. 85: 891.
VanRaden P.M., Van Tassell C.P., Wiggans G.R., Sonstegard T.S., Schnabel R.D., Taylor J.F. and Schenkel F.S. (2017) *J. Dairy. Sci*. 92: 16.
VanRaden P., Tooker M., O'Connell J., Cole J. and Bickhart D. (2017) *Genet. Sel. Evol*. 49: 32.
Walmsley B.J., Donoghue K.A., Johnston D.J., Clark S.A., Siddell J.P., Walkom S.F., Granleese T. and Arthur P.F. (2021) *Proc. Assoc. Advmt. Anim. Breed. Genet*. 24: 423.

Proc. Assoc. Advmt. Anim. Breed. Genet. 25:23 - 26

AN OPTIMAL CONTRIBUTION SELECTION TECHNIQUE THAT UTILISES NON-ADDITIVE GENETIC COMPONENTS

Z. Loh, J. H. J. van der Werf, S. Clark

School of Environmental & Rural Science, University of New England, Armidale, NSW, 2351 Australia

SUMMARY

This study proposes an optimal contribution selection method (OCS) that utilizes both additive and non-additive genetic components. Using a genetic algorithm, the contribution of sires toward a cohort of dams, along with their mate allocation, were optimized under a constraint of a 1% increment of inbreeding per generation. The inclusion of dominance into the OCS increases the total genetic gain in offspring initially by 30.5% improvement from +4.02 to +5.27 units compared to using additive genetic component alone for a trait with a 15% dominance-to-additive variance ratio. Compared to additive-only OCS, optimization of the dominance component resulted in one-off additional gains, with no additional merit thereafter, despite continued optimization. In conclusion, this inclusion of dominance in mate allocation can give a significant genetic lift in total genetic merit.

INTRODUCTION

While optimal contribution selection (OCS) has successfully optimized the additive genetic gain in livestock breeding systems within a constraint of inbreeding, it has only focused on estimated breeding values (EBVs) and generally not focused on optimizing the non-additive genetic component, such as dominance. Dominance could explain a significant proportion of the genetic variance for some traits, but it has been difficult to exploit due to its dependency on sire-dam mating configuration and the difficulty of predicting these specific effects. The advent of genomic information, however, allows direct prediction of the expected offspring heterozygosity, which could be used to predict dominance effects for mate allocation.

The aim of this study was to develop an OCS that could optimize both additive and non-additive genetic components, using information easily available to a breeding program. It is anticipated this OCS can be used in improving both additive and non-additive effects in a trait.

LAYOUT OF THE OPTIMAL CONTRIBUTION SELECTION METHOD

The OCS requires several inputs: sire and dam genotype arrays of size $N_m \times M$ and $N_f \times M$ respectively, with N_m, N_f and M be number of sires, dams and markers respectively; sire and dam phenotypic vector of length N_m and N_f respectively, and narrow sense heritability. The genotype, phenotype and heritability were used to calculate the sire EBVs ($\widehat{\boldsymbol{\beta}}_{\boldsymbol{m}}$) and sire GRM ($\boldsymbol{G}$) using method by VanRaden (2008). A targeted level of increment of consanguinity (ΔF_t) were also needed for the OCS.

This OCS has three phases: the first phase optimized additive and inbreeding components; the second phase optimized the non-additive genetic components only, and the final phase combined the results from both phases. Such a design was needed to improve the feasibility of the method from the significantly increased sample space when optimizing the dominance genetic components.

To initialize the GA, 1500 candidate solutions of length N_f, denoted as $\boldsymbol{s}$, that contain the indices of sires that paired with each dam, were generated, with the i-th entry of $\boldsymbol{s}$ contains which sire that would be paired with i-th dam. This formatting was required due to the mate-specific nature of the dominance component, which depends on the exact permutations of the sires. This set of $\boldsymbol{s}$ vectors were compiled into a sire index matrix of size $1500 \times N_f$, denoted as $\boldsymbol{S_1}$.

The first phase of this OCS optimized the additive and inbreeding coefficients, which were initialized by translating $\boldsymbol{S_1}$ into its corresponding sire proportion matrix $\boldsymbol{X_1}$, defined as a matrix of size $1500 \times N_m$ with its i-th row and j-column representing the proportion of j-th sire that would contribute into the next generation for the i-th solution. The objective function for this phase was defined as follows:

$$f_{obj}(\boldsymbol{X_1})_{AI} = \boldsymbol{X_1}\widehat{\boldsymbol{\beta}}_{\boldsymbol{m}}{}' - \lambda_i * diag(\boldsymbol{X_1 A X_1}') \quad [1]$$

where λ_i denoted the scalar weightage for the inbreeding component for this phase of OCS.

From this objective function, the top two $\boldsymbol{s}$s in term of $f_{obj}(\boldsymbol{X_1})_{AI}$ were chosen, which were propagated into a new $\boldsymbol{S_1}$. This new $\boldsymbol{S_1}$ was subjected to five genetic operators: mutation, where sires in $\boldsymbol{S_1}$ were replaced with new sires; vertical and horizontal recombination, where the part of $\boldsymbol{S_1}$ were exchanged, column-wise and row-wise, respectively, and vertical and horizontal inversions, where the orders of sires in $\boldsymbol{S_1}$ were reversed, column-wise and row-wise respectively. The hyper-parameters values for these operators were based on Srinivas and Patnaik (1994).

This phase was then iterated with the new $\boldsymbol{S_1}$. For each iteration, the λ_i was adjusted with the amount $100\left(average\left(diag(\boldsymbol{X_1 A X_1}')\right) - \Delta F_t\right)$. The mutation, recombination and inversion rate were also adjusted adaptively based on the method by Srinivas and Patnaik (1994). This process continued until convergence, defined as the point where the slope of the curve of $f_{obj}(\boldsymbol{X_1})_{AI}$ is less than 1×10^{-3} across the last 50 iterations. To reduce the chance of premature convergence for subsequent phases, this phase was repeated eight times, with the converged solutions from each repeat pooled into a new sire index array, $\boldsymbol{S_2}$. From each repeat, the average of the λ_i at the point of convergence, denoted by λ_{avg}, was also recorded.

The $\boldsymbol{S_2}$ was then used for Phase 2 optimization, which maximizes the offspring dominance component. From $\boldsymbol{S_2}$, 3000 solutions were resampled and altered using genetic operators. Only vertical recombination and horizontal inversion were used on $\boldsymbol{S_2}$, as they only affect the permutations of the sires within the $\boldsymbol{s}$s, thus with no effects on its additive and inbreeding scores, thus not affecting their Phase 1 optimality. The performance of each solution in $\boldsymbol{S_2}$ was tested, with the objective function for k-th solution defined as follows:

$$f_{obj}(\boldsymbol{S_2})_D = \sum_{i=1}^{N_f} \boldsymbol{H}_{S_2(k,i),i} \quad [2]$$

where $\boldsymbol{H}_{S_2(k,i),i}$ is defined as the expected heterozygosity for $\boldsymbol{S_2}(k,i)$-th sire and i-th dam, which $\boldsymbol{S_2}(k,i)$ is the k-th row and i-th column of $\boldsymbol{S_2}$. The top two $\boldsymbol{s}$s in terms of $f_{obj}(\boldsymbol{S_2})_D$ were extracted from $\boldsymbol{S_2}$ and used to generate a new $\boldsymbol{S_2}$ array, subjected to vertical recombination and horizontal inversion. This phase was iterated until convergence, defined as the point where the slope of the curve of $f_{obj}(\boldsymbol{S_2})_D$ is less than 1×10^{-4} across the last 200 iterations. To increase the chance of finding the global maximum, Phase 2 was repeated eight times, and the solutions pooled into $\boldsymbol{S_3}$.

In the final phase, the $\boldsymbol{S_3}$ was translated into its corresponding sire proportion array $\boldsymbol{X_3}$. The performance of each solutions was evaluated as follows:

$$f_{obj}(\boldsymbol{S_3}, \boldsymbol{X_3})_{ADI} = \boldsymbol{X_3}\widehat{\boldsymbol{\beta}}_{\boldsymbol{m}}{}' + \sum_{i=1}^{N_f} \boldsymbol{H}_{S_3(k,i),i} - \lambda_{avg} * \left(average\left(diag(\boldsymbol{X_3 A X_3'})\right) - \Delta I_t\right) \quad [3]$$

Equation [3] served as the final objective function for the OCS. The top $\boldsymbol{s}$ in terms of $f_{obj}(\boldsymbol{S_3}, \boldsymbol{X_3})_{ADI}$ were deemed as the optimized solution, and were the final output of the OCS.

TESTING THE OPTIMAL CONTRIBUTION SELECTION METHOD

The OCS was tested with simulated genotypic arrays generated using QMSim (Sargolzaei and Schenkel 2009). For the ancestral population, 5,000 animals and 20,000 loci across 10 chromosomes of 100 cM each were simulated. This population was gene-dropped for 1,000 generations, with the population size increasing up to 10,000 in the final generation. Either 500 or 1000 sires and dams were then randomly chosen for genotyping and these were selection candidates (Table 1).

From all loci, 500 of them were assigned as QTL, with both additive and dominance effects. Using these effect sizes, the phenotypes were calculated as follows:

$$\boldsymbol{y} = \boldsymbol{Z_a\beta} + \boldsymbol{Z_h\delta} + \boldsymbol{e} \quad [4]$$

where $\boldsymbol{y}$ is the phenotype vector; $\boldsymbol{Z_a}$ is the additive genotypic array encoded in the format of $\{0,1,2\}$; $\boldsymbol{Z_h}$ is the heterozygosity array with a value of 1 for heterozygotes and 0 otherwise; $\boldsymbol{\beta}$ and $\boldsymbol{\delta}$ are vectors with additive and dominance effect sizes for each QTL, respectively, and $\boldsymbol{e}$ is a vector with the residual component of the phenotypes. Both $\boldsymbol{\beta}$ and $\boldsymbol{\delta}$ were generated using a gamma distribution, with shape parameters set at 0.3 and scale parameters provided in Table 1. The vector $\boldsymbol{e}$ was generated using a normal distribution, with mean zero and variance $\frac{(1-h^2)var(G\beta)}{h^2}$, where h^2 is the narrow sense heritability. The h^2 was set at 0.3 for all simulations.

These genotypes and phenotypes were used in a four-generation selection program. Three selection regimes were tested: truncation genomic selection (denoted as TS), OCS with additive component (OCSA); and OCS with both additive and dominance components (OCSAD). The ΔF_t is set at 1% per generation for OCSA and OCSAD. To ensure validity of comparison for TS, the proportion of sires selected was determined by the number of selected top sires that would produce the same ΔF_t. A non-selected population (NSEL) was used to establish the offspring baseline performance. For each generation, the additive, dominance and total genetic merits (TGM) from each selection regime were recorded.

The parameters and values tested in this study were provided in Table 1. When a parameter was under study, default values were used for other parameters. When neither the additive and dominance genetic variances were under study, the default scale parameters of the effect size distributions were chosen such that the dominance genetic variance is 15% of the additive genetic variance. For each set of parameter values and selection regimes, 20 replicates were conducted. To test the performance between selection regimes, a two-sample Welch's t-test was used, with the performance deemed significantly different if the $logpval = -\log_{10}(p - value) \geq 3$.

Table 1. Parameters and values tested in this study

Parameters	Default values	Alternative values
Number of Sires and Dams	500	1000
Additive Effect Size Scale Parameter	1.0	3.0
Dominance Effect Size Scale Parameter	0.5	1.5

RESULTS

The additive, dominance and TGM across four generations for the different selection regimes were provided in Figure 1. The first-generation total genetic merit under different parameter values and selection regimes were provided in Table 2.

Compared to TS, both OCS methods significantly improved the additive genetic component of the offspring across all parameter values tested. The OCSAD method significantly improved the dominance component compared to OCSA from +0.17 to +1.51 ($logpval = 22.54$), and this led to a 30.5% additional improvement in TGM from +4.02 to +5.27 ($logpval = 8.48$) in the first generation of selection under the default parameter values. The additional gain from the dominance component is a one-off genetic lift, however, with no further additional increments in dominance genetic merit despite its continued optimization in the subsequent generations (Figure 1b).

The improvement in TGM in OCSAD compared to OCSA was observed for all parameter values tested, although these parameters affect the significance of improvement. For example, by increasing the scale parameter for additive QTL effect sizes from 1.0 to 3.0, which increases the additive genetic variance, the increment in TGM becomes less significant ($logpval = 1.59$). While this change of parameter value has decreased the dominance-to-additive variance ratio to 2.1%, the TGM for OCSAD is still 11.4% higher than OCSA after the first generation of selection, indicating the potential merit of mate allocation in exploiting dominance variation.

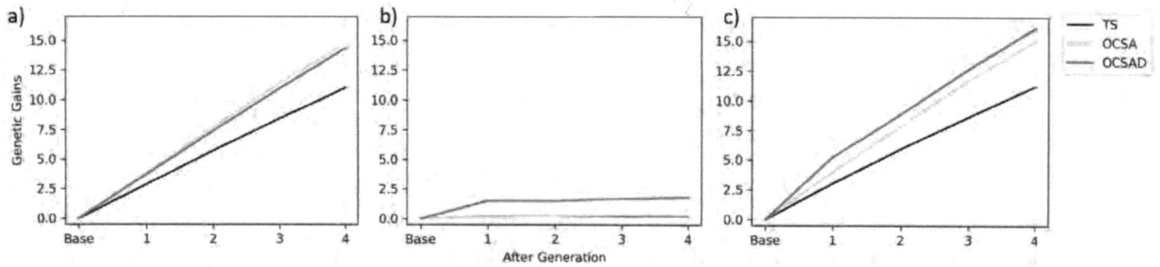

Figure 1. Plots for the base scenario showing (a) additive, (b) dominance and (c) total genetic merit of the offspring under truncation selection (TS), additive-inbreeding OCS (OCSA) and additive-dominance-inbreeding OCS (OCSAD) across four generations

Table 2. First generation total genetic merit with truncation selection (TS), additive-inbreeding OCS (OCSA) and additive-dominance-inbreeding OCS (OCSAD) under varying parameter values and selection regimes. Superscripts with different letters (row wise) denote significant differences between selection regimes

Parameter values	Value tested	Total genetic merit		
		TS	OCSA	OCSAD
Number of sires and dams (default)	500	3.045[a]	4.019[b]	5.247[c]
(alternative)	1000	4.010[a]	4.527[a]	5.827[b]
Additive effect size scale parameter	3.0	8.845[a]	11.328[b]	12.616[b]
Dominance effect size scale parameter	1.5	3.052[a]	3.953[b]	7.855[c]

DISCUSSION AND CONCLUSION

In this study, an OCS method that optimized the additive and dominance component was proposed. Using heterozygosity for all loci as a proxy for the optimization of dominance, with a 15% dominance-to-additive variance ratio, this method improved the initial TGM by 30.5% compared to only optimizing the additive component. The one-off lift from the dominance component optimization means that after the first generation both OCS would have the same rate of genetic gain despite the continued optimization of dominance. Some computational aspects of the proposed method could be further optimised.

In conclusion, an OCS that optimizes additive and dominance effects was proposed in this study, and gave a significant lift in total genetic merit of a selected trait. The method can be used to improve the within-population genetic merit for economically important traits in livestock.

REFERENCES

Sargolzaei M and Schenkel F.S. (2009). *Bioinformatics*. 25: 680.
Srinivas M. and Patnaik L.M. (1994). *IEEE Sys. Man. Cybern.* 32: 115.
VanRaden P.M. (2008). *J. Dairy Sci.* 75: 3136.

EWE GENOTYPE EFFECTS IN GENETIC EVALUATION OF MERINO FLEECE TRAITS ACROSS AGES

S.I. Mortimer[1], K.L. Egerton-Warburton[2], T.L. Bird-Gardiner[3] and A.A. Swan[4]

[1]NSW Department of Primary Industries, Armidale, NSW, 2351 Australia
[2]NSW Department of Primary Industries, Orange, NSW, 2800 Australia
[3]NSW Department of Primary Industries, Trangie, NSW, 2823 Australia
[4]Animal Genetics and Breeding Unit*, University of New England, NSW, 2351 Australia

SUMMARY

The significance of ewe genotype and its interactions with sire effects were examined for Merino fleece traits recorded between post weaning and adult stages of progeny at the Macquarie site of the Merino Lifetime Productivity project. Across post weaning, hogget and 3 adult expressions, ewe bloodline effects significantly influenced fleece traits, particularly fleece weight and fibre diameter. Sire X ewe genotype interactions on fleece traits across ages were generally unimportant, accounting for small amounts of the phenotypic variance (less than 1%). Correlations between sire progeny performances were generally greater than 0.70. These results support the methods used routinely to account for these effects in MERINOSELECT genetic evaluations.

INTRODUCTION

Atkins *et al.* (1999) concluded that for across-flock genetic evaluation of Merino sires, though sire X dam source interactions were small, the influence of heterosis on some Merino traits may also need to be included in evaluation models. This has been addressed by the fitting of a sire X flock-year interaction in the model; its component of a genotype X genotype interaction allows the MERINOSELECT genetic evaluation system to account for potential sire X ewe genotype interactions (Li *et al.* 2015).

In the context of central test sire evaluation and on-farm progeny testing, it is usual to have the sires mated to an even line of ewes selected to meet a breeding objective that differs from that of the majority of the sires. Anecdotally, some Merino breeders have concerns that these genetic benchmarking systems will be biased by heterosis as sires are mated to ewes of a different genetic background, with these concerns greater for assessments at later ages. Several recent studies have shown that these sire X ewe bloodline interactions are unimportant for measured fleece traits (Egerton-Warburton *et al.* 2019), visually assessed wool traits (Mortimer *et al.* 2021b) and body composition and reproduction traits (Mortimer *et al.* 2021a) recorded at post weaning, hogget and/or a first adult shearing. At most, the interaction effect in those studies accounted for less than 2% of the phenotypic variance. Using data recorded on progeny at the Macquarie site of the Merino Lifetime Productivity (MLP) project, this study extends the findings of Egerton-Warburton *et al.* (2019) by examining the influence of sire X ewe bloodline interaction on fleece traits recorded at hogget and 3 adult shearings.

MATERIALS AND METHODS

Data were available from the progeny born in 2017 and 2018 at the Macquarie MLP site. The site's establishment at the Trangie Agricultural Research Centre has been outlined by Egerton-Warburton *et al.* (2019), while the overall design of the MLP project has been described by Ramsay *et al.* (2019). Briefly, each drop was generated by AI matings of industry sires (30 sires in total, including a link sire) to foundation ewes, which had previously lambed, of 2 bloodlines sourced

* A joint venture of NSW Department of Primary Industries and the University of New England

from representative commercial flocks. Bloodline 1 (B1) was selected to meet a dual purpose objective, where selection aimed to improve wool, fertility and growth traits. The breeding objective of bloodline 2 (B2) was set to increase wool production and body size. The allocation of ewes to sires was balanced across the ewe flock sources. The progeny were assessed at post weaning (average age of 280.6 days), hogget (average age of 533.7 days), first adult (average age of 899.3 days), second adult (average age of 1,273.7 days) and third adult (average age of 1,659.2 days) shearings. Animals were shorn for assessment of greasy fleece weight (gfw, kg), clean fleece weight (cfw, kg), mean fibre diameter (fd, μm), coefficient of variation of fd (fdcv, %), fibre curvature (curv, degrees/mm), staple length (sl, mm) and staple strength (ss, N/ktex). Birth type and rearing type of the progeny were inferred from parentage determination based on DNA samples and the dam's pregnancy scan results. Fleece traits recorded on the ewe progeny have been used for this study, except for the post weaning assessment where the records were available for both drops on wethers only. From the post weaning assessment, records were available from 529 (52% of records) and 495 animals for B1 and B2 respectively, with these proportions maintained across the later assessments. Mean performances for each trait of each bloodline are shown in Table 1. Coefficients of variation were similar for each combination of bloodline and trait within an assessment (results not shown).

Table 1. Means (standard deviations) for fleece traits of bloodline 1 (B1) and bloodline 2 (B2) for post weaning to third adult assessments of fleece traits

		Post weaning	Hogget	First adult	Second adult	Third adult
gfw	B1	3.6 (0.72)	5.1 (0.72)	7.5 (1.30)	6.8 (1.12)	6.8 (1.17)
	B2	3.5 (0.69)	5.1 (0.68)	7.7 (1.29)	7.4 (1.11)	7.4 (1.19)
cfw	B1	2.4 (0.54)	2.8 (0.49)	4.5 (0.89)	4.7 (0.94)	4.8 (0.98)
	B2	2.4 (0.53)	2.9 (0.54)	4.9 (0.84)	5.4 (0.92)	5.6 (1.01)
fd	B1	16.5 (1.12)	18.0 (1.33)	18.3 (1.29)	18.9 (1.44)	19.5 (1.49)
	B2	17.5 (1.34)	19.0 (1.45)	19.5 (1.40)	20.5 (1.59)	21.2 (1.70)
fdcv	B1	18.3 (2.20)	15.8 (2.35)	16.5 (2.36)	15.8 (2.00)	15.3 (1.83)
	B2	19.2 (2.26)	17.2 (2.60)	17.8 (2.55)	16.9 (2.22)	16.0 (1.98)
curv	B1	60.1 (10.02)	63.1 (10.80)	61.6 (11.4)	62.5 (12.29)	62.9 (12.68)
	B2	60.4 (8.85)	63.1 (10.62)	61.5 (10.52)	60.5 (11.63)	60.0 (12.21)
sl	B1	77.3 (9.77)	81.3 (10.02)	115.3 (11.28)	115.8 (10.41)	108.6 (10.36)
	B2	77.1 (8.79)	79.4 (9.90)	112.6 (10.51)	114.7 (9.76)	108.6 (9.68)
ss	B1	26.9 (8.03)	44.3 (9.29)	28.7 (9.41)	34.9 (11.13)	38.5 (12.47)
	B2	26.6 (9.58)	48.3 (10.11)	30.6 (9.51)	37.2 (10.92)	40.0 (12.74)

Analyses were performed using ASReml (Gilmour et al. 2021). Initially, the significance of ewe bloodline fitted as a fixed effect was tested in univariate models that fitted other fixed effects and a random effect of sire. Those fixed effects included birth type (single, twin, triplet), rearing type (single, twin), dam age (3, 4, 5, 6 and 7 years old at mating), current reproduction (adult traits only) and contemporary group. Fixed effects were excluded from the model when not significant. Then, to this model a sire X ewe interaction was added to test if it increased significantly ($P < 0.05$) the log-likelihoods between models. Treating the performances of the progeny of each ewe bloodline as individual traits, a multivariate approach also was used to evaluate the correlation between predicted sire effects as a measure of the genetic correlation between performance in each bloodline.

RESULTS AND DISCUSSION

In general, the ewe bloodline effect was significant ($P < 0.001$) for the fleece weights and mean and variability of fd (Table 2). Bloodline 1 cut less clean wool than B2 at hogget (0.45 kg) and adult shearings (0.9 to 1.1 kg) and grew finer fleeces at all shearings (from 0.5 µm to 1.9 µm at post weaning and third adult shearings respectively) which were of more uniform diameter. The results were consistent with the breeding objectives described for the bloodlines and agreed with the findings for post weaning fibre diameter traits reported earlier from the Macquarie data (Egerton-Warburton *et al.* 2019). The ewe bloodline effect also influenced curv at later shearings, sl at hogget and first adult shearing and ss at the hogget and the first and second adult shearings.

In agreement with Egerton-Warburton *et al.* (2019), the sire X ewe bloodline interaction was significant for cfw at the post weaning shearing, accounting for 4.6% of the phenotypic variation (Table 2), versus 1.8% in the earlier study. Previously, Mortimer and Casey (2015) had found the interaction to be not significant for both clean and greasy fleece weights, accounting for negligible amounts of phenotypic variance. Otherwise, the interaction was unimportant for fleece traits recorded at later stages, where it tended to account for less than 1% of the phenotypic variation, particularly for traits recorded at hogget and later adult shearings. For yearling Merino fleece traits recorded on the Information Nucleus flock, sire by site interaction effects have been shown to be at most moderate, accounting for 6% of the phenotypic variation in cfw (Swan *et al.* 2016).

Table 2. Significant ewe bloodline effects[1], and their estimates (deviation from bloodline 2), for post weaning to third adult fleece traits and percentage variation accounted for by sire X ewe bloodline interaction

	Post weaning	Hogget	First adult	Second adult	Third adult
Ewe bloodline effect					
gfw	0.02 (0.13)***	ns	-0.83 (0.21)***	-0.92 (0.21)***	-1.07 (0.22)***
cfw	ns	-0.45 (0.09)***	-0.87 (0.16)***	-0.89 (0.17)***	-1.07 (0.18)***
fd	-0.48 (0.24)***	-1.56 (0.25)***	-1.46 (0.26)***	-1.82 (0.29)***	-1.89 (0.32)***
fdcv	-0.21 (0.46)***	-1.81 (0.48)***	-1.14 (0.47)***	-1.67 (0.42)***	-1.32 (0.40)***
curv	ns	ns	ns	2.29 (2.25)**	5.09 (2.42)***
sl	ns	2.10 (1.43)***	0.84 (2.07)***	ns	ns
ss	ns	-3.35 (1.79)***	-6.49 (1.95)**	-3.85 (2.17)**	ns
Sire X ewe bloodline (%)					
gfw	3.6	0	1.7	0.7	0
cfw	4.2*	0.5	2.6	0.7	0
fd	1.7	0	0	0.7	0.5
fdcv	0	0	2.0	0	0
curv	0	0	0	0	0
sl	2.7	0	0	0	0
ss	0.7	0	2.5	0	0.9

[1] ns, not significant; *, $P < 0.05$; **, $P < 0.01$; ***, $P < 0.001$.

Correlations between predicted sire progeny means across the stages for each bloodline for both greasy and clean fleece weights were generally around 0.80 (Table 3). While the sire X ewe bloodline interaction was not significant for post weaning gfw, the correlation was 0.36. At this assessment, progeny of B1 ewes produced more greasy wool than progeny of B2 ewes in contrast to the trend observed across later assessments. Most sire progeny means for post weaning gfw of B1 and B2 were not significantly different from each other, but significantly different means were detected where the sire progeny means were at the higher end of the range for B1. Reasonably high correlations, usually greater than 0.70, were also estimated for mean and variability of fd across assessments. For fd, there tended to be at least half the sire means of B1and B2 detected to be significantly different from each other.

Table 3. Correlations between predicted sire progeny means for post weaning to third adult fleece traits in 2 different ewe bloodlines

	Post weaning	Hogget	First adult	Second adult	Third adult
gfw	0.36	0.77	0.90	0.80	0.78
cfw	0.79	0.88	0.79	0.81	0.79
fd	0.77	0.90	0.86	0.79	0.71
fdcv	0.75	0.85	0.82	0.78	0.67
curv	0.72	0.87	0.82	0.90	0.86
sl	0.54	0.97	0.87	0.77	0.85
ss	0.96	0.91	0.35	0.93	0.61

CONCLUSION

Although ewe bloodline effects influenced the fleece traits across stages, sire X ewe genotype interactions were generally unimportant and accounted for minor amounts of phenotypic variation in Merino fleece traits. Rankings of sires for fleece traits would be reasonably consistent for evaluations conducted across different ewe genotypes. This study supports the methods adopted by MERINOSELECT genetic evaluations that routinely fit this interaction to account for ewe bloodline source effects (Li *et al.* 2015).

ACKNOWLEDGEMENTS

The Merino Lifetime Productivity Project is being undertaken in partnership between the Australian Merino Sire Evaluation Association Incorporated (AMSEA) and Australian Wool Innovation (AWI). AMSEA and AWI acknowledge those entities who also contribute funding: woolgrowers through sire evaluation entry fees, site hosts, site committee in-kind contributions, and sponsors of AMSEA. The MLP Macquarie site gratefully acknowledge the generous financial support provided by the site host NSW Department of Primary Industries. Staff and Site Committee volunteers are also thanked for their contributions to data collection. AWI's funding of the wether data through Project ON-00536 is acknowledged.

REFERENCES

Atkins K.D., Mortimer S.I. and Casey A.E. (1999) *Proc. Assoc. Advmt. Anim. Breed. Genet.* 13: 58.

Egerton-Warburton K.L., Mortimer S.I. and Swan A.A. (2019) *Proc. Assoc. Advmt. Anim. Breed. Genet.* 23: 520.

Gilmour A.R., Gogel B.J., Cullis B.R., Welham S.J. and Thompson, R. (2021) 'ASReml User Guide Release 4.1 Functional Specification'. VSN International Ltd, Hemel Hempstead, UK.

Li L., Swan A.A., Brown D.J. and van der Werf J.H.J. (2015) *Proc. Assoc. Advmt. Anim. Breed. Genet.* 21: 386.

Mortimer S.I. and Casey A.E. (2015) *Proc. Assoc. Advmt. Anim. Breed. Genet.* 21: 322.

Mortimer S.I. Egerton-Warburton K.L. and Swan A.A. (2021a) *Anim. Prod. in Aust.* 33: lxxix.

Mortimer S.I., Egerton-Warburton K.L. and Swan A.A. (2021b) *Anim. Prod. in Aust.* 33: lxxx.

Swan A.A., Brown D.J. and van der Werf J.H.J. (2016) *Anim. Prod.Sci.* 56: 87.

Ramsay A.M.M., Swan A.A. and Swain B.C. (2019) *Proc. Assoc. Advmt. Anim. Breed. Genet.* 23: 512.

Proc. Assoc. Advmt. Anim. Breed. Genet. 25: 31 - 34

EVOLUTION OF SHEEP BREEDS WITHIN LAMBPLAN AND THE RISE OF THE COMPOSITES

A.J. McMillan, S.F. Walkom and D.J. Brown

Animal Genetics Breeding Unit*, University of New England, Armidale, NSW, 2351 Australia

SUMMARY

The LAMPLAN – Terminal and Maternal analysis are large and contain multiple breeds. Over the last 20 years there has been an increase in cross breeding in seedstock flocks and thus the number of composite animals in these analyses, especially in the Maternal analysis and an overall reduction in the number of breed pure animals. The increase in crossbred animals will require some development to ensure that breed and heterosis effects are being modelled accurately. Further use and reporting of breed composition via genomics and pedigree methods should be considered. However, composite animals provide the comparisons needed for an accurate multibreed LAMBPLAN analysis allowing selection of animals across breeds for the industries diverse needs.

INTRODUCTION

Sheep Genetics (SG) has made significant advancements to the Australian national sheep genetic evaluation since its inception in 2005. Combining multiple database (for Merinos) and developing a uniform "language" to describe genetic evaluation for Australian sheep has proven extremely successful and allowed a more streamlined pipeline for delivery of genetic tools and extension activities (Collison *et al.* 2018). Much research has focused on the technical advancements to the genetic evaluation, with the main development work, first outlined by Brown *et al.* (2007), being completed and implemented into the current genetic evaluation. Advances in genomic technologies and development of resource populations (Brown *et al.* 2018) have seen changes to the analyses and these are incorporated as key component in the evaluation and in many breeding programs.

This paper examined the occurrence of the major contributing breeds and/or composites within the SG population and examine utilisation of pure animals and composite animals over time.

MATERIALS AND METHODS

Data for this analyses was obtained from the Terminal and Maternal LAMBPLAN database from the February 2023 routine analyses. Table 1 below shows a summary of pedigree related data for these two analyses. The breed of animals is assigned based on the flock of origin with animals within that flock being designated a breed. This included a number of breed codes specifically for composite animals (CT; Terminal, CM; Maternal).

As part of the routine analyses a pedigree-based breed composition matrix for all animals was calculated along with both generalised direct heterosis and maternal heterosis which was calculated across all breeds but not accounting for specific breed combinations (Brown et al., 2016). Flock and breed level trends were calculated for the following statistics breed purity (animals which have 90% or greater of assigned breed proportion), homebred (proportion of animals where the sire's flock code is the same as to progeny's flock code), outside sire breed (proportion of animals where sire type does not match the flocks breed code), direct heterosis and maternal heterosis.

* A joint venture of NSW Department of Primary Industries and University of New England.

Table1. Data summary of LAMBPLAN pedigree for the February 2023 Analyses

	Animals	Sires	Dams	Flocks	Breeds
Maternal	2,671,734	35,484	608,956	2,635	65
Terminal	3,726,242	54,788	958,806	4804	73

RESULTS AND DISCUSSION

The LAMBPLAN analysis contains many breeds however for this study we focus on the major breeds with minor breeds grouped into an "Other" category. The number of minor breeds contributing has declined from 39 breeds in 2000 to 19 breeds in 2020 for the Maternal analysis and from 42 in 2000 and to 28 in 2020 for the Terminal analysis (Table 2).

Within the Maternal analyses the major difference in breed contribution between 2000 and 2020 can be associated with a 7,000% increase in the number of CM animals due to large increases in the use of composite sires and dams as well as a substantial increase in number of composite stud flocks. Furthermore, it is worth noting that there was a significant proportion of CM animals in 2000 were missing pedigree, a likely by-product of the development of composites from non-SG sources. Excluding the CM breed all other breeds have exhibited a reduction in flock numbers since. The number of Border Leicester and Booroola animals in the analysis has increased while the Coopworth and White Suffolk have maintained similar number of animals. The Corriedales and minor breeds (other) have seen reductions in the number of animals and flocks.

In the Terminal analysis the use of a composite breed (CT) has significantly increased in both the number of animals and flocks, although not to the same extent as observed in Maternals. White Dorper, Dorper, White Suffolk, Poll Dorset, Suffolk and Southdown all show large increases in animals. With the Texel and the minor breeds (other) breeds having a reduction in numbers.

Table 2. Summary of the major breed contributions in the LAMBPLAN analyses in 2000 and 2020

Breed	Animals		Sires		Dams		Flocks	
	2000	2020	2000	2020	2000	2020	2000	2020
			Maternal Analysis					
Border Leicester	8,130	15,341	277	292	5,319	9,114	53	45
Corriedale	9,462	5,520	177	116	5,781	3,506	25	21
Coopworth	35,095	32,371	406	417	18,922	17,561	52	35
White Suffolk	2,282	2,444	212	94	1,618	1,597	81	17
East Friesian	885	798	47	12	454	446	12	1
Booroola	224	625	6	10	111	322	2	1
Composite	345	25,096	19	365	12	12,771	3	39
Other	1,914	684	595	64	1,559	490	229	22
			Terminal Analysis					
White Dorper	831	6,394	89	142	285	4,180	25	18
Dorper	494	4,056	46	108	113	2,271	15	14
White Suffolk	24,995	58,053	663	1,102	16,647	36,794	166	193
Suffolk	3,144	6,369	172	220	2,224	4,073	49	62
Texel	4,123	1,302	192	48	2,975	919	63	10
Poll Dorset	37,595	55,662	1,034	1,049	24,847	35,477	199	152
Southdown	483	2,204	33	85	362	1,345	8	15
Composite	1,164	8,791	51	311	683	5,292	21	34
Other	5,269	4,466	619	172	3,541	2,983	196	41

Figure 1 below summarises the trend in animal number for the two most populous breeds along with the composite breed for their respective analyses. The terminal analyses showed a rise in the composite animals, but they remain below the number of Poll Dorset (PD and White Suffolks (WS) within the analysis. In comparison the Maternal composites have had a marked rise in popularity since the mid-2000s. Equalling the top two breeds Border Leicester (BL) and Coopworth (CW) for animals born in 2015 and increasing rapidly to almost 25,000 animals in 2020.

Whilst the popularity of developing a composite line has increased the desire to maintain purity varies across breeds, most likely due to breed society convention and capacity to achieve desired genetic gains with the breed population. For example, the Poll Dorset and Border Leicester breeds which have remained largely pure with only small influence from outside breeds. This contrasts with the White Suffolk in Terminals and Coopworth in Maternal which show only a small number of animals born per year which could be considered pure. Unsurprisingly the composite animals in both analyses have considerable influence from outside breeds.

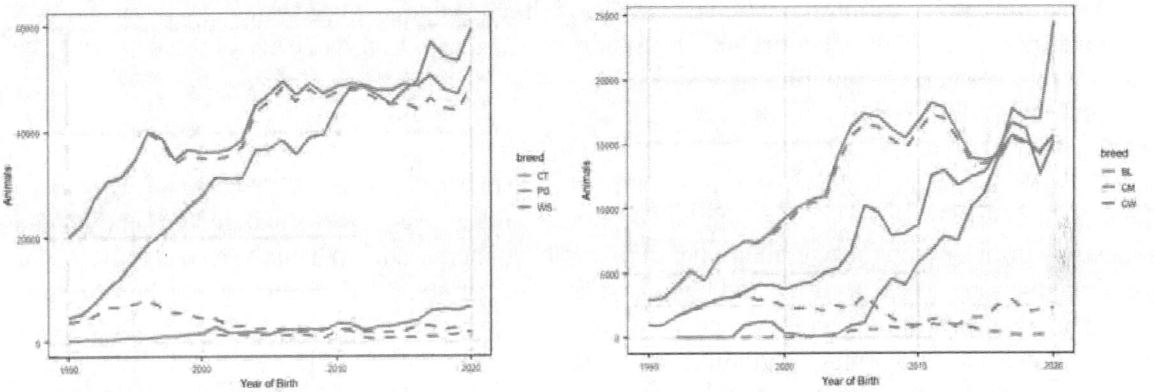

Figure 1. Breed contributions within the LAMBPLAN of the two largest breeds and composites, Maternal (left) and Terminal (right), databases from 1990-2020. Solid lines are counts of animals with dotted lines being counts of pure (>90%) animals

The LAMBPLAN Terminal and Maternal analysis are large multibreed analysis when we look at trends overtime for statistics relating to breed some interesting differences were observed between the two analyses. Figure 2 presents for both analyses the proportion of animals which are pure of designated breed or above 90 percent of that breed via the black and green lines, respectively. The reduction in purity across the analyses is greatest in the Maternal analysis, where there is a stronger willingness from breeders to look at individual animals from outside their breed rather than limiting sire uses to their breed. Thus, allowing maternal breeders to take advantage of across breed and within breed genetic variation and potential heterosis effects. The red lines (Figure 2) represent the proportion of animals which are the progeny of a homebred sire, this has increased overtime in both analyses and approaching 50% in Terminals and almost 70% in Maternal. The proportion of progeny born to outside breeds is significantly higher in Maternals compared to Terminals, suggesting an increased willingness from breeders to look to capitalise on high merit animals from outside breeds. However, the trend to use and outside breed remains proportionally relatively constant (Figure 2, blue line). Direct and Maternal Heterosis levels increase to around 25% in the Terminal analysis and approaching double that in the Maternal analysis however the level of heterosis looks to be stabilising in the Terminal analysis while the Maternal animals are continuing to trend towards higher levels of heterosis. Overall the trends across both analyses are for less pure and more cross bred animals with Maternal analysis showing this trend much more strongly than the Terminal analysis.

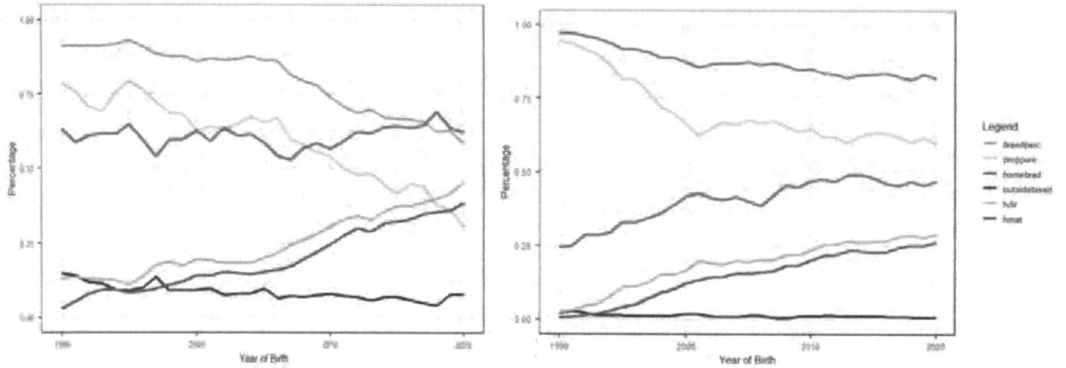

Figure 2. Breed purity, sire selection and Heterosis trends within the Sheep Genetics Maternal (left) and Terminal (right) databases from 1990-2020. The mean percentage of pure breed (grey), proportion of animals who are > 90% of their assigned breed (green), proportion of homebred animals (red), proportion of outside breed sires (blue), Direct Heterosis (orange) and Maternal Heterosis (purple)

CONCLUSIONS

This study showed the change overtime in the breed structure of the LAMBPLAN Terminal and Maternal analysis. In general, the number of flocks and breeds represented in the analysis has reduced overtime, while the overall animal numbers have increased. Both analyses have also had an overall reduction in breed purity and a consequent rise in composite animals, this is especially prominent in the Maternal analysis where composites are now the largest "breed" represented. These changes provide both opportunities and challenges for the evolution of the analysis. Future challenges included modelling of more heterosis effects and providing information around the breed proportion of these composites. Also, with the number of breed crosses and composites and large number of animals and pedigree in common could provide the possibility of a future joint Maternal and Terminal Analysis.

REFERENCES

Brown D.J., Huisman A., Swan A.A., Graser H.U., Woolaston R., Ball A.J., Atkins K.A. and Banks R.G. (2007) *Proc. Assoc. Advmt. Anim. Breed. Genet* 17: 187.

Brown D.J., Swan A.A., Gill J.S., Ball A.J. and Banks R.G. (2016) *Anim. Prod. Sci.* 56: 1442.

Brown D.J., Swan A.A., Boerner V., Li L., Gurman P.M., McMillan A.J., van der Werf J.H.J., Chandler H.R., Tier B., and Banks R.G. (2018) *Proc World Cong. on Genet. App. Livestock Prod.* 11: 460.

Collison C.E., Brown D.J., Gill J.S., Chandler H.R., Apps R., Swan A.A. and Banks R.G. (2018) *Proc World Cong. on Genet. App. to Livestock Prod.* 11: 661.

GENETIC AND ECONOMIC EFFECTS OF GENOMIC SELECTION AMONGST SURPLUS DAIRY HEIFERS IN AUSTRALIAN DAIRY HERDS

J.M. Morton[1] and G.J. Nieuwhof[2]

[1]Jemora Pty Ltd, Geelong, VIC, 3220 Australia
[2]DataGene Ltd, AgriBio, Bundoora, VIC, 3083 Australia

SUMMARY

When dairy farmers have reared surplus dairy heifers, a subset are selected for retention and entry to the milking group. We used stochastic simulation modelling followed by economic modelling to compare effects of 1) selection of heifers based on genomic Balanced Performance Index estimates to either 2) selection of early-born heifers in seasonal and split calving herds (the most common calving systems in Australia), or 3) random selection in year-round calving herds.

Based on those results, for Holsteins in seasonal and split calving herds, selection in a single birth-year group based on Balanced Performance Index typically delivers a small profit over the medium term (i.e. estimated net present value for the first 7 years $1,704 for a 300 cow herd) but larger profits over 12 ($5,354) and 17 ($6,942) years. These estimates did not include the additional costs incurred due to retaining later-born heifers when heifers are selected based on Balanced Performance Index estimates compared to selection of early-born heifers. Effects on profit were estimated as typically being larger in year-round calving herds ($4,897, $9,958 and $12,124, respectively). In Jerseys, effects would be expected to be a little less than these as there was typically less variation in Balanced Performance Index estimates within groups of Jersey heifers.

These results indicate that, under the model assumptions, the medium-term effects of using genomic selection to select replacements from surplus dairy heifers on herd profit are typically small in seasonal and split calving herds, but are larger in year-round calving herds. However, there was large stochastic variation between birth-year groups of 100 heifers in effects of selection strategy on true breeding values for Balanced Performance Index, indicating that effects would vary substantially between individual birth-year groups.

INTRODUCTION

When dairy farmers have reared surplus dairy heifers, a subset are selected for retention, calving and entry to the milking group and the remainder are sold before their first calving. Heifers can be selected based on various attributes. In seasonal and split calving herds (the most common calving systems in Australia), preferential selection of early-born heifers helps ensure all heifers are at target liveweight by yearling mating start date. Alternatively, selection can be based on genetic estimates for the heifers. With the availability of commercial genomic testing services, genetic estimates based jointly on animal pedigree and genomic results are readily available for heifers prior to their first calving.

The aim of this work was not to assess the profitability of rearing and selling surplus dairy-breed heifers relative to other management strategies available for the herd but rather, for herds that have an excess of AI-sired dairy heifers, to estimate the effects of selection of a subset of those heifers for retention based on genomic Balanced Performance Index estimates compared to each of a) selection of early-born heifers in seasonal and split calving herds or b) random selection in year-round calving herds (representing any selection strategy that is independent of both the heifers' true Balanced Performance Index values and birth date).

MATERIALS AND METHODS

We used stochastic simulation modelling followed by economic modelling to compare selection strategies in a single birth-year group in a commercial herd. To provide input parameter values for this modelling, we analysed genomic Balanced Performance Index estimates and reliabilities, and birth dates from 25,423 Holstein heifers born in 2019 or 2020 from 205 herds, and 61,631 Jersey heifers born in 2019, 2020, or 2021 from 396 herds. Only genomic estimates from before the heifers' first calving dates were used.

The steps in the simulation modelling were as follows:

1. Generate a simulated group of 100 heifers, each with a true breeding value for Balanced Performance Index and a birth date
2. Generate Balanced Performance Index estimates for each heifer
3. Select 50 heifers from the 100 heifers using each of three methods:
    a. Select the 50 heifers with highest Balanced Performance Index estimates
    b. Select the 50 earliest-born heifers
    c. Select 50 heifers independently of their genetic attributes and birth date (simulated by random selection)
4. Calculate mean true breeding value for Balanced Performance Index under each strategy, and calculate differences between means for strategy a versus strategy b and strategy a versus strategy c
5. Also calculate differences in distributions of birth dates under each strategy
6. Repeat steps 1 to 5 a further 9,999 times, and summarise differences

True breeding values for Balanced Performance Index were simulated using specified parameter values for 1) the standard deviation of Balanced Performance Index estimates in the source population from which the birth year group of heifers were drawn, 2) Balanced Performance Index estimate reliabilities and 3) the genetic correlation between Balanced Performance Index and birth date. Birth dates were selected from a log-normal distribution based on that observed in the study dairy heifers and the observed value for the (negative) correlation between Balanced Performance Index estimate and $\log_e$-transformed birth date in Holstein study heifers of -0.21 used as the genetic correlation parameter value. The Balanced Performance Index includes the daughter fertility estimated breeding value (Australian Breeding Value or ABV) so cows with higher Balanced Performance Index values on average, conceive and so calve earlier in the calving period in seasonal and split calving herds. As daughter fertility ABVs between dams and daughters are correlated, it was expected that Balanced Performance Index would be correlated with birth data, and this is what we found in our analyses.

Economic effects of selection based on Balanced Performance Index estimates relative to each of the other strategies in a single birth-year group were assessed for a 300-cow herd rearing 132 dairy heifers each year and retaining 66 to calve in the herd (i.e. 22% replacement rate). Herd replacement rates and age structures were held constant in every year under all three strategies. Differences between means for the strategies from the simulation modelling (Table 1) were estimated with the standard deviation of Balanced Performance Index values in the source population from which the birth year group of heifers were drawn of 85.9 units, with Balanced Performance Index estimate reliabilities of 0.64, with genomic testing costs of $53 per heifer ($50 testing cost plus $3 labour). We assumed that each 1 unit increase in herd average true breeding value for Balanced Performance Index in a particular year over the previous year causes a $1 increase in herd profit in that year, as inferred by Byrne (2016). Effects of higher Balanced Performance Index values of the selected birth-year group on true breeding value for Balanced Performance Index of their daughters, granddaughters etc were incorporated when calculating differences in herd average true breeding values for Balanced Performance Index by year. Economic effects were estimated for 17 years where year 1 is the year of birth of the selected group. Net present values were calculated using a 5% discount rate. The additional costs incurred due to retaining later-born heifers (additional feed costs to attain higher growth rates to the herd's yearling mating start date and/or economic costs of failing to achieve the same liveweights by then) when heifers are selected based on Balanced Performance Index estimates compared to selection of early-born heifers were not included in economic calculations.

RESULTS AND DISCUSSION

Means of true breeding values for Balanced Performance Index amongst 50 heifers selected from 100 heifers based on Balanced Performance Index estimates were, on average, modestly higher than for other selection methods but there was large stochastic variation between birth-year groups (Table 1 and Figure 1). The mean of differences relative to random selection of 54.6 was close to the expected value from the breeders' equation (Falconer 1989) of 54.8. Relative to selection of the earliest-born heifers, more of the heifers selected based on Balanced Performance Index estimates had been born after day 42 of the herd's calving period, (Table 1).

Table 1. Distribution of differences in means of true breeding values for Balanced Performance Index and percentages of heifers born after day 42 of the herd's calving period between 50 simulated Holstein heifers selected from 100 heifers as those with the highest Balanced Performance Index estimates and either the 50 earliest-born heifers or 50 randomly selected heifers selected from the same 100 heifers; distributions were from 10,000 replications

Outcome variable and statistic	Relative to selection of earliest-born heifers	Relative to random selection of heifers
Means of true breeding values for Balanced Performance Index		
Mean of differences	39.9	54.6
Standard deviation of differences	9.6	10.8
Range of differences	7.2 to 82.9	16.2 to 103.9
Percentages of heifers born after day 42 of the herd's calving period		
Mean of differences	17%[1]	-4%
Standard deviation of differences	5%	6%
Range of differences	0% to 40%	-26% to +18%

[1]For example, the percentage of the 50 heifers selected as those with the highest Balanced Performance Index estimates that had been born after day 42 of the herd's calving period was, on average, 17% more (range 0% to 40%) than that for the 50 earliest-born heifers

From the economic modelling, for seasonal and split calving herds, selection in a single birth year based on Balanced Performance Index typically delivers a small profit over the medium term (i.e. estimated net present value for the first 7 years $1,704 for a 300 cow herd) but larger profits over 12 ($5,354) and 17 ($6,942) years. The larger profits for 12 and 17 years relative to 7 years were mainly due to higher Balanced Performance Index values in the daughters, granddaughters etc of the selected heifers. Effects of selection on profit were estimated as typically being larger in year-round calving herds ($4,897, $9,958 and $12,124, respectively). Newton *et al* (2018) also reported small estimated profits from selected heifers based on genomic Balanced Performance Index estimates in seasonal calving herds compared to selection in the absence of genetic information. Due to stochastic variation in genetic effects between single birth-year groups under each heifer selection method, in seasonal and split calving herds, net present value for the first 7 years would be expected to be negative (i.e. the benefits being less than the costs of genomic testing) for 20.4% of birth-year groups but net present value for the first 17 years would be expected to be negative for only 1.2% of birth-year groups. In contrast, in year-round calving herds, negative net present values due to stochastic variation in genetic effects would be very unlikely (1.4% and 0.04% of birth-year groups, respectively).

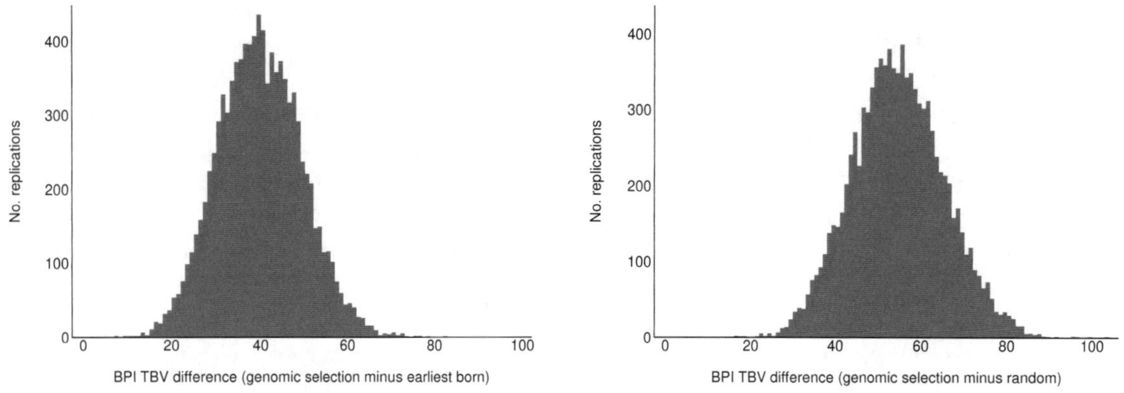

Figure 1. Distribution of differences in means of true breeding values (TBVs) for Balanced Performance Index ('BPI TBV difference') between 50 simulated Holstein heifers selected from 100 heifers as those with the highest Balanced Performance Index estimates and a) left-hand graph: selection of the 50 earliest-born heifer and b) right-hand graph: 50 heifers randomly selected from the same 100 heifers. Distributions were from 10,000 replications

There was typically less variation in Balanced Performance Index estimates for groups of Jersey heifers (median of the standard deviations 63.5 compared to 68.7 for Holsteins). Accordingly, economic effects of selection on Balanced Performance Index estimates would be expected to be a little less for groups of Jersey heifers than for Holsteins as reported above.

CONCLUSIONS

These results indicate that currently the medium-term effects of genomic selection from surplus dairy heifers on profit are typically small in seasonal and split calving herds but are larger in year-round calving herds, on average. Genomic testing can also assist in correcting pedigrees and reducing inbreeding, avoiding recessive lethal genes, and selecting for desired genetic variants, and the benefits of these effects were not included in the economic analyses.

ACKNOWLEDGEMENTS

The authors thank Dairy Australia for funding time inputs by the first-named author, Stephanie Bullen and Michelle Axford for ongoing inputs into study design and interpretation of results from this study, and DataGene for providing the heifer data used in this study.

REFERENCES

Byrne T.J., Santos B.F.S., Amer P.R., Martin-Collado D., Pryce J.E. and Axford M. (2016) *J. Dairy Sci.* 99: 8146.

Falconer D.S. (1989) Introduction to Quantitative Genetics (3rd ed.), Longman Scientific and Technical, New York

Newton J.E., Hayes B.J. and Pryce J.E. (2018) *J. Dairy Sci.* 101: 6159.

Proc. Assoc. Advmt. Anim. Breed. Genet. 25: 39 - 42

EFFECTS OF SELECTION AND DATA TRUNCATION ON ESTIMATES OF GENETIC PARAMETERS OBTAINED FITTING A SINGLE-STEP MODEL

Karin Meyer

Animal Genetics Breeding Unit*, University of New England, Armidale, NSW 2351 Australia

SUMMARY

Simulation was used to illustrate the effects of genomic selection on estimates of genetic parameters, comparing values when genomic relationships were ignored with those obtained accounting for the joint relationship matrix of genotyped and non-genotyped individuals. Analyses were carried out with increasing truncation of earlier records, pedigrees and genotype information. Results showed that estimates from pedigree only analyses could be markedly biased downwards as more historical data is ignored, especially with strong genomic selection, causing predicted breeding values for selection candidates in the last generation to be under-dispersed.

INTRODUCTION

Increasingly genetic evaluation schemes for livestock incorporate genomic information on a routine basis. To date, the most common method is single-step genomic best linear unbiased prediction (ssGBLUP) fitting a breeding value model. This replaces the pedigree-based inverse of the numerator relationship matrix with its counterpart which combines pedigree and genomic information (Aguilar *et al.* 2010). It is a conceptually simple extension of the classic prediction procedures using pedigree based relationships only (PBLUP). Like PBLUP, ssGBLUP requires appropriate values of genetic parameters as input. It is common practice to estimate these fitting the same – or at least a very similar – model as used for prediction of breeding values (EBV). Reviewing the status of genomic evaluation, Misztal *et al.* (2020) advocated inclusion of genomic relationships when estimating genetic parameters to counteract the bias due to genomic selection. The authors also recommended frequent re-estimation as genetic variances appeared to change quicker with ssGBLUP.

However, to date, estimates are mostly obtained considering pedigree based relationships only, and little is known about the impact of doing so on the efficacy of genomic selection. This paper presents a simple simulation study exploring the effects of accounting for genomic relationships on estimates of genetic parameters and the resulting accuracy of ssGBLUP based selection.

MATERIAL AND METHODS

Data were simulated for a trait with heritability of 0.3 and for individuals from 13 generations using the software package AlphaSim, version 1.05 (Faux *et al.* 2016). The data set contained records for 2100 and 3150 animals, respectively in generations 1 to 7 and 8 to 13, who were the progeny of 100 and 150 sires and 1000 and 1500 dams, respectively. To mimic a distribution over fixed effects subclasses, records were randomly assigned to 51 'contemporary groups' per generation. Genotypes were constructed by sampling 10 chromosomes with 4,000 single nucleotide polymorphism (SNP) and 50 quantitative trait nucleotide (QTN) each, randomly allowing for some QTN to be included among the SNP and assuming no mutation or recombination. Marker information for all individuals in generations 10 to 13 was retained, disregarding earlier genomic information.

AlphaSim provides the option to carry out selection in individual generations externally by allowing the user to select the parents and mating allocations of the next generation (Faux *et al.* 2016). This was utilised to implement three alternative selection schemes, combining random selection with selection on EBV obtained using pedigree relationships only and EBV from ssGBLUP analyses.

* A joint venture of NSW Department of Primary Industries and University of New England,

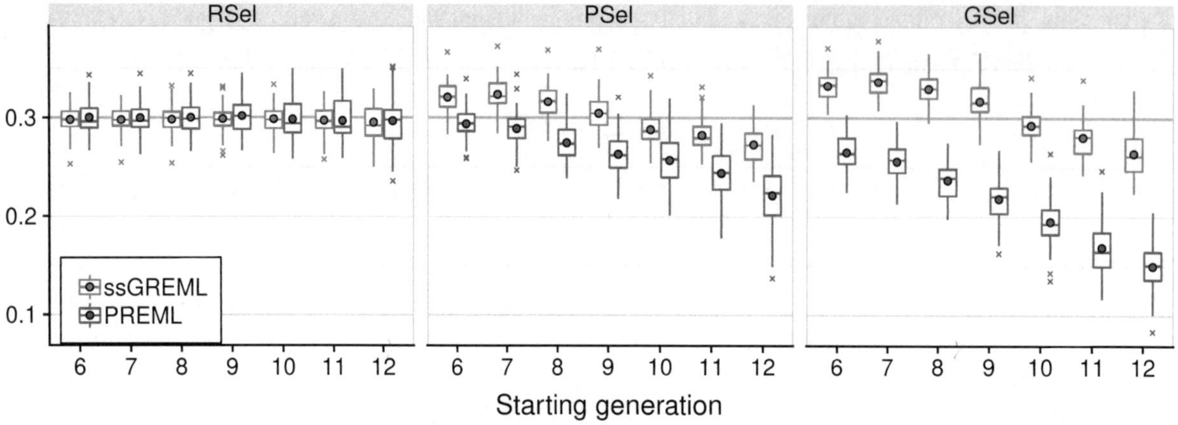

Figure 1. Distribution of heritability estimates over replicates (see text for definitions)

Discarding generations 1 to 4 as burn-in, parents of generations 5 to 7 were chosen at random. 1) For a genomic scenario (GSel) selection was on EBV from PBLUP in generations 8 to 10 and on EBV from ssGBLUP in generations 11 to 13. This was contrasted with 2) selection on EBV from PBLUP in generations 8 to 13 (PSel) and 3) random selection throughout (RSel). EBV were obtained from restricted maximum likelihood (REML) analyses at convergence. For generation i analyses utilised data and pedigree information from generation 6 to i (to select the parents of generation $i+1$) and, where applicable, all marker information from generation 10 to i.

To investigate the effects of selection bias and truncation of data on estimates of genetic parameters, analyses were carried out successively ignoring information from earlier generations, i.e. considering records, pedigrees and marker counts from generations i to 13 only where $i = 6,\ldots,12$. In the following, we refer to generation i as the 'starting generation' for an analysis. Accuracy and dispersion of EBV for selection candidates in generation 13 were measured as the correlation between and regression of true breeding values (TBV) on EBV. 50 replicates were carried out for each scenario.

REML analyses (for both the external selection steps and the data sets sampled) used either pedigree based relationships only (PREML) or pedigree and genomic relationships jointly (ssGREML), fitting a simple animal model with contemporary group as the only fixed effects. Genomic relationship matrices (**G**) were built using Method 1 of Van Raden (2008), eliminating SNP with minor allele frequencies less than 2% and centering allele counts using mean frequencies in the data. These were aligned to their pedigree based counterparts ($\mathbf{A}_{22}$) following Vitezica *et al.* (2011).

RESULTS

The distribution of heritability estimates over replicates for the three selection strategies is summarised in Figure 1. In all cases, means – depicted by circles – agreed closely with the median values. As expected, for RSel, estimates from ssGREML and PREML did not differ noticeably and showed no bias, though some differences in variability across replicates were evident. For PSel and GSel, however, estimates depended strongly on the subset of data utilised. Loosely described, REML can account for selection bias, provided the information that selection decisions were based on is included in the analysis. Hence, for data starting at generations $i = 6$ or 7, no selection bias was evident for PSel, while corresponding estimates from PREML analyses for GSel were somewhat lower. The latter could be attributed to stronger selection in the last three generations for GSel, together with the fact that PREML ignored the genomic information which facilitated it. Conversely, as more and more of the generations subject to selection were omitted from the data (i.e. as the 'starting generation' increased), estimates reflected the reduced genetic variation available in what was implicitly treated as the base generation in the truncated data set.

Estimates from ssGREML were consistently higher than those from PREML for both PSel and GSel and, for analyses including data from unselected generations, were somewhat higher than the population value of 0.3 simulated. Including pedigree information for individuals in starting generation *i*, we would expect estimates to reflect the genetic variance in base generation $i - 1$. Presumably the overestimates might be attributed, to some extent at least, to the effects of pedigree truncation – and thus underestimates of inbreeding – resulting in imperfect alignment of **G** to $\mathbf{A}_{22}$. Limited additional analyses for GSel using data from generations 3 to 13 yielded a mean heritability estimate closer to 0.3, suggesting so.

As shown in Figure 2 for GSel, the higher heritability estimates from ssGREML analyses were mainly due to higher genetic variance estimates. Interestingly, ssGREML estimates of the residual variance did not depend strongly on the amount of data truncation, while values for PREML exhibited distinct repartitioning of genetic into residual variation.

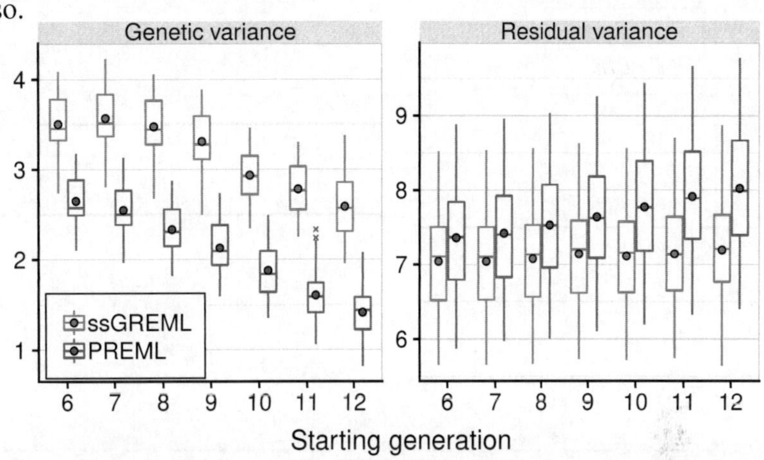

Figure 2. Distribution of variance component estimates over replicates for GSel (see text for definitions)

The distributions of correlations between TBV and EBV and regressions of TBV on EBV for selection candidates in generation 13 are shown in Figure 3. With the simulation involving strong selection and, for ssGREML, all individuals from generation 10 onward having genomic information, mean correlations for ssGREML analyses were very high and substantially exceeded those from PREML, in particular for GSel. For all three scenarios, values for PREML differed little between the subsets of data utilised. Robustness of such correlations, in particular for univariate analyses, is a well known phenomenon for PBLUP. In contrast, means for ssGREML and starting generations 11 and 12 dropped, due to the omission of marker information in these analyses. Mean regressions of TBV on EBV were close to their expected value of unity for all ssGREML analyses. Corresponding values for PREML and PSel or GSel, however, showed increasing underdispersion of EBV (i.e. regression coefficients greater than unity) with increasing starting generation, mirroring the underestimates of genetic variation reported above.

DISCUSSION

Simulation studies on ssGREML have been presented by Cesarani *et al.* (2019) and Junqueira *et al.* (2022) but involved different set-ups and questions considered. Our study attempted to mimic, in a simplified scheme, the progression from random selection to pedigree based and finally to genomic assisted selection which might occur in a livestock improvement programme. Clearly, results are at least partially specific to the scenario considered. In particular, for analyses using genomic information all individuals in the relevant generations were assumed to be genotyped. This yielded substantial differences in estimates of variance components from PREML and ssGREML. Additional ssGREML analyses retaining only genotypes for a proportion of randomly selected animals reduced estimates closer to values from PREML (not shown).

Truncation of data and pedigrees redefines the base generation. This implies that estimates of the genetic variance reflect the amount of 'usable' genetic variation in that generation. Consequently, when

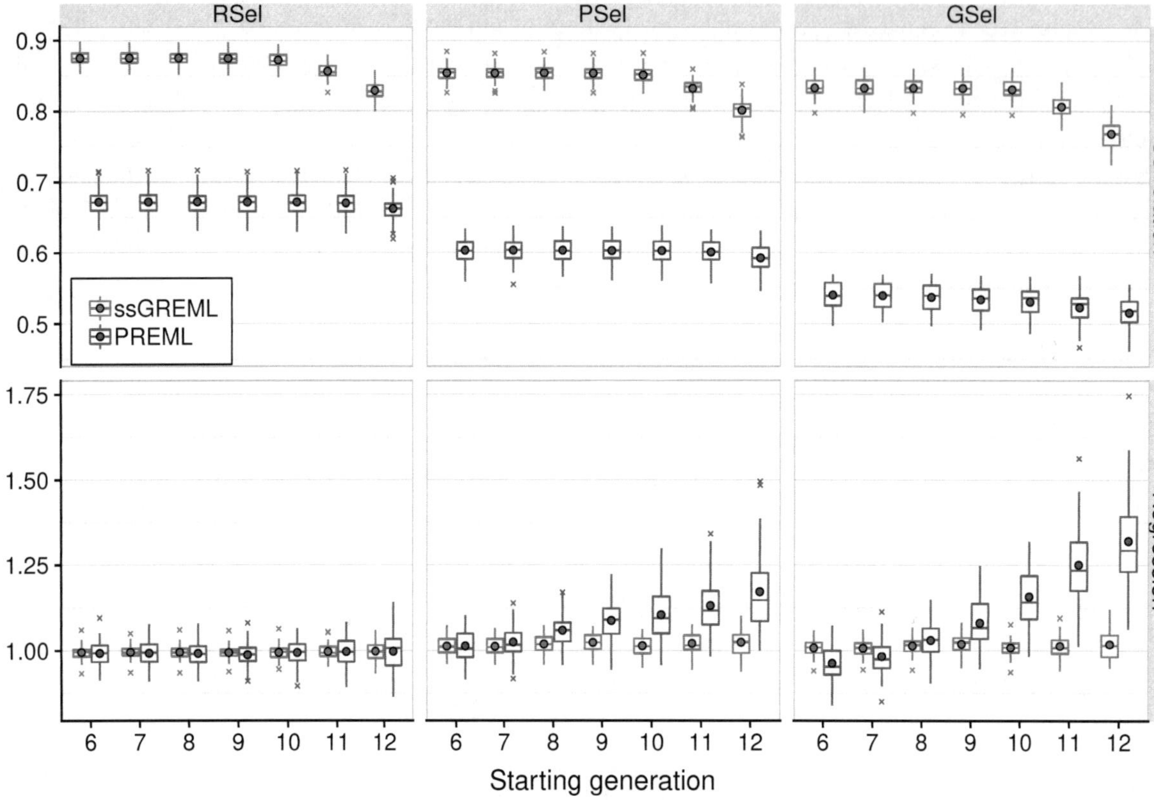

Figure 3. Distribution over replicates of correlations between true and predicted breeding values and regressions of true on predicted breeding values for animals in generation 13 (see text for definitions)

omitting information on which selection decisions have been based, estimates declined, especially for GSel. Implications thereof need to be considered when predicting response to selection or evaluating reliabilities of EBV (Gorjanc *et al.* 2015). As more and more animals are genotyped and as the emphasis on genomic selection increases, ssGREML estimation of genetic parameters will become a necessity.

ACKNOWLEDGEMENTS

This work was supported by Meat and Livestock Australia grant L.GEN.2204.

REFERENCES

Aguilar I., Misztal I., Johnson D.L., Legarra A., Tsuruta S. and Lawlor T.J. (2010) *J. Dairy Sci.* **93**:743.

Cesarani A., Pocrnic I., Macciotta N.P.P., Fragomeni B.O., Misztal I. and Lourenco D.A.L. (2019) *J. Anim. Breed. Genet.* **136**:40.

Faux A.M., Gorjanc G., Gaynor R.C., Battagin M., Edwards S.M., Wilson D.L., Hearne S.J., Gonen S. and Hickey J.M. (2016) *Plant Genome* **9**:1.

Gorjanc G., Bijma P. and Hickey J.M. (2015) *Genet. Sel. Evol.* **47**:65.

Junqueira V.S., Lourenco D., Masuda Y., Cardoso F.F., Lopes P.S., Silva F.F.E. and Misztal I. (2022) *J. Anim. Sci.* **100**:skac082.

Misztal I., Lourenco D. and Legarra A. (2020) *J. Anim. Sci.* **98**:skaa101.

Van Raden P.M. (2008) *J. Dairy Sci.* **91**:4414.

Vitezica Z.G., Aguilar I., Misztal I. and Legarra A. (2011) *Genet. Res.* **93**:357.

Proc. Assoc. Advmt. Anim. Breed. Genet. 25: 43 - 46

GENETIC BENCHMARKING OF MATERNAL SHEEP FLOCKS USING GENOMIC TESTING

D.J. Brown[1,2], P.M. Gurman[1] and A.A. Swan[1]

[1] Animal Genetics Breeding Unit*, University of New England, Armidale, NSW 2351 Australia
[2]Advanced Livestock Measurement Technologies project, Meat & Livestock Australia, North Sydney, NSW, 2060 Australia

SUMMARY

This study aimed to examine the predictive ability of the "Flock Profile" genomic benchmarking method in maternal sheep flocks, estimated from the Maternal LAMBPLAN analysis. Data from this analysis was used in a validation study to test the accuracy of predicting mean flock performance for reproductive traits. For each validation flock, the pedigree, genotypes and performance data were removed for the entire flock and then its Flock Profile result was estimated from genomic predictions based on estimated SNP marker effects from single step genomic BLUP analyses (ssGBLUP). The Flock Profile results were then compared to the original Australian Sheep Breeding Values (ASBVs) from the full analysis. The accuracy of ranking of mean flock performance was high (r>0.85) for all traits except ewe rearing ability. However, the Flock Profile results were generally over-dispersed and thus had more variation compared to their ASBVs. Genomic predictions for individual animals were also highly correlated to the full ASBVs. This initial study supports further investment into the development of these products, with the potential to offer commercial producers new genetic tools to foster ongoing improvement in on-farm profitability.

INTRODUCTION

The Flock Profile test is successfully used to genetically benchmark commercial Merino flocks (Swan *et al.* 2018). While the average Australian Sheep Breeding Value (ASBV) of rams purchased is often the most accurate metric of genetic merit, it is not available to commercial flocks from outside of Sheep Genetics when sourcing rams. Thus, Flock Profile tests are an important tool for those breeders without any knowledge of their current genetic benchmark. At present, these are only commercially available in purebred Merino flocks and does not include reproduction traits. However, dissemination of genetic gain made in the seedstock sector would be enhanced across industry if similar products were available for other breeds, and in particular for commercial crossbred flocks. Brown *et al.* (2022) conducted a preliminary validation in terminal sire breeds for carcase traits, which demonstrated that genomic flock profiling accurately predicted their true ASBV mean. Another application under consideration is a Flock Profile product to support the marketing of maternal replacements, allowing purchasers to value sale lots on more accurate genetic benchmarks for all the key traits rather than relying on visual appraisal alone. In addition to flock benchmarking, the methodology could also be used to perform genomic prediction on individual animals.

This study aimed to examine the predictive ability of the Flock Profile test for reproductive traits in maternal sheep breeds, estimated from the maternal LAMBPLAN analysis (Brown *et al.* 2007).

MATERIALS AND METHODS

Data from the reproductive component trait analysis for Maternal LAMBPLAN (as described by Bunter *et al.* 2019) were utilised for this study. Reproductive data and genotypes were identified for 14 selected seedstock flocks within this analysis, chosen based on volume and quality of data. The

* A joint venture of NSW Department of Primary Industries and University of New England

flocks consisted of Border Leicester, Coopworth and composite breeds. The data available for each trait are illustrated in Table 1. To generate an independent ASBV analysis, all the phenotype data, pedigree and genotypes from these 14 flocks were removed sequentially and 14 special Maternal LAMBPLAN analyses conducted. Using the genotypes only, ASBVs were estimated for each animal using the back-solve methods described by Swan et al. (2018) and these were then averaged for each flock to estimate the Flock Profile result. The Flock Profile results were then compared to their true ASBV means from the full LAMBPLAN analysis using all data. A key aspect of the Flock Profile method is the projection of genetic group effects using a regression of genetic group coefficients on genomic relationships between reference animals in the ssGBLUP and the animals targeted for prediction. Genetic groups in Maternal LAMBPLAN are defined at the breed level, and an important difference between Merino and Maternal LAMBPLAN ssGBLUP analyses is that the latter uses a genomic relationship matrix which accounts for breed structure (Gurman *et al.* 2019) while the Merino analysis does not. The current Maternal LAMBPLAN analysis uses a breed-adjusted genomic relationship matrix (G) and a lambda of 0.5. In an attempt to capture breed effects in the Flock Profile prediction in this study, back-solving was conducted using genomic relationships without accounting for breed structure. In addition, a "lambda" value of 1 was used i.e. variance fully explained by markers with no residual polygenic variation.

The component reproduction traits analysed included conception rate (CON: 0=failed to conceive, 1=conceived) litter size (LS: 1 to 4 lambs born) and ewe rearing ability (ERA: lambs surviving/lambs born for ewes which lambed). All three traits have yearling and adult expressions separated. Additional correlated traits included maternal behaviour score (MBS: from 1: good to 5: poor) and, pre-joining weight (AWT) and condition score (CS) recorded within the 30 days before joining. Body composition and development traits also included in the analysis were post-weaning body weight (PWT), carcase fat (PFAT) and eye muscle depth (PEMD), along with post-weaning (PSC) or yearling (YSC) scrotal circumferences.

Table 1. Descriptive statistics of the data used in the validation for each trait.

Trait	Flocks	Animals	Genotyped	Mean	SD
PWT	14	94,176	14,632	45.50	9.54
PCF	14	94,095	14,616	3.33	1.26
PEMD	14	94,145	14,628	26.52	3.95
AWT	13	65,016	19,790	68.80	11.58
PSC	10	31,029	1,449	28.88	4.44
YSC	7	15,075	269	28.49	2.90
MBS	11	26,845	19,320	1.82	0.92
CS	12	33,976	15,542	3.49	0.64
CON	14	102,562	17,447	0.92	0.27
LS	14	154,410	25,279	1.76	0.60
ERA	14	108,618	19,415	0.85	0.30
YCS	6	4,859	3,580	3.41	0.55
YCON	12	43,130	13,963	0.66	0.47
YLS	14	37,854	10,231	1.51	0.55
YERA	14	28,719	7,707	0.78	0.36

RESULTS AND DISCUSSION

The accuracy of ranking of flocks was high (r>0.85) for all traits except ewe rearing ability. However, the ASBV means and variation between flocks were significantly different between the full ASBV and Flock Profile results (Table 2). The difference in the variation between flocks and slope values significantly less than 1 indicate that the Flock Profile results were generally over-dispersed compared to their ASBVs. This over-dispersion maybe due to the use of the unadjusted G and lambda of 1.0 in this study and further research is required to study the impact of these factors.

Table 2. Relationship between Flock Profile (n=14) results and ASBV means from the full Maternal LAMBPLAN analysis

Trait	Flock Profile mean	ASBV mean (SD)	Slope	Corr	RMSE#
PWT	-0.50 (3.89)	2.76 (1.61)	0.39 (0.04)	0.95	0.49
PCF	0.27 (0.13)	0.35 (0.13)	0.85 (0.14)	0.87	0.06
PEMD	0.33 (2.09)	2.21 (1.00)	0.44 (0.06)	0.92	0.40
AWT	-1.31 (1.47)	0.80 (1.13)	0.70 (0.09)	0.91	0.47
PSC	-0.56 (1.97)	1.10 (0.93)	0.44 (0.05)	0.94	0.33
YSC	-1.05 (1.70)	0.71 (0.87)	0.49 (0.04)	0.96	0.25
MBS	-0.10 (0.04)	-0.11 (0.05)	0.92 (0.19)	0.81	0.03
CS	0.13 (0.08)	0.13 (0.05)	0.47 (0.12)	0.75	0.03
CON	0.00 (0.04)	0.03 (0.01)	0.33 (0.07)	0.83	0.01
LS	-0.18 (0.25)	0.08 (0.11)	0.41 (0.05)	0.92	0.04
ERA	0.01 (0.02)	0.02 (0.01)	-0.02 (0.14)	-0.03	0.01
YCS	0.15 (0.10)	0.24 (0.08)	0.73 (0.10)	0.91	0.03
YCON	0.01 (0.16)	0.19 (0.06)	0.36 (0.05)	0.91	0.03
YLS	-0.11 (0.20)	0.11 (0.08)	0.36 (0.07)	0.85	0.04
YERA	0.01 (0.03)	0.03 (0.01)	0.28 (0.05)	0.86	0.01

RMSE: Root mean square error

The results of the back-solved breeding values at the level of individual animal are shown in Table 3. These results highlight that the Flock Profile methodology could accurately predict the ranking of ASBVs within the flocks tested with correlations generally greater than 0.7 for most traits and regression slopes of close to 1.0. The relationships across all animals and flocks were lower with correlations ranging from 0.34 to 0.96. Further research is required to refine the methodology to more accurately partition flock and breed effects.

It should be noted that unlike most commercial flocks, the flocks used in this analysis were seedstock breeders with stronger genetic links to other breeding and reference flocks in the Maternal LAMBPLAN analysis and some descendants of these flocks would have existed in other flocks that remained in the analysis. Thus, the correlations observed here may be higher compared to those observed in less related commercial flocks in industry that are the target of Flock Profile products.

The longer-term challenge for the development of a commercial Flock Profile test for industry flocks is to accommodate their crossbred structure. Lamb production flocks generally incorporate breed components from the 3 major breed types of Merino, maternal and terminal, each of which are analysed separately by Sheep Genetics in their MERINOSELECT, Terminal LAMBPLAN and Maternal LAMBPLAN evaluations. Therefore, the results would need to be aligned relative to each of these 3 different ASBV analyses. One difficulty of alignment across analyses not covered in this study is the potential effects of heterosis in commercial crossbred ewes, which is one of the key benefits of using these maternal sheep, for example in the Border Leicester x Merino production system. This requires further consideration. Another technical challenge is that the LAMBPLAN analyses are multi-breed, with genomic information corrected for breed effects (Gurman *et al.* 2019). One of the motives of this study was to investigate this issue and ensure breeds effects could be accommodated in the Flock Profile method. Aside from addressing the technical challenges associated with breed structure and heterosis, Flock Profile testing should be expanded to cover more of the traits that influence profitability in sheep enterprises, including reproduction and ewe efficiency, product quality and disease resistance.

Table 3. Relationship of the genomic only animal level breeding values to the ASBV from the full Maternal LAMBPLAN analysis both across and within flocks

Trait	Across all flocks					Average of within flock	
	Flock Profile mean (SD)	ASBV mean (SD)	Slope	Corr	RMSE[#]	Corr	Slope
PWT	1.59 (3.34)	3.69 (2.02)	0.51 (0.00)	0.85	1.06	0.87 (0.03)	0.88
PCF	0.30 (0.25)	0.35 (0.26)	0.93 (0.00)	0.89	0.11	0.90 (0.03)	0.95
PEMD	1.49 (1.73)	2.69 (1.13)	0.54 (0.00)	0.83	0.64	0.86 (0.04)	0.90
AWT	-0.84 (2.88)	1.05 (2.84)	0.89 (0.00)	0.90	1.21	0.88 (0.03)	0.92
PSC	0.64 (1.63)	1.77 (1.03)	0.54 (0.00)	0.86	0.52	0.87 (0.06)	0.93
YSC	-0.01 (1.47)	1.30 (0.96)	0.58 (0.00)	0.89	0.44	0.86 (0.06)	0.92
MBS	-0.09 (0.13)	-0.12 (0.14)	0.96 (0.00)	0.87	0.07	0.90 (0.03)	1.00
CS	0.09 (0.12)	0.12 (0.12)	0.81 (0.00)	0.85	0.06	0.90 (0.03)	0.96
CON	0.02 (0.03)	0.04 (0.02)	0.34 (0.00)	0.49	0.02	0.58 (0.19)	0.76
LS	-0.04 (0.21)	0.16 (0.14)	0.55 (0.00)	0.83	0.08	0.68 (0.13)	0.79
ERA	0.01 (0.02)	0.01 (0.02)	0.39 (0.01)	0.35	0.02	0.64 (0.12)	0.92
YCS	0.20 (0.10)	0.27 (0.10)	0.86 (0.00)	0.85	0.05	0.87 (0.04)	0.98
YCON	0.10 (0.13)	0.23 (0.09)	0.46 (0.00)	0.65	0.07	0.65 (0.18)	0.75
YLS	-0.01 (0.17)	0.17 (0.13)	0.57 (0.00)	0.78	0.08	0.70 (0.14)	0.84
YERA	0.03 (0.03)	0.04 (0.03)	0.52 (0.00)	0.57	0.02	0.66 (0.15)	0.81

RMSE: Root mean square error

CONCLUSIONS

The results of this study demonstrate accurate ranking of flocks, but more work is required to produce accurate ASBV benchmarks for all traits. This initial study supports further investment into the development of Flock Profile products, which has the potential to expand the range of genetic tools available to the sheep industry to foster ongoing improvement in on-farm profitability.

ACKNOWLEDGEMENTS

This work was funded by the Australian Government Department of Agriculture, Water and the Environment, MLA (L.GEN.2204) and a range of industry partners throughout the supply chain. The authors would also like to thank the industry seedstock producers in LAMBPLAN who contributed to this work.

REFERENCES

Brown D.J., Gurman P.M. and Swan A.A. (2022). In 34th Biennial Conference of the Australian Association of Animal Sciences, Cairns, Queensland, 3-8 July 2022.

Brown D.J., Huisman A., Swan A.A., Graser H.U., Woolaston R., Ball A.J., Atkins K.A. and Banks R.G. (2007) *Proc. Assoc. Advmt. Anim. Breed. Genet* 17: 187.

Bunter K.L., Swan A.A., Gurman P.M., Boerner V., Mcmillan A. and Brown D.J. (2019) *Proc. Assoc. Advmt. Anim. Breed. Genet* 23: 560.

Gurman P.M., Bunter K.L., Boerner V., Swan A.A. and Brown D.J. (2019) *Proc. Assoc. Advmt. Anim. Breed. Genet* 23: 254.

Swan A.A., Gurman P.M., Boerner V., Brown D.J., Clark S., Gore K., Granleese T. and van der Werf J.H.J. (2018) In 'Proceedings of 11th World Congress on Genetics Applied to Livestock Production', 11–16 February 2018, Auckland, New Zealand. p. 346.

PRELIMINARY EVALUATION OF THE IMPACT OF VISUAL TRAITS ON LIFETIME EWE PERFORMANCE

P.K. Wahinya[1], D.J. Brown[1], S.F. Walkom[1], T. Bird-Gardiner[2], B.E. Clarke[3], J.L. Smith[4], A.A. Swan[1]

[1] Animal Genetics Breeding Unit [*], University of New England, Armidale, NSW 2351 Australia

[2] NSW Department of Primary Industries, Agricultural Research Centre, Trangie, NSW, 2823 Australia

[3] College of Science, Murdoch University, Murdoch, Western Australia, 6150 Australia

[4] CSIRO, Agriculture and Food, F.D. McMaster Laboratory, Armidale, NSW 2350 Australia

SUMMARY

Visual traits are considered valuable components within the breeding objectives of many Merino breeders. This paper aimed to estimate genetic and phenotypic correlations between visual traits and growth, body composition, reproduction and survival in adult ewes. The data were derived from Merino Lifetime Productivity (MLP) sites. Heritability estimates were high for body weight, eye muscle depth, fat depth, body wrinkle, breech wrinkle, breech cover and classer grade (0.32 – 0.64), moderate for urine stain (0.21) and legs score (0.23) and low for weaning rate (0.07) and ewe survival (0.06). Low to moderate negative (favourable) genetic correlations were estimated between the visual traits and body weight and composition, reproduction, and survival traits. Phenotypic correlations between the visual traits and adult body composition and weaning rate traits were negative and low. The genetic and phenotypic correlations estimated in this study were generally favourable hence consideration of visual traits in selection and classing may have beneficial effects on adult ewe performance.

INTRODUCTION

Merino sheep are often visually assessed for a range of traits that are not easily evaluated by quantitative measurements (Mortimer *et al.* 2009). These traits contribute to the cost of production, the value of wool and meat and the welfare of the sheep; hence, they are considered valuable components within the breeding objective of Australian Merino sheep. Professional sheep classers and trained technicians currently use standardised scoring systems to visually assess sheep for evaluations by Sheep Genetics and the Australian Merino Sire Evaluation Association (Brown *et al.* 2007; Australian Wool Innovation 2019; https://merinosuperiorsires.com.au/australian-sire-evaluation). Moderate heritabilities and low genetic correlations have been reported in the literature between some visual traits and body composition (Mortimer *et al.* 2009). Walkom and Brown (2016) estimated genetic parameters and relationships among some visual and production traits in the Sheep Cooperative Research Centre Information Nucleus Flocks. However, the association among early visual traits and ewe survival are largely unknown. This study utilised data from the Merino Lifetime Productivity (MLP) project (Ramsay *et al.* 2019) to estimate preliminary genetic relationships between visual classing traits recorded pre-selection and adult ewe measures of body composition, reproduction and survival.

MATERIALS AND METHODS

Data. Data were extracted for 5,916 Merino ewes from the Balmoral, MerinoLink, New England, Macquarie and Pingelly Merino Lifetime Productivity (MLP) project sites (Ramsay *et al.*

[*] A joint venture of NSW Department of Primary Industries and University of New England

2019). These first generation (F1) ewes were the progeny of 134 sires and 4,266 dams. All sites provided two cohorts of F1 ewes with lifetime data up to seven years of age. Additionally, the sires and sites represent the main wool growing regions in Australia and genotypes found in Australia (Ramsay *et al.* 2019). The data included lifetime records (all repeat records available) for weaning rate (WR), pre-joining adult body weight (AWT, kg), live ultrasound eye muscle depth (AEMD, mm) and live ultrasound fat at the C site (AFAT, mm). The visual traits included body wrinkle, breech wrinkle, breech cover, urine stain, legs score and classer grade, as defined in Table 1. The visual traits were scored on a scale of one to five except for grade, which was scored in categories of tops (1), flock (2) and culls (3). Ewe survival was defined as the ability of ewes to survive from yearling to beyond their fourth year of age (0 or 1). Individual ewes that missed consecutive adult reproduction, body and wool trait measurements due to involuntary culling or culling for welfare reasons were assumed to have been dispersed from the flock and assigned 0 for survival. Outlier measurements beyond four standard deviations across the dataset for body weight, fat and eye muscle measurements were dropped from the analysis.

Table 1. Visual trait descriptions, age stages considered and their standard scoring scale (Australian Wool Innovation 2019)

Trait	Description	Scores
Breech wrinkle (MBRWR)	Degree and quantity of wrinkle on the breech at marking (1 – no wrinkle and 5 – extensive wrinkle)	1 - 5
Body wrinkle (YBDWR)	Degree and quantity of wrinkle on the body at yearling (5 – extensive wrinkles and heavy folds of skin over the entire body)	1 - 5
Breech cover (MBCOV)	Amount of natural bare skin around the perineum and breech area at marking (5 – complete wool cover)	1 - 5
Urine (HURINE)	A score of the extent of breech, hind legs and tail wool stained by urine at hogget (5 – extensive urine)	1 - 5
Legs (PLEGS)	Overall soundness of the front and back leg and feet structure at post-weaning (5 – extreme angulation at the hocks and pasterns of the back legs)	1 - 5
Grade (HGRADE)	Standard of the sheep for visual performance relative to the flock breeding objective at hogget (1 – tops and 3 – culls)	1 – 3

Statistical analysis. Univariate and bivariate mixed linear models were used to estimate variance components and, genetic and phenotypic correlations between the visually assessed traits and body composition traits using the ASReml software package (Gilmour *et al.* 2015). Fixed effects in the models included contemporary group (flock, year of birth and management group, 97 levels), and the interaction between birth and rear type (8 levels). Additive genetic, permanent environmental and genetic group effects (182) were fitted as random. The permanent environmental effect was fitted for adult traits with repeated records. An extended pedigree with 10,546 animals from MERINOSELECT (Brown *et al.* 2007) was used to capture all known ancestors of the animals with data and their parents, and with genetic groups defined for base animals with unknown parents. Ultrasound fat and eye muscle depth records were adjusted for body weight (van der Werf 2004). Variance components and heritability for survival were based on a binomial univariate model. The correlations between muscle and fat and visual traits were post-adjusted for body weight as shown by van der Werf (2004).

RESULTS AND DISCUSSION

Low heritabilities were estimated for weaning rate (0.07) and survival (0.06) (Table 2). Low heritabilities for reproduction (Walkom and Brown 2016; Bunter *et al.* 2019) and survival traits

(Hatcher *et al.* 2009) have been commonly reported for Merino sheep. The heritability for fat, urine stain, leg score, classer grade, body weight, eye muscle depth, body wrinkle and breech wrinkle ranged from 0.21 to 0.64, indicating considerable genetic variation that could be exploited to improve these traits through selection. These parameters were within the ranges of estimates reported by Brown *et al.* (2010) and Walkom and Brown (2016). However, lower estimates for classer grade, body and breech wrinkle and legs scores for front and back legs have been estimated Mortimer *et al.* (2009). Future analysis using threshold models will be considered for the categorical traits.

Table 2. Data summary, genetic groups ($\boldsymbol{\sigma_{gg}}$), additive genetic ($\boldsymbol{\sigma_a}$), permanent environment ($\boldsymbol{\sigma}_{\mathrm{pe}}$) and phenotypic ($\boldsymbol{\sigma_p}$) variances and heritabilities for body composition, weaning rate (WR), visual traits and ewe survival traits (full trait definitions in Table 1 and in data section)

Trait	Records	Mean (std)	σ^2_{gg}	σ^2_a	σ^2_p	h^2
AWT	15,338	59.06 (11.21)	4.15	19.86	48.93	0.41 (0.04)
AEMD	15,337	24.21 (3.23)	0.30	2.05	5.22	0.39 (0.04)
AFAT	15,331	3.36 (1.70)	0.08	0.43	1.10	0.40 (0.04)
WR	15,298	1.09 (0.67)	0.03	0.03	0.42	0.07 (0.02)
YBDWR	3,318	2.28 (0.84)	0.20	0.14	0.32	0.44 (0.08)
MBRWR	5,771	2.53 (0.95)	0.21	0.48	0.76	0.64 (0.06)
MBCOV	5,771	3.60 (1.11)	0.08	0.17	0.49	0.34 (0.05)
HURINE	2,564	1.71 (0.75)	0.01	0.08	0.41	0.21 (0.06)
PLEGS	3,824	2.08 (0.77)	0.03	0.13	0.54	0.23 (0.05)
HGRADE	5,304	2.00 (0.65)	0.03	0.13	0.41	0.32 (0.05)
Ewe Survival	5,494	0.77 (0.42)	0.01	0.21	3.50	0.06 (0.03)

The genetic relationships between early visual traits and adult body weight (Table 3), indicated that lower wrinkle, barer breech cover, lower urine stain, better legs and/or classer grade scores were all associated with heavier ewes. This relationship supports previous findings by Mortimer *et al.* (2009) and Brown *et al.* (2010). The association between early body and breech wrinkle scores and adult muscle and fat was also favourable, implying that plainer ewes (less wrinkle) were genetically more likely to have higher body condition. Similar results were reported by Walkom and Brown (2016) between wrinkle and joining condition scores, who also observed high genetic correlations between condition scores and muscle and fat. Ewes with more breech cover were genetically likely to have higher body fat. The positive correlation between classer grade and adult muscle and fat implies that classers favour ewes with lower body condition. This may be related to relationships between these traits and others not included in this study (wool traits for example), which is an area for further investigation. Phenotypic correlations followed a similar trend to the genetic correlations except for the positive correlation between urine stain and adult body weight and fat. Bigger and heavier ewes with longer fleece could, therefore, tend to have more urine stain.

Low and favourable genetic relationships existed between weaning rate and body wrinkle, breech wrinkle and leg scores showing that ewes with lower wrinkle and good legs would tend to wean more lambs. The relationship between urine stain and leg scores with weaning rate should be treated cautiously due to the high standard errors. The genetic correlation between classer grade and weaning rate was also negative, indicating a favourable relationship between classing and reproduction. Low phenotypic correlations were estimated between the visually assessed traits and weaning rate. Negative and moderate genetic correlations were estimated between survival and the wrinkle traits. These results suggest that plain bodied ewes with low breech cover score (barer breech) at an early age are likely to survive longer in the flock. The genetic correlations estimated between survival and breech cover, urine stains, legs and classer grade were considered not significantly different to 0. The phenotypic correlations between survival and the visual traits were also close to zero or not significantly different from zero. Further analysis of survival is required to better understand the impact of other traits.

Table 3. Genetic and phenotypic correlations between welfare traits at yearling, post-weaning or hogget stage and lifetime adult production

	Trait	AWT	AEMD	AFAT	WR	Ewe survival
Genetic	YBDWR	-0.18 (0.05)	-0.06 (0.05)	-0.12 (0.06)	-0.15 (0.08)	-0.43 (0.26)
	MBRWR	-0.09 (0.03)	-0.11 (0.03)	-0.08 (0.03)	-0.22 (0.06)	-0.46 (0.19)
	MBCOV	-0.33 (0.04)	0.02 (0.04)	0.11 (0.04)	-0.05 (0.08)	-0.01 (0.21)
	HURINE	-0.35 (0.08)	-0.14 (0.08)	-0.03 (0.08)	-0.21 (0.13)	-0.03 (0.29)
	PLEGS	-0.36 (0.07)	-0.11 (0.06)	0.06 (0.06)	-0.23 (0.10)	0.04 (0.27)
	HGRADE	-0.53 (0.04)	0.16 (0.04)	0.21 (0.05)	-0.24 (0.08)	-0.01 (0.20)
Phenotypic	YBDWR	-0.12 (0.02)	-0.04 (0.02)	-0.17 (0.03)	-0.07 (0.02)	-0.01 (0.02)
	MBRWR	-0.06 (0.02)	-0.13 (0.02)	-0.15 (0.02)	-0.10 (0.02)	-0.03 (0.01)
	MBCOV	-0.26 (0.02)	0.05 (0.02)	0.01 (0.04)	-0.03 (0.02)	-0.03 (0.01)
	HURINE	0.21 (0.03)	-0.11 (0.03)	0.01 (0.03)	0.03 (0.03)	-0.04 (0.02)
	PLEGS	-0.16 (0.02)	-0.06 (0.02)	-0.01 (0.03)	-0.03 (0.02)	-0.01 (0.02)
	HGRADE	-0.42 (0.02)	0.05 (0.02)	0.01 (0.02)	-0.05 (0.02)	-0.00 (0.01)

CONCLUSION

The genetic and phenotypic correlations estimated in this study were generally favourable hence emphasis on visual traits prior to first selection of maiden ewes into the breeding flock may have beneficial effects on adult ewe performance. This was a preliminary analysis and after data collection is completed a more comprehensive analysis will be conducted.

ACKNOWLEDGEMENTS

The authors acknowledge Australian Wool Innovation, New South Wales Merino Stud Breeders and the Merino Lifetime Productivity project sites for supporting and funding this work. The authors thank all the site managers and committees for facilitating data collection. The Australian Government supports research, development and marketing of Australian wool through AWI.

REFERENCES

Australian Wool Innovation (2019) Visual sheep scores. Australian Wool Innovation: Sydney.

Brown D., Huisman A., Swan A., Graser H., Woolaston R., Ball A., Atkins K. and Banks R. (2007) *Proc. Proc. Assoc. Advmt. Anim. Breed. Genet.* 17*:*187.

Brown D.J., Swan A.A. and Gill J.S. (2010) *Anim. Prod. Sci.* 50: 1060.

Bunter K., Swan A., Gurman P., Boerner V., McMillan A. and Brown D. (2019) *Proc. Assoc. Advmt. Anim. Breed. Genet.* 23*:* 560.

Gilmour A., Gogel B., Cullis B., Welham S. and Thompson R. (2015) 'ASReml user guide release 4.1 structural specification.' (VSN international ltd: Hemel Hempstead)

Hatcher S., Atkins K. and Thornberry K. (2009) *Proc. Assoc. Advmt. Anim. Breed. Genet.* 18*:*580.

Mortimer S.I., Robinson D.L., Atkins K.D., Brien F.D., Swan A.A., Taylor P.J. and Fogarty N.M. (2009) *Anim. Prod. Sci.* 49: 32.

Ramsay A., Swan A. and Swain B. (2019) *Proc. Assoc. Advmt. Anim. Breed. Genet.* 23*:* 512.

van der Werf J.H.J. (2004) *Aust.J. Exp. Agric.* 44: 405.

Walkom S. and Brown D. (2016) *Anim. Prod. Sci.* 57: 20.

COMPARISON OF UDDER AND TEAT TRAITS IN MERINO EWES RECORDED AT LAMBING AND WEANING

E.G. Smith[1,2], G.A. Acton[1], A.M. Bell[1] and J.L. Smith[1]

[1]CSIRO, Agriculture and Food, F.D. McMaster Laboratory, Armidale, NSW, 2350 Australia
[2]School of Environmental and Rural Science, University of New England, Armidale, NSW, 2350 Australia

SUMMARY

In Australia, there is currently no standard system for assessing ewe udder traits for genetic improvement. The aim of this study was to provide preliminary genetic parameter estimates of four visually scored udder and teat traits recorded at lambing and weaning, to inform recommendations about how and when to record udder and teat traits. Udder depth, teat size and teat placement were moderately heritable at both lambing and weaning (0.23 ± 0.08 to 0.36 ± 0.09) and the traits recorded at the two stages showed high genetic correlations (udder depth 0.75 ± 0.14; teat size 0.79 ± 0.12; teat placement 0.70 ± 0.16). Udder cleft, showed lower heritability, and lower genetic correlation across the two stages, with increased phenotypic variance from lambing to weaning. These results suggest that either stage is appropriate for recording udder depth, teat size and teat placement for genetic improvement of Australian Merinos.

INTRODUCTION

Neonatal lamb mortality is the most significant health issue of Australian sheep with substantial economic, welfare and sustainability implications (Shephard *et al.* 2022). In Australia, neonatal mortalities are mainly caused by dystocia/birth injury and the starvation/mismothering/exposure (SME) complex, each contributing to approximately 40% of neonatal deaths (Hinch and Brien 2014). Starvation mortalities are typically regarded as multifactorial, however poor udder and teat conformation of the dam have been implicated (Jordan and Mayer 1989). Mortality rates in lambs born to ewes with defective udder function have been shown to be more than double that observed in lambs born to ewes with sound udder conformation (Hayman *et al.* 1955; Griffiths *et al.* 2019). Smith *et al.* (submitted) showed that udder and teat conformation traits of Australian Merino ewes are heritable, with estimates ranging from 0.09 to 0.56 across visually scored and measured traits, and that overall udder soundness was associated with lamb survival. Further, (Smith *et al.* submitted) observed that in some instances udder conformation issues noted at birth were not readily discernible at weaning, which was consistent with the findings of Griffiths *et al.* (2019). The objective of this study was to build on earlier work, providing preliminary genetic parameter estimates of udder and teat traits assessed at birth and weaning, to inform recommendations regarding the optimal time for their assessment.

MATERIALS AND METHODS

Data source. The study was conducted during 2022 using ewes from the New England Merino Lifetime Productivity (MLP) flock (Ramsay *et al.* 2019), maintained by CSIRO at the FD McMaster Laboratory, Chiswick, Uralla NSW, Australia, according to MLP project protocols (AMSEA 2020). The flock was generated by artificial insemination in 2017 and 2018 from 28 genetically diverse Merino and Poll Merino sires (15 sires per year with 2 sires used across years for genetic linkage). In 2022 ewe progeny per year-sire group ranged from 27 to 57 ewes. The flock comprised 619, 4 year old (yo) (born 2018) and 638 5 yo (born 2017) ewes, however, only those ewes that lambed in 2022 (number lambs born, NLB>0) and reared at least one lamb to weaning (number lambs weaned, NLW>0) were included in the statistical analysis (n=1,105). The ewes were natural syndicate mated within age groups for 35 days (d) commencing 28th March (d0). Lambing took place from d142-187. Lambs from the 4yo and 5yo ewes were weaned on d248 (median age 89d) and d252 (median age d93) respectively. Udder and teat traits were recorded on the ewes at lambing during lambing rounds (twice-daily), and on the days following weaning (d248-249 and d252-254 for the 4yo and 5yo ewes, respectively). Experimental procedures conducted on animals were approved by the CSIRO Armidale Animal Ethics Committee (Animal Research Authority no. 21/24).

Udder and teat appraisal. Ewes were visually scored (1-5) while in a standing position for 4 udder and teat traits at lambing (L) and weaning (W). Traits assessed were udder depth (UD, size of the udder in relation to the hock, 1=smallest to 5=largest, udder floor below hock); udder cleft (UC, reflects udder symmetry, strength of the medial ligament and attachment to the abdomen, 1=well defined cleft (strong medial ligament), 2=evident cleft, 3=flat udder floor or 'broken' (weak) ligament, 4=asymmetric but both halves functioning; 5=asymmetric with one half involuted); teat size (TS, combination of teat width and length, 1=smallest to 5=largest) and teat placement (TP, position of teat relative to horizontal, 1=high on udder, horizontal, 3=45° from vertical, 5=vertical). Score 3 is considered optimal in terms of productivity and ewe health for all traits, except UC where score 1 is optimal. At lambing, ewes were assessed by 1 of 4 trained operators during lambing rounds, and at weaning by a single operator (1 of the initial 4) in a classing crate.

Statistical analysis. Univariate mixed animal models were applied using ASReml software package (Gilmour *et al.* 2021) for determination of significant fixed effects and covariates on the udder and teat traits, and to estimate (co)variance components and heritability. The maternal environmental effect was tested, but determined by likelihood ratio testing to be non-significant and was not considered further. All traits approximated normality and no interactions among fixed effects were considered. Phenotypic and genetic correlations among the udder and teat traits were estimated from pairwise bivariate models. Fixed effects tested for both the lambing and weaning-stage traits were dam source (3 levels, reflecting the genetic background of the MLP Base ewes) and contemporary group (CG, 4 levels, combined ewe birth year and management group at/following lambing). Assessor of the lambing-stage traits was confounded with lambing management group. For the lambing-stage traits, number of lambs born in 2022 (NLB, 3 levels) and total NLB up to and including 2022 (TotNLB, 10 levels) were also tested, along with bodyweight and condition score pre-mating (pmWT and pmCS) and at late-pregnancy (lpWT and lpCS), all as linear covariates. For the weaning-stage traits, number of lambs weaned in 2022 (NLW, 2 levels), total NLW up to and including 2022 (TotNLW, 8 levels) and day of assessment (ie. from weaning, DoA, 3 levels) were also tested. Linear covariates tested on the weaning-stage traits were days of lactation (DoL), pmWT, pmCS, lpWT, lpCS, as well as weight and condition score at weaning (wWT and wCS).

RESULTS AND DISCUSSION

Phenotypes and heritabilities. The majority of ewes exhibited udders of moderate size with a defined udder cleft and moderately sized teats positioned at or near 45° from vertical (Table 1). The udder and teat trait heritabilities estimated here ranged from 0.09 ± 0.05 to 0.36 ± 0.09 among the two stages. These estimates were higher than those estimated previously by Smith *et al.* (submitted) in the same ewe population and for the same traits at weaning (0.01 to 0.17), but similar to those estimated by McLaren *et al.* (2018) in a terminal breed (0.14 to 0.35). The differences observed in the MLP flock across the different studies may be attributable to some refinements to the scoring system, exclusion from the current study of the ewes that were not lactating at weaning, and age of the ewes. The heritability estimates for UD and TP were consistent across the stages (Table 2). The heritability of TS was higher at weaning than lambing, and UC was lower at weaning than lambing. The phenotypic variance of UC doubled from lambing to weaning which suggests deterioration in UC during the lactation period, with increased expression of udder asymmetry at weaning.

Table 1. Descriptive statistics, significance of fixed effects and phenotypic variance (Vp) for ewe udder depth (UD), udder cleft (UC), teat size (TS) and teat placement (TP) at lambing (L) and weaning (W)

	LUD	LUC	LTS	LTP	WUD	WUC	WTS	WTP
Mean	2.99	2.21	2.70	2.87	3.17	1.97	2.75	3.08
Sd	0.68	0.69	0.68	0.55	0.48	1.00	0.54	0.37
Range	1 - 5	1 - 5	2 - 5	1 - 5	1 - 5	1 - 5	1 - 5	2 - 5
CG	***	***	***	***	ns	*	*	ns
NLB22	**	ns	*	*	-	-	-	-
NLW22	-	-	-	-	***	**	ns	***
DoA	-	-	-	-	***	***	**	***
Vp	0.35±0.02	0.45±0.02	0.42±0.02	0.27±0.01	0.21±0.01	0.96±0.04	0.29±0.01	0.13±0.01

n=1,105 for all traits; CG=contemporary group, NLB22=number lambs born 2022, NLW22=number lambs weaned 2022, DoA=day of assessment after weaning; *** P<0.001, ** P<0.01, * P<0.05, ns not significant, '-'=not tested; dam source, Total NLB, Total NLW, and days of lactation were ns effects on all traits, and ewe weight and condition scores pre-mating, late pregnancy and weaning were mostly ns (not reported here)

Table 2. Heritability (bold, diagonal), phenotypic correlations (above diagonal) and genetic correlations (below diagonal) (all ±s.e.) for ewe udder depth (UD), udder cleft (UC), teat size (TS) and teat placement (TP) at lambing (L) and weaning (W)

Trait	LUD	LUC	LTS	LTP	WUD	WUC	WTS	WTP
LUD	0.29±0.09	0.12±0.03	0.29±0.03	0.03±0.03	0.21±0.03	0.02±0.03	0.11±0.03	0.10±0.03
LUC	0.52±0.23	0.17±0.07	0.04±0.03	0.07±0.03	0.06±0.03	0.13±0.03	0.06±0.03	0.03±0.03
LTS	0.37±0.22	-0.24±0.25	0.24±0.08	0.42±0.03	0.11±0.03	0.26±0.03	0.31±0.03	0.28±0.03
LTP	-0.61±0.22	-0.59±0.23	0.55±0.18	0.23±0.08	-0.05±0.03	0.00±0.03	0.14±0.03	0.27±0.03
WUD	0.75±0.14	0.67±0.18	0.21±0.22	-0.60±0.19	0.28±0.08	-0.11±0.03	0.14±0.03	0.07±0.03
WUC	0.04±0.31	0.31±0.34	0.02±0.32	-0.03±0.34	0.00±0.30	0.09± 0.05	0.05±0.03	0.06±0.03
WTS	-0.04±0.21	0.22±0.25	0.79±0.12	0.34±0.21	0.19±0.20	-0.23±0.28	0.36±0.09	0.33±0.03
WTP	-0.16±0.24	-0.20±0.26	0.69±0.15	0.70±0.16	0.04±0.23	-0.65±0.23	0.60±0.16	0.24±0.08

Phenotypic and genetic correlations. Within stages, phenotypic correlations among the udder and teat traits were low to moderate (0.04 ± 0.03 to 0.42 ± 0.03 at lambing, and -0.11 ± 0.03 to 0.33 ± 0.03 at weaning). Phenotypic correlations between individual traits across the two stages were also generally moderate ranging from 0.13 ± 0.03 (UC) to 0.31± (0.03) (TS). At lambing, the genetic correlations between UD and UC (0.52 ± 0.23) and between UD and TP (-0.61 ±0.22) indicate that increasing UD is associated with deteriorating UC and high/horizontal TP, but at weaning those correlations were not different from zero. At both lambing and weaning UC and TP were unfavourably correlated genetically (-0.59 ± 0.23 and -0.65 ±0.23 respectively), indicating well-defined UC was associated with more vertical TP. Moderate positive genetic correlations between TS and TP at both lambing (0.55 ±0.18) and weaning (0.60 ± 0.16) imply that large teats tend to be placed vertically. In general, the genetic correlations estimated here are consistent with those of Fernandez *et al.* (1997). Scores for UD (0.75 ±0.14), TS (0.79 ± 0.12) and TP (0.70 ± 0.16) at lambing and weaning were highly correlated genetically. While these genetic parameter estimates have high associated errors and should be interpreted with caution, they do suggest that for UD, TS and TP there would be minimal re-ranking between lambing and weaning. UC at lambing and weaning was not significantly correlated genetically and the estimate had a high error (0.31 ± 0.34).

Implications. Ewe udder soundness, which encompasses aspects of udder and teat conformation has been shown to impact neonatal lamb survival (Hayman *et al.* 1955; Griffiths *et al.* 2019; Smith *et al.* submitted). Genetic improvement of ewe udder conformation may be a means of reducing lamb mortality. The phenotypic and genetic parameters estimated here suggest that for UD, TS and TP, there could be similar genetic gain through trait recording at either lambing or weaning. However, there are likely trade-offs relating to data collection logistics of scoring udder traits at lambing or weaning. Breeders who do not conduct birth records are likely to favour udder scoring at weaning. This may be advantageous for the UC trait, for which deleterious levels may not become evident until weaning. For those who already collect birth records, additional udder scores are likely a minor imposition and may offer better selection outcomes in terms of future lamb survival. Where a lamb dies as a neonate due to an udder issue of the dam, the problem would likely be identified if udder scoring were conducted at birth. If udders were not assessed until weaning, the issue may not be identifiable because the udder will go through involution returning to a dry state. In the current study, DoL was a non-significant effect on weaning udder scores, which is in contrast to Smith *et al.* (submitted), and therefore requires further investigation. However, DoL can only be accurately calculated with knowledge of the date of birth, adding support to udder scores at lambing. Further, assessor of udder traits at lambing was confounded with management group, and at weaning DoA was a significant effect, so both of these factors require consideration in udder trait data collection.

CONCLUSION

Udder depth, teat size and teat placement scored at lambing and weaning on Merino ewes was moderately heritable. For these three traits, the genetic correlations between records at lambing and weaning were high. This indicates that among ewes that have reared a lamb(s) to weaning, there would be minimal re-ranking of ewes across those stages. Udder cleft had lower heritability and lower genetic correlations from lambing to weaning than the other three traits. The phenotypic variance of udder cleft increased from lambing to weaning, suggesting that udder cleft issues develop during lactation and therefore may be more accurately assessed at weaning.

ACKNOWLEDGEMENTS

The New England MLP flock is co-funded by AWI and CSIRO with support from woolgrowers. The Australian Government supports research, development and marketing of Australian wool through AWI. The conduct of this study was funded by CSIRO. The authors thank Heather Brewer and Daniel Driscoll for their technical assistance.

REFERENCES

AMSEA (2020) https://merinosuperiorsires.com.au/mlp-project/.

Fernandez G., Baro G.A., de la Fuente L.F. and San Primitivo F. (1997) *J. Dairy Sci.* 80: 601.

Gilmour A., Gogel B., Cullis B., Welham S., Thompson R., Butler D., Cherry M., Collins D., Dutkowski G. and Harding S. (2021) *VSN International Ltd, Hempstead, UK.*

Griffiths K.J., Ridler A.L., Compton C.W.R., Corner-Thomas R.A. and Kenyon P.R. (2019) *N.Z. Vet. J.* 67: 172.

Hayman R.H., Turner H.N. and Turtont E. (1955) *Aust. J. Agric. Res.* 6: 446.

Hinch G.N. and Brien F. (2014) *Anim. Prod. Sci.* 54: 656.

Jordan D.J. and Mayer D.G. (1989) *Aust. J. Exp Agric.* 29: 31.

McLaren A., Kaseja K., Yates J., Mucha S., Lambe N.R. and Conington J. (2018) *Animal* 12: 2470.

Ramsay A., Swan A. and Swain B. (2019) *Proc. Assoc. Advmt. Anim. Breed. Genet* 23: 512.

Shephard R., Ware J.W., Blomfield B. and Niethe G. (2022) *MLA Limited, B.AHE.0327.*

Smith E.G., Hine B.C., Acton G.A., Bell A.M., Doyle E.K. and Smith J.L. (*sumbitted*). *Small Rumin. Res.*

Proc. Assoc. Advmt. Anim. Breed. Genet. 25: 55 - 58

AGE AT PUBERTY, DAYS TO CALVING AND FIRST PARITY RETURN TO OESTRUS IN AUSTRALIAN TEMPERATE BEEF BREEDS

K.A. Donoghue[1], R. Rippon[2], M. Wolcott[3], K.L. Moore[3], S.A. Clark[4] and B.J. Walmsley[3,5]

[1]NSW Department of Primary Industries, Agricultural Research Centre, Trangie, NSW, 2823 Australia

[2]NSW Department of Primary Industries, Primary Industries Institute, Grafton, NSW, 2460 Australia

[3]Animal Genetics and Breeding Unit*, University of New England, Armidale, NSW, 2351 Australia

[4]School of Environmental and Rural Science, University of New England, Armidale, NSW, 2351 Australia

[5]NSW Department of Primary Industries, Livestock Industries Centre, Armidale, NSW, 2351 Australia

SUMMARY

571 females from six beef breeds (Angus, Brahman, Charolais, Hereford, Shorthorn and Wagyu) from the first cohort of the Southern MultiBreed project were recorded for fertility traits at different physiological stages up until their second mating. Traits included age at puberty, days to calving and days to return to oestrus following first calving. Sire least-square means for these traits were used to examine relationships between traits. There was a strong positive relationship between age at puberty and days to calving, indicating that sires whose progeny reached puberty at a later age also conceived and calved later. There was a weaker positive relationship between age at puberty and return to oestrus indicating that sires whose progeny reached puberty at a later age also took longer to return to oestrus after the birth of their first calf. A weak negative relationship between days to calving and return to oestrus indicates that sires whose progeny calved later in the calving season exhibited a quicker return to oestrus. The nature of the relationship between these two traits was unexpected given previous studies, and further analyses once data from other years/cohorts is available will be required to gain confidence in the nature of the relationships between these three traits.

INTRODUCTION

Research in Australian northern beef cattle breeds has shown that fertility traits measured with a high degree of precision (such as serial ovarian scanning to detect age at puberty) are heritable, favourably genetically correlated with lifetime reproductive outcomes, and may be suitable to achieve genetic improvement in fertility (Johnston *et al.* 2009). Studies in temperate beef breeds have found moderate heritabilities (0.38-0.42) for age at puberty (Wolcott *et al.* 2019; 2021), highlighting the potential of these traits for improved fertility outcomes in the southern beef industry. While studies in tropically-adapted breeds have indicated that early-life fertility traits have a strong genetic relationship with later fertility traits (Johnston *et al.* 2014), these relationships have not yet been quantified in temperate breeds. This current study aimed to characterise fertility traits at various physiological stages in young beef females from several temperate beef breeds and gain an understanding of the relationships between these different fertility traits.

* A joint venture of NSW Department of Primary Industries and University of New England

MATERIALS AND METHODS

The Southern MultiBreed (SMB) project is being conducted across New South Wales Department of Primary Industries research facilities; Trangie Agricultural Research Centre, Trangie; Grafton Primary Industries Institute, Grafton; Tocal Agricultural Centre, Tocal; Glen Innes Agricultural Research and Advisory Station, Glen Innes; and Elizabeth MacArthur Agricultural Institute (EMAI). Animals from the six different breeds (Angus, Brahman, Charolais, Hereford, Shorthorn and Wagyu) are managed in mixed breed groups at all stages of the production cycle except for joining, which is undertaken in breed groups. See Walmsley *et al.* (2021; 2023) for further details on the SMB project.

Female progeny born in 2020 at the research sites were recorded for fertility traits from weaning through to post-calving in the first parity. Age at puberty (AP) was detected by serial ultrasound ovarian scanning to identify the animal's first *corpus luteum* (CL) (Johnston *et al.* 2009). Scanning commenced within the first month after weaning (approximately nine months of age) and was conducted at 4-5 week intervals. The decision to cease pubertal ovarian scanning was made on a within-breed basis at each site, and occurred once 100% (or extremely close to 100%) of heifers had reached puberty, such that there was little value in collecting additional records. Animals not observed as having reached puberty by the end of ovarian scanning that were pregnant were given an age of puberty value equal to their date of conception, calculated using foetal age at pregnancy test. Animals not observed as having reached puberty by the end of ovarian scanning that failed to fall pregnant were given a penalty value equal to the largest AP trait value within their site breed group + 21 days. Females were joined by natural mating at approximately 15 months of age for 60 days and commenced calving at approximately two years of age. Days to calving (DC) was calculated as the number of days from the start of the joining period until the date of calving. Animals that failed to calve were given a penalty value equal to the largest DC record within their contemporary group + 21 days. Return to oestrus interval (RO) was detected by serial ultrasound ovarian scanning to identify the animal's first *corpus luteum* post-calving. Only females that calved and were lactating were scanned to identify the first return to oestrus. Scanning commenced approximately 45 days after the first calf was born, and ceased once the percentage of females that had cycled post-calving within a breed was at 100% or extremely close to 100%, such that there was little value in collecting additional records. Animals not observed as having cycled post-calving by the end of ovarian scanning were given a penalty value equal to the largest RO trait value within their site breed group + 21 days (if not pregnant) or a trait value equal to their date of conception, calculated using foetal age at pregnancy test .

Statistical analyses. PROC MIXED in SAS (SAS Institute, Cary, NC, USA) was used to obtain least-square mean estimates for the effect of sire. The model fitted for AP included site, contemporary group and sire; for DC the model included site, joining group and sire; and the model for RO included site, joining group, sex of calf and sire.

RESULTS

Table 1 contains a statistical summary of the raw fertility phenotypes pooled across sites and breeds. There were 571 age at puberty (AP) records including 20 females whose first CL was not observed prior to the cessation of ovarian scanning, and received a penalty AP value. There was significant variation for age at puberty, with the first detected CL ranging from approximately 7 months to approximately 27 months with an average of 12 months of age. Results from the pubertal ovarian scanning showed that 72% of heifers were pubertal at joining, though this varied between sites and breeds. There were 542 days to calving records including 105 females that failed to calve and received a penalty DC value. DC ranged from 273 to 386 days, with an average of 316 days. There were 416 return to oestrus records, including 30 females whose first CL post-calving was not observed prior to the cessation of ovarian scanning, and received a penalty RO value. RO was only recorded on lactating females and ranged from approximately 1 month to 6 months with an average of approximately 3 months post-calving.

Table 1: Summary statistics of the raw unadjusted age at puberty (AP), days to calving (DC) and return to oestrus interval (RO) phenotypes across sites and breeds

Trait	Number	Mean	SD	Minimum	Maximum
AP (days)	571	355.4	81.2	207	816
DC (days)	542	316.1	34.3	273	386
RO (days)	416	98.3	25.3	23	180

Table 2 contains summary statistics by breed for the number of trait records, the number of sires represented in the data set, and the average number of progeny for these sires. There were 158 sires with progeny recorded for AP, with the average number of recorded progeny per sire ranging from 2.6 to 4.0. The number of sires with progeny recorded for DC was 151. The number of sires with progeny recorded for RO was 140 (average number of progeny ranged from 2.6-3.4), which was lower than AP and DC as only progeny that were lactating (i.e. successfully raised a calf) were recorded for this trait.

Table 2: Number of records, number of sires, average number of progeny (standard error) per sire by breed for age at puberty, days to calving and return to oestrus interval

	Age at puberty			Days to calving			Return to oestrus Interval		
Breed	No. records	No. sires	Av No. progeny	No. records	No. sires	Av No. progeny	No. records	No. sires	Av No. progeny
Angus	205	51	4.0 (3.5)	203	50	4.0 (3.5)	157	46	3.4 (2.8)
Brahman	13	5	2.6 (2.1)	13	5	2.6 (2.1)	1	1	-
Charolais	56	16	3.5 (1.8)	55	16	3.4 (1.7)	43	16	2.7 (1.5)
Hereford	133	38	3.5 (2.2)	111	34	3.3 (1.8)	82	31	2.6 (1.6)
Shorthorn	79	20	4.0 (1.9)	77	19	4.1 (1.8)	56	19	2.9 (1.4)
Wagyu	85	28	3.0 (2.0)	83	27	3.1 (2.1)	77	27	2.8 (1.7)

The relationships between the three fertility traits were examined by plotting the sire least-square means. There was a strong positive relationship between AP and DC (Figure 1a), indicating that, in general, sires whose progeny reached puberty at a later age also conceived later and hence calved later in the calving season. This result concords with previous studies which have reported strong genetic correlations (~0.80) between these traits (Johnston *et al.* 2014). However, there was variation observed, with some sires having progeny that reached puberty earlier than average but calved later (and hence conceived later) than average. There was a weaker positive relationship between AP and RO (Figure 1b), indicating that, in general, sires whose progeny reached puberty at a later age also took longer to return to oestrus after the birth of their first calf. Johnston *et al.* (2014) reported moderate to strong genetic correlations (0.31 to 0.72) between these traits. There was a weak negative relationship between DC and RO (Figure 1c), indicating that, in general, sires whose progeny calved later in the calving season exhibited a quicker return to oestrus. The nature of the relationship between these two traits is different to previous studies (Johnston *et al.* 2014) in tropically-adapted cattle, which reported a strong positive genetic correlation (0.75) between DC and lactation anoestrus interval. In this study, sires had relatively low numbers of progeny as only the first year/cohort of females born within the project were included. It is anticipated that future analyses including data from other years/cohorts will allow for greater confidence in the nature of the relationships between these three traits.

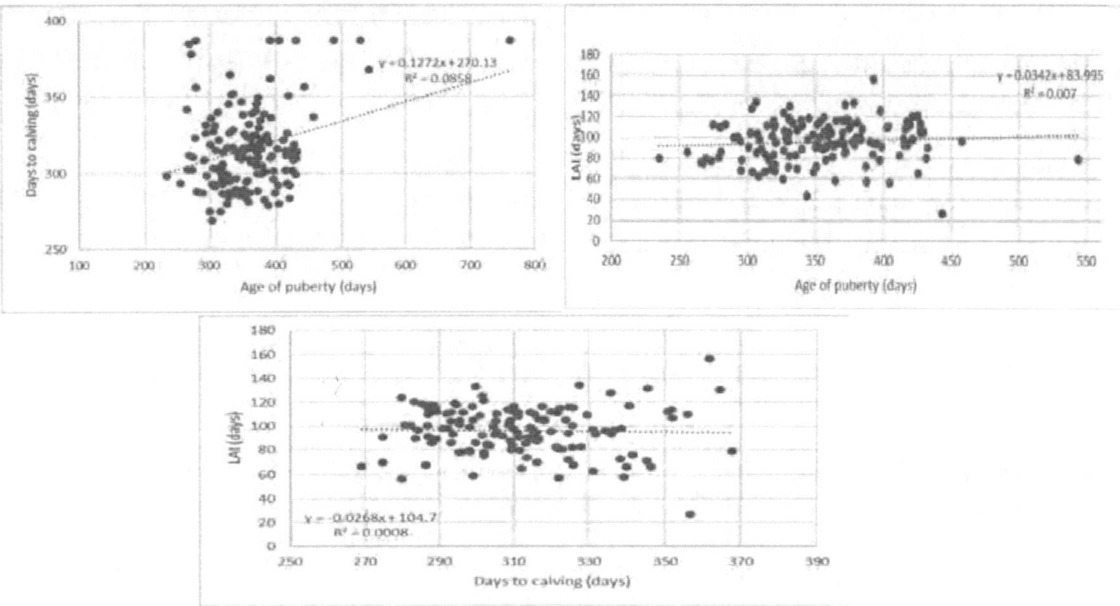

Figure 1: Plot of (1a) age at puberty and days to calving least-square sire means; (1b) age at puberty and return to oestrus interval least-square sire means; (1c) days to calving and return to oestrus interval least-square sire means

CONCLUSIONS

This study reports an initial investigation of female fertility traits in several temperate beef breeds that have been managed in mixed-breed groups. Results showed that phenotypic variation exists in age at puberty, days to calving and return to oestrus interval for these breeds. The next steps will investigate whether genetic variation is also present for these traits once sufficient records are available. Plots of least-square sire means indicated a positive relationship between age at puberty and days to calving and a weaker positive relationship between age at puberty and return to oestrus. A weak negative relationship was found between days to calving and return to oestrus interval, which was contrary to previous studies. Further analyses will be undertaken once more data is available to quantify these relationships with greater confidence.

ACKNOWLEDGEMENTS

The Southern Multi-Breed project is an initiative co-funded by the NSW Department of Primary Industries, the University of New England and the Meat and Livestock Australia Donor Company. The technical expertise provided by Shane Thomson and Jeremy Lim (Holbrook Veterinary Centre) and John Wilkins is gratefully acknowledged.

REFERENCES

Johnston D.J., Barwick S.A., Corbet N.J., Fordyce G., Holroyd R.G., Williams P.J. and Burrow H.M. (2009) *Anim. Prod. Sci.* 49: 399.

Johnston D.J., Corbet N.J., Barwick S.A., Wolcott M.L. and Holroyd R.G. (2014b) *Anim. Prod. Sci.* 54: 74.

Walmsley B.J., Donoghue K.A., Johnston D.J., Clark S.A., Siddell J.P., Walkom S.F., Granleese T. and Arthur P.F. (2021) *Proc. Assoc. Advmt. Anim. Breed. Genet.* 24: 423.

Walmsley B.J., Moore K.L., Walkom S.F., Clark S.A., Granleese T., and Donoghue K.A. (2023) *Proc. Assoc. Advmt. Anim. Breed. Genet.* 25: *These proceedings.*

Wolcott M.L., Banks R.G., Tweedie M. and Alder G. (2019) *Proc. Assoc. Advmt. Anim. Breed. Genet.* 23: 91.

Wolcott M.L., Banks R.G. and Alder G. (2021) *Proc. Assoc. Advmt. Anim. Breed. Genet.* 24: 439.

Proc. Assoc. Advmt. Anim. Breed. Genet. 25: 59 - 62

NEW MODULE FOR PREDICTION OF REPRODUCTIVE TRAITS EBV IN BREEDPLAN

M.G. Jeyaruban and D.J. Johnston

Animal Genetics Breeding Unit*, University of New England, Armidale, 2351, NSW Australia.

SUMMARY

BREEDPLAN publishes EBVs for days to calving (DTC) from natural mating (NAT) as the key measure of genetic merit for female reproduction. More recently, oestrus synchronization and artificial insemination (AI) have become more widely used in beef cattle in Australia to improve reproductive efficiency. The aim of this study was to develop a reproductive module to predict reproductive performance in beef cattle in Australia by using mating outcomes from AI, NAT from females and scrotal circumference (SC) in males. The study analysed mating and calving data collected on Angus cattle in Australia and New Zealand using the events-based recording system introduced in 2010. Genetic parameters for 1st, 2nd and 3rd parities conception rate (CR) to AI and DTC from NAT were estimated. Mean CR from 1st, 2nd and 3rd parities of AI were 51.5%, 56.2% and 70.4% and for DTC of NAT were 303, 308 and 305 days, respectively. Estimated heritability for CR from 1st, 2nd and 3rd parities were 0.15, 0.12 and 0.10 and for DTC were 0.05, 0.11 and 0.14, respectively. Moderate negative genetic correlations (-0.41 to -0.10) were estimated between CR and DTC of all three parities and were significantly lower than 1 suggesting that they were different traits. Therefore, there are benefits in genetic evaluation from including AI data and modelling parities as different traits.

INTRODUCTION

Cow reproductive efficiency is important for the productivity and profitability of beef cattle in Australia. DTC has been implemented in the BREEDPLAN genetic evaluation as the key measure of genetic merit for female reproduction in naturally mated (NAT) females (Schneeberger *et al.* 1991). DTC is calculated as the number of days between the first joining date for a cow and its subsequent calving. However, low heritability, low intensity of selection together with repeat observations accumulating relatively late in life limit the capacity to improve female fertility using DTC measures alone. More recently, oestrus synchronization and AI have become more widely used in beef cattle in Australia to improve reproductive efficiency and increase genetic gain, and has now become the dominant mating technique for seedstock breeders in temperate Australia. CR to AI has been proposed as an important trait to describe reproductive performance in heifers (Bormann *et al.* 2006). However, published heritabilities for CR were also low (Bormann *et al.* 2006) and relationship to natural joining traits unknown. Therefore, we combined several measures of male and female traits to increase the accuracy of EBVs for fertility traits in Angus cattle.

MATERIALS AND METHODS

An enhanced event-based recording system was introduced in 2010 for submission of records for genetic evaluation of reproductive efficiency in BREEDPLAN. This recording system includes all mating events such as mating and pregnancy test outcomes, in addition to culling and disposal dates as well as codes identifying all heifers and cows subjected to synchronization, AI and/or NAT. Mating and calving records for AI and NAT, along with the pedigree data, were obtained for heifers and cows in the BREEDPLAN evaluation for Angus cattle in May 2022. Initial examination of the

* A joint venture of NSW Department of Primary Industries and University of New England

data revealed incomplete data submission for animals from unsuccessful mating. Therefore, 400-days weight management groups were used to eliminate these selective records from AI or NAT. Only contemporary groups with more than 90% of heifers with 400-day weight records also having mating records (either AI or NAT) were included.

Currently in BREEDPLAN, DTC for NAT, was defined as the number of days from "bull in date" to the resultant calving and all non-calving cows were included by assigning a penalty DTC record as described by Johnston and Bunter (1996). Up to 6 DTC records per cow, in a repeatability model, are used to predict EBVs. In this study, for AI mating traits, CR was defined as a binary trait with females, who calved to the first AI sire used, recorded as 1 and those which failed to calve as 0. Records of heifers initially mated by AI, but conceived to subsequent AI or NAT, were analysed as failing (0) to calve to the first AI. Similarly, to DTC, CR to AI for heifers aged between 270 to 625 days at the time of first breeding were identified and their parity records between 270 to 625, 626 to 990 and 991 to 1340 days were defined as 1st (CR1), 2nd (CR2) and 3rd (CR3) CR records, respectively. Scrotal circumference (SC) records of males and their contemporaries measured for DTC and CR, were also included as an extra trait in the analyses. Number of records and descriptive statistics for CR using AI, DTC using NAT and SC are presented in Table 1.

Combined threshold and linear animal models were used to estimate genetic parameters for binary traits (CR) and linear traits (DTC and SC).

Model for CR was $l_{ijs} = cg_i + age_j + age^2{}_j + a_j + s_s + e_{ijs}$

Where l_{ijs} is the liability on the underlying scale for the CR score of animal j in a fixed contemporary group i (cg_i) and age_j and $age^2{}_j$ are linear and quadratic covariates for age at mating, a_j is the random additive genetic effect of female j. The s_s is an additional random effect of service sires. The random error variance was fixed at 1. The contemporary group included herd of birth, year of birth and date of AI. Bayesian analysis, using Gibbs sampling, was used to estimate the means of marginal posterior distributions for CR. The analysis was carried out using THRGIBBS1F90 (Misztal *et al.* 2002). Single chains of 100,000 iterations were sampled with the first 20,000 samples discarded. Every 20th sample was stored and a total of 4,000 were kept to compute posterior means and highest posterior density interval (95%) credible regions.

Model for DTC was $Y_{ij} = cg_i + age_j + age_j^2 + a_j + e_{ij}$

Where Y_{ij} is the DTC of female j in a fixed contemporary group i (cg_i), age_j and $age^2{}_j$ are linear and quadratic covariates for age at mating, a_j is the random additive genetic effect of female j and e_{ij} is the random error associated with this observation. The contemporary group included herd of birth, year of birth and service sire as defined in BREEDPLAN (Graser e*t al.* 2005). In order to account for the selection of data in the 2nd and 3rd AI and NAT, a tri-variate analysis using CR and DTC records from all parities was performed.

Model for SC was $Y_{ijk} = cg_i + age_j + age_j^2 + age_k + age_k^2 + a_j + e_{ijk}$

Where Y_{ijk} is the SC of male j in a fixed contemporary group i (cg_i), age_j and $age^2{}_j$ are linear and quadratic covariates for age at measurement, age_k and $age^2{}_k$ are linear and quadratic covariates for age of dam at birth in days, a_j is the random additive genetic effect of male j and e_{ijk} is the random error associated with this observation.

Estimates of (co)variance components and solutions for fixed effects of DTC for NAT were obtained by REML using an Average Information algorithm (AI algorithm) and the Expectation-maximisation algorithm (EM algorithm) in WOMBAT (Meyer 2007). Genetic correlations between CR, DTC and SC were estimated in a multivariate animal model by combining the threshold model for CR with linear models for DTC and SC. Models identified for the univariate analysis were used in the multi-variate analysis. Pedigree information from up to six generations was used. All multivariate animal model analyses were carried out using THRGIBBSF90 (Misztal *et al.* 2002).

RESULTS AND DISCUSSION

Descriptive statistics for the data used in this analysis are presented in Table 1. Mean ages of heifers at first mating were very similar for AI and NAT and were 434 and 439 days, respectively. However; there was a large difference in the mean calving rate for heifers mated by AI as compared to NAT (51.5% for CR1 and 93.4 for DTC1). Expertise of oestrous synchronization, heat detection and AI all influence the CR for AI heifers. Furthermore, NAT heifers may have had more than one exposure to bulls and more than one expression of heat in the natural mating period. Similar mean calving rate of 93% was observed for 1st, 2nd and 3rd NAT, as well as for their DTC for 1st, 2nd and 3rd parities (303, 308 and 305 days respectively). The percentage of females conceiving to first insemination using AI was similar to the value of 60% reported by Bormann *et al.* (2006) for American Angus heifers. Donoghue *et al.* (2004*a*) reported 79.3% for first CR of Angus cattle in Australia. The higher rate observed may be due to the incomplete submission of data analysed for that study where some reproductive data from animals with unsuccessful mating outcomes were not recorded or included. Furthermore, the data used in Donoghue *et al.* (2004*a*) were collected prior to 2003 and the data used in this analysis were collected after 2010.

Table 1. Descriptive statistics for 1st, 2nd and 3rd parity conception rate to artificial insemination (AI) and days to calving to natural mating (NAT) of heifers and cows and scrotal circumference (SC) of bulls

Variables	AI			NAT (days)			SC (cm)
	1st	2nd	3rd	1st	2nd	3rd	
Number of records	13233	5119	2717	25291	12196	3695	14516
Number of sires	1052	531	357	4014	2275	1112	1274
Number of dams	8805	3778	2148	20684	9357	3347	10083
Number of contemporary groups	165	168	141	2204	1155	850	710
Mean age (days)	434.1	811.8	1174.1	439.4	761.2	1173.7	394.8
(SD)	(22.9)	(31.8)	(24.9)	(37.8)	(38.4)	(34.4)	(45.7)
Mean conception rate (%)	51.5	56.24	70.4	93.4	92.8	93.0	
Trait means				303.4	308.3	305.2	36.6
(SD)				21.4	22.4	20.6	2.97

Estimated posterior means for additive genetic variances, heritabilities and genetic correlations are presented in Table 2. Estimated additive variance and heritabilities for DTC2 and DTC3 were higher than DTC1, supporting the need for splitting the DTC records based on parities. Similar low heritabilities were estimated for CR1, CR2 and CR3. Low to moderate negative genetic correlations were estimated between CR of first three parities using AI and DTC from the first three parities using NAT, illustrating that higher CR is associated with shorter DTC. Overall, the genetic correlations between CR and DTC were lower than 1, suggesting that the CR and DTC were different traits. This is expected for the reasons given in the previous paragraph and in addition, CR is a binary trait and DTC is a continuous trait. Low genetic correlations were estimated between SC with CR and DTC was a continuous trait. Therefore, the data from CR and SC could increase the accuracy of DTC EBVs, enhancing the scope for selection and genetic improvement of female reproduction in Angus heifers.

Additive variance for CR1 was 0.09 and was lower than the value of 0.11 reported by Bormann *et al.* (2006) for American Angus heifers. However, the estimated heritability was slightly higher than the value of 0.03 reported by Donoghue *et al.* (2004*b*) for Angus heifers in Australia. This may be due to the fact that the herds selected in this study, have minimised the incomplete submission of data for animals from unsuccessful mating than the above study. Estimated additive genetic variance and heritabilities for DTC1 of NAT heifers were 16.6 and 0.05 (±0.02) respectively. Estimated heritability was slightly lower than the value of 0.06 reported by Donoghue *et al.* (2004*b*). The genetic correlation between CR and DTC of 1st NAT was lower in magnitude than the value of -0.66 reported by Donoghue *et al.* (2004*b*).

Table 2. Estimated additive genetic variances (σ^2_a), heritabilities (h^2) and genetic correlations for 1st, 2nd and 3rd parity conception rate by artificial insemination (AI) and days to calving by natural mating (NAT) and scrotal circumference (SC) in bulls

	σ^2_a	h^2	Genetic correlations					
Traits[ab]			2nd AI	3rd AI	1st NAT	2nd NAT	3rd NAT	SC
1st AI	0.09	0.08	0.31	0.47	-0.29	-0.14	-0.10	0.10
2nd AI	0.17	0.13		0.44	-0.33	-0.37	-0.34	0.15
3rd AI	0.13	0.12			-0.22	-0.31	-0.41	0.23
1st NAT	16.62	0.05				0.61	0.52	-0.17
2nd NAT	38.21	0.11					0.44	-0.12
3rd NAT	43.73	0.14						-0.10
SC	2.71	0.53						

[a] standard deviation from 4000 iterations from threshold model ranged between 0.04 to 0.06
[b] approximate standard error from linear model evaluation ranged between 0.02 to 0.06.

CONCLUSIONS

This study has shown AI mating records could be included in genetic evaluation of reproduction traits. Higher estimated additive variance and heritabilities for DTC2 and DTC3 than that of DTC1, suggest that separation of DTC records based on parities also benefits the evaluation. Moderate to high non-zero genetic correlations were estimated between CR from the first three AI and DTC from the first three NATs, suggesting that both CRs from AI mating and DTC from NAT need to be included in the BREEDPLAN evaluation to enhance selection for higher heifer fertility. The CR and DTC traits reported here, together with SC, will form the core of a new reproduction trait analysis for BREEDPLAN.

ACKNOWLEDGEMENT

The authors would like to thank Meat and Livestock Australia (MLA) for their financial support through L.GEN.2204 and Angus Australia and Agricultural Business Research Institute (ABRI) for providing data for this study.

REFERENCES

Bormann J.M, Totir L.R, S.D. Kachman, R.L. Fernando and D.E. Wilson (2006). *J. Anim. Sci.* 84: 2022.

Donoghue K.A., Rekaya R., Bertrand J.K. and Misztal I. (2004a) *J. Anim. Sci.* 82: 362.

Donoghue K.A., Rekaya, R., Bertrand J.K. and Misztal I. (2004b) *J. Anim. Sci.* 82: 987.

Graser H-U., Tier B., Johnston D. J. and Barwick S.A. (2005) *Aus. J. Exp. Agr.*45: 913.

Johnston D.J. and Bunter K.L. (1996) *Livest. Prod. Sci.* 45: 13.

Meyer K. (2007) *J. Zhejiang Uni. Sci.* 8:
815

Misztal I., Tsuruta S., Strabel T., Auvray B., Druet T. and Lee D.H. (2002) *World Cong. Genet. Appl. Livest. Prod.* 7: 743.

Schneeberger M., Tier B. and Hammond K. (1991) Proc. *Assoc. Advmt. Anim. Breed. Genet.* 9: 194.

FIBRE DIAMETER VARIATION AS A MEASURE OF RESILIENCE IN SHEEP

E.G. Smith[1], S.F. Walkom[2] and S.A. Clark[1]

[1]School of Environmental and Rural Science, University of New England, Armidale, NSW, 2350 Australia
[2]Animal Genetics Breeding Unit*, University of New England, Armidale, NSW, 2350 Australia

SUMMARY

The ability to select sheep which have a greater capacity to overcome environmental fluctuations is topical given the severity of climatic events, labour shortages and increased productive demands (lamb, meat and wool). In this paper, we review the possibility of using variation in fibre diameter (FD), measured along the wool staple as an indicator of how sheep respond to the fluctuations of their environment.

INTRODUCTION

Production animals are exposed to significant fluctuations in their internal and external environments which can hinder their productive performance, health and well-being. Significant research efforts over the last decade have focused on quantifying the ability of animals to cope with these fluctuations and in turn, selecting those with a greater capacity to overcome them, particularly among intensively raised livestock (Berghof *et al.* 2019). The ability for animals to be minimally affected by environmental fluctuations or to promptly recover from them is referred to as resilience (Colditz and Hine 2016). Many recent studies of resilience are based on quantifying the variable rate of resource accretion into production tissues or products (muscle, milk and eggs) (Colditz *et al. in press*). In idealised form, these measures capture the animal's inherent success in maintaining homeostatic balance, as it modulates resource allocation between survival and production (Neville 1967). An animal undergoing a challenge from its internal or external environment is likely to temporarily divert resources away from non-essential functions (typically production variables), which can then be used to inform assessments of resilience. Genetically, animals with greater uniformity to these production variables have been associated with better current and future health outcomes (Berghof *et al.* 2018) and improved longevity (Adriaens *et al.* 2020; Poppe *et al.* 2020), qualities which are increasingly valued by consumers and producers alike.

Similar methodologies have yet to be investigated in extensive sheep populations, primarily due to a lack of appropriately structured data (frequent measures over long time periods). Assessment of fibre diameter variation along the wool staple is a promising avenue that offers frequent measurement intervals. Variation in FD observed along the wool staple is reflective of changes in nutrient supply and demand to the wool follicle in accordance with the sheep's interaction with its prevailing internal and external environment. In most production systems wool is harvested annually and therefore becomes an archive of these interactions over the previous 12 months. Importantly, wool is among the final stores of energy and protein in the body and unlike other tissue structures such as muscle and fat, resources accumulated in wool cannot be remobilised in times of nutritional deficits (Freer *et al.* 1997). This review will provide contextual background as to how FD variation along the staple has been measured and analysed in other applications. It will also discuss alternative methods of modelling and analysing FD variation. The paper concludes with a discussion of challenges and opportunities for refinement and validation of along-staple FD variation as a measure of resilience.

---

* A joint venture of NSW Department of Primary Industries and University of New England

MEASURING ALONG STAPLE VARIATION

Fibre diameter variations along staples can be assessed through a sampling technique called a fibre diameter profile (FDP). These profiles are created from repeat measures of average FD taken longitudinally and in sequence along the wool staple (Figure 1). The last 50 years have seen significant advances in the instrumentation used to measure FDP. Historically, FDPs were created by segmenting staples into snippets (2 or 5mm), individually measuring each snippet for FD and plotting the average FD against its relative position along the staple. The method was labour intensive which confined studies at the time to small numbers of animals or a reduced number of samples per staple (Brown *et al.* 2000). The commercialisation of OFDA2000 instrumentation in the early 2000's allowed FD variation along staples to be measured quickly and cheaply (Brims *et al.* 1999), and therefore on a scale sufficient to provide phenotypes for genetic evaluations. OFDA2000 generates profiles on entire, greasy or clean staples, typically at measurement increments of 5mm.

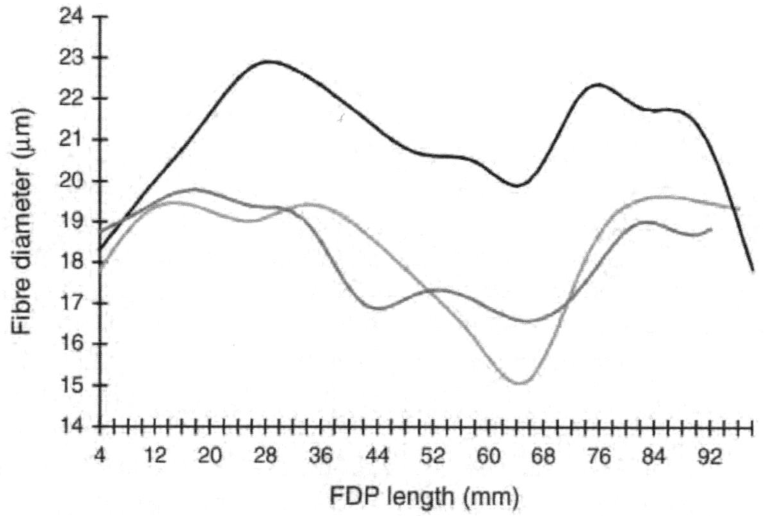

Figure 1. Example of fibre diameter profiles from three animals (Brown and Crook 2005)

GENETIC PARAMETERS OF FD VARIATION ALONG THE STAPLE

Traditionally, FD variation along the staple has been analysed as summary statistics including the minimum and maximum FD and along staple FD coefficient of variation (CV (%)) or standard deviation (SD (µm)), with the intent of investigating their relationships to staple strength. Minimum and maximum FD typically produces high to moderate heritabilities (0.47 to 0.68) (Greeff 2002; Preston and Hatcher 2013a). However, measures of variation (CV and SD) along the staple are inconsistent, ranging from 0.07 to 0.30 (Yamin *et al.* 1999; Greeff 2002; Preston and Hatcher 2013a). The latter two measures are most akin to a trait that reflects FD uniformity in response to environmental conditions throughout the year. However, both of these variation traits are potentially biased due to failures to account for the disparity of staple length between animals which influences the number of FD measures contained in the profile. Similarly, studies to date have not examined the phenotypic and genetic correlations between traits describing along staple FD variation and other important performance traits, with exception of wool quality characteristics (Greeff 2002; Preston and Hatcher 2013b). This is despite indications that reproduction, growth and health may account for some of the variation observed in the profile (Brown and Crook 2005; Gonzalez *et al.* 2020). Overall, these preliminary findings suggest that genetic variation exists for traits derived from FDP, however, for the purpose of examining resilience, further work should progress beyond summary characteristics from the FDP into more comprehensive measures of the variation over time.

POTENTIAL METHODS FOR ANALYSING ALONG STAPLE VARIATION

There are potentially many ways in which FD variation along the staple could be analysed for the purpose of examining resilience. FDP can be thought of as repeat records of FD made between two known time points. Animal breeding has several approaches for analysing longitudinal records, for instance the repeatability or multi-trait models. Perhaps the most appropriate method involves the fitting of curves to phenotypic values across time points and analysing the fitted parameters such as the slope and intercept. This may be considered optimal as it takes account of the genetic and environmental covariance structures between FD measured along the staple.

Random regression models (RRM) are among the most popular methods of analysing longitudinal data such as lactation or growth curves. RRM include a function nested inside the random effects which allows the variance components to vary along a trajectory (Schaeffer 2004). In animal breeding, this function is nested in the individual, thereby modelling the individual deviation from a fixed regression of the trait over time (Kolmodin 2002). RRM most commonly uses Legendre polynomials to fit the fixed and random regressions. The use of splines has also been advocated as an alternative to Legendre polynomials due to greater flexibility in fitting curves of arbitrary shapes (Meyer 2005), which is consistent with FDPs. The possibility remains to use the linear or curve components from RRM to determine the uniformity of animal performance across the trait trajectory, which may be interpreted as a greater ability to cope with environmental fluctuations.

Other studies on resilience have analysed deviations from longitudinal data in what is referred to as profile analysis, where deviations are calculated between reference and observed production curves (Colditz *et al. in press*). Reference curves are typically modelled based on individual or contemporary group means (Elgersma *et al.* 2018; Doekes *et al.* 2022). Statistical measures are then applied to describe the amount of deviation between the reference and observed curve including; natural log variance, skewness and lag-one autocorrelations of the deviation. Such indicators are typically shown to have heritability estimates ranging from 0.01 to 0.26 (Berghof *et al.* 2019; Poppe *et al.* 2020). Together, the performance of these methods of analysis of FD variation are yet to be determined, and each is likely to have merits and limitations.

DISCUSSION

Quantifying FD variation along the staple may offer a unique way of assessing resilience in Australian sheep. Many important questions however remain regarding the analysis and interpretation of such measures. Firstly, provided that genetic variation exists for traits describing along staple FD variation, what is a desirable amount of variation to select for, in regard to resilience? From the points raised above, it may seem that a uniform or relatively flat profile is desired, as the resilient animal is considered to defend the trait expression against the environment. However, conformity to this normative model may be explained by other factors such as inadequacy to perform other productive functions such as to rear a lamb, which is not necessarily desired. This highlights the necessity to examine both the phenotypic and genetic correlations between FD variation and other key performance traits, as well as the need to validate resilience indicators to ensure they are able to provide economic benefits in terms of better health, welfare or long-term productive outcomes.

It is also important to understand how the FD variation along staples performs both across life stages and under different environmental conditions (existence of G x E interactions). The absolute level of the FD variation shown in a contemporary group contains important information about the quality of the environment experienced. This information could not only be used to form an environmental gradient in a reaction norms analysis, but would also complement assessments of profile analysis. Other studies have shown that environmental conditions experienced during the development of young animals can have lasting consequences on the resilience of adult genotypes

(Parois *et al.* 2022). It would be extremely useful to be able to quantify such measures from FDP across years, in particular, with respect to the relationship between resilience measures and performance longevity.

Finally, it is important to remember that breeding for improved resilience to environmental fluctuations should not be interpreted as a means of "forcing" animals to endure less than optimal living or management conditions. This work merely seeks to form part of an integrated approach to helping both animals and producers achieve better production and health outcomes amidst the challenges of future farming systems.

CONCLUSION

The use of FD variation along staple as a way of quantifying the resilience of sheep remains under-explored and offers a research opportunity to inform whether genetic variation exists for such traits in Australian sheep populations. Further work is warranted to understand the most appropriate ways of analysing FDP data and the potential application of these measures in breeding programs.

REFERENCES

Adriaens I., Friggens N.C., Ouweltjes W., Scott H., Aernouts B. and Statham J. (2020) *J. Dairy. Sci.* 103: 7155.

Berghof T.V.L., Bovenhuis H. and Mulder H.A. (2019) *Front. Genet.* 10: 1216.

Berghof T.V.L., Poppe M. and Mulder H.A. (2018) *Front. Genet.* 9: 692.

Brims M., Peterson A. and Gherardi S. (1999) *Inter. Wool. Text. Organi.* 4: 1.

Brown D.J. and Crook B.J. (2005) *Aust. J. Agri. Res.* 56: 673.

Brown D.J., Crook B.J. and Purvis I.W. (2000) *Wool. Tech. Sheep. Breed.* 48: 1.

Colditz I.G. and Hine B.C. (2016) *Anim. Prod. Sci.* 56: 1.

Colditz I.G., Smith E.G., Ingham A.B. and Dominik S. (in press) *Anim. Prod. Sci.*

Doekes H.P., Bovenhuis H., Berghof T.V., Peeters K., Visscher J. and Mulder H.A. (2022) *Poul. Sci.* 102312.

Elgersma G.G., de Jong G., van der Linde R. and Mulder H.A. (2018) *J. Dairy. Sci.* 101: 1240.

Freer M., Moore A. and Donnelly J. (1997) *Agri. Syst.* 54: 77.

Gonzalez E.B., Sacchero D.M. and Easdale M.H. (2020) *J. Arid. Environ.* 181: 1.

Greeff J. (2002) *Wool. Tech. Sheep. Breed.* 50: 11.

Kolmodin R., Strandberg E., Madsen P., Jensen J. and Jorjani H. (2002) *Acta. Agric. Scand. - A: Anim. Sci. 52*: 11.

Meyer K. (2005) *Genet. Sel. Evol.* 37: 473.

Neville A. (1967) *Biol. Rev.* 42: 421.

Parois S., Van Der Zande L., Knol E., Kemp B., Rodenburg T. and Bolhuis J. (2022) *Sci. Rep.* 12: 6804.

Poppe M., Veerkamp R.F., van Pelt M.L. and Mulder H.A. (2020) *J. Dairy. Sci.* 103: 1667.

Preston J. and Hatcher S. (2013a) *Proc. Aust. Assoc. Anim. Breed. Genetics* 20: 118.

Preston J. and Hatcher S. (2013b) *Proc. Aust. Assoc. Anim. Breed. Genetics* 20: 106.

Schaeffer L.R. (2004) *Livest. Prod. Sci.* 86: 35.

Yamin M., Hynd P., Ponzoni R., Hill J., Pitchford W. and Hansford K. (1999) *Int. J. Sheep. Wool. Sci.* 47: 151.

DEVELOPMENT OF FEMALE FERTILITY INDICATOR TRAITS FOR THE ANGUS HeiferSELECT GENOMIC TOOL

P.A. Alexandre[1], L.R. Porto-Neto[1], B.C. Hine[2], A.M. Samaraweera[3], A.I. Byrne[3], A.B. Ingham[1], C.J. Duff[3] and A. Reverter[1]

[1]CSIRO, Agriculture and Food, Queensland Bioscience Precinct, St Lucia, QLD, Australia.
[2]CSIRO, Agriculture and Food, F.D. McMaster Laboratory, Armidale, NSW, Australia.
[3]Angus Australia, Armidale, NSW, Australia

SUMMARY

Angus HeiferSELECT is a genomic tool designed to inform the selection of replacement heifers by providing genomic estimated breeding values (GEBV) for traits related to cow-calf production, feedlot performance, carcase quality, and resilience. Here, we explore the incorporation of fertility indicator measures into the gamut of traits using data from 9,155 heifers in the Angus Australia database. The heritability of age at first calving (AFC), days to calving (DC), and pregnancy test measured in weeks (PREG) were 0.25, 0.26 and 0.32, respectively. The three traits were favourably correlated. AFC and DC presented a genetic correlation of 0.45, while PREG presented negative correlations to the other traits (-0.23 and -0.45, respectively). The accuracy of the GEBVs varied from 0.24 for DC to 0.34 for PREG. Although the three traits showed low to moderate heritability and prediction accuracy, phenotypic differences between animals at the top and bottom quartiles when ranking animals based on GEBV demonstrate the positive impact that could be achieved by selecting for improved female fertility in commercial enterprises. The findings from this study have demonstrated that DC, AFC and PREG would all be suitable traits for inclusion in the Angus HeiferSELECT tool.

INTRODUCTION

Targeted selection of replacement females is crucial for optimising genetic gain in commercial beef enterprises. The decision of which heifers to keep in the operation, and which to sell, potentially affects the profitability of the herd for years to come (Wathes *et al.* 2014). Angus HeiferSELECT is an advanced genomic tool developed to inform the selection of replacement heifers in commercial beef breeding operations. It includes GEBV for thirteen maternal, growth, feed intake, carcase, and resilience traits. Recently, genomic predictions for birth weight, weaning weight, yearling weight and mature cow weight have been validated based on the animal's self-performance as well as the average performance of their progeny (Alexandre *et al.* 2022). However, worldwide there is an increasing effort to include fertility traits in genetic evaluations (Brzáková *et al.* 2020).

Fertility traits are notorious for having low heritability and some, such as the result of a pregnancy test (PREG), are particularly difficult to measure in beef cattle since it requires a qualified technician. In addition, traits such as age at first calving (AFC) and days to calving (DC) are complex because they involve the steps required to conceive, gestate, and deliver a calf (Minick Bormann and Wilson 2010). Yet, these traits not only allow the identification of animals that are more likely to conceive, but also those who will conceive early in the breeding season, which has implications on calf performance, the heifer's successive re-breeding, and overall herd productivity (Moorey and Biase 2020). For instance, shortening the AFC has been shown to decrease replacement rates, decrease production costs and consequently increase profit (López-Paredes *et al.* 2018).

In the present study, we investigate an opportunity to include fertility indicator traits in the Angus HeiferSELECT trait repertoire. Using data from heifers in the Angus Australia database we investigate the heritability of AFC, DC and PREG, the accuracy of genomic predictions and the possible phenotypic impacts of selecting for these traits.

MATERIALS AND METHODS

Data for the 9,155 heifers were retrieved from the Angus Australia database. It included genomic information for 45,364 autosomal SNPs and three fertility indicator traits (Figure 1A): AFC (n=6,806, 734.4±50.9 days), DC calculated from the start of the joining period (n=2,883, 364.4±197.4 days), and PREG (n=6,070, 13±7.4 weeks). Heifers that failed to calve were penalized with a DC value of 980 days. Records for PREG included N for "non-pregnant" (n=819), P for "pregnant" if the number of weeks pregnant was unknown or over 20 weeks (n=1,668), or a number between 3-20 for the number of weeks pregnant at the time of assessment as advised by a qualified technician (n=3,583). To transform PREG into a numerical trait, we assigned a random 0 to 1 to the "N"s and a random 21 to 25 to the "P"s. We reached this decision after comparing the average age of the heifers at the time of assessment for animals with an N, a P, and four groups based on the number of weeks pregnant (Figure 1B).

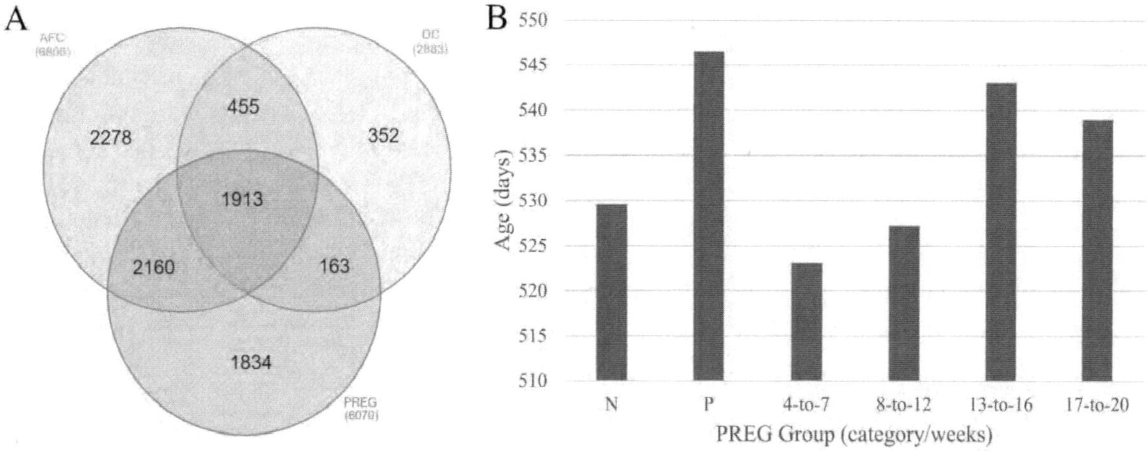

Figure 1. Number of animals with records for age at first calving (AFC), days to calving (DC) and pregnancy test (PREG) (A) and the average age at pregnancy test per category (B)

Heritabilities and genetic correlations were estimated using Qxpak5 (Pérez-Enciso and Misztal 2011). The linear mixed model used to analyse all traits (n=9,155) contained the fixed effects of contemporary group (CG), including mating program type and a minimum CG size of five, and the linear covariate of age at measurement for DC and PREG. The random additive polygenic and residual effects were fitted with assumed distributions $N(0, G \otimes V_G)$ and $N(0, I \otimes V_R)$, respectively, where G represents the genomic relationship matrix (GRM) generated using the first method of VanRaden (2008), V_G is the genetic covariance matrix, I is an identity matrix, V_R is the residual covariance matrix and $\otimes$ represents the Kronecker product.

To ascertain the quality of the resulting GEBVs we used the LR Method following Legarra and Reverter (2018). The method compares predictions based on partial and whole data, resulting in accuracy, dispersion, and bias estimates. For that, a series of univariate analyses were undertaken using adjusted phenotypes, first using the whole dataset (calibration), and then using a partial dataset in which data from a random 20% of records were treated as missing (validation). Finally, animals in the validation population were ranked based on their GEBVs from the analyses of the partial dataset and the difference between the average adjusted phenotype of animals in the top and the bottom quartile was calculated (Q1Q4 measure).

RESULTS AND DISCUSSION

Among the fertility indicators, AFC is the most studied trait. Our estimate of heritability for AFC (0.25, Table 1) is well within the values reported in the literature for black Angus, which range from 0.17 to 0.35 (Brzáková *et al.* 2020; Minick Bormann and Wilson 2010). In contrast, studies report a lower heritability for DC compared to our results (0.26), varying between 0.06 and 0.12 for Angus and Nellore (Donoghue *et al.* 2004; Ferreira Júnior *et al.* 2018). The literature is scarce for PREG, particularly when recorded as a continuous trait. When recorded as a binary trait, pregnancy shows low heritability, around 0.13 to 0.17 for heifers (Bormann *et al.* 2006; Buddenberg *et al.* 1989). In this study, PREG showed the highest heritability (0.32) suggesting that our strategy to transform PREG records into a continuous trait was reasonable and perhaps more suitable for genomic selection than binary pregnancy.

As expected, we found a positive genetic correlation between AFC and DC (0.45, Table 1) although not as high as reported in the literature for Nellore (Forni and Albuquerque 2005). While lower values for AFC and DC are indicative of early conception and are therefore desirable, the opposite is true for PREG. This is reflected in the negative genetic correlation between PREG and the other traits, which was stronger for DC (-0.45).

Table 1. Heritabilities (diagonal), genetic correlations (above diagonal) and residual correlations (below diagonal)

	AFC	DC	PREG
AFC	0.25±0.04	0.45±0.31	-0.23±0.36
DC	0.88±0.01	0.26±0.03	-0.45±0.39
PREG	-0.80±0.02	-0.85±0.01	0.32±0.02

The metrics of GEBV quality are presented in Table 2. The GEBV accuracy varied from 0.24 for AFC to 0.34 for PREG. Indeed, increased accuracy is expected for traits with a higher heritability (Fernandes Júnior *et al.* 2016). There were no signs of bias given the high standard errors, but there could be an indication of overdispersion, particularly for AFC, which is not uncommon (Legarra and Reverter 2018) and can be related to the low heritability of the traits.

Table 2. Method LR accuracy, bias, and dispersion of GEBV for age at first calving (AFC), days to calving (DC) and pregnancy test (PREG)

	AFC	DC	PREG
Accuracy	0.27	0.24	0.34
Bias	-0.17±0.20	-0.64±0.48	0.03±0.02
Dispersion	0.53±0.01	0.06±0.04	0.26±0.02

The Q1Q4 measure for AFC, DC and PREG were respectively 11.4 days, 25.0 days, and 1.7 weeks. Although one can expect the low to moderate heritabilities and GEBV accuracies to be reflected in the size of phenotypic differences between animals in the highest and lowest GEBV quartile, there are still gains that can be anticipated based on genomic selection.

CONCLUSION

This study has demonstrated that DC, AFC and PREG would all be suitable traits for inclusion in the Angus HeiferSELECT tool, with selection based on either trait resulting in gains in female sexual precocity. The phenotypic differences between animals at the top and bottom of the ranks demonstrate the positive impact that could be achieved by selecting for improved female fertility in commercial enterprises using the Angus HeiferSELECT tool.

REFERENCES

Alexandre P., Porto-Neto L., Hine B., Ingham A., Samaraweera M., Duff C., and Reverter A. (2022) *12th World Congress on Genetics Applied to Livestock Production*. 1: 1930.

Bormann J.M., Totir L.R., Kachman S.D., Fernando R.L. and Wilson D. E. (2006) *J. Anim. Sci.* 84: 2022.

Brzáková M., Čítek J., Svitáková A., Veselá Z. and Vostrý L. (2020) *Animals* 10: 2122.

Buddenberg B.J., Brown C.J., Johnson Z.B., Dunn J.E. and Peterson H.P. (1989) *J. Anim. Sci.* 67: 2589.

Donoghue K.A., Rekaya R., Bertrand J.K., and Misztal I. (2004) *J. Anim. Sci.* 82: 987.

Fernandes Júnior G.A., Rosa G.J.M. Valente B.D., Carvalheiro R., Baldi F., Garcia D.A., Gordo D. G.M., Espigolan R., Takada L., Tonussi R.L., de Andrade W.B.F., Magalhães A.F.B., Chardulo L.A.L., Tonhati H. and de Albuquerque L.G. (2016) *Genet. Sel. Evol.* 48: 7.

Ferreira Júnior R.J., Bonilha S.F.M., Monteiro F.M., Cyrillo J.N.S.G., Branco R.H., Silva J.A.I. and Mercadante M.E.Z. (2018) *J. Anim. Sci.* 96: 4035.

Forni S. and Albuquerque L. G. (2005) *J. Anim. Sci.* 83: 1511.

Legarra A., and Reverter A. (2018) *Genet. Sel. Evol.* 50: 53.

López-Paredes J., Angeles Pérez-Cabal M., Jiménez-Montero J. A., and Alenda R. (2018) *J. Anim. Sci.* 96: 4015.

Minick Bormann J., & Wilson D.E. (2010) *J. Anim. Sci.* 88: 1947.

Moorey S.E. and Biase F.H. (2020) *J. Anim. Sci. and Biotechno.* 11: 97.

Pérez-Enciso M. and Misztal I. (2011) *BMC Bioinformatics* 12: 202.

VanRaden P.M. (2008) *J. Dairy Sci.* 91: 4414.

Wathes D.C., Pollott G.E., Johnson K.F., Richardson H. and Cooke J.S. (2014) *Animal* 8: 91.

LIBRARY PREPARATION METHOD AFFECTS OBSERVED MICROBIOME VARIATION WHEN USING OXFORD NANOPORE SEQUENCING

E.M. Ross[1], Z. Chen[1], L.T. Nguyen[1], S. Meale[2], C. T. Ong[1]

[1]Queensland Alliance for Agriculture and Food Innovation, University of Queensland, St. Lucia, QLD, 4072 Australia
[2]School of Agriculture and Food Sciences University of Queensland, Gatton, QLD, 4343 Australia

SUMMARY

New sequencing technologies are opening up new opportunities to explore microbiome variation; however, the technical effects of the molecular methods used have not been characterized. In this study, we aimed to investigate the potential impact of different library preparation methods and base calling algorithms on the observed microbiome variation when using Oxford Nanopore Technologies sequencing. To achieve this, we sequenced technical replicates of a single rumen fluid sample from a cannulated Bos taurus. Our results showed that the use of higher accuracy base calling methods led to a significant increase in the number of classified reads, resulting in more usable data. We did not observe any alteration in the microbial profile due to the use of different base calling algorithms. We also found that the rapid library preparation sequencing kit, which uses an enzymatic method to cut the DNA and ligate the adapter, resulted in shorter sequence lengths and lower numbers of classified reads compared to the Ligation library preparation kit, which does not cut the DNA during library preparation. Importantly, we observed significant differences in the proportion of microbial species within the data generated using the Ligation versus the rapid library preparation kit. Our study suggests that the library preparation method used can impact the observed microbiome and is therefore important to consider in any downstream analysis.

INTRODUCTION

Metagenomics is a popular method to describe microbiome variation, with one important application being the investigation of the relationships between microbiome variation and host phenotype (e.g. Ross *et al.* 2013). Accurate representation of microbiome variation is essential to detect these associations. While short-read sequencing has been the primary method for microbiome analysis to date, the declining cost of long-read sequencing has made it a potential alternative (e.g. Ong *et al.* 2023). To confidently adopt long-read sequencing, specifically using Oxford Nanopore Technologies (ONT), it is crucial to investigate the technical effects of the molecular methods used, as well as the algorithms used to analyse the raw output signal. In this study, we aimed to test the hypothesis that the library preparation method used for generating the ONT sequencing library significantly affects the observed microbiome. Additionally, we tested the hypothesis that the base calling algorithm significantly affected the observed microbiome.

MATERIALS AND METHODS

Sample. This study used technical replicates from a single rumen fluid sample taken from a single 3-year-old cannulated cow (*Bos taurus*) under animal ethics number 2021/AE000991. The animal was fed with hay as a regular diet. Rumen fluid collection was performed by restraining the animal in a crush, removing the cannula, and collecting rumen contents. The rumen fluid was squeezed from the rumen contents and then sieved to remove large particulate matter. The rumen fluid was distributed into 1.5 mL tubes after homogenization and stored at -20°C until samples were processed.

DNA extraction. Thawed 1.5 mL rumen fluid samples were centrifuged at 14,000 rpm for 5 min at 4°C, followed by the removal of the supernatants. Multiple DNA extraction methods were performed to characterise microbiome differences between extraction kits compared to sequencing methods. DNA extraction was performed on the cellular pellet in triplicate for each method. The DNeasy Plant Mini Kit (QIAGEN, Germany) was performed following the manufacturer's protocol. The PowerFecal Pro DNA Kit (QIAGEN, Germany) was used according to the instruction from the manufacturer. The Puregene Blood Core Kit (QIAGEN, Germany) extraction was performed by following the Gram-positive bacteria protocol provided by the manufacturer. Chemical cell lysis was performed in the DNeasy Plant Mini Kit and Puregene Blood Core Kit, while the PowerFecal Pro DNA Kit was combining chemical and mechanical processes. The extracted DNA was stored at -20°C for subsequent use.

Sequencing. Two sequencing kits, the ligation kit (SQK-LSK109) and the rapid kit (SQK-RBK110.96), were used in this study. The Ligation Kit was used for the library preparation for all extraction methods. Exclusively, the rapid kit was used with the PowerFecal Pro DNA kit (Table 1). Barcoding during the library preparation of DNA samples from the Puregene Blood Core Kit was performed using EXP-NBD104. Library preparations were conducted according to the manufacturer's instructions with some modifications as previously described (Hayes et al., 2021). Sequencing was performed on the PromethION P24 (ONT, UK) with the MinKNOW v.22.03.4 software using FLO-PRO002 (R9.4.1) flow cells. Samples were sequenced for 24 hours. Three basecall models, named Fast basecalling (FA), High accuracy basecalling (HAC), and Super Accurate basecalling (SUP), as well as the barcode demultiplexing, were operated by Guppy v.6.0.7. The adapter and barcode trimming functions were not selected during the sequencing.

Bioinformatics. Porechop v.0.2.4 (Wick *et al.* 2017) was performed for the trimming of adapters and barcodes. Minimum Q scores for reads generated from FA, HAC, and SUP basecall models were 8, 9, and 10, respectively. Reads under the minimum Q scores of corresponding basecall methods and less than 100 bp were filtered by Nanofilt v.2.8.0 (De Coster *et al.* 2018). Read-based taxonomic classification was performed by Kraken2 v.2.1.2 (Wood *et al.* 2019) with a customized Kraken2 database. A customized Kraken2 database was used in this study to increase the taxonomic classification efficiency. The complete genomes of bacteria, fungi, archaea, and protozoa from the NCBI RefSeq were downloaded to construct the customized database, with the low-complexity sequences masked. The Vegan v.2.6-2 (Dixon 2003) and phyloseq v.1.40.0 (McMurdie and Holmes 2013) package implemented in R, were used for the calculation of alpha diversity (Shannon index). A linear model with the DNA preparation method and/or sequencing kit as covariates was employed to assess significance.

RESULTS AND DISCUSSION

The sequencing process generated a total of 49,917,517 raw reads. Following trimming and filtering, 2,096,033 reads (4.2%) were excluded, leaving 47,821,484 reads that passed quality control. These reads were subsequently classified using the Kraken2 tool (Figure 1). The N50 values for sequence data generated from the Ligation Kit were higher (6,558 to 7,941) than for the Rapid Kit (4,662 to 4,952) with Powerfecal kit extraction (Table 1). The N50 value was positively correlated with the proportion of classified reads ($r = 0.88$, $P < 0.001$). Increasing the basecalling accuracy led to an increase in the proportion of classified reads (Figure 1A), rising from a mean of 29.71 (FA) to 38.40 (SUP). Notably, within the Powerfecal excitation kit data the ligation library preparation kit resulted in a greater proportion of reads assigned to a taxon than the rapid library preparation kit (Figure 1B).

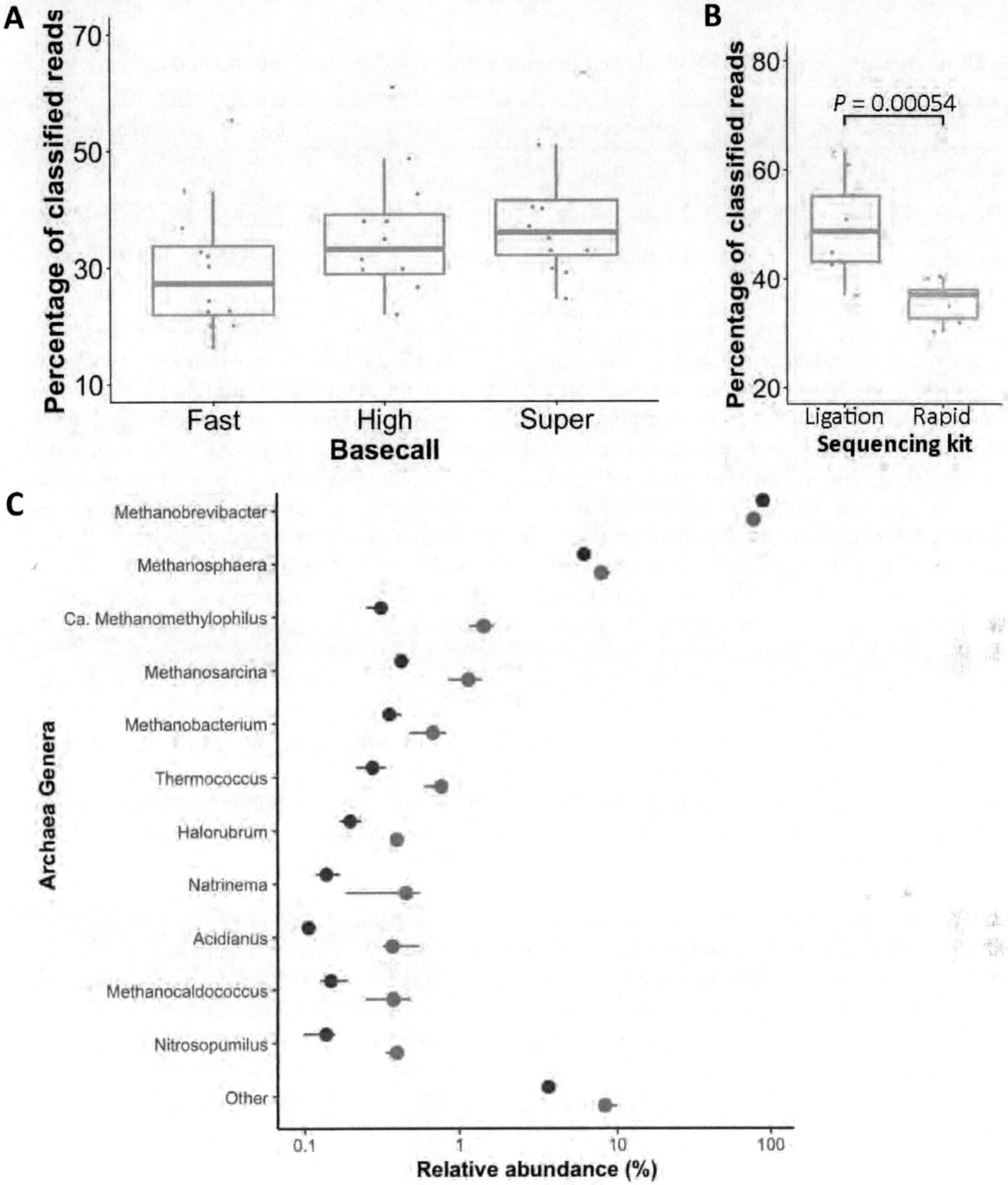

Figure 1. A) Percentage of reads assigned to a taxon for each base calling accuracy level. B) Percentage of reads assigned to a taxon for each library preparation kit (Powerfecal DNA extraction method only). A linear model was used to assess statistical significance. C) Within Archaea, the proportion of reads assigned to each genera from the ligation (Red) and rapid (green) sequencing kits

Table 1. Average lengths (N50) of the sequencing reads from each of the molecular methods

Extraction_Method	Sequencing_Kit	Mean	Sd	Median
DNeasy	Ligation_Kit	1532.22	587.87	1382.00
PowerFecal	Ligation_Kit	7460.89	659.32	7875.00
Puregene	Ligation_Kit	1431.44	149.80	1418.00
PowerFecal	Rapid_Kit	4792.89	115.46	4731.00

Microbial abundances at the Kingdom level were affected by DNA extraction ($P < 0.01$) and library methods ($P < 0.05$), but not basecall models ($P > 0.05$). Bacteria dominated the rumen microbial community (> 90.90%) for both extraction and sequencing kits. Significant effects were observed for the abundance of archaea genera (Figure 1C) based on both extraction and sequencing kits ($P < 0.05$), but not basecall models ($P > 0.05$). The Ligation Kit had a higher Shannon index (H=2.56) than the Rapid Kit (H=2.08). Conversely, the Rapid Kit had greater bacterial species diversity (H=6.13) than the Ligation Kit (H=6.11). The fungal diversity was slightly higher in the Rapid Kit than in the Ligation Kit (H=4.40 versus H=4.46, $P < 0.05$). Notably, DNeasy and Puregene extracted samples had less archaea abundance, but higher archaeal Shannon index compared to PowerFecal extracted samples. Basecall models did not affect the archaeal richness and evenness ($P > 0.05$).

CONCLUSION

Base calling accuracy in ONT sequencing of microbiome samples affects the proportion of reads that can be classified, but not species ratios, thereby impacting data acquisition costs. The choice of library preparation kit has a significant influence on the observed distribution of microbial species. Therefore, it is crucial to record the library preparation kit information in the metadata of public sequence repositories and account for it in statistical models.

ACKNOWLEDGEMENTS

The authors would like to acknowledge the invaluable support of their colleagues at the Centre for Animal Science, within the Queensland Alliance for Agriculture and Food Innovation, and the staff at the Research Computing Centre of The University of Queensland. This research was funded by Meat and Livestock Australia (project P.PSH.2010) and Research Strategic Package 3 funding from the University of Queensland. The authors also acknowledge the use of ChatGTP, an AI algorithm from OpenAI, to assist with editing the final manuscript on February 16, 2023. A small number of corrections were made to the suggested edits to preserve the accuracy and nuance.

REFERENCES

De Coster W., D'Hert S., Schultz D.T., Cruts M. and Van Broeckhoven C. (2018) *Bioinformatics* 34: 2666.

Dixon P. (2003) *J Veg Sci.* 14: 927.

Hayes et al. (2021). *Frontiers in Genetics* 12: 760450

McMurdie P.J. and Holmes S. (2013) *PLOS One* 8: e61217.

Ong C.T., Boe-Hansen G., Ross E.M., Blackall P.J., Turni C., Hayes B.J. and Tabor A.E. (2022) *Microbiology Spectrum.* 10: e00412.

Ross E.M., Moate P.J., Marett L.C., Cocks B.G. and Hayes B.J. (2013) *PLOS One* 8: e73056.

Wick R.R., Judd L.M., Gorrie C.L. and Holt K.E. (2017) *Microb Genom.* 3: e000132.

Wood D.E., Lu J. and Langmead B. (2019) *Genome Biol.* 20: 257.

Proc. Assoc. Advmt. Anim. Breed. Genet. 25: 75 - 78

SIMULTANEOUS INVESTIGATION OF GENOMIC REGIONS OF INTEREST – THE USE OF ADAPTIVE SAMPLING

R.M. Clarke[1], A. Hess[2], A, Caulton[1], R. Brauning[1], K.M. McRae[1], A. Chen[3] and S.M. Clarke[1]

[1]AgResearch, Invermay Agricultural Centre, Mosgiel, Otago, NZ
[2]University of Nevada, Reno, Nevada, USA
[3]University of Otago, Dunedin, Otago, NZ

SUMMARY

Utilization of more complex genetic variation present within a population can help to address the challenge of identifying animals that perform optimally in their environment while reducing their environmental impact. The objective of this study was to determine if Oxford Nanopore sequencing technology provides a potential solution to capture these data types in a cost-effective and high-throughput method. Adaptive sampling was used to investigate regions of interest surrounding known genome-wide association studies peaks and copy number variation regions in sheep with higher enrichment achieved in targeted areas. Multiplexing of three animals was achieved, but further work is needed to determine cost-effectiveness of this tool for the animal industry.

INTRODUCTION

Providing energy-rich protein to the world while reducing the environmental impact is one of the largest challenges facing the animal industry today. To face this challenge, novel tools need to be adopted for methods to identify animals that perform optimally in their environment. This includes, but is not limited to, utilizing more complex variation, epigenetics, and microbial communities present within the host. A major hurdle in utilizing these data lies in the development of cost-effective and high-throughput methods for data capture. We propose the sequencing platform developed by Oxford Nanopore Technologies (ONT) as a potential solution.

Adaptive sampling, a software-controlled enrichment unique to the nanopore sequencing platform, enables targeted sequencing of specific regions of a genome or species of interest at higher coverage than non-selected regions of the genome. Adaptive sampling allows the sequencing of particular regions of DNA to be enriched through the comparison of the first 400bp of a strand of DNA to a provided sequence list. If the 400bp match the sequence list then the strand continues to be sequenced, if there is no match then the stand is ejected, and the pore is available for the next strand (Payne *et al* 2021). This enrichment approach not only circumvents the need for upfront sample manipulation but also enables simultaneous capture of multiple sources of information such as methylation, mutations, and structural variances in a single run, with the aim of reducing costs (Payne *et al.* 2021).

MATERIALS AND METHODS

DNA was extracted from 19 sheep (Montgomery and Sise 1990) and libraries were prepared using SQK-LSK109 with the native barcode expansion pack EXP-NBD104 (ONT) as per ONT protocols. Adaptive sampling was done on 52 regions of interest (ROI) surrounding previously identified GWAS peaks (unpublished data) and known structural variations such as the Haemoglobin region. High molecular weight (>60kb) DNA and fragmented DNA samples (10-20kb) were compared, as well single verseuse multiplexed samples to determine the optimal output for adaptive sampling in sheep.

Retained reads were analysed using Nanoplot (De Coster *et al.* 2018) to check quality and mapped to the Oarv3.1 (Jiang *et al.* 2014) and ARS-UI_Ramb_v2.0 (Rambv2.0; Davenport *et al.* 2022) sheep genomes using minimap2 (Li, 2018). Mosdepth (Pedersen and Quinlan 2018) was used to determine the mean and median read depth across the ROI and the whole genome. Coverage results were displayed using R studio and Samplot (Belyeu *et al.* 2021).

Faecal samples were taken from sheep challenged with a single isolate of the gastrointestinal nematode *Haemonchus contortus*. DNA was extracted using the QIAamp PowerFecal Pro DNA kit (QIAGEN), and sequenced using random Genotyping by Sequencing (GBS; Dodds *et al.* 2015) to determine the percentage of DNA mapping to the *Haemonchus* (Doyle *et al.* 2020) and sheep (Oarv3.1) genomes. To determine if parasite and host DNA can be detected from faecal samples using ONT, adaptive sequencing was completed with enrichment for *Haemonchus* genome sequence. The passed reads were mapped to the *Haemonchus* genome using minimap2 and the percent of accepted reads and reads mapped to the genome were calculated. The passed reads and the failed reads, which are the first 400bp that are sequenced then rejected as not matching, were mapped against the Oarv3.1 genome using minimap2. The reads that passed were also BLAST searched using BLASTn to determine the possible source of the DNA. The percent of reads mapped to *Haemonchus* and the sheep genome were compared between techniques.

RESULTS AND DISCUSSION

Multiplexing of the 52 ROI showed that higher enrichment is seen in the selected regions versus non-selected regions and that the median coverage of these regions ranges from 1-3x coverage when three samples are multiplexed together in the same run (Figure 1).

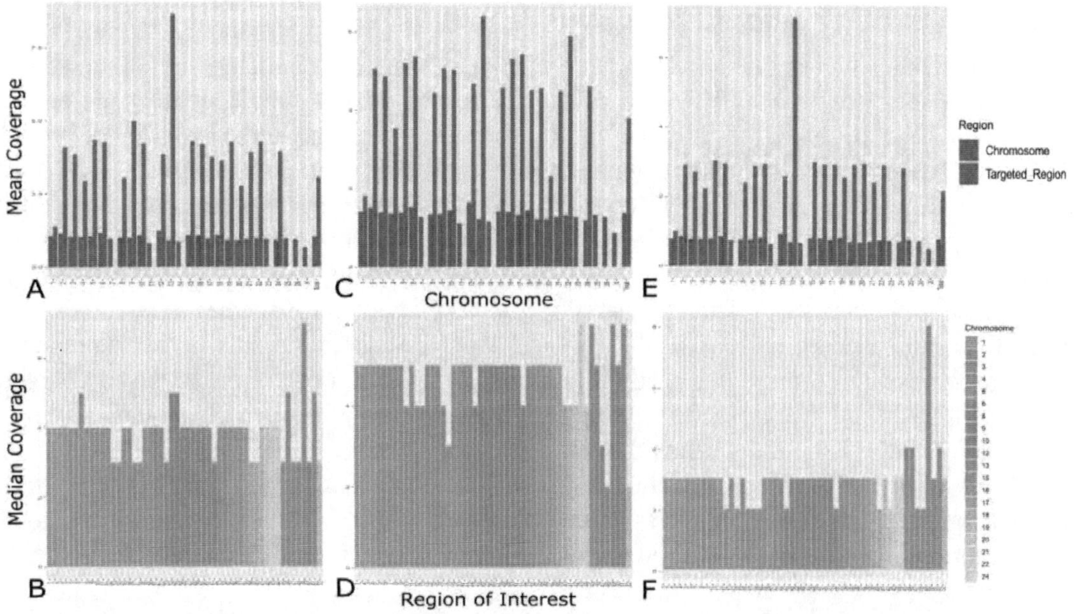

Figure 1. Mean coverage of the target regions compared to the chromosomes and the median coverage per target region for three multiplexed samples. The top panel (A, C, E) shows the mean coverage of each chromosome (red) and the mean coverage of the target regions per chromosome (blue). The x-axis labels the chromosomes. The bottom panel (B, D, F) shows the mean coverage of each of the 52 ROI, x-axis labels the regions of interest (1-52) with the colours indicating the chromosome on which the region is located. Barcode 1 (A-B), 2 (C-D) and 3 (E-F) are three individual animals that were multiplexed.

The β-globin locus of the ovine genome was chosen as an exemplar in this study, which is a region on chromosome 15 formed by the duplication of an ancestral four-gene set consisting of two embryonic-like genes, a pseudogene, and a β globin gene. Each set contains a different form of the β-globin gene, which is synthesised to make different forms of haemoglobin during development; in the foetus (F), pre-adult (C) and adult (A). There are two haplotypes, a long one comprising of all three gene sets (haplotype A) and a short one (haplotype B), where the juvenile set is missing (Figure 2) (McRae *et al.* 2022). These two haplotypes are tagged by two SNPs and the individuals are genotyped allowing a comparison between the SNP genotype and the adaptive sampling targeting the entire haemoglobin locus. The adaptive sampling shows a corresponding genotype as the SNPs with the long version of the locus Hb-A having reads across the area and the short Hb-B haplotype not having reads across the juvenile gene set (Figure 2).

Figure 2. A) Schmetic of the Haemoglobin locus with both the short (HapB), the long (HapA)

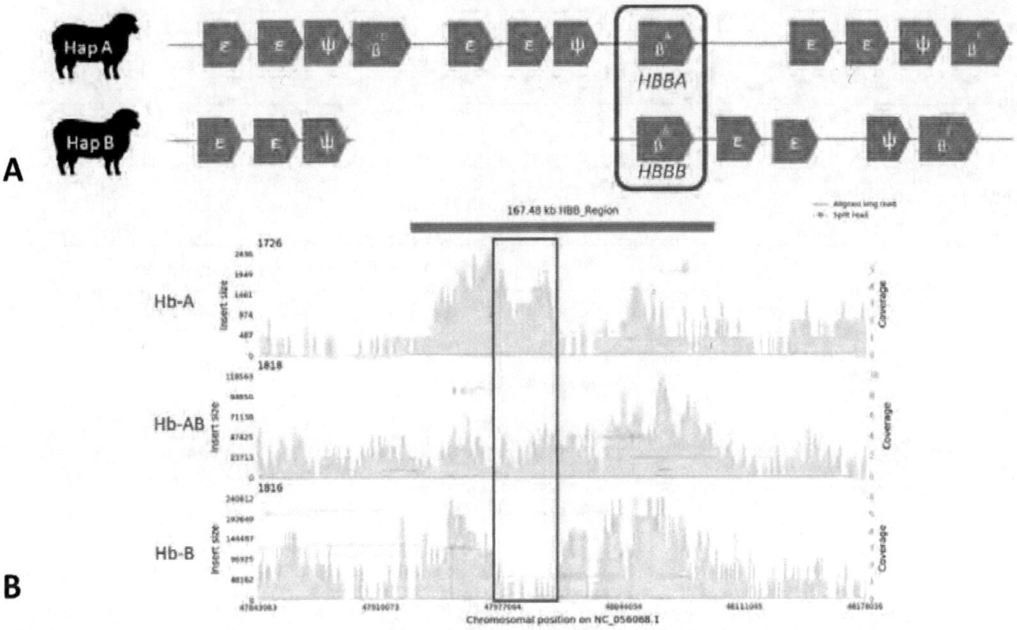

haplotype. B) Mapped examples of the long (Hb-A), short (Hb-B) and the heterozygous (Hb-AB) aligned to the Rambv2.0 (long haplotype) genome. Left axis shows the aligments scaled by insert size (distance between pair ends) and the right axis shows the per base coverage

The comparison between the GBS and adaptive sampling shows that when targeting the entire *Haemonchus* genome, a similar percentage of reads mapped to the genome using either GBS or adaptive sequencing (Table 1). When the adaptive sequencing reads were mapped against the *Haemonchus* genome only 0.38% of the reads were mapped. These reads were BLASTn searched to determine the source and the top 5 hits are shown in table 2. This suggests that the first 400bp of the read which adaptive sampling uses to make its decision to accept, or reject is matching a common sequence in the genome that is present in other species. To make this more specific to parasite DNA in faecal samples, the ITS2 region could be provided as a target region. The percentage of host DNA that is detected in adaptive sampling from both the accepted and failed reads combined shows a higher percentage of host DNA than is detected in the GBS (Table 1).

CONCLUSIONS

We have utilized adaptive sampling to investigate ROI surrounding known GWAS peaks and CNV regions with ~2-15x higher enrichment in selected areas versus non-selected areas. Enrichment of both host and parasite DNA from faecal samples shows that this technique can be utilized for different sample types and has flexibility in the information acquired. The results show that the use of the whole genome of a single parasite as the target sequence resulted in reads being accepted from a range of sources and not only the intended targets. To overcome this, adaptive sampling targeting the ITS2 region may provide a better sequencing enrichment of parasite DNA from faecal samples.

Table 1. Reads mapped to the *Haemonchus* and sheep genomes from both GBS and adaptive sampling on the same DNA faecal samples

	GBS (% mapped)	Adaptive sequencing (% reads accepted)	Accepted reads mapped to corresponding genome (%)	Accepted and failed reads mapped to corresponding genome (%)
Haemonchus	0.14	0.15	0.38	
Sheep	0.01			0.14

Table 2. Top 5 BLASTn matches for reads that were accepted as matching to the *Haemonchus* genome

BLASTn match	Accepted reads matched (%)
Haemonchus contortus	22
Plasmodium berghei ANKA	18
Chrysodeixis includens	13
Heterocephalus glaber	11
Bos taurus	3

We have also shown multiplexing can be achieved in conjunction with adaptive sequencing, but the current level of multiplexing that can be achieved to still provide the required coverage suggests that while this tool is useful for discovery and validation, it is not at the point of moving through to industry uptake. If the number of samples that can be run in one multiplexing run can be increased by, for example, using the R.10.4 flow cells on the PromethION, where current predictions are 15 samples for multiplexing, this would provide a more cost-effective option for the industry.

ACKNOWLEDGEMENTS

Thanks must go to Tania Waghorn for providing the ovine faecal samples. Funding provided by Genomics Aotearoa and AgResearch's Ministry of Business, Innovation and Employment (MBIE) Strategic Science Investment Funding (SSIF).

REFERENCES

Belyeu J.R., Chowdhury M. ,Brown J., … and Layer R.M. (2021) *Genome Biol.* 22:1.
Davenport K.M., Bickhart D.M., Worley K., … & Rosen B.D. (2022) *Gigascience*, 11.
De Coster W., D'hert S., Schultz D.T., Cruts M., and Van Broeckhoven C. (2018) *Bioinformatics*, 34: 2666.
Dodds K.G., McEwan J.C., Brauning R., … & Clarke S.M. (2015) *BMC Genomics* 16: 1.
Doyle S.R., Tracey A., Laing R., … & Cotton J.A. (2020) *Commun.Biol.* 3: 656.
Jiang Y., Xie M., Chen,W., Talbot R., … and Dalrymple B. (2014) *Science*, 344: 1168.
Li H. (2018) *Bioinformatics* 34: 3094.
McRae K.M., Rowe S.J., Johnson P.L., … and Clarke S.M. (2021) *Genes* 12: 1560.
Montgomery G.W., & Sise J. (1990) *New Zeal. J.Agr. Res.* 33: 437.
Payne A., Holmes N., Clarke T., … and Loose M. (2021). *Nat. Biotechnol.* 39: 442.
Pedersen B.S., and Quinlan A.R. (2018). *Bioinformatics*, 34:867-868.

Proc. Assoc. Advmt. Anim. Breed. Genet. 25: 79 - 82

GENE NETWORK PREDICTION FOR BULL FERTILITY TRAITS

W.L.A Tan[1], N. Hudson[2], L.R. Porto-Neto[3], A. Reverter[3], J. Afonso[4], M.R.S. Fortes[1]

[1]The University of Queensland, School of Chemistry and Molecular Bioscience, St Lucia, QLD 4072 Australia

[2]The University of Queensland, School of Agriculture and Food Sciences, Gatton, QLD 4343 Australia

[3]CSIRO Agriculture & Food, St Lucia, QLD 4067 Australia

[4]Empresa Brasileira de Pesquisa Agropecuária, Pecuária Sudeste, São Carlos, São Paulo, 13560-970 Brazil

SUMMARY

Most beef breeding herds globally still use natural mating, and therefore, conception rates are influenced by bull fertility. Many indicator traits are captured in the Bull Breeding Soundness Evaluation (BBSE). This paper uses a set of BBSE phenotypes subjected to Genome-Wide Association Studies (GWAS) to predict a gene co-association network. Gene networks can be used to mine the genetic basis of complex traits, thereby deriving a better biological understanding of the underlying mechanisms and informing genomic predictions. Here we described how a dataset of BBSE traits in a multibreed population resulted in a network of 537 connected genes whose topology and prediction will serve as the starting point for future work.

INTRODUCTION

The standardised Bull Breeding Soundness Examination (BBSE) intends to evaluate bulls' traits relevant to fertility (Entwistle and Fordyce 2003). The quantitative traits of BBSE are heritable (0.17 to 0.57) (Corbet *et al.* 2013; Porto-Neto *et al.* 2023) and possibly suitable for improvement via genomic selection. Previously, we have performed a multibreed sequence level GWAS (~ 13 million SNPs), which includes data from 6,422 beef bulls. As a result, we identified 179440 variants associated with one or more of the seven BBSE traits tested (unpublished results). The traits were body weight, condition score, scrotal circumference, sheath score, and semen morphology. In an effort to take these results beyond simple associations with our phenotypes of interest and explore underlying biology, this study utilises an Association Weight Matrix (AWM) (Fortes *et al.* 2010) approach to identify co-associations between SNPs and build a gene network. SNP selection through the AWM could highlight genes that potentially explain a key fertility phenotype, giving us insight into the genetics of bull fertility.

MATERIALS AND METHODS

Animals and phenotypes. BBSE records from 6,422 bulls comprising six different breeds were included in this study. Two breeds were research herds from the Cooperative Research Centre for Beef Genetic Technologies (Beef CRC) consisting of 1,051 Brahman (BRH) and 1,819 Tropical Composite bulls (TRC). The remaining four breeds were obtained from industry, which consists of 1,288 Santa Gertrudis (SGT), 760 Droughtmasters (DMT), 844 Ultra blacks (UBK), and 660 Belmont Tropical Composite (BTC). Descriptive statistics of BBSE records obtained for these six populations are shown in Table 1. Phenotypes include body weight (Weight), body condition score (CS), scrotal circumference (SC), sheath score (Sheath), percent normal sperm (PNS), proximal droplets (PD) and mid-piece abnormalities (MP).

Genotypes. Most animals were genotyped at ~ 50K. A reference panel that utilised BeefCRC and industry animals that were at higher density (~700K) and sequence level (~25 million) were used to impute animals to higher density and, subsequently, to sequence level. The animals used in the reference population was representative of the bulls used in this study (Porto-Neto *et al.* 2021). This was conducted using a phased reference generated by Eagle 2 (v2.4.1) and then imputed using Minimac3 for autosomes and Minimac 4 for Chromosome X. Imputation $r^2 > 0.8$, a call rate > 0.9 and a minor allele frequency > 0.01 were kept for further analysis. After quality control, a total of 13,398,171 SNPs, including 92,134 SNPs mapped onto the X chromosome. After running a Leave One Chromosome Out (LOCO) GWAS in GCTA (Yang *et al.* 2011), a total of 179,440 variants were significant ($P < 5 \times 10^{-8}$) for at least one trait. 19, 337 variants were significant for two or more traits.

Table 1. The number of records and descriptive statistics of the observed traits*

	N^A	$Mean^B$	SD^C	Min^D	Max^E
Weight, kg	6014	391.59	98.65	124.00	810.00
CS, score	5917	2.96	0.37	2.00	4.00
SC, cm	6235	30.82	4.26	15.50	52.50
Sheath, score	6417	3.19	1.77	1.00	9.00
PNS, %	6055	61.76	27.53	0.00	100.00
PD, %	6052	13.50	19.96	0.00	96.00
MP, %	6052	11.39	11.04	0.00	83.00

[A] Number of records available for a trait. [B] Mean of a trait. [C] Standard deviation of a trait. [D] Minimum value of the trait. [E] Maximum value of the trait.

AWM-PCIT methodology. The AWM was constructed using the procedure described by (Fortes *et al.* 2010). This method applies a series of selection steps to choose relevant SNPs from the 179440 significant variants base on our previous GWAS study (Figure 1). Firstly, we only considered significant SNPs that mapped to genes expressed in the testis, which were previously reported by de Lima *et al.* (2021). PNS was chosen as the key phenotype for the AWM as sperm morphology is an important aspect of bull fertility that is heritable (0.24) and correlated with commonly used bull fertility indices (Attia *et al.* 2016; Butler *et al.* 2019; Porto-Neto *et al.* 2023). We selected SNPs that were associated with PNS ($P < 0.05$). If SNPs were not associated with PNS but with at least three other traits ($P < 0.05$), these SNPs were also kept. The final selection step for the AWM chose SNPs that map to coding regions or was within 2,500 bp of known genes. SNP-to-gene mapping was done using the Map2NCBI package (Hulsman Hanna and Riley 2014) in R. SNPs were grouped by gene to map one representative SNP per gene. This was achieved by selecting the SNP within each gene group associated with the highest number of phenotypes. Next, SNPs within each group were chosen using the most significant average *p*-value across traits. The result is a matrix with rows representing genes (I) and columns representing phenotypes (J). Each element (I, J) contains the association of the SNP to the phenotype. We applied the partial correlation and information theory (PCIT) algorithm described by Reverter and Chan (2008) to the AWM. This algorithm assigns zero for non-significant correlations and retains significant correlations to establish edges in the network (Reverter and Chan 2008). The PCIT algorithm allows for a less stringent threshold ($P < 0.05$) to be used, because SNPs are highlighted based on a number of features and not just it's association to the phenotype (Reverter and Chan 2008). The correlation values can be used as input for Cytoscape (Shannon *et al.* 2003) to establish gene interactions in the gene network analyses.

RESULTS AND DISCUSSION

The gene network constructed using the AWM is shown in Figure 2. The network contains 537 genes forming two distinct clusters with 279 genes on the left and 237 genes on the right. Among these genes, 21 are transcription factors (TF). This network can serve as a starting point for further downstream analysis that can serve two aims: biological discovery and genomic prediction.

For example, biological discovery with the STRING database (Szklarczyk *et al.* 2020) will perform functional enrichment analysis to derive biological information from the gene network. Recent efforts have shown that biological data and the discovery of causal variants can positively impact genomic prediction (Xiang *et al.* 2021). Botelho *et al.* (2021) proposed AWM weighted single step genomic best linear unbiased prediction (AWM-WssGBLUP) as a method to derive weights when building the genomic relationship matrix (G). However, this method did not significantly increase the predictive ability of genomic predictions in their dataset of boar taint compounds. Nonetheless, biological information can still be useful in genomic predictions. Tahir *et al.* (2022) showed that slight improvements in predictive accuracy could be attained using biologically informed SNPs in heifer fertility traits. The SNPs that underpin the network described here are leads for causal variants that could be used to improve predictions of bull fertility traits.

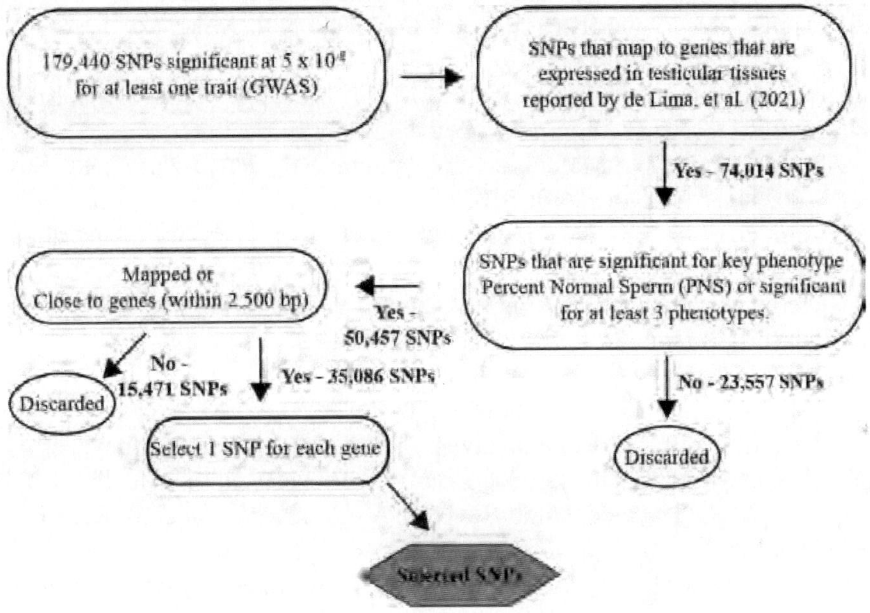

Figure 1. AWM SNP selection flow chart

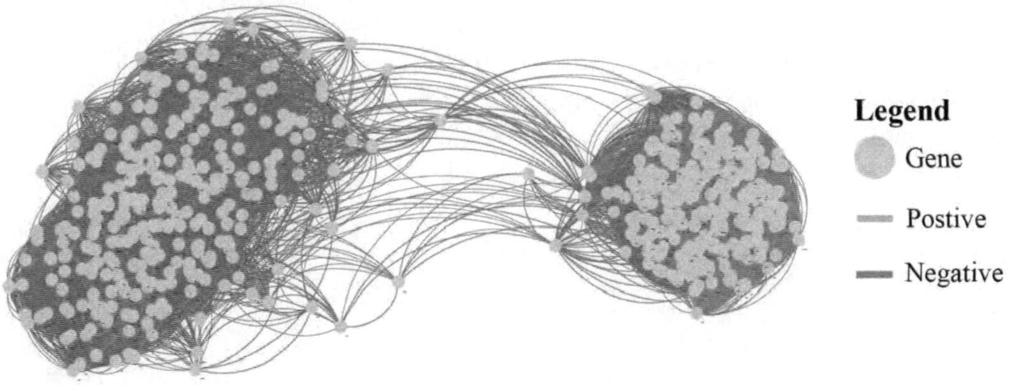

Figure 2. Gene network derived from the Association Weight Matrix (AWM)

CONCLUSION

The gene network created using the AWM highlights several genes and TFs associated with bull fertility traits. These genes and TFs, together with the significant SNP in our sequence-level GWAS, are promising leads to discover causal variants important for bull fertility. This network can be a starting point for further downstream analysis, giving insight into important molecular mechanisms for bull fertility traits.

ACKNOWLEDGMENTS

This work was supported by the Bull Fertility update project, an initiative from CSIRO, the University of Queensland, and Meat & Livestock Australia. The authors are thankful to the beef producers who took part in this collaborative effort and Meat & Livestock Australia that co-funded the project (L.GEN.1818).

REFERENCES

Attia S., Katila T. and Andersson M. (2016) *Reprod. Domest. Anim.* 51: 54.

Botelho M.E., Lopes M.S., Mathur P.K., Knol E.F., Guimarães S.E.F., Marques D.B.D., Lopes P. S., Silva F.F. and Veroneze R. (2021) *J. Anim. Breed. Genet.* 138: 442.

Butler M.L., Bormann J.M., Weaber R.L., Grieger D.M. and Rolf M.M. (2019) *Transl. Anim. Sci.* 4: 423.

Corbet N.J., Burns B.M., Johnston D.J., Wolcott M.L., Corbet D.H., Venus B.K., Li Y., McGowan M.R. and Holroyd R.G. (2013) *Anim. Prod. Sci.* 53: 101.

de Lima A.O., Afonso J., Edson J., Marcellin E., Palfreyman R., Porto-Neto L.R., Reverter A. and Fortes M.R.S. (2021) *Front. Genet.* 12: 610116.

Entwistle K. and Fordyce G. (2003) 'Evaluating and reporting bull fertility'. Australian Association of Cattle Veterinarians.

Fortes M.R.S., Reverter A., Zhang Y., Collis E., Nagaraj S.H., Jonsson N.N., Prayaga K.C., Barris W., Hawken R.J. (2010) *Proc. Natl. Acad. Sci.* 107: 13642.

Hulsman Hanna L.L. and Riley D.G. (2014) *Livest. Sci.* 162: 59.

Porto-Neto L.R., McWilliam S.M., Alexandre P.A., Reverter A., McGowan M., Fortes M.R.S., Hayes B. and Betram J. (2021) Meat and Livestock Australia, North Sydney.

Porto-Neto L.R., Alexandre P.A., Hudson N.J., Bertram J., McWilliam S.M., Tan A.W.L., Fortes M.R.S., McGowan M.R., Hayes B.J. and Reverter A. (2023) *PloS one* 18: e0279398.

Reverter A. and Chan E.K.F. (2008) *Bioinformatics* 24: 2491.

Shannon P., Markiel A., Ozier O., Baliga N.S., Wang J.T., Ramage D., Amin N., Schwikowski B. and Ideker T. (2003) *Genome Res.* 13: 2498.

Szklarczyk D., Gable A.L., Nastou K.C., Lyon D., Kirsch R., Pyysalo S., Doncheva N.T., Legeay M., Fang T., Bork P., Jensen L.J. and von Mering C. (2020) *Nucleic Acids Res.* 49: D605.

Tahir M.S., Porto-Neto L.R., Reverter-Gomez T., Olasege B.S., Sajid M.R., Wockner K.B., Tan A.W.L. and Fortes M.R.S. (2022) *J. Anim. Sci.* 100: 340.

Xiang R., MacLeod I.M., Daetwyler H.D., de Jong G., O'Connor E., Schrooten C., Chamberlain A. J. and Goddard M.E. (2021) *Nat. Commun.* 12: 860.

Yang J., Lee S.H., Goddard M.E. and Visscher P.M. (2011) *Am. J. Hum. Genet.* 88: 76.

METHYLOME PROFILING IN RESPONSE TO STRESS: MYCOTOXIN (SPORIDESMIN) EXPOSURE IN SHEEP

A. Caulton[1], K.M. McRae[1], K.G. Dodds[1], R. Brauning[1], N.K. Pickering[2], P.L. Johnson[1] and S. M. Clarke[1]

[1]AgResearch, Invermay Agricultural Centre, Mosgiel, New Zealand
[2]Focus Genetics, Napier, New Zealand

SUMMARY

Epigenetic modifications, including DNA methylation, alter gene expression without changing the DNA sequence, allowing for immediate and reversible modulation of physiological responses to abiotic/biotic stress. Facial eczema (FE) is a metabolic disease, which causes liver damage in affected animals. It occurs as a result of ingestion of the mycotoxin sporidesmin, which is found in the spores of the pasture-dwelling fungus *Pseudopithomyces chartarum.* This pilot study investigated DNA methylation changes that occurred as a result of sporidesmin exposure and identified a number of differentially methylation genomic regions in animals with liver stress. Of note, the *HBA* gene showed differential methylation in the promoter region; the *HBA* co-subunit of haemoglobin *HBB* has previously been identified as a QTL for the disease in sheep. There may be potential for DNA methylation markers to be used as a diagnosis proxy for FE or as a selection marker for resilient animals in the future.

INTRODUCTION

Facial eczema (FE) is a metabolic disease responsible for major economic losses and animal welfare concerns in New Zealand. The disease is caused by ingestion of the mycotoxin sporidesmin, causing liver damage and leading to decreased productivity and reproduction in clinically and sub-clinically affected animals. Current strategies to reduce the severity of FE outbreaks include dosing animals with zinc, spraying pastures with fungicides, managing pastures, alternative feeds, and breeding for animals with increased tolerance to the disease.

Currently, ram breeders in NZ use an ethical dosing strategy using sporidesmin from laboratory-cultured *Pse. chartarum* to predict FE tolerance (RamGuard™; Aymes & Hawkes 2014). The physiological effects of the disease are assessed by measuring serum gamma-glutamyltransferase (GGT) at 21 days post-challenge (GGT21), which is recorded in the national genetic evaluation ($h^2 = 0.44 \pm 0.03$) (McRae *et al.* 2021). While the underpinning genomics continues to be assessed, we have extended our investigation to include epigenetics profiles. Advances in "omics" technologies have fuelled investigation into the epigenome as a tool to enhance livestock selection and breeding practices. DNA methylation is an important epigenetic mark that is essential for genomic stability and maintenance throughout development and serves as a biomarker of chronological age and a biological fingerprint of a stress response (Clarke *et al.* 2021). A pilot study was conducted investigating changes in the methylome in response to a sporidesmin challenge to assess the potential application of methylation profiling for livestock breeding.

MATERIALS AND METHODS

Animals were managed following the provisions of the New Zealand Animal Welfare Act 1999, and the New Zealand Codes of Welfare developed under sections 68-79 of the Act. All work was undertaken with the approval of the AgResearch Ruakura Animal Ethics Committee (Approval number: 15059). Reduced-representation bisulphite sequencing (RRBS; Smith *et al.* 2009) was used to profile a cohort of sheep exposed to a controlled FE disease challenge with the identified stress-imposed changes to DNA methylation across two timepoints, day 0 (pre-challenge), day 21 (post-challenge) presented. The animals used in this study consisted of 70 nine-month-old ram lambs from two breed groups. A total of 50 ram lambs (5 progeny per sire, 5 sires per breed) were challenged through the Ramguard™ (Amyes & Hawkes 2014) program. The remaining 20 rams (10 per breed) were selected from remaining unchallenged animals and were from a mixture of sires (1-2 progeny per sire). The Ramguard™ programme uses *Pse. chartarum* that is cultured in a laboratory to produce the toxic forms of mycotoxin sporidesmin, specifically sporidesmin A, B and E, with a >90% predominance of sporidesmin A. Animals were dosed with precise amounts of sporidesmin by intra-ruminal intubation at a volume that is dependent on the animal's live weight (mg per kg live weight). Blood samples were taken for GGT testing before dosing (for a base activity; d0) and at 21 days after dosing (d21) and processed for serum GGT activity (IU/L) through a commercial laboratory (IDEXX, Hamilton, New Zealand) (Johnson & Amyes 2021). Ear tissue punches (Allflex Tissue Sampling Unit; TSU samples) were also collected at d0 and d21 and genomic DNA was extracted from tissue samples using a high-salt method (Clarke *et al.* 2014).

The Zymo-seq RRBS Library Kit (Zymo Research, Irvine, CA, United States) was used for bisulphite conversion and library construction as per the manufacturer's instructions, using 500 ng of input DNA, and sequenced at AgResearch on a NovaSeq 6000 (Illumina Inc, San Diego, CA, USA.), yielding 101 bp single-end reads (minimum 5 x coverage). Data were processed and analysed as follows. Briefly, TrimGalore v.0.5.010 was used to trim raw reads to remove adapter oligos and poor-quality bases (Phred score < 20) with the flags: --non_directional --rrbs -q 20 (https://www.bioinformatics.babraham.ac.uk/projects/trim_galore/). Trimmed reads were aligned to the reference sheep genome ARS-UI_Ramb_v2.0 (Davenport *et al.* 2022) using the BSSeeker23 script bs_seeker2-align.py and Bowtie2, with RRBS settings and allowing four mismatches (-m 4) (Guo *et al.* 2013). Methylation levels were called using the "bam2cgmap" function within CGmaptools with default options (Guo *et al* 2018).

After sequencing 7 samples were excluded due to low sequencing depth leaving 63 animals in the analysis. The challenged rams were categorized into high and low susceptibility groups based on their d21 GGT, which indicates liver/bile duct damage, with GGT<=300 IU/L classified as having a low GGT score (n=29) and GGT >=301 IU/L classified as having a high GGT score (n=14) (Johnson & Aymes 2021). Differentially methylated regions (DMR) were identified between groups and time points with MethylKit v1.12.0261 in R, which applies a sliding-window approach with a window of 1,000 bp and a step size of 500 bp (Akalin *et al.* 2012). The data were filtered for potential PCR duplicate reads by excluding bases with more than the 99.9th percentile of coverage in each sample (hi.perc=99.9). Read coverage distributions between samples were normalised using a scaling factor derived from differences between the median of the coverage distributions to avoid oversampling of reads from more highly sequenced individuals in downstream statistical analyses.

Methylated sites common between samples were identified and combined into a single R object for further analysis. To calculate differential methylation, groups were compared via a logistic regression model. P-values were adjusted to q-values using the Sliding Linear Model (SLIM) method (Wang *et al.* 2011). DMR were defined as regions with at least a 15% difference between the group being tested and the remaining samples and a Q-value $\leq$ 0.05, controlling for false discovery rate based on the SLIM method. DMRs were overlapped with annotated genes and gene promoters, defined as 1kb upstream flanks of genes in the ARS-UI_Ramb_v2.0 genome.

RESULTS AND DISCUSSION

Day 21 serum GGT activity ranged from 50 UI/L to 1056 UI/L (Figure 1). A total of 6 rams showed clinical signs of FE and had to be euthanised due to the severity of their symptoms, TSUs were collected from these animals on the day of euthanasia and used in this study. A total of 14 rams had GGT activity >=301 UI/L and were ranked as high GGT animals including the 6 euthanised animals. A series of group comparisons were performed to identify DMRs between time points and

GGT score groupings. These are summarised in Figure 2A. The number of DMRs between groups differed greatly. There was one DMR identified between the challenged and control groups at d0 which was also seen at d21, this DMR was located in the *FBXL17*gene which is involved in the mitotic cell cycle. There were a large number of DMRs (1330) identified in the comparison between d0 and d21 of the High GGT group.

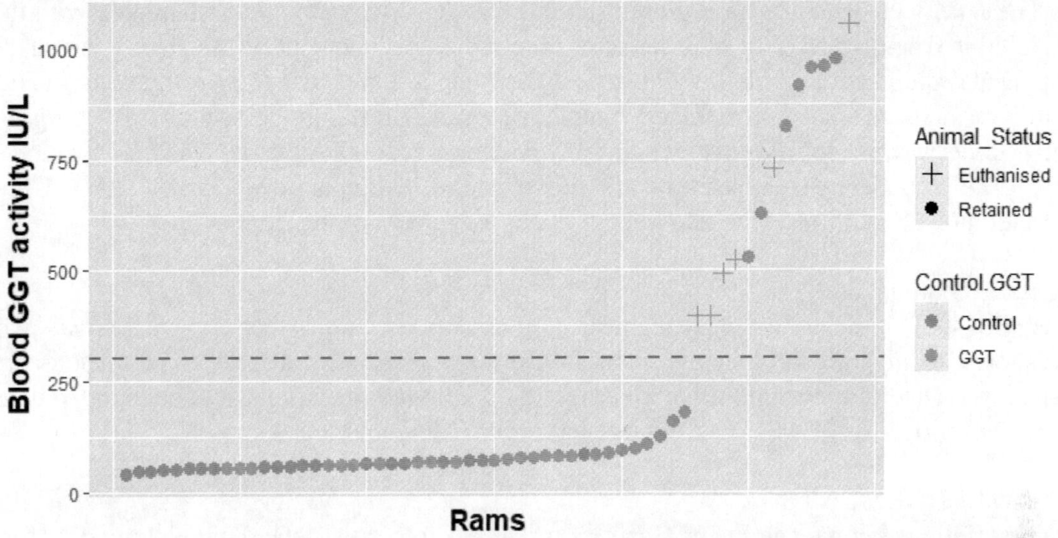

Figure 1. Blood GGT activity d21 post- Ramguard™ challenge, the red dashed line indicates the threshold for high GGT rank >=301 UI/L

A

Group 1	Group 2	# DMRs	# Genes
Control d0	Case d0	1	1
Control d0	Control d21	3	0
Case d0	Case d21	1	1
Control d21	Case d21	4	2
Low GGT d0	Low GGT d21	0	0
Low GGT d21	High GGT d21	66	45
High GGT d0	High GGT d21	1330	811

B

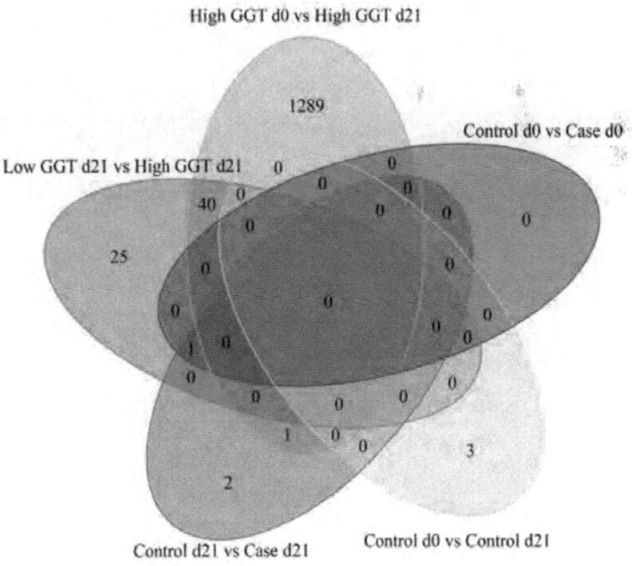

Figure 2. DMR identified between groupings; (A) number of DMRs identified between group 1 and group 2 in each comparison and number of gene regions (gene + 1KB upstream) identified in the DMRs. (B) common DMRs identified in each comparison

Interestingly there was only one DMR identified between d0 and d21 in the challenged group, however, when subsetted into low and high GGT groups a large number of DMRs were identified between d0 and d21 in the high GGT group, suggesting that the low GGT responders have a different response mechanism compared to the high GGT responders. Two genes of note were identified as having DMRs in the promoter regions when comparing the high GGT d0 and d21 groups, *HBA* and *CARD11*. The β-globin gene, *HBB* has previously been identified as a QTL for FE susceptibility in a previous GWAS, explaining 5% of the phenotypic variance of resistance to FE (McRae *et al.*, 2022). The *HBA* locus encodes an α-globin subunit that forms a haemoglobin complex with the β-globin subunit. The discovery of a DMR across the regulatory element of the *HBA* gene supports the notion that the haem complex is linked to FE susceptibility. Further investigations including proteomic mass spectrometry from blood samples of animals through the Ramguard™ program is currently underway to fortify this hypothesis. Another gene of note is *CARD11,* which has previously been associated with severe atopic dermatitis in humans (Ma *et al.* 2017). Clinical manifestation of FE in animals include flaking of exposed areas of the skin.

CONCLUSIONS

This pilot study indicates there are significant methylation changes that occur in animals with a poor response to mycotoxin challenge. The DMRs associated with a high GGT response have biological relevance and warrant further investigation. Methylation markers hold potential to be used as a diagnosis proxy for FE or as a selection marker for resilient animals.

ACKNOWLEDGEMENTS

We gratefully acknowledge Focus Genetics for access to ear tissue and phenotypic data for the animals used in this study. This work was supported by the New Zealand Ministry of Business, Innovation and Employment: C10X1906 "Beyond the genome: Exploiting methylomes to accelerate adaptation to a changing environment"

REFERENCES

Akalin A., Kormaksson M., Li S., Garrett-Bakelman F.E., Figueroa M.E., Melnick A. and Mason C.E. (2012) *Genome Biology*, 13: R87.

Amyes N.C. and Hawkes A.D. (2014) *Proc.N.Z. Soc.Anim. Prod.* 74: 154.

Clarke S., Caulton A., McRae K., Brauning R., Couldrey C. and Dodds, K.D. (2021). *Animal Frontiers*, 11:,90.

Clarke S.M., Henry H.M., Dodds K.G., Jowett T.W.D., Manley T.R. Anderson R.M. and McEwan J.C. (2014) *PLOS ONE*, 9: e93392.

Davenport K., Bickhart D., Worley K., Murali S., Salavati M., Clark E., Cockett N., Heaton M., Smith T., Murdoch B. and Rosen, B. (2022) *GigaScience* 11.

Guo W., Fiziev P., Yan W., Cokus S., Sun X., Zhang M.Q., Chen P.-Y. and Pellegrini, M. (2013) *BMC Genomics*, 14: 774.

Guo W., Zhu P., Pellegrini M., Zhang M.Q., Wang X. and Ni Z. (2018) *Bioinformatics* 34: 381..

Johnson P. and Amyes N. (2021) *N.Z. J. Anim. Sci. Prod.* 80: 34.

Ma C.A., Stinson J.R., Zhang Y., … Milner J.D. (2017) *Nature Genetics* 49: 1192.

McRae K.M., Rowe S.J., Johnson P.L., Baird, … Brauning R., Amyes N.C., Dodds K.G., Newman S.-A. N., McEwan J.C. and Clarke S.M. (2021) *Genes* 12: 1560.

Smith Z.D., Gu H., Bock C., Gnirke A. and Meissner (2009) *Methods* 48: 226.

Wang H.-Q., Tuominen L.K. and Tsai C.-J. (2011) *Bioinformatics* 27: 225.

RESOURCES ASSESSMENT FOR DETERMINATION OF ANCESTRAL ALLELES IN CATTLE

J. Dorji[1], L.R. Porto-Neto[1], A.J. Chamberlain[2], C.J. Vander-Jagt[2], J. Kijas[1] and A. Reverter[1]

[1]CSIRO Agriculture and Food, St. Lucia, QLD, 4067 Australia
[2]AgriBio, Centre for AgriBioscience, Bundoora, VIC, 3083 Australia

SUMMARY

Selection in animals whether natural or artificial leaves imprints on the genome also known as selection signatures. Such signals can pinpoint genomic regions that have undergone fixation during the selection for production, adaptation, and other domestication events. Few approaches to identifying selection signals require designation of ancestral alleles. Selection signature studies particularly at a sequence level are seldom undertaken possibly for the lack of a comprehensive list of ancestral alleles. Therefore, we reviewed the published lists of ancestral alleles in cattle and other resources of potential use in the derivation of an exhaustive list of ancestral alleles in cattle. Our results suggest the current list of ancestral alleles in cattle are few, incomplete and has low coverage on the genome. We also report on the publicly available resources particularly raw sequence reads from non-cattle *Bos* species and the 1000 Bull Genomes as readily usable resources to determine ancestral allele in cattle. Altogether, the use of genomic variants from the 1000 Bull Genomes is expected to help determine ancestral allele for about 73 million genomic positions in cattle.

INTRODUCTION

Selection signatures are genetic imprints resulting from selection, adaptation or domestication process. The identification of selection signature is increasingly used to mine genomic regions influencing complex production and adaptation traits in cattle (Stella *et al.* 2010; Zhao *et al.* 2015; Cheruiyot *et al.* 2018). Several tools employing iHS, XP-EHH, iSAFE parameters for the detection of selection signatures require designation of ancestral alleles in the input. The ancestral alleles are alleles that persisted prior to selection and are commonly determined by comparing alleles at orthologous sites to a closely related species (Naji *et al.* 2021). However, in the absence of a list of ancestral alleles in the cattle, most if not all selection signature studies in cattle using ancestral allele dependent parameters assume the major allele as the ancestral allele in some tools (e.g., rehh package) (Gautier and Vitalis 2012). As such the major alleles are not always the ancestral allele. The minor alleles constituted more than 13 and 19% of ancestral alleles in Xiang *et al.* (2021) and Naji *et al.* (2021) respectively and such assumption can have significant influence the inferences.

In this study, we reviewed published lists of ancestral alleles and assessed genomic resources of cattle and out-species with potential for the determination of ancestral alleles. We examined the coherence of genomic position among the published lists of ancestral alleles and drew insight on the population structure of previously unused out-species.

MATERIALS AND METHODS

We reviewed the published list of ancestral alleles in cattle for their coverage or the number of sites, associated reference genomes and the closely related species used for its determination. Wherever comparable, the number of sites in common and concordance of ancestral alleles were estimated. In terms of resource for determination of ancestral alleles in cattle, we queried the raw sequences of non-cattle *Bos* species in the NCBI-SRA. Further, we use raw sequence reads of *Bos* species that were not previously used in the determination of ancestral alleles to draw insights on the population structure in relation to cattle. The sequence reads were processed following the pipelines used for processing the 1000 Bull Genome. Further, we explored the coincidence of variant position in the 1000 Bull Genomes with previous studies and estimated its potential contribution to the existing ancestral allele list.

RESULTS AND DISCUSSION

Ancestral alleles. Up to now, at least four studies have determined the lists of ancestral alleles in cattle (Table 1). They were determined by comparing allele of cattle species to non-cattle *Bos* species and other non-Bos species. The earlier two lists of ancestral alleles were based on the older bovine reference genome (UMD3.1) and Bovine SNP panels. The third and fourth studies were based on the reference genome, ARS-UCD1.2 and independently determined ancestral allele for 32 and 40 million positions corresponding to about 25% of the total variants detected in cattle to date.

Table 1. Summary of ancestral alleles in cattle published

Coverage	Reference genome	Cattle spp[†]	Non-cattle *Bos* spp	Out-group spp	No. of AA* determined	References
BovineSNP50	UMD3.1	Taurus Indicine Composite	Gaur Banteng Yak	Bison Low-land anoa Cape buffalo	50.1 thousand	Matukumalli *et al.* 2009
BovineSNPs (19.5 million variants)	UMD3.1	Taurine	Yak	Sheep Water buffalo	14.4 million	Rocha *et al.* 2014
Whole genome sequence	ARS-UCD1.2	Taurine Indicine	Yak Banteng Gayal Gaur Auroch	Bison	32.4 million	Naji *et al.* 2021
Whole genome sequence	ARS-UCD1.2	Taurine Indicine Composite	Yak	Sheep, camel	39.9 million	Xiang *et al.* 2021

*AA: Ancestral allele

Between the two recent lists of ancestral alleles, about 9 million positions were in common. The coincidence of the ancestral alleles in the common positions were very high (99.8%). Altogether, these two lists presented ancestral alleles for about 60 million positions.

Query of raw sequence reads of non-cattle *Bos* species showed *Bos mutus* (N=4) and *Bos sauveli* (N=2) have not been previously used in the determination of ancestral allele in cattle. The inclusion of these species is likely to improve the reliability of some of the current ambiguous and low probability sites in the lists. The population structure of cattle and out-species (Figure 1) showed less prominent segregation among the out-species group compared to cattle. This is expected because the variant positions in out-species were ortholog of cattle which not necessarily segregated in the out-species. The PC1 largely separated out cattle and out-species groups while PC2 segregated *Bos indicus* and *Bos taurus*.

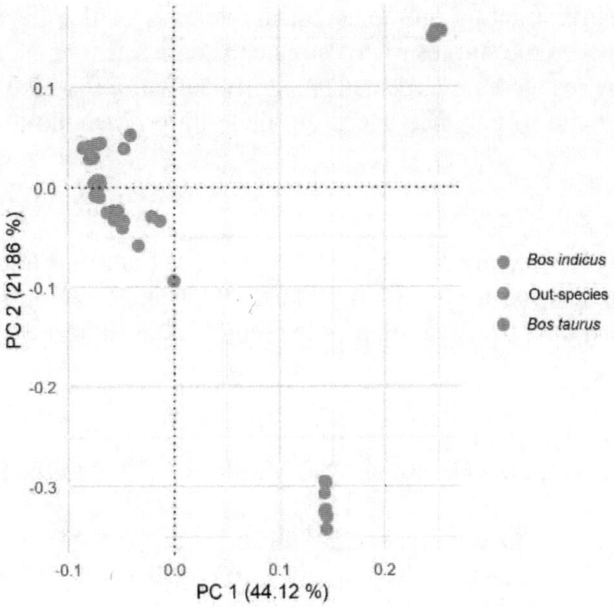

Figure 1. PC plot (PC1 and PC2) of cattle (*Bos indicus* and *Bos Taurus*) and out-species

1000 Bull Genome. It is a massive genomic resource collating variants for genomic imputation and genome wide association studies in cattle. The dataset has ~32 million high confidence (i.e., PASS) biallelic variant positions for cattle. It also provided genomic variants for out-species (five non-cattle *Bos* species including bison). This is a readily usable resource for the determination of ancestral allele by comparing alleles in cattle with alleles in orthologous positions in the out-species. The coherence of this genomic positions with two previous studies combined were more than 65% (Figure 2). This dataset would add another 11 million genomic positions for ancestral alleles to the existing list to reach 73 million. Further, considering the next best confidence category of variants (i.e, Tranche90to99) which is about 40 million positions can substantially increase the positions of ancestral alleles in cattle up to 100 million.

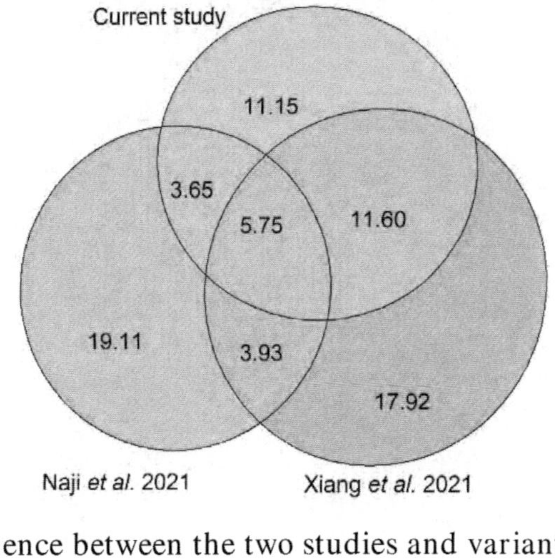

Figure 2. Positional coherence between the two studies and variant positions in the 1000 Bull Genomes (in million)

CONCLUSION

Despite limited studies investigating on the ancestral alleles in cattle, there is a high proportion of positions in common between the studies to investigate the coherence of ancestral alleles. Further, the use of available genomic resources is expected to significantly improve the coverage of ancestral allele on the cattle genome and to enhance ancestral allele-based detection of selection signature studies in cattle.

ACKNOWLEDGEMENTS

Authors sincerely thank Agriculture Victoria, Department of Energy, Environment and Climate Action (DEECA), Victoria for sharing the semi-processed sequences of out-species. We are also thankful to the 1000 Bull Genome Consortium for the meta-data and summary statistics of the 1000 Bull Genomes.

REFERENCES

Cheruiyot E.K., Bett R.C., Amimo J.O., Zhang Y., Mrode R., and Mujibi F.D.N. (2018) *Front. Genet.* 9: 607.

Gautier M. and Vitalis R. (2012) *Bioinformatics.* 28: 1176.

Matukumalli L.K., Lawley C.T., Schnabel R.D., Taylor J.F., Allan M.F., Heaton M.P., O'Connell J., Moore S.S., Smith T.P.L., Sonstegard T.S. and van Tassell C.P. (2009) *PLOS ONE.* 4: e5350

Naji M.M., Utsunomiya Y.T., Sölkner J., Rosen B.D. and Mészáros G. (2021) *BMC Genomics.* 22: 108.

Rocha D., Billerey C., Samson F., Boichard D. and Boussaha M. (2014) *J. Anim. Breed. Genet.* 131: 483.

Stella A., Ajmone-Marsan P., Lazzari B. and Boettcher P. (2010) *Genetics.* 185: 1451.

Xiang R., Breen E.J., Bolormaa S., Vander-Jagt C.J., Chamberlain A.J., Macleod I.M. and Goddard M.E. (2021) *Comm. Biol.* 4: 1353.

Zhao F., McParland S., Kearney F., Du L. and Berry D.P. (2015) *Genet. Sel. Evol.* 47: 49.

Proc. Assoc. Advmt. Anim. Breed. Genet. 25: 91 - 94

INCORPORATING GENOMIC INFORMATION IN EVALUATIONS FOR FARMED DEER IN NEW ZEALAND

R. Costilla[1], R. Brauning[2], J. Ward[2], S. Newman[2], A. Hess[3], S. Hickey[1], T. van Stijn[2], J. McEwan[2], K. Dodds[2], S. Clarke[2] and S. Rowe[2]

[1] AgResearch Ltd, Ruakura Research Centre, Hamilton, New Zealand.
[2] AgResearch Ltd, Invermay Agricultural Centre, Mosgiel, New Zealand.
[3] University of Nevada, Reno, Nevada, USA

SUMMARY

The New Zealand deer industry has made notable genetic progress in the last decades. Initially based around velvet antler and venison, deer are now selected based on several traits including carcass composition, reproduction, and disease resistance within DEERSelect, the industry performance recording system in New Zealand. Due to its low cost and manageable logistics for deer, the genotyping-by-sequencing (GBS) technology was chosen to replace microsatellites for parentage assignment and has allowed a national genetic evaluation across flocks since 2015. Genomic information, however, is not yet fully exploited as evaluations currently only use a traditional pedigree-based, best linear unbiased prediction (BLUP) approach, to estimate the genetic merit of an animal. To assess the benefits of using genomic information in the evaluations, here we compare BLUP, genomic BLUP (GBLUP) and single-step genomic BLUP (SSGBLUP) evaluations for several production traits in the NZ deer industry. Using forward-validation, we estimate the prediction accuracy and bias for these three approaches in 19,863 red animals born between 2018 and 2020. We show that regardless of the approach, GBLUP or SSGBLUP, incorporating genomic information explicitly improves prediction accuracy and reduces bias. Across all traits, we estimate gains in accuracy of 16% for GBLUP and 18% for SSGBLUP on average for red deer. We therefore recommend the incorporation of genomic information in the evaluations performed by DEERSelect and propose a computational pipeline to support the medium to long-term growth of this dynamic livestock industry in New Zealand.

INTRODUCTION

The New Zealand deer industry has made substantial genetic progress in the last decades (Ward *et al.* 2016). Initially based around velvet antler and venison, deer are now selected based on several traits including carcass composition, reproduction, and disease resistance within DEERSelect, the deer industry performance recording system in New Zealand (Gudex *et al.* 2013; Ward *et al.* 2016).

Since 2015, the AgResearch deer genomics program has developed and implemented genotyping-by-sequencing (GBS) methods for deer parentage (Dodds *et al.* 2019; Rowe *et al.* 2018), gender and breed assignment (Bilton *et al.* 2019). Deer are routinely genotyped by GenomNZ (https://www.agresearch.co.nz/genomnz) using these GBS methods (Rowe *et al.* 2018). Genomic information, however, is not yet fully exploited in current evaluations as they only use the theoretical relationships between animals contained in the pedigree to estimate their genetic merit (A matrix). This approach to generate breeding values (BVS) is also known as best linear unbiased prediction (BLUP) approach. In contrast to that, the genomic BLUP (GBLUP) approach uses the realized relationships between animals (Genomic Relationship Matrix - GRM) and thus allows a more accurate estimation of their breeding values. Under the GBLUP approach, the actual relatedness for between animals is estimated using genomic markers (realized relationships).

The SSGBLUP approach is intermediate between the latter two, as it weights both genomic (GRM) and pedigree (A) contributions to construct a unified relationship matrix known as H matrix. The SSGBLUP approaches thus requires an extra parameter α ($0 \leq \alpha \leq 1$) to weight the GRM and A matrices. The main advantage of the SSGBLUP approach is to allow the direct incorporation of animals with either pedigree or genomic information in the genetic evaluation.

Here we investigate ways to incorporate genomic information in routine deer evaluations of genetic merit. We use production traits related to growth, meat and carcass yield, health, and reproduction to compare the performance of genomic prediction using pedigree (BLUP), genomic BLUP (GBLUP) and single-step genomic (SSGBLUP) relationships among animals.

MATERIALS AND METHODS

Animals and Phenotypes. We used phenotypes for red animals born between 2018 and 2020. Phenotypic data from six production traits related to growth, meat and carcass yield, health, and reproduction was retrieved from DEERSelect. Production traits analysed were weaning live weight (WWT), 12-month live weight (W12), carbohydrate larval antigen-specific immunoglobulin A levels at 10 months of age (CARLA10), ultrasound measured eye muscle area at 10 months of age (EMA), velvet weight at 2 years (VW2), and conception date at 2 years (CD2). Table 1 details the number of records per trait and their corresponding summary statistics.

Table 1. Number of records and summary statistics by trait for New Zealand red deer

	n records	Mean	SD	Min.	Max.
WWT	18,649	55	7	22.2	89.4
W12	15,894	90	8.5	51.4	130.3
CARLA10	6,962	2.4	1.2	-2.3	6.6
EMA	6,205	25.9	2.9	12.7	36.3
VW2	1,031	3.1	1	0.7	8.1
CD2	2,163	103.7	12.4	73	130

GBS Genotypes. We used genotypic information from 19,863 animals and 55,784 SNPs mapped using GBS data. Genomic relationship matrices (GRM) were constructed using the KGD software (Dodds *et al.* 2015). Principal components (PCs) were obtained from the GRM in GCTA (Yang *et al.* 2011).

Population structure. The deer population in New Zealand is composed of several crosses from two species of the genus Cervus: C. elaphus (red deer) and C. canadensis (wapiti deer/elk), which have notable phenotypic differences. Given that the inclusion of genetically divergent breeds can reduce prediction accuracy in genomic evaluations (Calus *et al.* 2014; Makgahlela *et al.* 2013), analyses are conducted separately for each breed. Birth herd codes were used as a proxy for breed. Only red analyses are presented here.

Statistical Analyses. Phenotypes were modelled using a univariate genetic model for each trait. Genetic models include relevant covariates for each trait, including contemporary group, age of dam, birth date deviation, and breed proportions measured with PCs from the GRM (PC1 to PC3).

Genetic parameters and breeding values were estimated using pedigree (BLUP), genomic BLUP (GBLUP) (VanRaden 2008) and single-step genomic BLUP (SSGBLUP) (Misztal *et al.*, 2009) relationships among animals in MTG2 (Lee and van der Werf 2016). This software tool was also used to construct the H matrix. In absence of any prior information, we used α=0.5, equal weights for pedigree and genomic relationships, to construct H.

Genomic prediction accuracy. We assessed the prediction quality of the three relationship matrices using a forward-validation scheme. This was done by removing the phenotypes of the last cohort (target population, animals born in 2020), estimating the genetic models again using the older cohorts (training population, animals born 2018 and 2019), and comparing the breeding value predictions with actual phenotypes for the last cohort of animals. Prediction quality was assessed using two measures: prediction accuracy and prediction bias. Prediction accuracy (acc) is defined as correlation between predicted breeding values and phenotypes adjusted by fixed effects divided by the squared root of the heritability (SSGBLUP model). Bias (bias) as the regression's slope between predicted breeding values and phenotypes adjusted by fixed effects.

RESULTS AND DISCUSSION

Population structure. Figure 1 shows the first two PCs by birth year (A) and birth herd (B). In the figure A, we observe that animals are spread out evenly across the plot, suggesting that there is little change in the genetic composition of the animals over time. In contrast to that, figure B shows that birth herds form clearly defined clusters in specific regions which do not mix with each other. We can thus conclude that there is more variation between birth herds than across time.

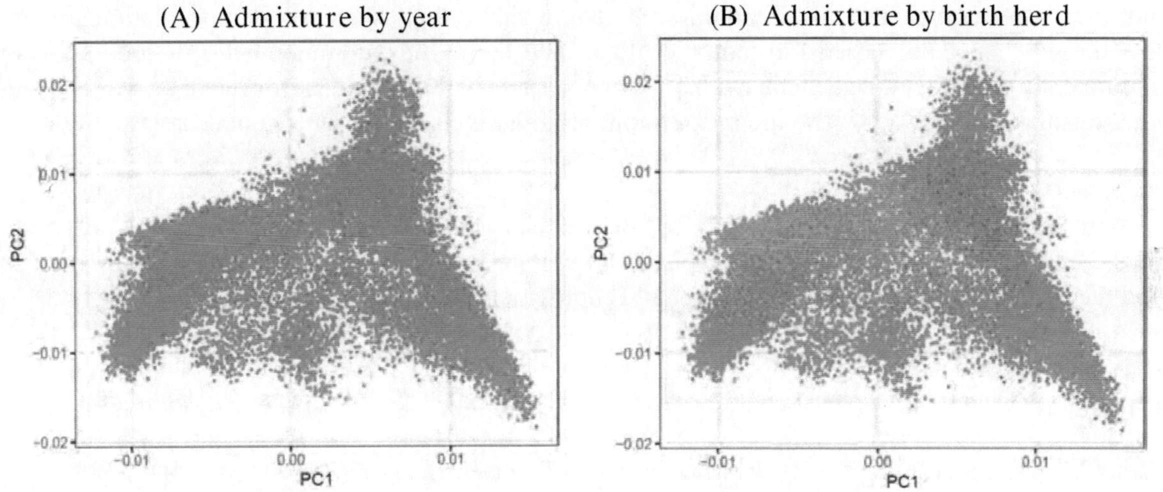

Figure 1. Genetic admixture by birth year (A) and birth herd (B) for deer born between 2018-2020. Principal components (PC) 1 and 2 are shown in the x and y-axis

Table 2. Accuracy and bias of genomic prediction using pedigree (BLUP), genomic (GBLUP) and single-step SSGBLUP) approaches for red deer

		BLUP		GBLUP		SSGBLUP	
	n target	acc	bias	acc	bias	acc	bias
WWT	6,476	0.33	0.41	0.34	0.57	0.28	0.87
W12	5,382	0.36	0.50	0.47	0.65	0.49	0.62
CARLA10	2,548	0.29	0.46	0.47	0.71	0.46	0.64
EMA	2,152	0.34	0.72	0.42	0.77	0.45	0.78
VW2	343	0.40	1.42	0.43	1.02	0.46	1.11
CD2	733	0.31	1.20	0.22	0.62	0.27	0.78
	Average	0.34	0.79	0.39	0.72	0.40	0.80

Genomic Prediction accuracy. The prediction accuracy and bias for the six production traits in red deer are presented in Table 2. Across all traits the prediction accuracy from BLUP, GBLUP and SSGBLUP approaches are 0.34, 0.39 and 0.40. This implies that incorporating genotypic information in the genetic evaluation could provide much more accurate breeding values, on average 16% and 18% more accurate for GBLUP and SSGBLUP, respectively. Similarly, bias is also reduced when using genotypic information (SSGBLUP) H, although variation by trait is still present.

Our analyses have some caveats. First, for computational easiness we focus only on red animals born between 2018 and 2020. This strategy reduces the computational time for the genetic evaluation but also limits the number of phenotypic records included for velvet weight (VW2) and conception date (CD2) as these are recorded at two years of age. Secondly, this strategy also reduces the number of deer with genotypes entering the evaluation. Despite these caveats, the prediction accuracy of the breeding values is improved, and the bias is consistently smaller across all traits as shown in the validation (Table 2).

CONCLUSIONS

We show that incorporating genomic information explicitly, either by using GBLUP or SSGBLUP, improves prediction accuracy and reduces bias. Across all traits, we estimate average gains in accuracy of 16% for GBLUP and 18% for SSGBLUP for red deer, the breed with the highest numbers of phenotypic records. In addition, breeding values that use genotype information are also less biased than those based on pedigree alone. We therefore recommend the incorporation of genomic information in the genetic evaluations performed by DEERSelect for red deer. Methods for joint evaluations for wapiti, red and red x wapiti cross animals are currently under investigation.

ACKNOWLEDGEMENTS

We are grateful to all deer breeders in New Zealand and the AgResearch Invermay deer farm who contributed phenotypic and genotypic information. This project was funded by DEEResearch and the Ministry of Business, Innovation and Employment (NZ).

REFERENCES

Bilton T.P., Chappell A.J., Clarke S.M., Brauning R., Dodds K.G., *et al.* (2019). *Anim. Genet.* 50: 307.

Calus M., Huang H., Wientjes Y., Ten Napel J., Bastiaansen J., *et al.* (2014). *Proc. WCGALP.*

Dodds K., Newman S., Clarke S., Brauning R., Hess A., *et al.* (2019) *Proc. Assoc. Advmt. Anim. Breed. Genet.* 23: 298.

Dodds K.G., McEwan J.C., Brauning R., Anderson R.M., van Stijn T.C., *et al.* (2015). *BMC Genomics* 16: 1047.

Gudex B., Johnson D., Ford D., Benton T. and Chardon J. (2013) *Proc. Assoc. Advmt. Anim. Breed. Genet.* 20: 336.

Lee S.H. and van der Werf, J.H.J. (2016). *Bioinformatics* 32: 1420.

Makgahlela M., Strandén I., Nielsen U., Sillanpää M. and Mäntysaari, E. (2013). *J. Dairy Sci.* 96: 5364.

Misztal I., Aguilar I., Johnson D., Legarra A., Tsuruta S. and Lawlor T. (2009). *Interbull Bulletin* 40: 240.

Rowe S., Dodds K., Ward J., Asher G., Brauning R., *et al.* (2018). *Proc. WCGALP.*

VanRaden P.M. (2008). *J. Dairy Sci.* 91: 4414.

Ward J., McIntyre S., Newman S. and Asher G. (2016) *Proc. NZ Soc. Anim. Prod.* 76: 54.

Yang J., Lee S.H., Goddard M.E. and Visscher P.M. (2011). *A. J. Hum. Genet.* 88: 76.

Proc. Assoc. Advmt. Anim. Breed. Genet. 25: 95 - 98

UNDERSTANDING TRAIT PREFERENCES AND VIEWS ON GENETIC TOOLS FOR THE NEW ZEALAND BEEF INDUSTRY

L. Kok[1], S. Harburg[1], B. Santos[1], P. Amer[1], J. Archer[2], A. Boyd[2] and G. Jenkins[2]

[1]AbacusBio Limited, PO Box 5585, Dunedin 9058, New Zealand
[2]Beef + Lamb New Zealand, PO Box 5501, Dunedin 9054, New Zealand

SUMMARY

Beef + Lamb New Zealand (B+LNZ) has implemented the Informing New Zealand Beef (INZB) programme, a 7-year Sustainable Food and Fibre Futures partnership with the Ministry for Primary Industries, which aims to boost the sector's profits by $460 million over the next 25 years. The programme includes the development of new traits for integration into future genetic evaluations. An online survey of beef producers was conducted. The survey consisted of a demographic section, asking questions about the respondent's production system and their views and attitudes on cattle selection and bull purchases. The survey also included a trait preference section for respondents to indicate their preferences and the relative importance of these traits to them. Overall, respondents had a preference for maternal traits, particularly cow fertility, cow functionality, and calving ease. In general, NZ beef producer survey respondents strongly support the use of estimated breeding values (EBVs) and indexes as tools to inform their selection and bull purchasing decisions. Almost 75% of respondents agree that NZ farm systems require specialized selection indexes. The survey also highlighted the importance of ongoing extension programs to improve adoption and understanding of genetic tools. There is strong priority for structural soundness and functionality traits, feed efficiency, and fertility traits, which are seen as important for new trait development.

INTRODUCTION

INZB is seeking to establish a beef genetic improvement system that can best support the needs and priorities of the NZ beef industry. One of the INZB projects aims to develop new traits for integration into industry phenotyping programs for future inclusion in industry genetic evaluations. An industry survey was undertaken between July and August 2022, to capture NZ beef industry perspectives on traits, trait priorities and selection index requirements to support genetic improvement within the NZ beef industry. Insights from the survey are being used to identify opportunities to develop new breeding traits, as well as to understand views and perceptions that might influence the scope and relevance of the genetic evaluation systems utilized by the NZ beef industry. Further, identification of beef farmer trait preferences (among current and future traits) and factors driving these preferences, will inform development of custom selection indexes that reflect the priorities of NZ beef industry stakeholders.

MATERIALS AND METHODS

A voluntary online survey was conducted from 1st of July to 12th of August 2022 and was distributed by B+LNZ to beef farmers and other industry stakeholders. The survey predominantly targeted beef farmer respondents (bull breeders, commercial breeders, and finishers), rural professionals, and other key stakeholders. Two distinct approaches were used to allow respondents to identify trait priorities, a direct ranking question (within the demographic component of the survey), and a Conjoint Analysis approach (Hensher *et al.* 2005) in the trait preference component. Firstly, the preference for traits was asked on a scale from 1 to 100 with the question displaying all traits jointly from which answers were converted to a relative weight of importance of each trait by dividing the score given to each trait by the sum of scores across all traits. Second, trait priorities were also captured through 1000minds (Hansen and Ombler 2009) which contrasts trade-off choices between pairs of traits, assuming the level of the trade-off has broadly equivalent economic values. This pairwise comparison is practical and requires less effort from participants than other methods, making choice decisions simpler and nearer to respondents' "true" preferences.

The demographics component of the survey received 439 complete responses and 290 partial responses, whilst the 1000minds component received 311 complete responses and 169 partial responses. A broad sample of the NZ beef industry participants from a variety of beef business activities responded. There was strong engagement from commercial breeders (44.5% of respondents) and finishers (24.9% of respondents). Survey respondents also represented a broad cross-section of farming regions across the North and South Islands of New Zealand.

A principal components analysis (PCA) and cluster analysis (CA) of trait preferences was performed to investigate groupings of respondents with similar preferences. Subsequent analysis was undertaken to characterise these clusters and understand whether differences in trait preferences reflected potential demographic differences or differences in breeding philosophy.

RESULTS AND DISCUSSION

Respondents were asked a series of questions about their breeding system to inform development of new maternal traits (e.g., cow body condition score). Approximately 82% of respondents mate their heifers targeting a first calving as a 2YO, with over half of the respondents first mating heifers between 300-350kg live weight. The overall frequency of calving difficulty averaged 3.71% (standard deviation of 8.12%) and is broadly consistent with published literature (e.g. Faucitano *et al.* 2012). Respondents estimated that the average weight of cows at weaning was 548kg (SD ± 94kg) versus an optimum weight of 555kg (SD ± 81kg). Generally, respondents agree that cow size and composition is important and that more descriptive traits (beyond cow weight) could be beneficial. This could provide a case for the development of additional traits for describing cow size and composition such as cow body condition score and cow height (as an indicator of frame size).

When asked a series of questions associated with general views on the use of genetic tools, most respondents believe that EBVs (64%) and economic selection indexes (54%) are useful tools for representing animal genetic merit and improving herd performance. Many respondents considered genetic tools to be important for bull purchase decisions, although a subset (30%) rated visual appraisal, structural soundness, and horn/poll status as being sufficient to predict performance and genetic merit. Similar levels of importance were placed on genetic tools and raw performance information (animal live weight, and other phenotypic measurements). This highlights the need for ongoing investment in extension activities to improve understanding and drive adoption of better genetics and tools across the industry.

Respondents generally supported a simplified portfolio of selection indexes covering maternal, terminal, and dairy-beef systems (69% somewhat agree/agree/totally agree). Almost 75% of respondents selected somewhat agree, agree, or totally agree that NZ farm systems require specialised selection indexes. There is support for development of sub-indexes to summarise animal merit (83%) and customisability to adapt indexes to specific requirements (70%). Respondents also believed (69% somewhat agree/agree/totally agree) that maternal selection indexes should include emphasis on carcass and eating quality traits alongside maternal traits. There is also very strong support for inclusion of functional traits (structure, docility, etc.) within selection indexes. These traits are currently omitted from most industry indexes.

Trait preferences. In the sociodemographic part of the survey, respondents placed greatest emphasis on maternal traits, particularly cow fertility, cow functionality (foot/leg structure, teat and udder scores, docility), and calving ease. This indicates an opportunity to improve understanding of existing trait EBVs and encourages greater effort on data collection for docility and calving ease.

Growth traits, carcass weight, and feed efficiency represent the next highest priority traits. Feed efficiency represents the only novel/new trait among the highest priority traits.

Trait preferences from the 1000minds part of the survey broadly were consistent with the answers on trait preferences based on relative rankings. The exception was cow fertility which ranked 5th highest (out of 11 traits). This could reflect either an overestimation by respondents of the relative importance of fertility, or the nominated trade-off for fertility within the 1000minds survey (3 less cows per 100 culled due to low fertility) did not adequately reflect an appropriate value for cow fertility relative to the other trait trade-offs.

Trait preferences tend to hold across all respondent demographic categories. There were some interesting insights in this demographic breakdown. For instance, the dairy farmer segment (3% of respondents) was the most clearly differentiated segment within the primary beef activity group, with a greater priority placed on methane emissions, marbling and weaning weight, and lower priority placed on docility, cow fertility and calving ease. Bull breeders and commercial breeders were generally very closely aligned, except for marbling (higher priority for bull breeders) and cow body condition score (higher priority for commercial breeders) as the key areas of divergence. The tendency for bull breeders to place higher priority on marbling is a likely reflection of sourcing genetics from overseas (particularly US and Australia) and the use of the combined TransTasman Angus Cattle Evaluation (TACE) with Angus Australia (Angus Australia, 2021).

Cluster analysis. Two distinct groups of respondents with similar preferences were identified. The 'production focus' cluster, which comprises 136 respondents, showed higher preference for both growth and carcass traits (i.e., feed efficiency, weaning weight and carcass weight) alongside calving ease. The 'maternal focus' cluster (comprising 168 respondents) had a stronger focus on maternal and functional traits (i.e., calving ease, cow fertility and docility). The average trait preferences (% weighting) of both clusters are presented in Figure 1. Interestingly, the PCA analysis indicated that a sub-cluster of novel, progressive traits, namely methane emissions, feed efficiency, and marbling could be formed from the 'production focus' group if clusters were to be further separated out. This reveals the existence of diversity of thought even within groups of similar preferences.

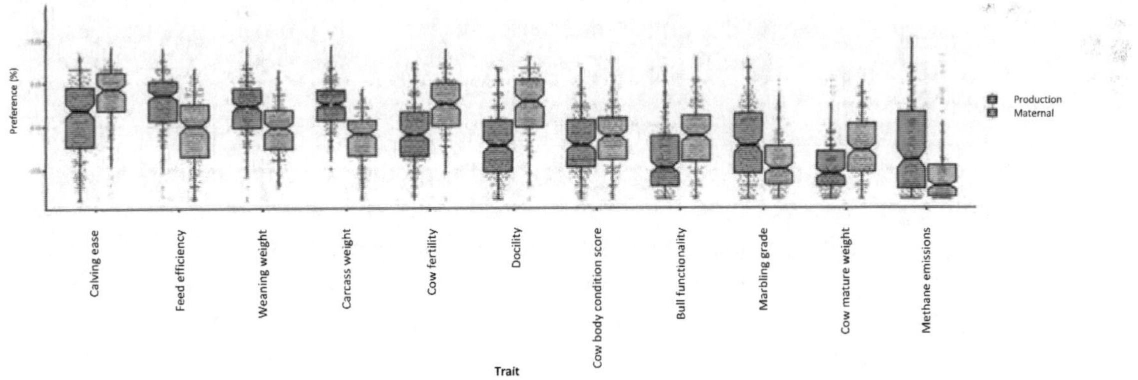

Figure 1. Cluster analysis of trait preferences from New Zealand beef industry stakeholders

Views and attitudes. Demographics only had a small influence on trait preferences. This is consistent with many similar studies, whereby breeding philosophy is intrinsic to the respondent, rather than reflecting their demographic situation or background. In addition to trait preferences, there are subtle differences in views and attitudes between the clusters. Respondents from the Production Cluster generally placed greater importance on genetic tools than those in the Maternal Cluster, however, overall patterns of response were quite similar between the two clusters (Figure

2). Whilst industry does consider genetic tools to be important for bull purchase decisions, their importance is lower than visual appraisal (structural soundness and overall appearance).

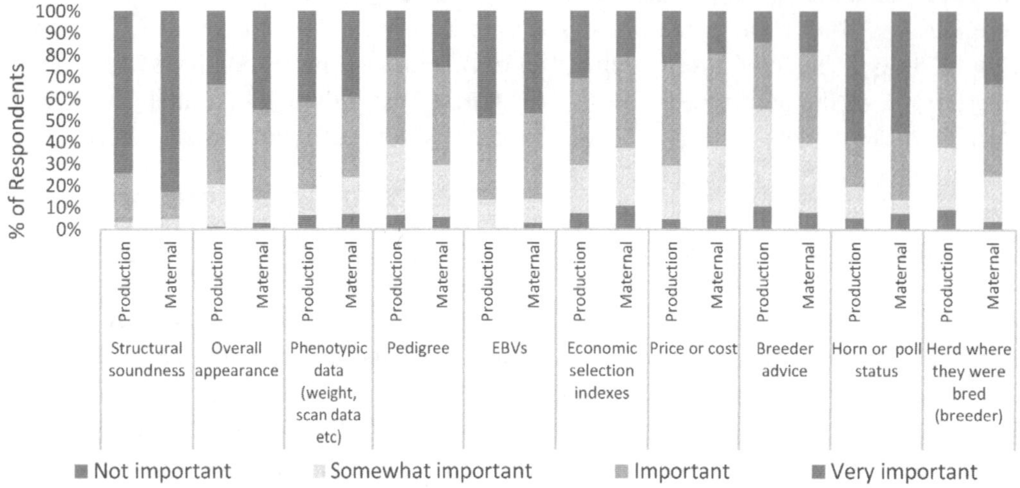

Figure 2. Respondent views represented by level of importance on bull purchase criteria

Indexes and index development. Survey results highlight several key opportunities to enhance the use and relevance of economic indexes to the NZ beef industry. The industry strongly supports the inclusion of functional traits (docility and structural traits) to indexes. This represents a key area of potential collaboration between B+LNZ and the breed societies to evaluate options to incorporate these traits within industry selection indexes. The implementation of a narrower range of selection indexes includes several key challenges to ensure these indexes are relevant to the trait preferences and breeding objectives of as many users as possible. These key challenges comprise 1) the relative importance of maternal versus production traits reflects a key area of divergence between key segments/clusters of the industry; 2) Whilst industry supports a simplified portfolio of indexes, there is strong interest in the ability to customise indexes to meet individual breeding objectives.

CONCLUSIONS

This industry survey underlines the importance of maternal and functional traits to most segments of the industry. It has identified key areas for B+LNZ and other key stakeholders to improve the scope and delivery of genetic tools. The importance of ongoing extension programs to improve adoption and understanding of these tools is critical for the success of the INZB programme. In addition to scoping the feasibility of collecting phenotypic data, B+LNZ should also engage with breed societies to assess the feasibility of implementing existing and developing new traits in a NZ genetic evaluation system. This engagement will secure collaboration and identify preferred approaches for implementation of future selection indexes for the New Zealand beef industry.

ACKNOWLEGEMENTS

The authors would like to acknowledge the support from the Informing New Zealand Beef (INZB) programme to develop this work.

REFERENCES

Angus Australia (2021) Introduction to TACE. https://angusaustralia.com.au/

Faucitano L. And Cassandro, M. (2012) *Animal* 6: 1780.

Hansen P. and Ombler, F. (2009) *J. Multi-Crit. Decis. Anal.* 15: 87.

Hensher D.A., Rose J.M. and Greene W.H. (2005) Cambridge University Press.

THE ECONOMIC VALUE OF GESTATION LENGTH

E. Ooi[1], N. Howes[1], F. Hely[1], M. Stephen[2], P. Amer[1] and C. Quinton[1]

[1]AbacusBio Limited, PO Box 5585, Dunedin 9058, New Zealand
[2]DairyNZ Limited, Private Bag 3221, Hamilton 3240, New Zealand

SUMMARY

Calving pattern is one of the most important factors affecting the overall profitability and reproductive performance of pasture-based dairy herds. Genetic selection for fertility typically incorporates a variety of traits, including those that use calving dates as part of their definition. Early calving can also be achieved through shortened gestation length (GL). Although not an official component of fertility evaluations in New Zealand, GL is an unavoidable contributor to calving season day (CSD), which is an official component. This is because CSD is the combined result of the conception date and GL of the foetus.

GL is not a true reflection of fertility, which typically comprises oestrus, fertilisation, and maintenance of pregnancy, and it also has comparatively high heritability, allowing it to dominate genetic progress over conventional fertility traits. Therefore, there is growing interest in separating the influence of GL from cow fertility evaluations. In this paper we outline an approach to derive the direct economic value (EV) for GL for a situation where it would be included in an index containing a conception date-based fertility date. Even though GL is not a true fertility trait, we find a high EV for GL through its indirect effect on fertility when farmers respond to a shorter GL population by delaying mating to achieve an identical seasonal calving pattern. Cows that have had a longer period between calving and first mating conceive at higher rates. This research facilitates revisions to the way fertility traits are included in national selection indices for seasonal dairy cows, allowing the development of non-linear index functions to avoid favouring selection for excessively short GLs challenging the welfare, viability, and productive performance of the resulting calves.

INTRODUCTION

Calving pattern is one of the most important factors affecting the overall profitability and reproductive performance of pasture-based dairy herds (Macdonald and Roche 2023). Genetic selection for fertility traits based on calving dates is used in New Zealand (e.g., calving season day; CSD) and many other countries to improve fertility with the aim of tightening calving patterns and reducing non-pregnant rates (Bowley *et al.* 2015).

Bias and censoring caused by this approach in pasture-based systems are usually addressed through the addition of other fertility traits, including cyclicity (e.g., PM21 in NZ) and conception (e.g., non-return rate in Australia). More recent research, however, has emphasized that early calving can be achieved not only through improved submission and conception rates, but also through shortened gestation length (GL). Although not an official component of fertility evaluations in NZ, GL is an unavoidable contributor to the CSD phenotype, as the latter is the result of conception date and GL of the foetus.

However, GL is not a true reflection of fertility – i.e., a cow's ability to resume and express oestrus, or to achieve fertilisation and maintain pregnancy. Further, the comparatively high heritability of GL dominates genetic progress over conventional fertility traits. Finally, short GL may have adverse consequences on calf health which must be carefully managed as part of a responsible approach to breeding. Most countries are likely to be indirectly selecting for short GL in their dairy cattle, and the impact of this on health, performance, and management systems is a current topic of research activity.

There is growing interest in separating the GL component of the EBVs for fertility into performance that is influenced by conception rate and performance that is due to GL. However, this requires an understanding of the direct economic value (EV) of GL. Therefore, the objective of this study was to use a stochastic fertility model to calculate the EV of GL for NZ dairy herds.

MATERIALS AND METHODS

Stochastic fertility model. To estimate the EV of GL, we adapted a stochastic fertility model (SFM) developed by Dennis *et al.* (2018) to simulate the performance of high and low fertility cattle. Briefly, the model simulates a cohort of 200,000 heifers' reproductive lifetime through 5 lactations, including genetic, physiological and management factors. It has the capacity to adjust genetic merit, oestrus duration, and fertility-related breeding values, but also incorporates events such as pregnancy diagnosis, oestrus detection, and embryonic loss.

The original SFM was tuned to reflect the phenotypic performance of heifers that had been divergently bred for low and high genetic merit for fertility in an experimental setting. Therefore, to model the more-realistic effects of GL changes on low- and high-fertility performance in commercial NZ dairy herds, we adjusted the base fertility traits underlying the model. For the analysis we defined low, average, and high fertility herds as having differences in their breeding values for postpartum anoestrus interval (-3, +3), number of services per conception (-0.03, +0.03), and oestrus duration (-1.6, +1.6).

To calculate the EV of GL, we reduced the mean GL (GL-3) and delayed planned start of mating (PSM) by 3 days in total (PSM+3). This reflects management changes currently being implemented by NZ dairy farmers in conjunction with the use of short GL semen. For each of the six runs (i.e., no intervention and GL-3/PSM+3 for low, average, and high fertility herds), 100 iterations were completed for a herd size of 200,000 animals, resulting in a dataset of 20 million lactations. Although Table 1 shows the effects of 3 days fewer GL for high and low fertility performance herds, these values are mainly provided to compare the effect of GL under a range of conditions. For index development, the following EV calculations are based on the performance of average herds.

EV calculation. The EV of GL is built from four component EVs that influence profitability: milk production, empty rate, value of artificially bred (AB) and beef calves sold, and number of natural mate (NM) bulls required. Component EVs were calculated as the change in $ profitability per lactation, per day change in GL, independent of changes in other New Zealand dairy breeding objective traits. Key parameters are summarised in Table 2. See Amer *et al.* (2013), Santos *et al.* (2022) and https://www.dairynz.co.nz/animal/animal-evaluation/interpreting-the-info/economic-values/) for details on the breeding objective and Breeding Worth index.

The milk component EV ($EVGL_{milk}$) was calculated as

$$EVGL_{milk} = \left[\left(\sum_{g=1}^{3} Pmilk_g \times \rho_{g_{GL-3}}\right) - \left(\sum_{g=1}^{3} Pmilk_g \times \rho_{g\_base}\right)\right] / -3,$$

where $Pmilk_g$ is the average milk profit per lactation in groups (denoted by *g*) of early (first 21 days of season), mid (21-42 d) and late (>42 d) calving cows. For each group, $Pmilk_g$ (\$/lactation) was calculated from weighted averages of milk production from Holstein-Friesian, Jersey and crossbred breeds in upper and lower North and South Islands, and average feed costs per lactation based on energy requirements and industry feed prices. In each group, $\rho_{g_{GL-3}}$ and ρ_{g_base} are the proportions of cows in the GL-3 and base herds, respectively (Table 1). -3 is the days change in GL in the base herd compared to the GL-3 herd.

The empty rate component EV ($EVGL_{empty}$) was calculated as

$$EVGL_{empty} = \left(\sum_{p=1}^{4} \Delta e_p \times \pi_p \times \rho_p\right) / -3,$$

where Δe_p is the change in the proportion of empty cows in the GL-3 herd at parity (p = 1, 2, 3 and 4+) compared to the base; π_p is the average value of an empty cow culled following parity p based on milk income, replacement rates, feed/purchase costs, and salvage value; and ρ_p is a weighting factor aggregating effects for parities based on the proportion of cows finishing lactation p.

The AB and beef calves sold component EV ($EVGL_{calf}$) was calculated as

$$EVGL_{calf} = \left[\left(\sum_{p=1}^{4} \Delta 6WICR_p \times \rho_p\right) / -3\right] \times \tau,$$

where $\Delta 6WICR_p$ is the change in 6-week in-calf rate (6WICR) from the base herd compared with the GL-3 herd at parity p (Table 1). τ is the average benefit of AB and high-value beef-sired calves per unit change in 6WICR ($/calf per proportion), based on industry data of proportions and prices for the range of AB and natural mating sired dairy and beef crossbred calves.

The NM bulls component EV ($EVGL_{NM}$) was calculated as

$$EVGL_{NM} = \left[\left(\sum_{p=1}^{4} \Delta 6WICR_p \times \rho_p\right) / -3\right] \times \nu,$$

where ν is the cost of NM bull per cow not-in-calf at the end of the 6-week AI mating season ($/cow) based on industry data of average yearly bull lease rate and ratio of cows to NM bull.

The total GL EV ($EVGL$) was calculated as

$$EVGL = \left(EVGL_{milk} + EVGL_{empty} + EVGL_{calf} + EVGL\_nm\right) \times \rho_{mp},$$

where ρ_{mp} is the industry average herd proportion of multiparous cows modified slightly to allow for delayed expression.

RESULTS AND DISCUSSION

SFM results. The mean reproductive performance metrics, produced by the SFM for each of the six scenarios are shown in Table 1. These do not include the performance of nulliparous heifers, which are often managed separately to the milking herd. The effect of GL-3/PSM+3 differs depending on the herd's existing reproductive performance. This is likely because a greater proportion of cows in a high fertility herd are already calving early at an optimal time.

Table 1. Reproductive performance metrics produced by the stochastic fertility model, adjusted for no intervention (base) and short gestation length (GL-3/PSM+3) scenarios

	Average fertility			Low fertility			High fertility		
	Base	GL-3/PSM+3	Δ	Base	GL-3/PSM+3	Δ	Base	GL-3/PSM+3	Δ
6-week in-calf rate (6WICR)	65.0%	67.0%	2.0%	41.3%	43.5%	2.3%	73.8%	75.3%	1.5%
Empty rate	12.3%	11.2%	-1.1%	27.9%	26.0%	-1.9%	7.9%	7.2%	-0.6%
% Calved by 21 days	48.6%	55.3%	6.7%	32.8%	38.3%	5.5%	55.5%	62.3%	6.8%
% Calved by 42 days	74.5%	78.3%	3.9%	58.5%	63.3%	4.8%	80.3%	83.5%	3.2%

EV results. The EVs for GL (Table 2) show that reducing mean GL by 3 days and delaying planned start of mating can have a significant impact on dairy farm profitability, especially in average and low fertility herds. Most of this is due to the indirect contribution of GL to improved fertility, with early calving allowing greater time for uterine involution and resumption of cyclicity.

The largest contributor to GL EVs was a reduction in empty rate at -$5.31 per extra day of gestation. Improving cow longevity not only reduces the need for more heifer replacements – which are costly to rear to maturity and take multiple seasons to achieve peak lactation – but is also consistent with societal expectations around animal welfare and survival. The second-most significant contributor was lactation profit at -$4.11 per extra day of gestation, which came from having a lower proportion of cows calving late in the season (i.e., a tighter calving pattern). Note that early in the calving season, the pattern of calving is relatively unchanged when farmers delay PSM as the average GL EBV of their herd shortens. Having more cows at peak lactation in conjunction with peak pasture availability is a key driver of profitability for pasture-based dairy herds. Finally, the contributions of having more high genetic merit artificially bred heifers (-$0.20 per extra day of gestation) and fewer natural follow-up bulls (-$0.13 per extra day of gestation) were smaller but still significant.

Table 2. Key gestation length EV model parameters and resultant component EVs

	Description	Values and unit
Model parameter	Milk profit for early, mid, and late calving cows ($Pmilk_g$)	$2435, $2372, $2162 / lactation
	Empty cow culled value following parity 1, 2, 3 and 4+ (π_p)	$1704, $1710, $1614, $1350 / cow
	Weighting factor for parity 1, 2, 3 and 4+ proportions (ρ_p)	0.25, 0.20, 0.16, 0.39
	AB and high value beef calves benefit value (τ)	$30 / calf per unit 6WICR
	Natural mating cost (ν)	$20 / cow
	Proportion of multiparous cows in herd (ρ_{mp})	0.73
Economic value	Milk profit component EV ($EVGL_{milk}$)	-$4.11 / lactation per d GL
	Empty rate component EV ($EVGL_{empty}$)	-$5.31 / lactation per d GL
	AB and beef calf component EV ($EVGL_{calf}$)	-$0.20 / lactation per d GL
	Natural mating component EV ($EVGL_{NM}$)	-$0.13 / lactation per d GL
	Total EV adjusted for ρ_{mp} ($EVGL$)	-$7.12 / lactation per d GL

CONCLUSION

The economic value of gestation length includes 1) increased milk income from having a greater proportion of early-calving cows at peak lactation when there is also peak pasture availability, 2) reduced involuntary culling due to decreased empty rates, 3) the value of more artificially bred high genetic merit heifers, and 4) a reduction in natural follow-up bulls required for the herd.

Our results show that reducing mean gestation length by 3 days and delaying PSM can have a significant impact on dairy farm profitability, especially in average and low fertility herds.

REFERENCES

Amer P.R., Santos B. Byrne T.J., Ludemann C. Visser B., Harris B.L. and Bryant, J. (2013) *Interbull Bulletin* 47: 95.

Bowley F.E., Green R.E., Amer P.R. and Meier S. (2015) *J. Dairy Sci.* 98: 2005.

Dennis N.A., Stachowicz K., Visser B., Hely F.S., Berg D.K., Friggens N.C., Amer P.R., Meier S. and Burke C.R. (2018) *J. Dairy Sci.* 101: 3176.

Jenkins G.M., Amer P., Stachowicz K. and Meier S. (2016) *J. Dairy Sci.* 99: 418.

Macdonald K.A. and Roche J.R. (2023). *N. Z. J. Agric. Res.* 1–118.

Santos B.F.S., Jenkins G.M., Kok J., Amer P.R. and Stachowicz K. (2022) *Proc. WCGALP* 12: 1782.

AUSTRALIAN BEEF CATTLE BREEDING OBJECTIVES

B.W. Gudex[1], P.J. Williams[1] and B.J. Walmsley[2,3]

[1]Agricultural Business Research Institute, Armidale, NSW, 2351 Australia
[2]Animal Genetics and Breeding Unit*, University of New England, Armidale, NSW, 2351 Australia
[3]NSW Department of Primary Industries, Livestock Industries Centre, Armidale, NSW, 2351 Australia

SUMMARY

BreedObject is the software used to formalise breeding objectives and create the selection indexes produced by BREEDPLAN. The BreedObject breeding objectives and selection indexes allow cattle producers to identify the most profitable cattle genetics for the beef production system modelled by each selection index. Since the release of the latest version (6.2) of the BreedObject software, eight Australian beef cattle breed organisations have implemented 29 new or revised selection indexes. This paper discusses the process by which the selection indexes were developed in conjunction with the relevant breed societies, summarises the EBV emphases applied in these new selection indexes, and discusses the breeder feedback and implications of the selection indexes in the greater industry.

INTRODUCTION

Selection indexes provide an overall estimate of an animal's genetic value for profit for a specified production system. Selection indexes are calculated by placing weightings on individual traits, with these weightings derived from the economic importance of the trait. As such, selection indexes reflect both the short-term profit generated by a bull through the sale of his progeny and the longer-term profit generated by his daughters if they are retained in the herd. The costs of production, including feed, are also accounted for. The selection indexes published by BREEDPLAN are calculated using BreedObject software (www.breedobject.com) and are reported in units of net profitability per cow mated ($) for the production system/market scenario that they represent.

MATERIALS AND METHODS

The development process for constructing a BreedObject selection index starts with individual breed organisations determining which production systems are the most relevant for their membership. This decision is influenced by the types of production systems that each breed organisation's genetics are currently used in or are expected to be used in the future. Once ready, each selection index is only made available for animals recorded on the relevant breed organisation's database. This allows the selection index definitions and inputs (including genetic parameters) to be specific to each breed organisation's recorded population.

Once the desired production systems were identified, a detailed description of the input costs and value generation of the commercial herd/production system was required for the BreedObject software. This process involved approximately 180 questions with the actual number varying between the selection indexes as the presence of some questions were reliant on prior answers. These questions included details of typical levels of production, herd population structure, prices received, costs of production etc in commercial herds.

Once the target production system was described, the BreedObject software assessed what emphasis needed to be applied to each trait to achieve profitability increases in the production system

* A joint venture of NSW Department of Primary Industries and University of New England

and market end point for which each selection index was designed. This step included evaluating the selection response expected from direct selection on the individual EBVs and the correlated responses expected from selection on related EBVs. Nonlinear effects (e.g. penalties for both under and over fat specifications) are also accounted for. Details of each selection index are available via the Help Centre on the BREEDPLAN website (https://breedplan.une.edu.au/help-centre/).

This paper summarised the EBV emphases applied in each of the new selection indexes implemented since 2018 using version 6.2 of the BreedObject software for Australian breed organisations. The selection indexes were grouped according to whether they were designed for replacement heifers to be retained in the herd (self-replacing) or not (terminal).

RESULTS AND DISCUSSION

A total of 29 new or updated selection indexes using version 6.2 of the BreedObject software have been made available since 2018 via the Australasian Charolais, Belmont Red, Brahman, Hereford, Performance Herds Australia, Southern Limousin, Trans-Tasman Angus, and Wagyu BREEDPLAN analyses. Of these selection indexes, 21 were self-replacing and 8 were terminal selection indexes (Table 1). Beyond the self-replacing/terminal differentiation, there was considerable variation in the target markets and production environments represented by the selection indexes. The target slaughter ages varied from 15 to 32 months, which in turn contributed to the corresponding variation in the target slaughter weights (Table 1) and emphasis on carcase EBVs. Part of this variation was due to the wide variety of production environments present across Australia from the tropical conditions in the north to the temperate regions in the south of the country. In addition, most breed organisations also had international members (predominantly from New Zealand) to consider at some level when developing their selection indexes. The presence of genotype by environment interactions in the resulting breeding objectives is consistent with the findings of Walmsley & Barwick (2018).

Table 1. Summary of the Selection Indexes and their corresponding market endpoints that are analysed in this paper

	Self-Replacing	Terminal
Number of Selection Indexes Analysed	21	8
Number of Breed Associations	8	6
Target Steer Slaughter Age Range (months)	15 to 32	12 to 29
Target Heifer Slaughter Age Range (months)	15 to 29	12 to 27
Target Steer Carcase Weight Range (kg)	250 to 460	205 to 360
Target Heifer Carcase Weight Range (kg)	230 to 410	190 to 300

Figures 1 and 2 show the range of, and the average EBV emphasis in the self-replacing and terminal selection indexes. With no daughters retained for future breeding, the maternal EBVs received no emphasis in the terminal selection indexes, thus allowing the emphasis applied to the calving, growth and carcase EBVs to be greater than in the self-replacing selection indexes. It should be noted that only one terminal selection index for a *Bos indicus* breed type was developed and implemented in the timeframe of this study. Therefore, some of the observed differences between the self-replacing and terminal indexes are likely to be due to the resulting variation between the breed types and the environments where they are typically (but not exclusively) run in Australia (*Bos indicus* breed types in the northern part of the country and *Bos taurus* in the south).

Within the self-replacing selection indexes, there were noticeable differences between the *Bos indicus* and *taurus* breed types. The selection indexes developed for the *Bos taurus* breed types typically had a higher emphasis on calving ease, earlier growth, and less emphasis on fertility than their *Bos indicus* counterparts. The higher emphasis on fertility (the Days to Calving EBV) in *Bos indicus* is to address the lower levels of fertility typically observed in Northern Australia (McCosker *et al.* 2010). The Calving Ease EBV emphasis was one of the more variable due to variation in the age of heifers at first calving, the existing levels of calving ease within each breed, and/or the breed with which the bulls were mated to. Additionally, in the breeds that run the majority of their cattle in Northern Australia, there was a desire to maintain or raise birthweights to improve calf vigour and post birth survival.

Regarding the emphasis applied to the three growth EBVs, the majority of the emphasis was applied to the weight EBV that matched the target slaughter age. Therefore, the other, non-target, growth EBVs can have a low or even negative emphasis, particularly if they occur after the target slaughter age. It should be noted that the expected selection response of the Growth EBVs with low or negative emphasis would still typically be positive due to the high genetic correlations between these traits. The Wagyu and any production system involving *Bos indicus* breed type genetics (including *Bos taurus* bulls over *Bos indicus* or *Bos indicus* cross cows) had target slaughter ages greater than 2 years of age, while the *Bos taurus x Bos taurus* selection indexes all targeted slaughter at 2 years or less.

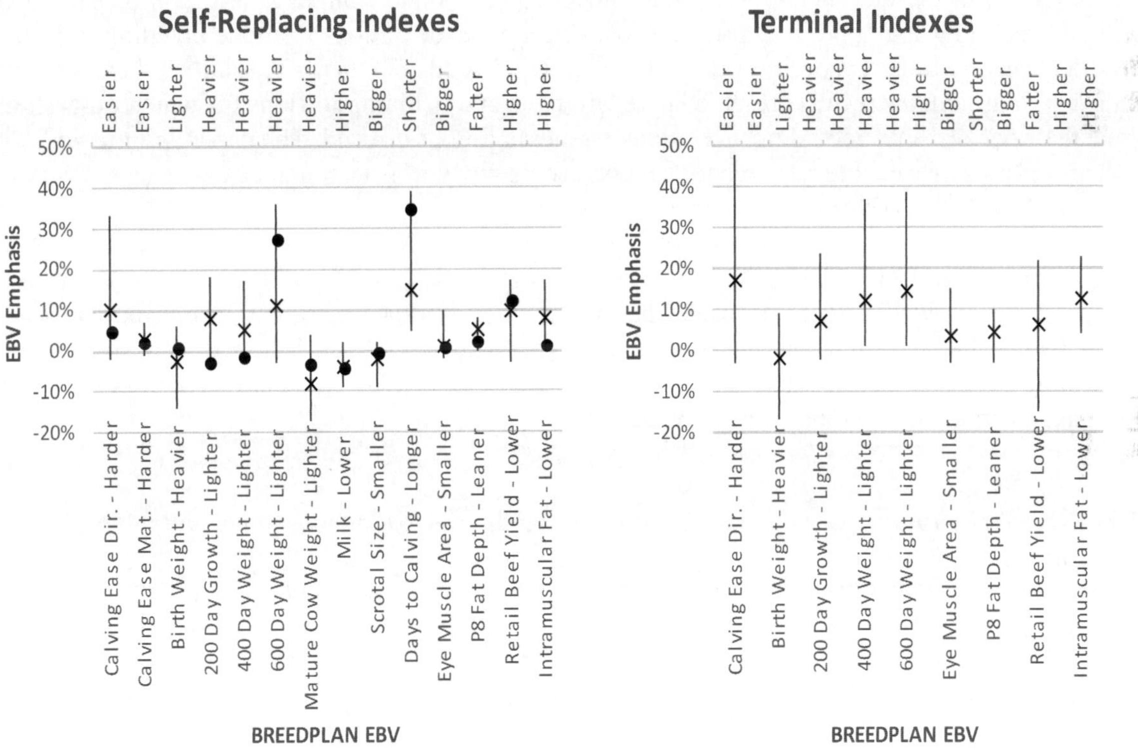

Figures 1 and 2. Range (line) and average (• for *Bos indicus,* × *for Bos taurus* breed types) of the EBV emphasis in 29 Australian Self-Replacing and Terminal beef cattle selection indexes

As part of the selection index development process, breeder input was sought and there were some examples where breeder expectation did not match the emphasis applied by the BreedObject software. There was considerable variation between breeders in their attitude towards the emphasis applied to calving ease, mature cow weight and the carcase traits in the selection indexes. As a consequence, a number of breeds implemented multiple selection indexes where one or more allowed mature cow weight to increase, and the other(s) held or reduced it. Further feedback on

Mature Cow Weight and Days to Calving, centred around the importance of these EBVs to the selection indexes and their relatively low levels of recording (Gudex & Millen 2019). This feedback places emphasis on the need for further extension efforts to promote the recording of these traits. For Days to Calving, this concern was compounded in the four breeds where this EBV was not reported and its emphasis in the selection index is applied through correlated traits. The Milk EBV typically received a low or negative emphasis in the selection indexes which caused concern for some breeders who assumed that increasing the weaning weight of the progeny was always desirable without considering the whole picture (e.g. the effect on cow BCS and her subsequent fertility and health). This assumption by breeders does not completely align with the standard BREEDPLAN advice which advocates selecting for a Milk EBV level appropriate for the environment where the cows are to be run (BREEDPLAN 2023). Environmental impacts were also discussed by some breed organisations, though none chose to add additional emphasis beyond the concept that animals with better production system efficiency will be better for the environment (and profitability).

CONCLUSION

Production systems vary between breeders and breeds, and therefore the corresponding trait emphasis in the selection indexes presented in this paper was variable. While this study summarised the breeding objectives, it is important to acknowledge that other sources of information should and will be used in most animal selection decisions. Therefore, deviations from the breeding objectives described here will be expected in the commercial and seedstock herds that utilise these selection indexes. That said, the results presented here will have practical implications for which traits should have their performance recording and extension messaging prioritised. The paper and methodology will also provide a valuable resource for benchmarking any new selection indexes that are developed or revised for BREEDPLAN.

ACKNOWLEDGEMENTS

The authors would like to thank the breed organisations who have contributed to and implemented the selection indexes that were analysed in this paper. There were also numerous contributions from the staff at the Animal Genetics and Breeding Unit, and the Agricultural Business Research Institute that should be acknowledged.

REFERENCES

BREEDPLAN https://breedplan.une.edu.au/understanding-ebvs/understanding-milk-ebvs/. Retrieved 13/01/2023

Gudex B.W. and Millen C.A. (2019) *Proc. Assoc. Advmt. Anim. Breed Genet*. 23: 488.

McCosker T., McLean D. and Holmes P (2010) B.NBP.0518. Meat and Livestock Australia, Sydney.

Walmsley B.J. and Barwick S.A. (2018) *Proc. Wld. Cong. Genet. Appl. Livest. Prod*. 11: 866.

Proc. Assoc. Advmt. Anim. Breed. Genet. 25: 107 - 110

SIGNATURES OF POSITIVE SELECTION FOR SCROTAL CIRCUMFERENCE IN THREE BEEF CATTLE BREEDS

Z. Manzari, D.J. Johnston, N.K. Connors and M.H. Ferdosi

Animal Genetics Breeding Unit*, University of New England, Armidale, NSW, 2351 Australia

SUMMARY

This study aimed to detect genomic regions associated with scrotal circumference in Australian Brahman, Hereford, and Wagyu beef cattle breeds. The presence of selection signatures was based on the F_{ST} test, using data on the genotype and BREEDPLAN estimated breeding values for SC of 100,990 animals. Signals of selection for scrotal circumference were identified in genomic regions on several chromosomes, especially chromosome 14 in Brahman with most candidate genes under selection associated with male fertility or growth. The findings of this study may be applicable to breeding programs using more informative markers and assigning higher weights to them to increase the accuracy of genomic predictions and improve the reproductive performance of beef cattle.

INTRODUCTION

The unique genetic patterns left on the genome by natural and artificial selection in livestock populations are known as "selection signatures" (Gouveia *et al.* 2014). Detecting these signatures may help explain the selection history, adaptation, and genetic advancement of traits that may be economically important. Scrotal circumference (SC) is commonly employed as a selection criterion for breeding bulls because it is easily measured and correlated with a number of favourable reproductive traits, such as sperm motility, morphology, and concentration (Ferreira *et al.* 2021). Identifying the genomic regions and genes associated with SC could improve future animal breeding programs by improving the genomic predictions used for selection. Various statistical tests have been developed to identify selection signatures, including the fixation index (F_{ST}), which can be used to infer genetic relationships between populations based on allele frequencies. This study used the F_{ST} index to detect selection signatures associated with the SC trait in three Australian beef cattle breeds: Brahman, Hereford, and Wagyu. These breeds represent the Indicus, Taurus, and East Asian Taurus lineages, respectively, and possess economically important traits that set them apart.

MATERIALS AND METHODS

Estimated breeding values (EBVs) for SC estimated independently in three Australian beef cattle breeds (20,312 Brahman, 27,356 Hereford, and 53,322 Wagyu) were based on a single-step genomic best linear unbiased prediction (ssGTBLUP) model that was extracted from BREEDPLAN along with their genomic data. The SC EBVs for genotyped animals were split into quartiles, and animals in the first and fourth quartiles were used to represent extremes for further analysis. The BREEDPLAN genomic pipeline quality control was applied to the genotypes (Connors *et al.* 2017), and imputation was performed using FImpute v3 (Sargolzaei *et al.* 2014). Additionally, PLINK v1.9 (http://www.cog-genomics.org/plink/1.9/) was used to remove single-nucleotide polymorphisms (SNPs) with minor allele frequency (MAF) $<$ 5% after imputation. The SNPs after quality control for Brahman, Hereford, and Wagyu were approximately 72K, 77K, and 67K, respectively. The GCTA v1.94.1 software (Yang *et al.* 2011) was used to calculate F_{ST} values, which indicate genetic differences between populations. A sliding window of five SNPs for F_{ST} values was applied to reduce noise and consider linkage disequilibrium between SNPs. Two distinct strategies were implemented to represent selection signals. The initial strategy utilised a standard F_{ST} value range

* A joint venture of NSW Department of Primary Industries and University of New England

(F_{ST} > 0.25 as very high differentiation, 0.15 - 0.25 as high, 0.05 - 0.15 as moderate, and < 0.05 as low). The second strategy involved considering only the top 0.1% of high windowed F_{ST} outliers as representative of selection signals. The BiomaRt R package (Durinck *et al.* 2009) was used to identify genes (ARS-UCD1.2 cow genome) within genomic regions under selection extending 250 kilobases (kb) upstream and downstream of significant SNPs. Functional enrichment analysis was also performed on the gene set using the DAVID Bioinformatics Resources (https://david.ncifcrf.gov/) to identify biological processes associated with SC.

RESULTS AND DISCUSSION

The distribution of high and low estimated breeding values (EBVs) for the trait of interest in each population (SC) is presented on the right side of Figure 1. Based on the genetic differences between the Q1 and Q4 groups, the EBVs for SC in the Brahman, Wagyu, and Hereford breeds were high, moderate, and low, respectively. In Brahman, the results of the top 0.1 % windowed F_{ST} values ($F_{ST} \geq 0.15$) showed that only chromosome 14 was under strong directional selection (Figure 1A). In agreement with our findings, some genome-wide association studies detected genomic regions for SC on chromosome 14 at 20–25 Mb for the Brahman breed (Fortes *et al.* 2012a; Fortes *et al.* 2012b; Soares *et al.* 2017). Genomic regions on chromosome 14 play an important role for both reproductive and growth traits across various cattle breeds through their pleiotropic effects. Candidate genes included *RP1*, *XKR4*, *TOX*, *PLAG1*, *PENK*, *RPS20*, *NSMAF*, *SNTG1*, and *MOS*. For example, *TOX* is a transcription factor that controls the development of puberty in tropically-adapted Brahman and Nellore beef cattle (Fortes *et al.* 2012a; de Camargo *et al.* 2015). From the set of 26 candidate genes, some significant Gene Ontology (GO) biological processes (P < 0.05) were found, including animal organ morphogenesis (GO: 0009887), protein metabolic process (GO: 0019538), establishment of spindle orientation (GO: 0051294), organic hydroxy compound catabolic process (GO: 1901616), and metabolic process (GO: 0008152).

In Wagyu, signals of selection were detected within several genomic regions distributed across seven chromosomes (BTA2, BTA3, BTA6, BTA7, BTA8, BTA14, and BTA20; Figure 1B). Chromosome 6 exhibited the signals of selection at 32 – 41 Mb, harbouring several genes involved in beef cattle growth, such as *SLIT2* and *CCSER1* (Smith *et al.* 2019). The *CATSPER3* gene on BTA7 encodes a specific ion channel in sperm and has been found to be exclusively expressed in the bovine testis. It has been related to male fertility in cattle (Johnson *et al.* 2017; Nani and Peñagaricano 2020). The most significant biological processes identified from 54 candidate genes on all chromosomes were associated with genitalia development (GO: 0048806), reproductive structure development (GO: 0048608), reproductive process (GO: 0022414), regulation of multicellular organismal development (GO: 2000026), sex differentiation (GO: 0007548), and positive regulation of nitrogen compound metabolic process (GO: 0051173).

In the Hereford cattle genome, several regions under selection pressure were detected across eight chromosomes (BTA1, BTA4, BTA5, BTA6, BTA8, BTA10, BTA11, and BTA15) that contained 92 candidate genes (Figure 1C). The region on BTA5 (105 Mb) was localized close to genes related to cattle growth (e.g. *FGF6*, *FGF23*, and *CCND2*) (Bernard *et al.* 2009; Bolormaa *et al.* 2014; Yin and König 2019; Fang *et al.* 2020). For instance, genes *FGF6* and *FGF23*, both members of the fibroblast growth factor family, play a role in various biological processes such as angiogenesis, tissue regeneration, oncogenesis, and morphogenesis (Yin and König 2019). *AKAP3* gene on BTA5 plays roles in spermatozoa, including acrosome reaction and sperm capacitation/motility (Han and Peñagaricano 2016; Selvaraju *et al.* 2018). On BTA11, there were two significant genomic regions at 71–75 Mb and 29 Mb. The 29 Mb region overlaps with regions identified by Irano *et al.* (2016), which were associated with the SC trait based on genome-wide association studies (GWAS) in Nellore. Xu *et al.* (2022) reported that the *CIB4* gene is positively associated with testis size. The following biological processes from all candidate genes related to the SC were identified: animal organ morphogenesis (GO: 0009887), animal organ development (GO: 0048513), regulation of nitrogen compound metabolic process (GO: 0051171), regulation of muscle contraction (GO: 0006937), tissue development (GO: 0009888), and regulation of metabolic process (GO: 0019222).

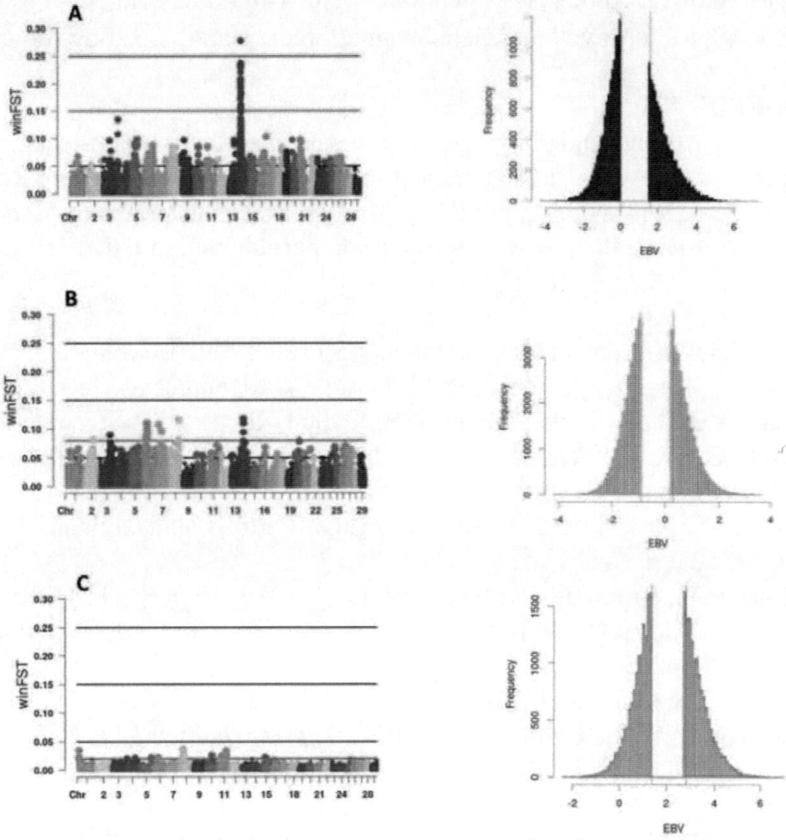

Figure 1. Manhattan plots of selection signatures for scrotal circumference using F_{ST} values with plots of the distribution of EBVs for (A) Brahman (B) Wagyu (C) Hereford. Windowed F_{ST} values are on the y-axis, chromosomal positions are on the x-axis, and the threshold lines represent the 0.1% (red) and standard F_{ST} value range (black) in the Manhattan plots

Selection signatures based on SC EBVs highlighted genes under selection in three Australian beef cattle breeds. The Brahman breed has lower reproductive rates (Reverter and Boe-Hansen 2011) than Wagyu and Hereford breeds, and improving fertility is a breeding program focus for many Brahman breeders. Putative signals of divergence within the Brahman breed had the strongest F_{ST} values compared to other breeds, suggesting higher differentiation in the breed for this trait. The Wagyu and Hereford breeds exhibited moderate and low levels of regional genetic differentiation, respectively. These results show that even though these breeds have different estimated breeding values (EBVs) for the SC in their populations, these diversities are not concentrated in certain genomic regions. GWAS can be used to confirm the relationships between the SC phenotype and genotype. Then, combining information from both selection signatures and GWAS could help in the validation of the informative SNPs. These SNPs can be used to improve the accuracy of genomic predictions for SC in the future. For example, combining these SNPs as a fixed effect or giving them greater weight in the genomic relationship matrix can potentially lead to improved accuracy in genomic predictions. In addition, the identification of informative SNPs improves our understanding of the biological mechanisms regulating the reproductive performance of beef cattle breeds.

CONCLUSIONS

Genotype data, along with EBV information, can provide insights into selection events for traits of interest in breeding programs. In this study, candidate genomic regions and genes associated with SC were detected using this method. These genomic regions could be confirmed by other validation studies, such as GWAS, to improve the genetic evaluation of animal breeding programs.

ACKNOWLEDGEMENTS

This research was partially funded by Meat and Livestock Australia project L.GEN.2204. ZM acknowledges receipt of an AGBU PhD scholarship, the Agricultural Business Research Institute provided access to BREEDPLAN data, and the Australian Brahman Breeders Association Limited, Herefords Australia, and Australian Wagyu Association participated in BREEDPLAN.

REFERENCES

Bernard C., Cassar-Malek I., Renand G. and Hocquette J.-F. (2009) *Meat Sci.* 82: 205.

Bolormaa S., Pryce J.E., Reverter A., Zhang Y., Barendse W., Kemper K., Tier B., Savin K., Hayes B.J. and Goddard M.E. (2014) *PLoS genetics* 10: e1004198.

Connors N., Cook J., Girard C., Tier B., Gore K., Johnston D. and Ferdosi M. (2017) *Proc Assoc Advmt Anim Breed Genet.* 22: 317.

de Camargo G.M., Costa R.B., Lucia G., Regitano L.C., Baldi F. and Tonhati H. (2015) *Reprod. Fert. Develop.* 27: 523.

Durinck S., Spellman P.T., Birney E. and Huber W. (2009) *Nat Protoc* 4: 1184.

Fang L., Cai W., Liu S., Canela-Xandri O., Gao Y., Jiang J., Rawlik K., Li B., Schroeder S.G. and Rosen B.D. (2020) *Genome Res.* 30: 790.

Ferreira C.E., Campos G.S., Schmidt P.I., Sollero B.P., Goularte K.L., Corcini C.D., Gasperin B.G., Lucia Jr T., Boligon A.A. and Cardoso F.F. (2021) *Theriogenology* 172: 268.

Fortes M., Lehnert S., Bolormaa S., Reich C., Fordyce G., Corbet N., Whan V., Hawken R. and Reverter A. (2012a) *Anim. Prod. Sci.* 52: 143.

Fortes M.R., Reverter A., Hawken R.J., Bolormaa S. and Lehnert S.A. (2012b) *Biol. Reprod.* 87: 58.

Gouveia J.J.d.S., Silva M.V.G.B.d., Paiva S.R. and Oliveira S.M.P.d. (2014) *Genet. Mol. Biol.* 37: 330.

Han Y., Peñagaricano F. (2016) *BMC Genet* 17: 1.

Irano N., de Camargo G.M.F., Costa R.B., Terakado A.P.N., Magalhães A.F.B., Silva R.M.d.O., Dias M.M., Bignardi A.B., Baldi F. and Carvalheiro R. (2016) *PLoS One* 11: e0159502.

Johnson G.P., English A.-M., Cronin S., Hoey D.A., Meade K.G., Fair S. (2017) *Biol. Reprod.* 97: 302.

Nani J.P. and Peñagaricano F. (2020) *BMC Genomics* 21: 1.

Reverter T. and Boe-Hansen G. (2011) Meat and Livestock Australia.

Sargolzaei M., Chesnais J.P. and Schenkel F.S. (2014) *BMC Genomics* 15: 1.

Selvaraju S., Parthipan S., Somashekar L., Binsila B.K., Kolte A.P., Arangasamy A., Ravindra J.P. and Krawetz S.A. (2018) *Syst. Biol.Reprod. Med.* 64: 484.

Smith J.L., Wilson M.L., Nilson S.M., Rowan T.N., Oldeschulte D.L., Schnabel R.D., Decker J.E. and Seabury C.M. (2019) *BMC Genomics* 20: 1.

Soares A., Guimarães S., Kelly M., Fortes M., e Silva F., Verardo L., Mota R. and Moore S. (2017) *J Anim. Sci.* 95: 3331.

Xu H., Sun W., Pei S., Li W., Li F. and Yue X. (2022) *Front. Genet.* 12: 2883.

Yang J., Lee S.H., Goddard M.E. and Visscher P.M. (2011) *Am. J. Hum. Genet.* 88: 76.

Yin T. and König S. (2019) *Genet. Sel. Evol.* 51: 1.

Proc. Assoc. Advmt. Anim. Breed. Genet. 25: 111 - 114

BREEDPLAN SINGLE-STEP GENOMIC EVALUATIONS DELIVER INCREASED ACCURACIES ACROSS ALL BREEDS AND EBVS

D.J. Johnston, M.H. Ferdosi, N.K. Connors, J. Cook, C.J. Girard and A.A. Swan

Animal Genetics Breeding Unit*, University of New England, Armidale, NSW 2351 Australia

SUMMARY

Forward cross-validation analyses were used to quantify the changes in BREEDPLAN EBVs from single-step genetic evaluations compared to traditional pedigree-based evaluations for Angus, Brahman, Hereford, Santa Gertrudis and Wagyu breeds. EBVs were generated from full multi-trait evaluations for each breed and compared to EBVs from an evaluation where all the phenotypic records were removed from the last four year drops of animals (termed Validation). Results for the sub-set of validation animals that were SNP genotyped showed the population-based accuracy of single-step EBVs were higher than pedigree-based accuracies for all breeds and traits. However, the magnitudes of the accuracy increases differed across breeds and traits, and generally reflected differences in the size of the training populations for each trait. The largest increase in accuracy, averaged across all traits in a breed, was observed for Angus (24%) and the smallest for Santa Gertrudis (5%). Across breeds, the largest increases in accuracy occurred for the growth trait EBVs compared to smaller increases for abattoir carcase, female reproduction and NFI EBVs. This study has shown the benefits of single-step genomic evaluations, and the opportunity to increase rates of genetic progress, through the increased accuracy generated. The study also highlighted breeds and traits which could benefit from additional recording to increase accuracies from single-step.

INTRODUCTION

Inclusion of SNP-based data is now routine in most livestock genetic evaluations worldwide. BREEDPLAN has included DNA marker data since 2010 and in 2017 implemented single-step SNP-based genomic evaluations (Johnston *et al.* 2018). Increasing rates of industry genotyping, coupled with the development of breed reference populations has resulted in increased accuracy of EBVs, especially for genotyped animals (Jeyaruban *et al.* 2019). Quantifying the benefits from single-step evaluations is not straightforward from large evaluations however Legarra and Reverter (2018) proposed using forward validation and semi-parametric estimates (called Method LR) as a relatively simple method to quantify the changes in accuracy, bias, and dispersion between two evaluations. The aim of this study was to use forward cross-validation and the Method LR to assess the changes in accuracies of single-step versus pedigree-based evaluations for a range of breeds that differ in the numbers of genotyped animals and the size and structure of their genomic reference populations.

MATERIAL AND METHODS

Data. The dataset used in this study included performance, pedigrees and genotypic data extracts from each of the five breed's databases from December 2022. Full BREEDPLAN multi-trait evaluations including maternal effects and genetic groups were run for each breed. Single-step evaluations were performed according to the procedures of Connors *et al.* (2017) and Johnston *et al.* (2018). Traits included: birth weight (BW), gestation length (GL), 200d weight (WW), 400d weight (YW), 600d weight (FW), cow weight (MCW), bull ultrasound rib fat (BRF), P8 fat (BP8), eye muscle area (BEMA), intramuscular fat percent (BIMF), heifer ultrasound rib fat (HRF), P8 fat (HP8), eye muscle area (HEMA), intramuscular fat percent (HIMF), days to calving (DC), abattoir

* A joint venture of NSW Department of Primary Industries and the University of New England

carcase weight (CWT), rib fat (CRIB), P8 fat (CP8), eye muscle area (CEMA), intramuscular fat (CIMF), marble score (CMS), marble fineness score (CMF), retail beef yield (CMY), shear force (CSF), net feed intake-postweaning (NFIP), net feed intake-finishing (NFIF), scrotal circumference (SC), heifer age at puberty (AP), flight time (FT) and percent normal sperm (PNS).

Forward validation. To perform the cross-validation animals were defined as "*validation*" animals or '*training*' based on their year of birth. Animals born after 2018 (except Wagyu, where 2019 was a better split in the data of genotyped animals) comprised the validation sub-set. BREEDPLAN evaluations were performed using pedigree-based BLUP analyses (PED) and single-step analyses (S-S). These runs included all phenotypes, with the resulting EBVs for the validation subset of animals denoted as $\hat{\boldsymbol{u}}_w$. Phenotypes for the validation animals were then removed and the analyses repeated, with the resulting EBVs denoted as $\hat{\boldsymbol{u}}_p$. The subscripts "w" and "p" refer to "whole" and "part" analyses respectively, with the part EBVs of validation animals ($\hat{\boldsymbol{u}}_p$) informed by their pedigree and genomic relationships with the training animals, respectively. A series of metrics were computed using the Method LR (Legarra and Reverter 2018) to compare EBVs from part versus whole subsets from each of the PED and S-S evaluations. A population-based accuracy (acc) of the evaluations was computed using the approximation below:

$$acc = \sqrt{\frac{cov(\hat{\boldsymbol{u}}_w, \hat{\boldsymbol{u}}_p)}{(diag(\boldsymbol{K}) - \bar{\boldsymbol{K}})\sigma^2_{u,\infty}}}$$

where, $\boldsymbol{K}$ is the relationship matrix (i.e. NRM for PED and GRM for S-S) for the validation animals with phenotypes for each trait and $\sigma^2_{u,\infty}$ is assumed to be the genetic variance. The dispersion (i.e. slope) was estimated by $disp = cov(\hat{\boldsymbol{u}}_w, \hat{\boldsymbol{u}}_p)/var(\hat{\boldsymbol{u}}_p)$ and the bias was estimated as $bias = (\overline{\hat{\boldsymbol{u}}_p} - \overline{\hat{\boldsymbol{u}}_w})$. While the validation animals included both genotyped and pedigree-only animals, metrics reported in this paper are for genotyped animals only. Bias and dispersion metrics were computed but only summary results are reported here.

RESULTS AND DISCUSSION

Table 1 presents population estimates of accuracies for genotyped validation animals for the five breeds across the range of BREEDPLAN EBVs. Results show accuracies from S-S were higher than PED across all breeds and all trait EBVs. This demonstrates that the validation animal's EBVs were on average more highly correlated from the S-S evaluation compared to the PED evaluation from the part versus whole runs. Across breeds S-S accuracies were generally highest for the growth traits, with an extreme value for Brahman birth weight (also evident in PED accuracy) suggesting the additive variance assumed is smaller than the true variance or is being influenced by maternal effects.

On average the increase in S-S accuracy was 0.05, 0.12, 0.16, 0.18, 0.23, for Santa Gertrudis, Brahman, Hereford, Wagyu and Angus, respectively (Figure 1). For Angus and Hereford the increased accuracies were in general agreement with earlier analyses of Jeyaruban *et al.* (2019) that used a subset of these data from previous years. Recently, Moore *et al.* (2023) presented an alternative approach applied to the Brahman data, and while their results were based on different edits and data subsets, the changes in accuracies from single-step across the traits were generally in agreement but not always, suggesting the different methods are possibly sensitive to assumptions and need further scrutiny.

The magnitude of the increases in accuracies generally reflected the size of the training populations. Figure 2 illustrates that pooled across traits and breeds, accuracies observed in validation animals tended to increase with the size of the training set. The plots plateau for S-S at 0.80 accuracy with more than 10,000 animals with records and genotypes in the training dataset compared to about 0.50 for pedigree evaluations of the same size. In general, for traits with greater than 5,000 animals in the training populations the S-S accuracies were above 0.60.

Estimates of dispersion (not shown here) were generally close to the expected value of unity, indicating little evidence of under- or over prediction of S-S EBVs. Bias estimates (not shown) were mostly small but further analyses are required to compare ungenotyped and genotyped contemporaries.

Table 1. Population accuracy estimates for genotyped validation animals from pedigree (PED) and Single-step (S-S) BREEDPLAN evaluations for five breeds across all EBVs

EBV*	Angus		Brahman		Hereford		Santa Gertrudis		Wagyu	
	PED	S-S	PED	S-S	PED	S-S	PED	S-S	PED	S-S
BW	0.44	0.81	0.79	0.99	0.63	0.85	0.33	0.38	0.49	0.77
WW	0.53	0.81	0.54	0.70	0.57	0.76	0.37	0.41	0.50	0.76
YW	0.53	0.81	0.43	0.57	0.64	0.83	0.41	0.45	0.53	0.85
FW	0.54	0.82	0.46	0.61	0.56	0.79	0.41	0.46	0.51	0.81
MCW	0.52	0.84	0.42	0.64	0.58	0.79	0.50	0.53	.	.
BEMA	0.58	0.81	0.41	0.49	0.54	0.66	0.43	0.52	0.38	0.56
HEMA	0.58	0.79	0.39	0.48	0.63	0.74	0.50	0.58	0.34	0.55
BIMF	0.67	0.84	.	.	0.65	0.78	0.34	0.39	0.33	0.42
HIMF	0.71	0.89	.	.	0.71	0.85	0.33	0.39	0.35	0.44
BP8	0.46	0.75	0.35	0.49	0.52	0.70	0.44	0.51	0.33	0.48
HP8	0.45	0.76	0.43	0.55	0.63	0.81	0.38	0.47	0.36	0.52
BRF	0.50	0.78	0.34	0.48	0.54	0.70	0.45	0.52	0.34	0.48
HRF	0.50	0.78	0.50	0.61	0.69	0.87	0.38	0.47	0.35	0.48
CWT	0.47	0.66	0.35	0.43	0.32	0.55	0.33	0.37	0.52	0.77
CEMA	0.39	0.69	.	.	0.34	0.46	.	.	0.33	0.44
CIMF	0.50	0.73	0.40	0.46	0.35	0.51	0.31	0.34	0.45	0.61
CMY	0.44	0.60	.	.	.	.	.	.	.	.
CP8	0.35	0.64	0.35	0.47	0.24	0.47	0.35	0.38	0.40	0.61
CRF	0.39	0.66	0.37	0.43	0.29	0.45	0.38	0.41		
CMF	.	.	.	.	.	.	.	.	0.27	0.37
CMS	0.40	0.62	.	.	.	.	.	.	0.53	0.63
CSF	.		0.36	0.42	.	.	0.28	0.32	.	.
DTC	0.47	0.54	0.36	0.60	0.25	0.31	0.55	0.67	.	.
AP	.	.	0.41	0.47	.	.			.	.
PNS	.	.	0.21	0.26	.	.	0.26	0.29	.	.
SC	0.44	0.72	0.59	0.76	0.49	0.61	0.40	0.46	.	.
GL	0.42	0.67	0.32	0.33	0.60	0.68	0.24	0.25	0.30	0.41
FT	.	.	0.41	0.46	.	.	0.51	0.57		
NFIF	0.22	0.40	.	.	0.17	0.17	.	.	.	.
NFIP	0.24	0.36	.	.	.	.	.	.	.	.

*see text for trait names; "." indicates trait not recorded or too few validation animals for the breed

CONCLUSIONS

This study has shown that single-step BREEDPLAN evaluations are delivering increased accuracies of EBVs across the full range of EBVs and breeds. This improvement in accuracies allow more genetic progress, particularly for economically important traits that are generally low accuracy at the time of selection. Improvements in accuracy from single-step will also benefit the commercial beef sector through better genetic description across a broader range of EBVs and allow more precise matching of genetics to specific production systems and markets. Further increases in accuracies are possible from single-step evaluations of particular breeds and traits by increasing the size of their reference populations.

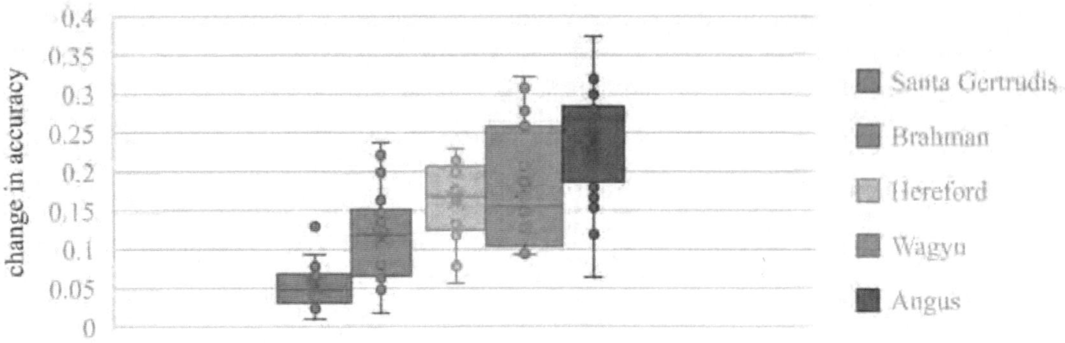

Figure 1. Average change in accuracy of validation animals across all traits by breed

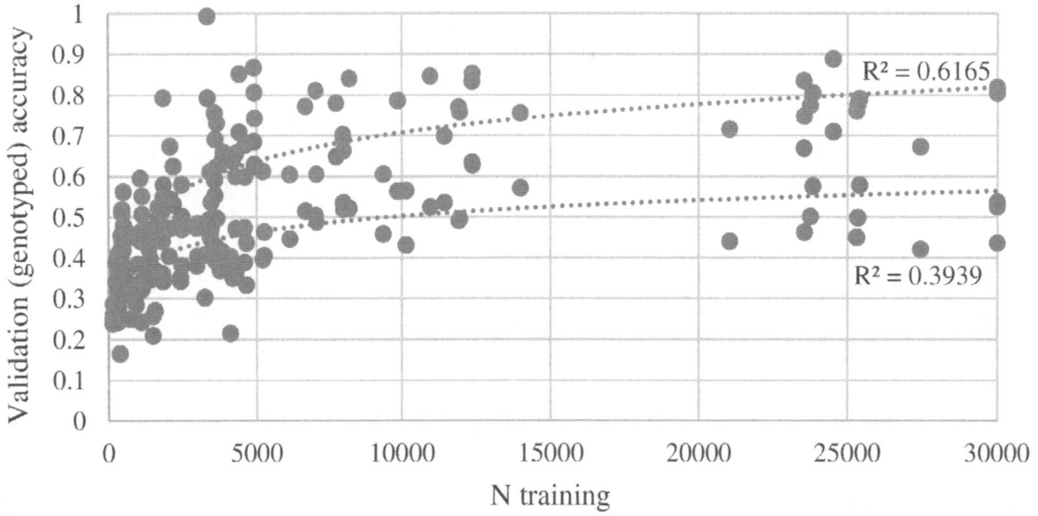

Figure 2. Accuracy of validation animals versus size of training populations for all traits and breeds for single-step (orange) and pedigree (blue) evaluations.

ACKNOWLEDGEMENT

This research was funded by Meat and Livestock Australia project L.GEN.2204. The authors acknowledge each of the five breed societies and their members for access to their data, and to the Agricultural Business Research Institute for providing phenotypic and genotypic data extracts.

REFERENCES

Connors N.K., Cook J, Girard C., Tier B, Gore K, Johnston D.J. and Ferdosi M.H. (2017) *Proc. Assoc. Advmt. Anim. Breed. Genet.* 22: 317.

Legarra A. and Reverter A. (2018) *Genet. Sel. Evol.* 50: 53.

Jeyaruban M.G., Gurman P.M., Johnston D.J., Swan A.A., Banks R.G. and Girard C.J. (2019) *Proc. Assoc. Advmt. Anim. Breed. Genet.* 23: 79.

Johnston D.J, Ferdosi M., Connors N., Boerner V., Cook J., Girard C., Swan A. and Tier B. (2018). *Proc. 11th World Cong. Genet. Appl. Livest. Prod.* 11: 269.

Moore K.J., Gurman P.M. and Johnston D.J. (2023). *Proc. Assoc. Advmt. Anim. Breed. Genet.* 25: *These proceedings.*

Proc. Assoc. Advmt. Anim. Breed. Genet. 25: 115 - 118

GENOTYPE BY TRAIT-SPECIFIC SEASON INTERACTIONS FOR FARROWING RATE AND AVERAGE PIGLET BIRTHWEIGHT

A.M.G Bunz[1,2], K.L. Bunter[2], J. Harper[1,2], R. S. Morrison[1] , B.G. Luxford[1] and S. Hermesch[2]

[1] Rivalea Australia (Pty Ltd), Corowa, NSW, 2646 Australia
[2] Animal Genetics Breeding Unit*, University of New England, Armidale, NSW, 2351 Australia

SUMMARY

The aim of this study was to investigate if there are interactions between genotype and trait-specific seasons (GxTrS) for average piglet birth weight and farrowing rate from sow litters. A series of bivariate animal models were used to estimate genetic parameters. The current study found GxTrS for farrowing rate (genetic correlation <0.4), but not for average piglet birth weight (genetic correlation >0.9). Farrowing rate recorded in the least stressful season (low temperature and positive change in day length) was genetically different to farrowing rate recorded in the two most stressful seasons (high temperature and increasing day length or high temperature and decreasing day length). The results of this study showed that seasonal infertility in sows can genetically be improved by using trait-specific seasons. However, the heritability of farrowing rate was very low (h^2=0.02) and improving the temperature control in the sow's housing environment and developing effective strategies to minimise the effects of changes in day length on the sows may be more effective to improve seasonal infertility.

INTRODUCTION

The reproductive efficiency of sows is economically important in pig production and has been observed to be affected by seasonal variation; in particular poor reproductive performance in summer (Love *et al.* 1993, Auvigne *et al.* 2010). Previous studies have defined seasons using calendar months or temperature and photoperiod information fitted separately at a single time point to investigate genotype by season interactions (Lewis and Bunter 2011, Sevillano *et al.* 2016). A novel methodology has been developed to define trait-specific seasons (modified from Bunz *et al.* 2019), which accounts for both temperature and photoperiod information simultaneously across multiple important time points. This study hypothesised that interactions between genotype and trait-specific seasons (GxTrS) exist for the reproductive traits of average piglet birthweight and farrowing rate recorded in mature sows.

MATERIALS AND METHODS

The traits investigated were farrowing rate calculated from the first insemination event within each mating cycle (FR1: 0=fail due to reproductive reasons, 1=pregnant) and average piglet birth weight (PWT) using multiparous-sow records. The data from two maternal lines (Large White: LW, and Landrace: LD) and one terminal line (Duroc: DC) were collected from a single farm in southern New South Wales, Australia. Data included 42,248 FR1 records from 14,667 sows (daughters of 1,161sires) and a subset of these sows (N=9,402 sows; daughters of 1,077 sires) with 20,293 PWT records collected between 2013 and 2019. All mating events were performed using artificial insemination with each sow receiving two inseminations from the same boar. Sows were housed in naturally ventilated sheds, during gestation and lactation.

The two steps used to define trait-specific seasons were: 1) A series of single-day models were applied to identify the most informative days ($P < 0.05$) for FR1 and PWT regarding maximum

* A joint venture of NSW Department of Primary Industries and University of New England

temperature (tmax) and change in day length (dl) in a trait-specific time period. The single-day linear mixed models were fitted using the *lme4* procedure in R (R Development Core Team 2022), represented as:

$$y_{ijklm} = \mu + \text{matingtype}_i + \beta(\text{tmax day x}) + \beta(\text{dl day x}) + \text{paritygr}_j + \text{lactlength}_k + pe_l + \text{quarteryear}_m + e_{ijklm} \ (Model\ 1)$$

where y_{ijklm} are the observations for FR1 or PWT, μ is the overall mean effect, matingtype_i is the fixed effect of the *i*th mating type (5 levels, LW x LW; LD x LD; LW x LD; LD x LW; DC x DC), paritygr_j is the fixed effect of the *j*th parity grouping (4 levels, p1;p2;p3; p4), covariate of tmax day x and dl day x, where x is one day in the investigated trait-specific time period, quarteryear_m is the fixed effect of the *m*th quarter year of trait recording (32 levels over the complete time periods) and the permanent environment of the sow (pe_l) was fitted to account for repeated records for FR1 and PWT. The fixed effect of previous lactlength_k, representing the *k*th lactation length grouping (4 levels, quartiles), was significant for FR1 only. For PWT, the time period considered was from 115 days before the farrowing date to the farrowing date and for FR1 the time period was from 45 days before to 18 days post the first mating date. Fitting a generalised linear model with a logit link function for FR1 did not converge; therefore, a linear mixed model was applied to FR1; 2) A cluster procedure (R Development Core Team 2022) was then used to group tmax and dl patterns based on the most informative days for every mating date (FR1) or farrowing date (PWT) which resulted in the definition of four clusters to represent trait-specific seasons (Table 1).

Parameter estimates for each trait were then obtained using an animal model in ASReml (Gilmour *et al.* 2015). Additional to Model 1, the random additive genetic effect of the *n*th sow (animal). Further, the permanent environmental effect of the *o*th service sire (s_n) was fitted for FR1. Covariate of tmax and dl was not fitted in the animal model. Random variables were included in models if significant ($P < 0.05$) based on a log-likelihood ratio test. Effects were distributed as var(a)=$\boldsymbol{A}\sigma_a^2$, where *A* is a matrix describing the relationships between animals (i.e., a numerator relationship matrix), and for the remaining effects: Var(pe)=$I\sigma_{pe}^2$, Var(s)=$I\sigma_s^2$ and Var(e)=$I\sigma_e^2$, where *I* is an Identity matrix. For each trait-specific season, genetic parameters were obtained. A series of bivariate animal models was applied to estimate genetic correlations between the trait-specific seasons to measure the magnitude of GxTrS for each trait fitting the same fixed and random effects that were fitted in the univariate analyses.

RESULTS AND DISCUSSION

Defining Seasons. Seasons were trait-specific, varied across years and were not the same as the standard four calendar seasons (Figure 1) due to the different informative days for tmax and dl between traits, and the variation in temperature across years. For FR1, high tmax and negative change in dl around the time of mating had the largest reduction in performance, which was Season 2. For PWT, high tmax and negative change in dl during early gestation and low tmax at late gestation had the lowest mean, which was Season 3 (Table 1).

Univariate Analysis. This study found low heritabilities for FR1 (Table 2), similar to those reported by Sevillano *et al.* (2016) and no additive variance was found in Season 3 (Table 2). However, heritability estimates for FR1 differed only marginally between seasons. Moderate heritabilities were found for PWT (Table 2), which were lower than previously reported by (Lewis and Bunter 2011). Season 1 of PWT had a lower heritability (h^2=0.17) than the other seasons due to a lower additive genetic variance, however the phenotypic variance was similar across Seasons.

Bivariate Analysis. Estimates of genetic correlations between the same trait recorded in different trait-specific seasons are shown in Table 2. The standard errors for genetic correlations were high for FR1 due to the low heritability. The genetic correlations were low between Season 4 and the first two seasons, suggesting that FR1 in Season 4 was genetically a different trait than in

Season 1 and Season 2. The first two seasons of FR1 had opposite tmax characteristics compared to Season 4, which could explain the low genetic correlation found in this study. Sevillano *et al.* (2016) found a higher genetic correlation (0.46±0.13) of farrowing rate between opposite environments (stressful and non-stressful) using a bivariate model. Better environmental control for the sows, during lactation, wean to service period and early gestation could reduce the magnitude of the GxTrS. The current study shows that the combined effects of high tmax and dl (Season 2) had the largest GxTrS and therefore it is important to account for both tmax and dl for defining the presence of genotype by season interactions for FR1 outcomes.

Table 1. Data characteristics for farrowing rate (FR1) and average piglet birthweight (PWT) recorded in sows according to trait-specific seasons (Sn), maximum temperature (tmax) and change in daylight length(dl) characteristics

Trait	Sn	Trait-specific season characteristics	N records	Mean(sd)
FR1	1	tmax from medium to high; dl positive	8,130	0.87(0.33)
	2	tmax from high to medium; dl negative	14,497	0.83(0.38)
	3	tmax low; dl negative	6,562	0.88(0.33)
	4	tmax low; dl positive	13,059	0.90(0.30)
PWT	1	tmax from low to high through gestation; dl positive	6,827	1.61(0.25)
	2	tmax from medium to high through gestation; dl negative	5,012	1.59(0.25)
	3	tmax from high to low through gestation; dl negative	5,137	1.56(0.25)
	4	tmax low through gestation; dl positive	3,317	1.59(0.26)

Abbreviations: Mean tmax: high >28°C; medium 22-28°C; low <22°C
Mean dl: positive = daylength is increasing; negative = day length is decreasing

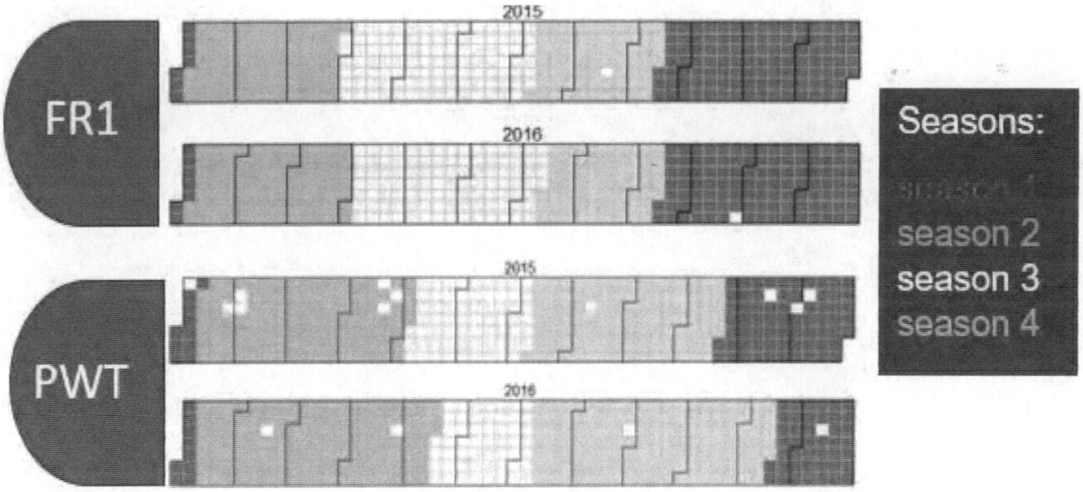

Figure 1. Distribution of calendar days according to the four trait-specific seasons for farrowing rate (FR1) and average piglet birthweight (PWT) over two years 2015 & 2016. Days with missing records were coloured white

Further, there was a high genetic correlation between Season 1 and Season 2 for FR1, indicating that as temperature tolerance is improved, seasonal tolerance will also genetically improve. These results are supported by Sevillano *et al.* (2016), who found that pigs tolerant to decreasing dl are also more tolerant to high tmax. The standard errors for genetic correlations were high for FR1 due to the low heritabilities in all seasons. It was not possible to estimate genetic correlations between Season 3 and other Seasons for FR1 due to the additive genetic variance not being estimable in Season 3.

Genetic correlations between PWT recorded in different Season were high (genetic correlation > 0.9) supporting the result of Lewis and Bunter (2011), which found that PWT was the same trait across all calendar seasons.

Table 2. Heritability estimates (in bold on the diagonal), with genetic (above diagonal) and phenotypic correlations (below diagonal) for farrowing rate (FR1) and average piglet birthweight (PWT) between trait-specific seasons

Trait	Season	1	2	3	4
FR1	1	0.02±0.01	0.80±0.36	ne	0.38±0.37
	2	0.04±0.00	0.02±0.01	ne	0.22±0.29
	3	ne	ne	0.00±0.00	ne
	4	0.04(0.02)	0.03(0.01)	ne	0.02±0.01
PWT	1	0.17±0.03	0.99±0.03	0.94±0.04	0.97±0.04
	2	0.31±0.02	0.27±0.03	1.00±0.04	1.00±0.06
	3	0.35±0.01	0.32±0.02	0.21±0.04	1.00±0.03
	4	0.33±0.02	0.34±0.02	0.34±0.02	0.24±0.03

CONCLUSION

This study showed that season defined by trait informative days for tmax and dl differed for FR1 and PWT traits, and was accompanied by seasonal differences in mean performances. The study also showed that genotype by season interactions existed for FR1 but not for PWT. Farrowing rates observed in Season 4 versus 1 or 2, which were characterised by opposite mean temperature patterns around mating events, were genetically different traits. The results of this study show that using trait-specific seasons can provide an opportunity to improve seasonal infertility in pigs genetically.

ACKNOWLEDGEMENT

Funded by Australian Pork Limited (Project 2017/2208) and student scholarship (2018-007).

REFERENCES

Auvigne V., Leneveu P., Jehannin C., Peltoniemi O. and Salle E. (2010) *Theriogenology* 74: 60.

Bunz A., Bunter K., Morrison R.S., Luxford B. and Hermesch, S.(2019) *Proc. Assoc. Advmt. Anim. Breed. Genet.* 23: 155.

Gilmour A., Gogel B., Cullis B., Welham S. and Thompson R. (2015) *Hemel Hempstead: VSN International Ltd*

Lewis C.R. and Bunter K.L. (2011) *Anim. Prod. Sci.* 51: 615.

Love R.J., Evans G. and Klupiec C. (1993) *J. Reprod. Fert.* 48: 191.

R Development Core Team, 2022.R: A language and environment for statistical

Sevillano C.A., Mulder H.A., Rashidi H., Mathur P.K. and Knol E.F. (2016) *J. Anim. Sci.* 94: 3185.

Proc. Assoc. Advmt. Anim. Breed. Genet. 25: 119 - 121

TRANSLATING MULTIBREED GENOMIC PREDICTION OF BULL FERTILITY TRAITS INTO ON-FARM SELECTION OF BULLS

L.R. Porto-Neto[1], P. Alexandre[1], J. Dorji[1], M.R.S. Fortes[2] and A. Reverter[1]

[1] CSIRO Agriculture & Food, St Lucia, QLD 4067 Australia
[2] The University of Queensland, School of Chemistry and Molecular Bioscience, St Lucia, QLD 4072 Australia

SUMMARY

Genomically enhanced estimated breeding values (GEBV) have been used by many protein-producing industries for several years. However, its use to improve herd fertility has been limited in beef cattle, especially in Northern Australia. The recording of fertility-related traits like those measured in a bull breeding soundness examination (BBSE) following a standard protocol offers an opportunity for improving those traits via genomic selection strategies. Here we describe analyses performed using a multibreed dataset comprising around 8,000 bulls of six tropical breed types and with BBSE data. The heritability estimates varied from low (0.168) for the percentage of proximal droplets to high (0.547) for the sheath score. The GEBV were unbiased and not over-dispersed. The overall accuracies of the GEBV varied from moderate (0.321, proximal droplet, %) to high (0.549, scrotal circumference, cm). These accuracies varied depending on the population. The phenotypic differences between animal quartiles ranked by the GEBV demonstrated the usefulness of those estimates. For example, 25kg of body weight and 2.5 cm in scrotal circumference were observed between quartiles one and four, demonstrating the value of those GEBV.

INTRODUCTION

The use of genomically enhanced estimated breeding values (GEBV) has been implemented in several animal production systems aiming at genetically improving a diverse range of traits. In cattle, the dairy industry leads the adoption by far, possibly followed by some of the Angus breed programs. In tropical cattle, there is limited adoption of the technology, especially when considering hard-to-measure traits like fertility. The use of the standardized bull breeding soundness examination (BBSE) (Entwistle and Fordyce 2003), known to have heritable components (Corbet *et al.* 2013), creates an opportunity to explore its use for genetic evaluation in a multibreed scenario. It might be hard to collect enough records within a single breed to build a breed-specific reference population. Therefore, the multibreed option becomes attractive. Here we tested the feasibility of a multibreed reference population for bull traits and evaluated if the accuracies obtained could be translated into a useful selection tool for on-farm selection of groups of bulls.

MATERIALS AND METHODS

We assembled a reference dataset of genotypes and trait observations on 6,063 bulls (Porto-Neto *et al.* 2023) which has now grown to more than 8,000. These comprise six tropical breed types, Brahman (n=1,817), Santa Gertrudis (n=1,314), Droughtmaster (n=1,008), Ultra-Black/Brangus (n=1,286) and different tropical composite populations (n=2,663) to which a BBSE was recorded. Here we present results for body weight (WT), scrotal circumference (SC), sheath score (SHEATH, 1 - tight to 9 - pendulous), percentage of normal sperm (PNS) and percentage of the most common sperm cell defect, proximal droplets (PD). Table 1 presents the number of records and descriptive statistics of the traits.

Most animals were genotyped using a commercial SNP array with around 50K markers. Genotypes were imputed to ~700K SNP using a reference population that encompassed Beef CRC, and industry cattle genotyped using the high-density Illumina array (BovineHD). Imputation was performed in two steps; first, genotypes were phased using Eagle software (Loh *et al.* 2016) and then imputed using either Minimac3 or 4 (Das *et al.* 2016).

The genomic analyses were performed with pre-adjusted phenotypes. The model for adjustment ran in SAS 9.4 (www.sas.com) included the fixed effects of the population (one per farm), year of birth and management group (within the farm). Additionally, it also fitted the covariates of age at measurement and the first two principal components derived from the genomic relationship matrix constructed following Van Raden's method 1 (VanRaden 2008). Univariate GBLUP models were run using QXPAK (Perez-Enciso and Misztal 2011). The accuracies of the GEBV were calculated as their correlation with adjusted phenotypes divided by the square root of the heritability and the LD method (Legarra and Reverter 2019), both following a five-fold cross validation where a random 20% of the traits data were set to missing. To evaluate the phenotypic potential of those GEBV, we ranked the animals using the GEBV, then calculated the average phenotypic difference between the quartile 1 to 4 (referred as Q1-Q4).

Table 1. The number of records and descriptive statistics of the observed traits *

Trait	N	Mean	SD	Min	Max
WT, Kg	7,730	383.05	93.82	109.50	810.00
SC, score	7,869	30.93	4.29	15.50	52.50
SHEATH, score	7,749	3.13	1.67	1.00	9.00
PNS, %	7,240	62.34	27.37	0.00	100.00
PD, %	7,214	13.13	19.78	0.00	96.00

* WT – body weight, SC – scrotal circumference, SHEATH – sheath score, PNS – the percentage of normal sperm, PD – the percentage of proximal droplets in sperm cells, N – number of observations, SD – standard deviation, Min – minimum value, Max – maximum value observed.

RESULTS AND DISCUSSION

Using our assembled multibreed reference population, we estimated the heritabilities varying from 0.168 (PD) and 0.547 (SHEATH) (Table 2). The moderate to high heritability estimates agreed with previously estimated values for those traits (Corbet *et al.* 2013), giving us confidence the dataset is sound and the traits amenable to improvement via selection.

The GEBV were unbiased and, with the possible exception of SHEATH, not over-dispersed (Table 2). Additionally, using a five-fold cross-validation approach, we obtained reasonably high accuracies (ACC LR, 0.321 to 0.549). The accuracies within populations varied (result not shown), in line with previous analyses using a partial dataset (Porto Neto *et al.* 2021).

Aiming to translate the observed accuracies into phenotypic differences between the validation bulls, we first ranked the bulls by their GEBV, split them into quartiles, and then observed their adjusted trait record (Table 2, Q1-Q4) within their quartile groups. The accuracy of 0.549 for SC translated into a 2.59 cm difference in scrotal size between to top and bottom quartile of bulls. Similarly, the Q1-Q4 analyses for PNS resulted in a 9.49% difference in sperm cells that passed the morphology test. These analyses resulted in group means with large SD and variation within breed types (result not shown). Nonetheless, the translation of the observed accuracies into phenotypic differences was encouraging and demonstrated the potential for using such a tool for on-farm selection of a group of bulls.

Table 2. Results summary. Heritability estimates for observed traits, bias, dispersion, accuracies of estimated breeding values, and the phenotypic difference between animal quartiles ranked by GEBV *

Trait	h2	Bias Mean (SE)	Dispersion Mean (SE)	ACC LR	ACC Trad	Q1-Q4
WT, Kg	0.310	-0.058 (0.365)	-0.042 (0.032)	0.531	0.460	25.14
SC, score	0.436	0.007 (0.027)	-0.001 (0.026)	0.549	0.565	2.59
SHEATH, score	0.547	-0.009 (0.008)	0.165 (0.020)	0.472	0.525	0.67
PNS, %	0.270	-0.112 (0.175)	0.089 (0.033)	0.365	0.300	9.49
PD, %	0.168	0.038 (0.073)	0.043 (0.028)	0.321	0.339	5.19

* WT – body weight, SC – scrotal circumference, SHEATH – sheath score, PNS – the percentage of normal sperm, PD – the percentage of proximal droplets in sperm cells, h2 – heritability estimated, SE – standard error, ACC LR – estimated accuracy of GEBV calculated using the method LR, ACC Trad – estimated accuracy of GEBV calculated using the correlation method, Q1-Q4 – the phenotypic difference between animal quartiles ranked by GEBV, where Q1 is the quartile of animals with highest GEBV and Q4 the quartile with lowest GEBV.

CONCLUSIONS

This study shows that it is feasible to assemble a multibreed reference population for fertility-related traits of tropical bulls. The reasonable to high heritability estimates confirm the quality of the dataset and encourage its adoption in selection breeding programs. The GEBV were mostly unbiased, and although variation within cattle type and population existed, the accuracies of GEBV could be translated into a useful tool for on-farm selection.

ACKNOWLEDGEMENTS

The author would like to acknowledge the contributions of J. Bertram, M. McGowan, B.J. Hayes to the Bull Fertility update project. The authors also would like to thank the cattle breeders that contributed data to that project and MLA for co-funding it (L.GEN.1818). Additionally, the authors are thankful to the CRC for Beef Genetic Technologies and all its participants that contributed data to this project via its legacy database.

REFERENCES

Corbet N.J., Burns B.M., Johnston D.J., Wolcott M.L., Corbet D.H. *et al.* (2013) *Anim. Prod. Sci.* 53: 101.

Das S., Forer L., Schonherr S., Sidore C., Locke A.E. *et al.* (2016) *Nat. Genet.* 48: 1284.

Entwistle K., and Fordyce G. (2003). Australian Association of Cattle Veterinarians, Australia.

Legarra A. and Reverter A. (2018) *Genet. Sel. Evol.* 50: 53.

Loh P.R., Danecek P., Palamara P.F., Fuchsberger C., Reshef Y.A. *et al.* (2016) *Nat. Genet.* 48: 1443.

Perez-Enciso M. and Misztal I. (2011) *BMC Bioinformatics* 12: 202.

Porto-Neto L.R., Alexandre P.A., Hudson N.J., Bertram J., McWilliam S.M. *et al.* (2023) *PLoS One* 18: e0279398.

Porto Neto L.R., Bertram J., McWilliam S., Fortes M.R.S., Alexandre P.A. *et al.* (2021) *Proc. Assoc. Advmt. Anim. Breed. Genet.* 24: 300.

VanRaden P.M. (2008) *J. Dairy Sci.* 91: 4414.

THE HIGHS AND LOWS OF RECORDING CALF AND CALVING TRAITS

Michelle Axford [1,2,3*], Majid Khansefid [1,2] Erica Jewell[2,3], Amanda Chamberlain [1,2] and Jennie E. Pryce [1,2]

[1]School of Applied Systems Biology, La Trobe University, Bundoora, Victoria 3083, Australia
[2]Agriculture Victoria, AgriBio, Centre for AgriBioscience, Bundoora, Victoria 3083, Australia
[3] DataGene Ltd, Bundoora,Victoria, 3083, Australia

SUMMARY

Centralised recording of calf identity, pedigree, treatments and health events are essential to the development of evaluations for calf traits so that farmers can breed for lower morbidity and mortality. Additionally, these same records help to inform decisions that 1) improve calf management protocols, 2) provide access to premium markets for surplus stock that demand evidence to support raising claims and 3) provide industry with greater insight of the health and welfare of young animals that is important in sustainability reporting activities. Farmers have recorded calf and calving traits for more than 30 years, but the frequency of recording is changing based on an analysis of almost 4M Australian calving records. In general, about 20% of herds in Australia and 10% of calvings are recorded with calving traits in the national database. The number of herds as well as recorded herds has declined over time but the number of herds with good calving records has declined at a slower rate. There are opportunities to improve the quality and quantity of calf trait data by lifting recording practices, improving connectivity between on-farm software and populating missing pedigree through genotyping.

INTRODUCTION

Producers and consumers share a deep interest in the welfare of calves. Calves that are born alive and with ease contribute to a sustainable dairy industry. Industry standards, quality assurance programs and animal raising label claims increase requirements for farmers to record and analyse animal health data (Animal Health Australia 2016; Saputo Dairy Australia 2022/23) of which calving traits are a component. Calf records are a valuable resource to farmers as they inform changes to herd management practices that lift productivity and sustainability. Dairy farmers have recorded calf health and calving traits for more than 30 years in Australia on a voluntary basis. Farmers are obliged to record the movement of an animal when it leaves their property through the National Livestock Identification System's (NLIS) central system, but NLIS devices simply record the property where the animal originated without any description of the animal's identity, breed or age (NLIS 2016). Current stores of calf data are the direct result of farmers willingly recording and sharing calf data, rather than through legal obligation. The aim of this study was to report trends in calf and calving traits over time and suggest opportunities for improved practices that will increase the monitoring and genetic improvement for calf survival and other traits affecting calf welfare.

MATERIALS AND METHODS

Farmer recorded calving and pedigree records were obtained for calvings with a calving ease observation that occurred from 1/1/1990 to 3/2/2023 as well as sire identity information from DataGene Ltd, Melbourne, Australia. These records can be defined as 'well-recorded' calving records which differ from more numerous calving records where only the dam's identity and calving date are recorded. Well-recorded calving records include dam identity, breed, pedigree, calf fate, calf sex, calving ease, calf size and litter size. Nonsensical calving records were removed e.g. if the calving date occurred before the dam's birth date. Using R Studio, data was grouped by calving year and then total counts, percentages and means were calculated to better understand trends over time (RStudio Team 2021).

RESULTS AND DISCUSSION

In 2022, 1 in 5 dairy herds recorded traits such as calf fate, calf size and calving ease using a system that supplies data to Australia's centralised data repository but 1 in 10 Australian dairy cows had a well-recorded calving event in the same year. The number and percentage of herds with well-recorded calvings between 1990 and 2022 is shown in Figure 1. At its peak in 2000, 1,926 herds with 128,821 calvings were well-recorded, which was 49% of recorded herds and 23% of all dairy herds in Australia (Australian Bureau of Statistics 2004). Twenty years later, 134,513 cows from 875 herds are well recorded, representing 71% of recorded herds and 20% of all dairy herds. This increase in the proportion of well-recorded calvings could be the result of 1) large scale genotyping projects, such as Ginfo, that have encouraged recording (Pryce *et al.* 2018) 2) those that remain in herd recording are more committed data recorders or 3) the increased use of technology on farm has made it easier to electronically record data through apps.

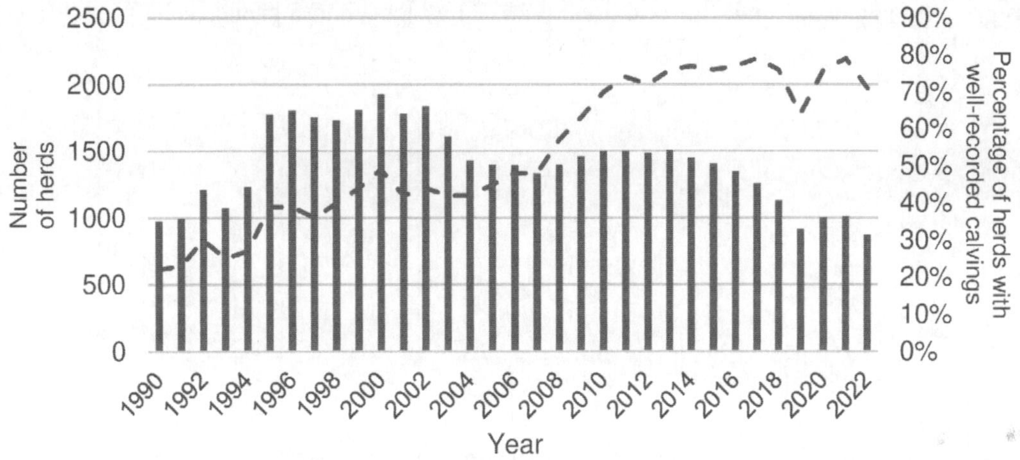

Figure 1: Number (column) and percentage (line) of herds with well-recorded calvings

Changes to Australia's dairy herd size, farm numbers and trends in herd recording practices, as described by Newton *et al.* (2021) are having a major impact on the availability of data, including calf data. Despite the improved proportion of herds recording calvings at a high standard, the number of records per year has declined. Over the past two decades, the size of the National milking herd has declined by about one third (Australian Bureau of Statistics 2004; Dairy Australia 2022). Additionally, herds that participate in official herd recording have a smaller average herd size compared to all herds (266 compared to 303 in 2022) and not all calvings in each herd are well-recorded (DataGene 2022). In summary, the number of herds as well as recorded herds has declined over time but the number of herds with good calving records has declined at a slower rate.

Based on the national herd identification codes, 69% of herds with well-recorded calvings are likely to be located in Victoria, followed by New South Wales (15%), South Australia (6%), Queensland and Western Australia (both 4%) and Tasmania (1%). This is inconsistent with the national distribution of herds where Victoria, New South Wales and Tasmania are the most populous states, specifically highlighting an under-representation of herds located in Tasmania. Regional differences provide an opportunity to tailor activities to specific groups for larger benefit.

Data quality, as well as quantity, is important for the evaluation of new traits. One way to characterise quality is to look for missing values, such as breed of the sire. In the past decade to 2022, the percentage of dams with well recorded calving observations but unknown breed of sire has almost doubled to 9% after being stable at around 5% throughout the previous decade. This presents an opportunity to recover reasonable quantities of data through better recording practices as well as using genotyping to populate pedigree and breed so that more animals can be evaluated.

Pleasingly, in the period 2012-2020, the percentage of calvings producing a female calf was close to 50% as expected. The percentage varied between 51-53% suggesting that calvings producing males and females are being similarly recorded. In 2021-22, the percentage of calvings producing females has increased to 55-57% which is in line with the increased use of sexed semen in dairy herds reported by the National Herd Improvement Association (2022).

The number of sires of dams is consistently more than double the number of calf sires in the dataset with 2385 sires of calves and 6594 sires of dams in 2022. Within a herd, both old and young cows produce calves so a larger number of sires of dams is expected. On average 56 progeny per sire were born in 2022, as illustrated in Figure 2. This has increased by 20 progeny after 2008 and the beginning of the genomic era. It will be important for researchers to carefully consider the minimum progeny per sire as one-third of sires in 2022 had less than 5 progeny.

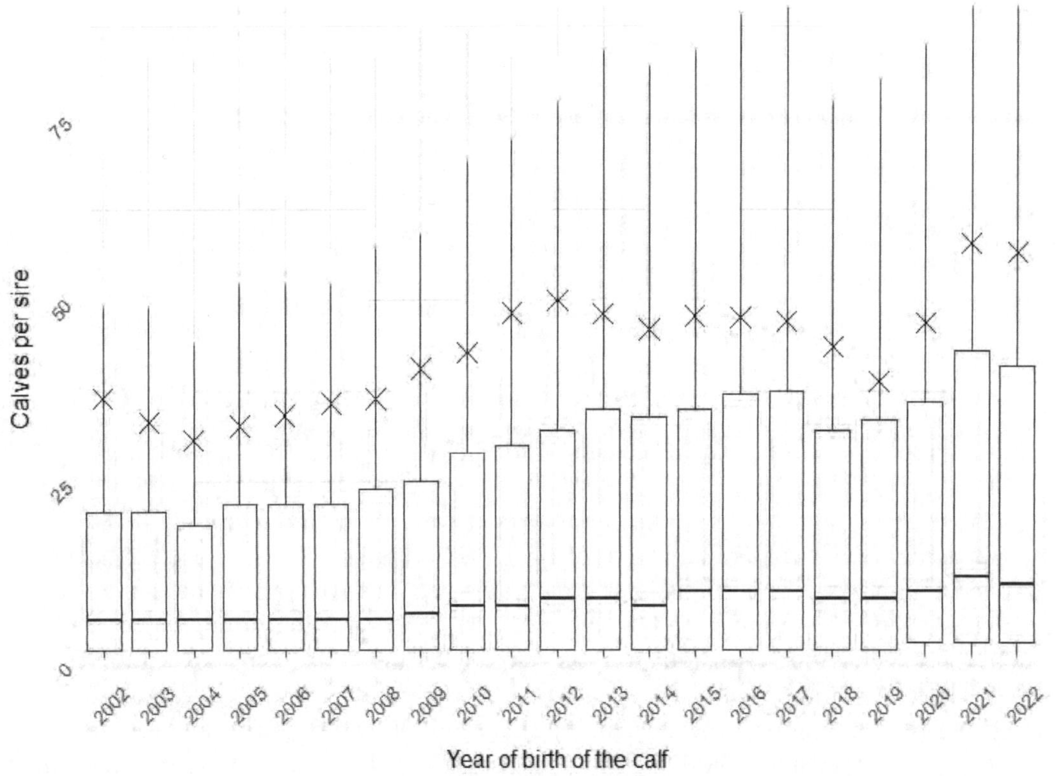

Figure 2: Average number of progeny by year of calving observation
Boxplots represent the mean (blue cross), median (solid line), first and third quartiles (contained in the boxes), outliers not shown.

A closer look at the characteristics of sires of calves revealed that only half of sires with at least 5 progeny had Estimated Breeding Values of sufficient merit to meet the minimum industry standards set by DataGene's Good Bulls Guide and are advised for use in all dairy matings. This group of bulls sired 73% of calves born in 2022. The remaining calves were sired by AI bulls that don't meet the Good Bulls Guide criteria (50% remaining calves), recorded herd bulls (28%), beef

AI or natural bulls (6%), cross-bred bulls (6%) and other groups that cannot be easily characterised. This data suggests that further improvements can be made to sire selection to ensure high quality AI sires are used for every joining to optimise the value of the resulting calves, however, a more complete dataset of well-recorded calvings would verify this suggestion.

According to the National Herd Improvement Association, 17% of semen sold is beef and an increasing proportion is used in lower merit cows in dairy herds (National Herd Improvement Association 2022). At 2% of all 2022 born calves, the recording of beef sired calves is under-reported in this dataset. This is likely the consequence of software, systems and protocols that were designed for dairy sire over dairy cow matings rather than beef sire over dairy cow matings.

CONCLUSIONS

Improvements to calf trait recording are of benefit to farmers, the industry that supports them and the broader community. This analysis reveals that there is a long history of good recording practices and that there are opportunities for continuous improvement. We conclude that improvements to calf trait recording may come through: 1) understanding the hurdles that prevent the recording of calving ease, calf size, calf fate and sire of most calvings, 2) generating pathways to participation for farmers not enrolled in conventional herd recording services, 3) targeting activities to regions and groups where small changes will return large quantities of new data and 4) the use of technology to make high quality data collection more efficient, 5) using genotyping to complete missing pedigree and 6) continuing to monitor for emerging trends in data recording practices.

ACKNOWLEDGEMENTS

The authors acknowledge the contribution of farmers who routinely collect data as well as the herd recording, genotyping and software services that participate in data collation. In particular, the authors recognise the late Mr Geoff Potts who was a driving force in Australia's herd improvement industry. We also thank DataGene Ltd and its staff for their ongoing contribution. This research has been funded by DairyBio, an initiative of Dairy Australia, Agriculture Victoria and The Gardiner Foundation and contributed to a PhD project at La Trobe University.

REFERENCES

Animal Health Australia (2016) Australian Animal Welfare Standards and Guidelines for Cattle. Animal Health Australia, Deakin, ACT, Australia.

Australian Bureau of Statistics (2004) 1310.0 - Year Book Australia, 2004. Australian Bureau of Statistics, Canberra, Australia.

Dairy Australia (2022) In Focus. Melbourne, Australia.[Accessed 04/01/2023].

DataGene (2022) Herd recording statistics 2021/2022 year. DataGene, Bundoora,Australia.

National Herd Improvement Association (2022) Market forces drive sexed and beef semen sales.

Newton J.E., Axford M.M., Ho P.N. and Pryce J.E. (2021) *Anim. Prod. Sci.* 61: 220.

National Livestock Identification System (2016). NLIS (Cattle) Traceability Standards. NLIS Ltd, North Sydney, NSW, Australia.

Pryce J.E., Nguyen T.T.T., Axford M., Nieuwhof G. and Shaffer M. (2018) *J Dairy Sci.* 101: 3702.

RStudio Team (2021) 'RStudio: Integrated Development Environment for R.' (PBC, Boston, USA:

Saputo Dairy Australia (2022/23) Milk Supply Agreement & Supplier Handbook.

Proc. Assoc. Advmt. Anim. Breed. Genet. 25: 126 - 129

USING FOETAL AGE ESTIMATES TO SUBSTITUTE BIRTH DATE RECORDING IN BEEF CATTLE EVALUATIONS

J. Kang[1], F. Weik[1], N. Sanderson[2], D. Robertson[3] and J.A. Archer[1]

[1]Beef + Lamb New Zealand, Dunedin, Otago, 9016 New Zealand
[2] Fossil Creek Angus, Oamaru, Otago, New Zealand
[3] Oamaru Veterinary Services, 311 Thames Street, Oamaru, 9400, New Zealand

SUMMARY

We investigated whether foetal age estimates can be used in beef cattle evaluations, primarily as the substitute for birth dates when applying pre-adjustment for the analyses of growth traits. By comparing different models that involve age calculated with conception dates (i.e., inferred based on the foetal age) and birth dates, we found that foetal age estimates can be used to adjust weaning weights without undermining the goodness of fit of statistical models.

INTRODUCTION

Birth dates are used in beef cattle evaluations for several purposes, including as fixed-effect age adjustments for early life traits (e.g., weights, carcase scans) and as part of the definition of traits for fertility (e.g., days to calving/calving date) and gestation length (Graser *et al.* 2005). Due to logistical challenges and labour requirements, data for birth dates are often unavailable from commercial (non-seedstock) herds. This, together with the lack of pedigree information, is one of the key limiting factors preventing wider utilisation of data from commercial herds in genetic evaluations.

An alternative is to use ultrasound scans to determine foetal age during early pregnancy, to estimate conception date (Beal *et al.* 1992). This can be combined with pedigree information (specifically dam-calf match, obtained using genomics) to provide an estimate of age of the calf from conception (rather than from birth) without any observations at calving. An argument can be made that date of conception, if known with sufficient accuracy, could potentially be an alternative to account for variation in weight due to age rather than birth date. The calf has a growth trajectory from conception, and birth date represents the switch from pre-natal to post-natal growth which may or may not be a significant point of inflexion on the growth curve. An immediate question is whether conception dates can be used to replace birth dates in genetic analyses, especially in genetic evaluations of growth.

Animals that were bred by artificial insemination (AI) may have different birth dates even though their conception dates are known to be the same. Consequently, these animals should have the same age from conception (AfC) but slightly different age from birth (AfB). In practice, one common approach is to pre-adjust weaning weights (WWs) based on AfB, where the WWs of calves with different AfB are projected onto the same linear model. Now, provided that AI-sired animals were conceived on the same dates (with identical AfC), it is questionable whether applying such pre-adjustment still makes sense.

To address these two questions, we analysed a small data set with both birth dates and foetal ages available and modelled WWs using age adjustments calculated from birth and conception dates, respectively.

MATERIALS AND METHODS

Data. Foetal age data was collected on 1,151 beef cattle from two New Zealand farms (Table 1). An experienced operator used rectal ultra-sound to age foetuses to approximately 5-day increments, with foetal aging conducted within a window of 42 to 140 days. A subset of 223 calves were bred by AI, and so true conceptions dates of these AI-sired animals are known. Otherwise, for those animals that were bred by natural service (NS), their conception dates were estimated by subtracting the foetal ages from the scan date. Additionally, birth and weaning dates of calves, as well as their WWs, were recorded. Subsequently, age from conception (AfC) and age from birth (AfB) at weaning were determined based on the estimated (or known for AI animals) conception and observed birth dates. Both estimated and observed AfC are highly correlated with AfB ($r = 0.81$ and 0.92). Gestation length (GL), effectively the difference between AfB and AfC, for animals that were bred by AI and NS were almost identical, with means equal to 280.4 and 280.8. Meanwhile, dam information, such as age and management group, was also available. Note that management groups are generally confounded with the age of dam for younger cows (i.e., management groups represented yearling heifers, two-year-old heifers and mixed age older cows).

Table 1. Sample size, mean and standard deviation of foetal age and weaning data collected from animals that were bred by artificial insemination (AI) and natural service (NS)

Calf Year of Birth	Sample Size		Age From Birth	Age From Conception		Gestation Length		Weaning Weight
	AI	NS		AI	NS	AI	NS	
2015	39	39	189.3 (±13.5)	481 (±0)	470.7 (±13.2)	281.8 (±6.3)	281.3 (±5.1)	217.9 (±30.8)
2018	37	207	202.6 (±15.2)	493.3 (±2.5)	484.1 (±13.7)	280.7 (±10.7)	281.5 (±5.5)	258.4 (±33.0)
2019	35	241	196.4 (±14.5)	485.1 (±1.9)	476.4 (±14.2)	281.4 (±3.9)	280.1 (±5.9)	257.5 (±33.2)
2020	82	183	183.6 (±13.0)	472.9 (±3.1)	462.7 (±11.1)	279.6 (±3.4)	279.2 (±5.7)	254.8 (±32.4)
2021	30	260	188.8 (±14.9)	482.3 (±7.6)	471.1 (±13.9)	280.0 (±4.4)	282.3 (±6.8)	265.6 (±32.8)
Total	223	928	192.3 (±15.9)	480.9 (±8.0)	473.2 (±15.2)	280.4 (±5.8)	280.8 (±6.1)	256.4 (±34.5)

Analyses. Two set of statistical analyses were carried out to study the effects of both age from conception (AfC) and age from birth (AfB) on the scaled WWs (WW_{sc}), by fitting a group of three nested linear models using the data from 1. AI animals only 2. all animals (both AI and NS), such that:

$$WW_{sc} = \beta_0 + \beta_1 CG_{WW} + \beta_2 AGE_{DAM} + \beta_3 AfC + \epsilon \quad (1)$$

$$WW_{sc} = \beta_0 + \beta_1 CG_{WW} + \beta_2 AGE_{DAM} + \beta_3 AfB + \epsilon \quad (2)$$

$$WW_{sc} = \beta_0 + \beta_1 CG_{WW} + \beta_2 AGE_{DAM} + \beta_3 AfC + \beta_4 AfB + \epsilon \quad (3)$$

where CG_{WW} represents the weaning contemporary group (CG_{WW}), defined by the combination of birth year, birth contemporary group ($CG_{BW} = \text{calf year of birth} \times \text{dam herd} \times \text{sex}$), weaning management group (i.e., farm A or B) and the sex of calves (i.e., male and female); scaled weaning weight (WW_{sc}) of each animal is calculated by multiplying raw weaning record by the population average (i.e., 256 kg) and then dividing it by its own contemporary mean; and AGE_{DAM} is the age of dam, with dam aged ten years and above were combined into the same class (i.e., "10+"). Subsequently, hypothesis testing was performed to determine whether there is any significant contribution by each factor, all models were further compared based on the adjusted R-squared values and residual standard errors (RSE).

RESULTS AND DISCUSSION

Results showed that fitting age from birth (AfB) and fitting age from conception (AfC) provided very similar results (Table 2). Including AfB (model 2) explained more variation than including AfC (model 1), but the difference was small. The estimated coefficients associated with AfC from model 1 and AfB from model 2 were very close (1.02 and 0.94). Interestingly, estimated coefficients of AfC and AfB from model 3 seemed to partition the coefficient provided in model 1, even though its standard errors were much higher. In fact, all models performed similarly, with the adjusted R-squared values range from 0.433 to 0.442 and the residual standard error range from 25.93 to 26.13. To summarise, there is only subtle difference between fitting AfC and AfB into the model when analysing WWs, and if AfC is already fitted into the model, little benefit was observed after adding AfB.

For AI animals that were conceived on a known and uniform date within a contemporary group, our results suggest that adding AfC and/or AfB failed to improve the fitted models when analysing WWs, with insignificant p-values associated with both terms (Table 3). Note that the standard errors of these estimated coefficients, especially for AfC (1.16 and 118), are relatively high; also indicating a lack-of-fit. Besides, all three models yielded very similar adjusted R-squared values (0.036 - 0.037) and residual standard error (26.03 - 26.07). In this case, neither AfC nor AfB was significantly contributing towards the predictions of WWs. In this case, applying pre-adjustment based on birth date is likely to introduce bias in the analyses. However, this needs to be further investigated once more data become available.

Although estimated conception dates are prone to measurement errors, its impact on breeding value (BV) predictions should only be noticeable at per individual level. In practice, animals that have an error of ± 5 days within their estimated AfC are likely to receive an estimated BV inflated (or deflated) by approximately half unit (obtained by multiplying the errors within AfC (5 days) to the coefficient of AfC (1.02 from Table 2) and the heritability of weaning weight (0.14, Weik *et al.* 2021). However, such impact is expected to be minimal when predicting sire BVs as errors in AfC are averaged out across multiple progenies. Overall, AfC should be considered as a practical alternative to AfB for pre-adjustment in genetic evaluations for beef cattle.

Table 2. Analyses of weaning weights (all animals) using different models that incorporate age from conception (AfC) and age from birth (AfB)

Model	Coefficient		*p*-value		Adjusted R-squared	Residual Standard Error
	AfC	AfB	AfC	AfB		
1	1.02 (± 0.06)	-	< 2e-16	-	0.221	26.13
2	-	0.94 (± 0.06)	-	< 2e-16	0.227	26.02
3	0.45 (± 0.15)	0.57 (± 0.13)	0.0027	1.61e-5	0.232	25.93

Table 3. Analyses of weaning weights (AI animals only) using different models that incorporate age from conception (AfC) and age from birth (AfB)

Model	Coefficient		*p*-value		Adjusted R-squared	Residual Standard Error
	AfC	AfB	AfC	AfB		
1	0.56 (± 1.16)	-	0.626	-	0.036	26.06
2	-	-0.25 (± 0.32)	-	0.438	0.037	26.03
3	0.77 (± 1.18)	-0.29 (± 0.33)	0.513	0.373	0.036	26.07

CONCLUSION

In this study, we compared different models to investigate whether it is feasible to use foetal age estimates in beef cattle evaluations. Our results showed that conception dates, inferred from foetal age data, could effectively substitute birth dates in the analyses of weaning weights. Moreover, careful consideration should be given if using birth dates to pre-adjust traits where animals are conceived on the same day (e.g., from fixed time AI programs), as applying pre-adjustments may introduce rather than reduce unwanted variation.

ACKNOWLEDGEMENTS

We are grateful to Blair and Jane Smith for the permission to use Fossil Creek Angus data in this analysis.

REFERENCES

Beal W.E., Perry R.C. and Corah L.R. (1992) *J. Anim. Sci.*70: 924.

Graser H.U., Tier B, Johnston D.J. and Barwick S.A. (2005). *Aust. J. Exp. Agric.* 45: 913.

Weik F., Hickson R.E., Morris S.T., Garrick D.J. and Archer J.A. (2021). *Animals*, 11: 2509.

PREDICTION OF GESTATION LENGTH AND DAYS TO CONCEPTION FROM FOETAL AGE SCANS FOR GENETIC IMPROVEMENT OF MATERNAL FERTILITY IN BEEF COW HERDS: A SIMULATION STUDY

F. Weik, J. Kang and J.A. Archer

Beef + Lamb New Zealand Genetics, Dunedin 9016, Otago, New Zealand

SUMMARY

A stochastic simulation model has been used to investigate the potential for combining foetal aging and birth date information to distinguish days to conception (DTCon) from gestation length (GL) under natural mating scenarios. The use of this data in genetic evaluations was assessed, with allowance for the error arising from an allocation to 5- or 10-day increments. The model for genetic evaluation included random additive genetic, permanent environmental and residual effects. Introducing an error associated with foetal aging of animals increased the error variances for DTCon and GL but had little effect on the additive genetic variances, resulting in overall lower heritabilities for both traits. Accounting for the foetal aging error, however, reduced the accuracies for estimated breeding values (EBV) only slightly, and this was true for sires, dams and progeny. The simulation outcomes indicate that foetal aging can be used as a tool to accurately predict the genetic merit of different classes of animals (sires, dams and progeny) for DTCon and to predict GL EBV with improved accuracies due to a larger number of phenotypes from naturally mated beef cow herds so long as the prediction of foetal aging can be done as accurately as 5, or 10 days.

INTRODUCTION

Genetic improvement of cow fertility within New Zealand beef cow herds is mainly achieved by selecting on traits such as days to calving (DTC) or gestation length (GL). Currently, DTC phenotypes require knowledge of the start date of mating and calving date records based on natural (unsynchronised) mating, while GL is based on the number of days between artificial insemination (AI) of cows and subsequent calving date (Graser *et al.* 2005). While the main variation in DTC is due to differences in the conception date following bull exposure, the use of birth date data alone does not allow for the separation of effects of conception date and GL under natural mating scenarios. Although evaluating GL includes information on the conception date, this is based on AI rather than natural mating (Graser *et al.* 2005) and is therefore only available on a restricted subset of animals. The use of foetal aging at pregnancy scanning may be a useful tool to separate the effects of days to conception (DTCon) and GL for naturally mated cows, providing an improved estimate of cow fertility (ability to conceive) and enabling GL estimated breeding values (EBV) to be assessed much more accurately on a larger sample of bulls. Foetal aging can be accurate to 5-day increments (Tweedie *et al.* 2019) such that an element of error is associated with the measurement of conception date from foetal age scans (White *et al.* 1985). The aim of this study was to develop a stochastic simulation to evaluate the impact of error associated with DTCon and GL estimated from foetal age scans and the applicability of foetal aging as a tool to separate the effect of DTCon and GL from DTC for the use in genetic evaluations.

MATERIALS AND METHODS

Simulation model overview. The simulation model was built using R version 4.2.1 (R Core Team 2019). The model simulated a beef cow herd and their progeny from mating to their subsequent calving for twenty consecutive years under New Zealand hill country conditions. The model ran simulations on an individual animal level and produced key production outcomes for the traits DTCon, GL and DTC. The cows simulated in the model were managed in an individual herd within the same mob, assuming the same environmental conditions and management strategies. The size of the cow herd at the start of the simulation was set to 1,000 animals.

Annual production cycle. Cows were mated annually via natural mating for a total of 63 days, allowing them to cycle 3 times based on a 21-day cycling interval. The percentage of bulls used for mating was set to 2 percent of the herd size aligning with typical farming practices within New Zealand beef breeding herds. Each year 50 percent of the bulls were replaced randomly. Cull cows were removed from the breeding herd after weaning of their calf at foot. Culling was conducted firstly due to cows not getting in calf and secondly due to age (cows older than 10 years of age were removed from the breeding herd). The annual replacement rate was set to 20 percent of the herd, such that additional cull cows were selected randomly from the remaining herd if the number of cull cows due to failure to conceive or age were below the threshold to maintain the herd size of 1,000 cows across multiple years. The model simulated a self-replacing herd where female progeny were retained and first mated at 2 years of age. The average number of progeny per sire was 81 (±66).

Simulated phenotypes. The true phenotype for DTCon (trait of the cow) for each animal i was calculated annually at time t as

$$DTCon_{it}=\mu+TBV_i+pe_i+e_{it}$$

and GL (trait of the calf) was calculated as

$$GL_i=\mu+TBV_i+e_i$$

where μ was the overall mean of the population for each trait; TBV_i was the true breeding value of animal i for DTCon or GL; pe was a permanent environmental effect due to repeated records of the animal and e a temporary environmental effect. True breeding values of calves were calculated as $TBV=0.5(TBV_{sire}+TBV_{dam})+ms$ where ms was a mendelian sampling component. True breeding values of the base population (i.e., sires (TBV_{sire}) and dams (TBV_{dam})), mendelian sampling components, permanent and temporary environmental effects were simulated from a Log-normal distribution for DTCon or Normal distribution for GL with zero means and variances equal to σ^2_a, $0.5\sigma^2_a$, σ^2_{pe} and σ^2_e, respectively (Table 1). Genetic correlations between DTCon and GL were set to zero. Phenotypes for DTC were obtained as the sum of DTCon and GL for each cow.

Table 1. Simulation input parameters for days to conception (DTCon) and gestation length (GL): phenotypic means (μ), heritabilities (h^2), additive genetic (σ^2_a), permanent (σ^2_{pe}) and temporary (σ^2_e) environmental variances

Trait	μ	h^2	σ^2_a	σ^2_{pe}	σ^2_e	References
DTCon[1]	2.87	0.21	0.15	0.28	0.28	Weik *et al.* (2021)
GL	282	0.64	11.83	0.00	6.66	Crews (2006)

[1]Values on the logarithmic scale.

Simulation of foetal aging error. A foetal aging error was added to the simulated DTCon phenotypes by rounding values to the nearest 5 ($DTCon_5$) or 10 increments ($DTCon_{10}$). This error was introduced to reflect the error associated with foetal aging, which is generally only accurate to 5-day (or 10-day) increments (White *et al.* 1985). Similarly, GL phenotypes were adjusted, aligning with the foetal aging error to obtain GL_5 and GL_{10}.

Genetic evaluation. Genetic evaluation was performed in ASReml (Gilmour *et al.* 2009) using univariate animal models. The following equation was used for genetic evaluation:

$$y_{ij}=\mu_i+a_{ij}+pe_{ij}+e_{ij}$$

for DTCon, $DTCon_5$, $DTCon_{10}$ and DTC due to repeated records on the same animal, or

$$y_{ij}=\mu_i+a_{ij}+e_{ij}$$

for GL, GL_5 and GL_{10} with no repeated records on the same animal, where y_{ij} was the phenotype of animal i for trait j; μ_j was the mean for trait j; a_{ij} was the random additive genetic effect of animal i for trait j; pe_{ij} the permanent environmental effect of animal i for trait j and e_{ij} the random residual effect unique to each y_{ij}. Cows that failed to conceive were included in the analysis by assigning a penalty of 21 days to DTCon, $DTCon_5$ and $DTCon_{10}$ from the last conception date within the herd (Meyer *et al.* 1990).

Model application. A total of 10 replicates were simulated, and key outcomes were averaged across replicates to determine the mean and SD for DTCon, GL and DTC EBV. Estimated breeding values were obtained for 3 classes of animals which were sires, dams and animals without progeny of their own (i.e., animals born in the last year of the model run). The accuracies of EBV were assessed within each grouping as the correlations between simulated TBV and EBV.

RESULTS AND DISCUSSION

Variances and heritabilities from univariate models for each trait are presented in Table 2. For both DTCon and GL the additive genetic variances were similar with or without the error associated with foetal aging, whereas the phenotypic variance increased. This resulted in a decline in heritability estimates from 0.20 to 0.16 for DTCon and from 0.63 to 0.43 for GL.

Table 2. Simulated additive genetic (σ^2_a), permanent (σ^2_{pe}) and temporary (σ^2_e) environmental variances, heritabilities (h^2) and repeatabilities (t) with standard errors shown in brackets for all traits considered in the analysis

	σ^2_a	σ^2_{pe}	σ^2_e	h^2	t
DTCon[1]	0.14 (±0.02)	0.28 (±0.01)	0.28 (±0.003)	0.20 (±0.02)	0.60 (±0.01)
$DTCon_5$[1,2]	0.13 (±0.01)	0.24 (±0.01)	0.28 (±0.003)	0.19 (±0.02)	0.57 (±0.01)
$DTCon_{10}$[1,2]	0.18 (±0.02)	0.37 (±0.02)	0.59 (±0.01)	0.16 (±0.02)	0.48 (±0.01)
GL	11.65 (±0.43)	-	6.77 (±0.23)	0.63 (±0.02)	-
GL_5[2]	11.60 (±0.47)	-	8.81 (±0.27)	0.57 (±0.02)	-
GL_{10}[2]	11.83 (±0.58)	-	15.46 (±0.37)	0.43 (±0.02)	-
DTC[1]	0.00032 (±0.00004)	0.00062 (±0.00004)	0.00113 (±0.00001)	0.15 (±0.02)	0.45 (±0.01)

[1]Values on the logarithmic scale.

[2]Error of 5 or 10 days associated with DTCon and GL records due to foetal aging.

Overall, EBV accuracies decreased with an increase in error associated with foetal age scanning of cows, and this was true for both DTCon and GL across all animals considered in the analysis (Table 3). However, the reduction in accuracy was small and decreased for the prediction of sire EBV from 0.73 to 0.71 for DTCon and from 0.97 to 0.95 for GL. Results indicated that the genetic merit of each class of animals may be assessed reasonably accurately for DTCon and GL using foetal age scanning, irrespective of the error associated with the actual measurement.

Although the error arising from foetal age scanning had a more prominent impact on the accuracy of GL EBV compared to DTCon EBV (especially for animals with less information on relatives), outcomes are likely to provide a suitable estimate for the duration of gestation from naturally mated beef cow herds. This would increase the number of phenotypes available independent of AI information, leading to an increase in accuracies for GL EBV. This has potential implications for beef on dairy herds such that a larger number of beef bulls may be identified with shorter GL to use over dairy cows to increase days in milk (Coleman *et al.* 2021).

Outcomes from this study indicate that foetal aging may be used as a tool to determine the ability of cows to conceive following natural mating and may provide a better estimate of cow fertility compared to DTC due to overall higher accuracies for each class of animal. Currently, foetal aging using transrectal ultrasonography is the most common method in New Zealand for estimating conception date under extensive farming systems (Brownlie *et al.* 2016). The highest accuracy may be obtained when cows are scanned between 42 and 90 days of gestation (White *et al.* 1985). The restricted mating season with seasonal calvings and high pregnancy rates in New Zealand farming systems would allow foetal aging to be estimated across the entire herd on a single day (Brownlie *et al.* 2016). This has the potential for wider use across the New Zealand beef population and may be implemented in a cost-effective and efficient way at pregnancy scanning when animals are yarded together. Future research may consider using other technologies, such as neck collars to measure cycling activity more accurately and provide a prediction of the actual conception day without error.

Table 3. Distribution of true (TBV) and estimated (EBV) breeding values and their accuracies for days to conception (DTCon), gestation length (GL) and days to calving (DTC) for 3 different classes of animals

Trait	Estimate	Sires			Dams			Progeny		
		Mean	SD	Acc	Mean	SD	Acc	Mean	SD	Acc
DTCon	TBV	0.00	0.37		-0.03	0.37		-0.04	0.37	
	EBV	0.01	0.27	0.73	-0.03	0.23	0.62	-0.04	0.13	0.32
$DTCon_5$[1]	EBV	0.01	0.26	0.73	-0.03	0.22	0.62	-0.04	0.12	0.32
$DTCon_{10}$[1]	EBV	0.01	0.30	0.71	-0.02	0.25	0.60	-0.03	0.14	0.30
GL	TBV	0.00	3.27		0.04	3.40		0.15	3.36	
	EBV	0.03	3.18	0.97	0.05	2.82	0.83	0.16	2.83	0.84
GL_5[1]	EBV	0.01	3.17	0.96	0.03	2.75	0.81	0.15	2.74	0.81
GL_{10}[1]	EBV	-0.01	3.14	0.95	0.02	2.62	0.76	0.11	2.56	0.75
DTC	TBV	0.00	3.29		0.00	3.43		0.11	3.39	
	EBV	0.00	0.01	0.30	0.00	0.01	0.29	0.00	0.01	0.19

[1]Error of 5 or 10 days associated with DTCon and GL records due to foetal aging.

CONCLUSIONS

The simulation study has demonstrated the value of foetal aging as a tool to separate the effects of DTCon and GL from DTC records for naturally mated beef cows. Foetal aging has the potential to add value to future genetic evaluations by providing an improved estimate of fertility based on the cows' ability to conceive and allowing more bulls from naturally mated beef cow herds to be evaluated for GL EBV with higher accuracies.

ACKNOWLEDGEMENTS

This research was conducted as part of the Informing New Zealand Beef programme, which was co-funded by the Ministry of Primary Industries and Beef + Lamb New Zealand Genetics.

REFERENCES

Brownlie T.S., Morton J.M. and McDougall S. (2016) *N. Z. Vet. J.* 64: 324.

Coleman L., Back P., Blair H. Lopez-Villalobos N. and Hickson R.H. (2021) *Dairy*. 2: 385.

Crews D.H. (2006) *J. Anim. Sci.* 84: 25.

Gilmour A.R., Gogel B.J, Cullis B.R. and Thompson R. (2009) ASReml User Guide Release 3.0. VSN International Ltd, Hemel Hempsted, UK.

Graser H.-U., Tier B., Johnston D.J. and Barwick S.A. (2005) *Aust. J. Exp. Agric.* 45: 913.

Meyer K., Hammond K., Parnell P.F., MacKinnon M.J. and Sivarajasingam S. (1990) *Livest. Prod. Sci.* 25: 15.

R Core Team. (2019). R: A language and environment for statistical computing. R Foundation for Statistical Computing. Vienna, Austria

Tweedie M., Archer J. and Proctor L. (2019) *Proc. Assoc. Advmt. Anim. Breed. Genet.* 23: 532.

Weik F., Hickson R.H., Morris S.T., Garrick D.J. and Archer J.A. (2021) *Animals.* 11: 2509.

White I.R., Russel A.J., Wright I.A. and Whyte T.K. (1985) *Vet Rec.* 117: 5.

MATCH BIRTHWEIGHT ASBVS TO FLOCK FECUNDITY FOR LAMB SURVIVAL

S. Hatcher[1,2], S. Robertson[2], D.J. Brown[3] and K.L. Bunter[3]

[1] Makin Outcomes, PO Box 8358, Orange East, NSW, 2800 Australia
[2] Charles Sturt University, Wagga Wagga, NSW, 2678 Australia
[3] Animal Genetics Breeding Unit*, University of New England, Armidale, NSW, 2351 Australia

SUMMARY

Selection of sires with high growth rates may unintentionally reduce lamb survival via dystocia due to the genetic relationships between high growth rates and birthweight. A range of Australian Sheep Breeding Values (ASBVs), including birthweight, lambing ease and gestation length, can be used as selection criteria to genetically increase lamb survival. However, their impact on lamb survival is likely to vary between birth types. Relationships between lambing ease scores, birth weights, gestation length and lamb survival of crossbred lambs born to Merino ewes from the MLA Resource Flock were quantified. Across all birth types, lamb survival was greatest for unassisted lambs; assisted lambs were of low incidence and above average birthweight. Increased lambing ease scores (i.e. more lambing difficulty) were associated with longer gestation length, higher birthweight and poorer lamb survival. Higher birthweight ASBVs were associated with increased lamb survival, but this was dependent on litter size and the lamb surviving parturition. Less fecund commercial flocks that experience dystocia related issues should place an upper limit on birthweight ASBVs and include lambing ease and gestation length ASBVs in their ram selection decisions. These flocks will also need to management ewe nutrition during late pregnancy, to ensure their single bearing ewes do not produce heavy lambs.

INTRODUCTION

Lamb survival is a key component of reproductive efficiency in sheep flocks (Hinch and Brien 2014) particularly in enterprises where the incidence of multiple births (twins and triplets) is relatively high. In extensively managed flocks, lamb mortality is highly variable between farms and years, but averages 10% for single born lambs and 30% for twins with the cause of death affected by the fecundity of the flock (Hinch and Brien 2014) as well as ewe nutrition during pregnancy, maternal behaviour and enviromental conditions at lambing. Dystocia has been implicated in up to 67% of lamb mortality and up to 41% of ewe mortality (Jacobson *et al.* 2020) with the risk of dystocia increasing at both high or low lamb birthweights (Horton *et al.* 2018). Variation in birthweight explains a large proportion of variation in lamb survival. The optimum birthweight for survival ranges between 4.5 and 5.5 kg (Hatcher *et al.* 2009), although birth type, breed and ewe age can shift the optimum range. Lambing ease scores have been genetically associated with all causes of lamb death (Brown *et al.* 2014).

Direct selection for lamb survival is problematic as both sires and dams must have survived as lambs and lamb mortality can occur more than 7 days post-partum (Hatcher *et al.* 2009). Birthweight and lambing ease scores have been identified as potential selection criteria to improve lamb survival (Brien *et al.* 2014) and Australian Sheep Breeding Values (ASBVs) for both traits are available through Sheep Genetics. Genetic gain in lamb survival can be slow due to it's low heritabilility, however the availability of ASBVs for a range of reproduction traits provide key tools for producers to utilise in their breeding flocks. Robertson *et al.* (2022) noted that high ASBVs for post-weaning

---

*A joint venture of NSW Department of Primary Industries and the University of New England

weight and birthweight can lead to unacceptably high levels of dystocia in single-bearing Merino ewes mated to Composite or Poll Dorset rams, which reduced lamb marking rates. This preliminary paper explores the relationships between Australian Sheep Breeding Values (ASBVs) for birthweight (BWT), lambing ease (LE) and gestation length (GL) with lamb survival amongst lambs born to Terminal sires over Merino ewes, when lambed in a common environment.

MATERIALS AND METHODS

Data were extracted from the MLA Resource Flock Katanning (van der Werf *et al.* 2010). LE scores were measured at birth, scoring each lamb on a 5-point scale consisting of: 1 for no assistance, 2 for some assistance, 3 for hard assistance, 4 for abnormal presentation and 5 for other (such as veterinary assistance). Individual lambs could be unobserved and receive no LE score. In this study, animals with LE scores greater than 3 were discarded, as these scores reflect problems which were at low incidence and considered non-genetic in origin (Sheep Genetics 2014). BWTs and GL were also extracted from the database for all animals. All the ewes were joined via artificial insemination, so the gestation length was known. Lamb survival to weaning was calculated using the rearing type records in the database. Least squares means were estimated in R (R Core Team 2022) with year of birth and lamb sex fitted as fixed effects. The predictability of BWT ASBVs was examined by regressing the (un)adjusted BWTs from this study on breeding values obtained from an independent analysis with the data from the resource flocks excluded.

RESULTS AND DISCUSSION

Birthweight and lamb survival. A curvilinear relationship between lamb survival and birth weight (Figure 1) was evident for all birth types. The 'optimal' birthweight for survival was similar for single and twin born lambs, but lower for triplets. The slope of the curve around the optimum was relatively flat for singles but steeper for both twins and triplets, although the latter do not typically extend across the same range of BWTs as single born lambs.

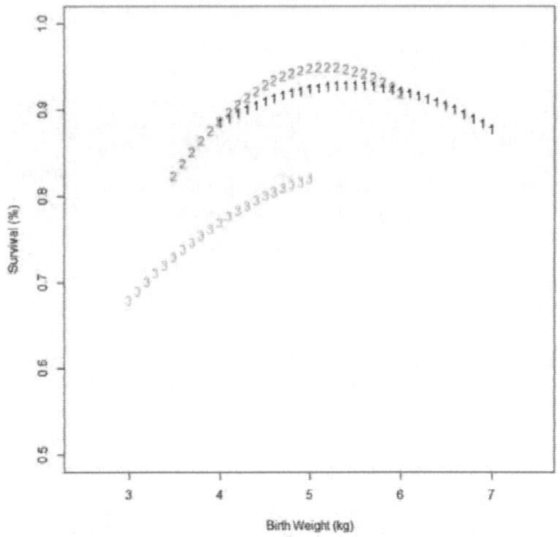

Figure 1. The relationship with BWT (kg) and lamb survival to weaning for single (1), twins (2) and triplet (3) born lambs in the MLA Resource Flock Katanning

Mean survival was highest for lambs born without assistance and, these lambs tended to have lower BWT than those requiring assistance – lambs requiring assistance were typically well above average BWT (Table 1). It is worth noting that most of the lambs that were assisted were close to death when they were assisted (i.e. no dead-at-birth lambs were recorded in the data) and our results are consistent with this observation. In many commercial sheep flocks, ewes typically lamb unassisted and would be expected to have higher rates of lamb mortality than this resource flock.

Longer gestation length was associated with increased assistance at lambing, higher BWT and poorer lamb survival (Table 2). However, gestation length is rarely known in commercial flocks with natural joining. Across all birth types, lambing ease score increased (i.e. increased birthing difficulty) when birthweight approached 5.5 kg (Figure 2). Bunter *et al.* (2023) reported that birthweight and lambing ease are antagonistic traits especially for single born lambs. Lambing ease scores are infrequently measured in commercial or stud flocks unless dystocia is a significant issue, so the lambing ease data available may be represented by the flocks with higher birthweight lambs.

Table 1. Least squares means (standard error) for survival (%) and birthweight (kg) for single, twin and triplet born lambs whose dams were either not assisted or assisted during parturition

Birth type	Assistance	Survival	Birthweight	n
Single	No assistance	0.91 (0.01)	5.30 (0.03)	894
Single	Assisted	0.60 (0.05)	5.74 (0.13)	25
Twin	No assistance	0.91 (0.01)	4.49 (0.02)	1,789
Twin	Assisted	0.59 (0.05)	4.93 (0.13)	18
Triplet	No assistance	0.78 (0.02)	3.83 (0.06)	245
Triplet	Assisted	0.46 (0.05)	4.27 (0.04)	8

Table 2. Least squares means for lambing ease scores, lamb survival (%) and birthweight (kg) by gestation length group (standard errors in brackets)

Gestation length	Lambing ease score	Survival	BWT	n
< 144	1.01 (0.03)	0.79 (0.06)	3.50 (0.17)	26
145 – 150	1.03 (0.01)	0.87 (0.01)	4.28 (0.04)	1,011
>150	1.15 (0.02)	0.79 (0.04)	4.83 (0.11)	62

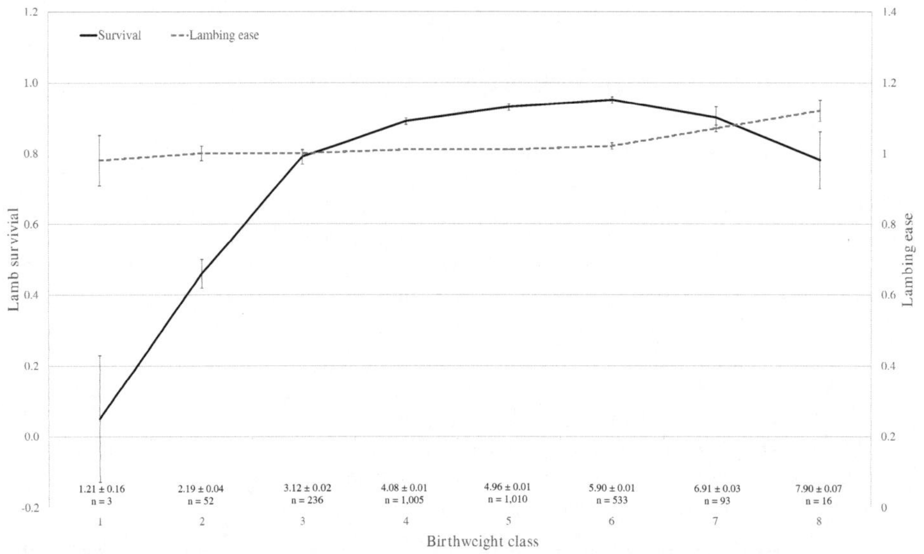

Figure 2. Across all birth types, lambing difficulty increased as birthweight approached 5.5 kg, at which point lamb survival decreases. Data are least squares means by birthweight class (± standard error) along with the number of lambs in that class

BWT ASBV predictive ability. The regression coefficients for BWT of progeny on their sire ASBVs were 0.955 for single-born lambs, 0.790 for twins and 0.478 for triplets. Higher ASBVs for BWT increased lamb survival, but this was contingent on litter size (affecting BWT) and the lamb surviving parturition (LE). Higher sire ASBVs for BWT were also associated with longer gestation length and higher LE scores, but this did not always translate into lower lamb survival. LE ASBVs were also significant predictors (P<0.001) of progeny LE outcomes.

CONCLUSIONS

Genetic improvement of lamb survival is complicated due to the direct genetic effects of the dam and lamb (i.e. half sire genes) and the mediating impacts of flock management and the lambing environment. Therefore, the current reproductive rate of a flock will have an impact. Commercial producers with a high proportion of single born lambs and evidence of lambing ease problems should consider placing an upper limit on BWT ASBVs, include some emphasis on both LE-dir and GL when choosing their rams and carefully manage the nutrition of their single-bearing ewes during late pregnancy. This is especially true for those flocks that place a high emphasis on post weaning growth rates as this trait is genetically associated with higher BWT.

Producers with more fecund flocks can afford to select rams with higher BWT ASBVs, because average lamb birth weight is lower in twin litters, and there is a positive relationship between birth weight and lamb survival overall.

REFERENCES

Brien F.D., Cloete S.W.P., Fogarty N.M., Greeff J.C., Hebart M.L., Hiendleder S., Edwards J.E.H., Kelly J.M., Kind K.L., Kleemann D.O., Plush K.L. and Miller D.R. (2014) *Anim. Prod. Sci.* 54: 667.

Brown D.J., Jones R.M. and Hinch G.N. (2014) *Anim. Prod. Sci.* 54: 736.

Bunter K.L., Brown D.J., Gurman P.M., Li L. and Swan A.A. (2023) *Anim. Prod. Sci.* 63: *In press.*

Hatcher S, Atkins KD. and Safari E. (2009) *J. Anim. Sci.* 87: 2781.

Hinch G.N. and Brien F.D. (2014) *Anim.Prod. Sci.* 54: 656.

Horton B.J., Corkrey R and Hinch G.N. (2018) *Anim. Prod. Sci.*58: 1125.

Jacobson C, Bruce M, Kenyon P.R., Lockwood A, Miller D, Refshauge P.G. and Masters D.G. (2020) *Small Rumin. Res.* 192: 106209.

R Core Team (2022). R: A language and environment for statiatical computing. R Foundation for Statistical Computing. Vienna, Austria. URL https://www.R-project.org/

Robertson S.M., Broster J.C., King B.J. and Friend M.A. (2022) *Anim. Sci. in Aust.*34: cvix.

Sheep Genetics (2014) Understanding Lambs Ease ASBVs. Available at http://www.sheepgenetics.org.au/files/58ed7ad1-e2e8-492f-9bd3-a08500b37fc3/Understanding-Lambing-Ease-ASBVs-041109.pdf

van der Werf J.H.J., Kinghorn B,P, and Banks R.G. (2010) *Anim. Prod. Sci.* 50: 998.

FecB CARRIER DECCANI CROSSBRED EWES IN MAHARASHTRA, INDIA HAVE MODERATELY HIGHER LITTER SIZES THAN NON-CARRIER EWES

C. Nimbkar[1], P.M. Ghalsasi[1], S.W. Walkden-Brown[2] and J.H.J. van der Werf[2]

[1]Nimbkar Agricultural Research Institute, Animal Husbandry Division,
Phaltan 415 523, Maharashtra, India
[2]University of New England, School of Environmental and Rural Science,
Armidale, NSW 2351, Australia

SUMMARY

Litter size records of Deccani crossbred sheep with 0, 1 or 2 copies of the *FecB* mutation were analysed. *FecB* carrier ewes had higher live litter size per ewe lambed, at birth and at the age of 3 months than non-carrier ewes. Live litter size per ewe conceived, at birth and at 3 months age, of both $FecB^{B+}$ and $FecB^{BB}$ ewes was the same, weaning 50% more lambs per ewe conceived than non-carrier ewes. The dampened expression of the *FecB* mutation thus leads to less lamb losses and increased lamb production in the Deccan plateau production system.

INTRODUCTION

The Deccani is an indigenous Indian sheep breed reared on the semi-arid Deccan plateau with a total population of about 1.4 million (GOI 2019). Deccani sheep are grazed in smallholder flocks of 25 to 200 breeding ewes and mostly have single lambs. Most income is earned from the sale of 4-5 months old unweaned lambs weighing 10 to 15 kg. Lambs are reared with personal attention including cross-fostering. In meat producing species, the reproductive rate of breeding females is a key determinant of productivity. There is high local demand for lambs and sheep meat. A (cross)breeding program was established at the Nimbkar Agricultural Research Institute (NARI) at Phaltan in Maharashtra state of India in 1996 to develop a more prolific and productive sheep adapted to the Deccan plateau environment and local sheep owners' management. Nimbkar (2005) found the economic value of litter size in Deccani sheep to be positive after accounting for feed cost mainly because of the personal care of ewes and lambs and the practice of selling lambs early.

The *FecB* or Booroola mutation is an autosomal mutation that has a large additive effect on ovulation rate and is partially dominant for litter size (Davis *et al.* 1982). $FecB^{B}$ is the allele at this locus promoting higher fecundity while $FecB^{+}$ is the wild type allele. The breeding program at NARI introduced the *FecB* mutation from the prolific Garole (Bengal) sheep of West Bengal, India into the local Lonand strain of Deccani sheep. Two strains of *FecB* carrier sheep were developed: the NARI Suwarna with contribution from only Deccani and Garole breeds; and the NARI Composite with additional infusion of the indigenous Bannur, the improved Awassi from Israel and the taller and heavier indigenous Madgyal sheep. The proportion of the Garole breed was reduced deliberately in the cross as its small size, low growth rate and poor mothering ability were not found desirable by local sheep owners. The nucleus breeding flock is still maintained at NARI although with a reduction in the number of breeding ewes from >350 to around 250 in 2020 due to a labour shortage. Ewe and ram lambs are first selected at 4 months age based on their *FecB* carrier status determined by DNA test, their body weight and the reproductive performance and mothering ability of their dams. Ewes are culled for old age or poor reproductive performance. These *FecB* carrier sheep have become popular with sheep owners in Karnataka, Telangana and Maharashtra states and 900 $FecB^{BB}$ breeding rams and 1400 *FecB* carrier ewes have so far been supplied from NARI. This paper compares the reproductive performance of *FecB* carrier and non-carrier ewes in the nucleus at NARI from 2009 to 2022.

MATERIALS AND METHODS

Location, climate and animal management. Phaltan is situated at 18^0 N latitude and 74^0 E longitude and has a dry monsoonal climate with an average annual rainfall of 500 mm. Ewes in the nucleus breeding flock were grazed on crop residues, seasonal grasses, weeds and fallows and housed in open-sided sheds at night, similar to the management of local shepherds. They were given cut-and-carry fodder in the evenings during severe shortage of grazing. Ewes were supplemented with a concentrate feed containing 18% crude protein from 2 months before lambing until weaning of lambs at 15 kg weight at 3-4 months' age. The quantity of concentrate given was 200 g/day/head during 2009-16 which was increased to 300 g from 2017 as the average weight of breeding ewes increased from 28 to 32 kg. Ewes were divided into three flocks and each flock of about 100 ewes was bred every 8 to 10 months. Ewes which did not exhibit oestrus during a particular breeding period were moved to the flock next-in-line for breeding. During the breeding period which lasted one month, oestrus detection was done every morning with vasectomized rams. Ewes found to be in oestrus were artificially inseminated cervically using fresh, diluted semen (so that accurate pedigrees could be maintained). One ram was used only for 5 to 10 ewes and rams were used for a maximum of two years to limit inbreeding and shorten the generation interval. Ewes were ultrasound scanned on average 55 days after insemination. Lambs of dams which did not secrete sufficient milk were cross-fostered to ewes which had lost their lambs or had ample milk supply. All ewes and lambs were genotyped at the *FecB* locus using a forced PCR-RFLP direct DNA test.

Breed proportions. The range of proportions of different breeds in the ewes with records was 30 to 100% Deccani, 0 to 25% Garole and Bannur and 0 to 50% Awassi and Madgyal. The data from both NARI Suwarna and Composite strains were analysed together.

Description of data. Table 1 shows the number of lambing/abortion records of 1235 ewes (3.3 records per ewe on average) and Table 2 shows the number of lambs born alive for each *FecB* genotype of the dam.

Table 1. Classification of lambings/abortions according to *FecB* genotype of ewe

Records where ewes	*FecB* genotype of ewe			Total
	$FecB^{BB}$	$FecB^{B+}$	$FecB^{++}$	
lambed with at least one live lamb	859	1859	578	3296
had stillborn lambs at completion of gestation	119	150	17	286
aborted before term	181	249	37	467
Total records	1159	2258	632	4049

Table 2. Distribution of live litter size according to *FecB* genotype of ewe

Number of lambs born alive	*FecB* genotype of ewe			Total
	$FecB^{BB}$	$FecB^{B+}$	$FecB^{++}$	
1	386	802	540	1728
2	414	1008	38	1460
3	54	49	0	103
4	5	0	0	5
Total	859	1859	578	3296

Traits analysed. The following traits were analysed.

i. LBTOT/EL: Number of live and dead lambs born per ewe lambed (includes lambs which died soon after birth)

ii. LBA/ELA: Number of live lambs born per ewe giving birth to at least one live lamb

iii. LS3m/ELA: Number of live lambs at 3 months per ewe giving birth to at least one live lamb
iv. LBTOT/EC: Number of live and dead lambs born per ewe conceived (i.e. all records used for trait (i) above and zeros for ewes which aborted before completion of pregnancy. Abortions were recorded at visual signs in pregnant ewes such as vaginal discharge for early term abortions or aborted fetuses later in the term.)
v. LBA/EC: Number of live lambs born per ewe conceived (i.e. all records used for trait (ii) above and additionally zeros for ewes which aborted before completion of pregnancy.)
vi. LS3m/EC: Number of live lambs at 3 months per ewe conceived

Model of analysis. All traits were analysed using the Echidna mixed model software (Gilmour 2023) both as Poisson variables using a square root link and as normal variables. The Poisson analysis was used only for testing fixed effects and comparison with the linear model. The repeated observations in this data for all traits are expected to reconcile the relationship between the mean and variance for the Poisson-distributed traits.

The linear model used for all traits analyzed separately in single trait models was as follows.

$$y = \mu + Xb + Z_1a + Z_2pe + e$$

where y is a vector of observations on the ewe, μ is the overall mean, b is a vector of fixed effects (year of insemination and *FecB* genotype of the ewe as fixed effects and age and weight of the ewe at insemination as covariates), a is a vector of random additive genetic effects of the ewe, pe is a vector of the permanent environmental effects of the ewe and e is a vector of residuals. X is the incidence matrix of fixed effects. Z_1 and Z_2 are incidence matrices relating observations to the associated random effects. Breed proportions were not fitted as they were confounded with ewe weights. All available pedigree relationships were used in the analysis mainly in order to obtain accurate predictions of the *FecB* genotype effect.

RESULTS AND DISCUSSION

FecB genotype of the ewe, year of insemination and weight of the ewe were highly significant (P<0.001) for all five litter size traits analysed (results not shown). The weight of the ewe had a positive influence on all traits while age of the ewe did not have any influence. The predicted means for the year of insemination showed a downward trend for all analysed traits (0.02 to 0.04 lambs per year) from the year 2016. Inadvertent culling of ewes with larger litter sizes because of consistently higher lamb losses could be one of the reasons for this reduction.

Table 3. Predicted means (pmean) and standard errors (s.e.) for ewe's *FecB* genotype

Trait	Ewe's *FecB* genotype								
	FecB^{BB}			*FecB*$^{B+}$			*FecB*$^{++}$		
	records	pmean	s.e.	records	pmean	s.e.	records	pmean	s.e.
LBTOT/EL	978	2.02	0.03	2009	1.72	0.02	595	1.01	0.03
LBA/ELA	859	1.64	0.02	1859	1.58	0.02	578	1.00	0.03
LS3m/ELA	859	1.57	0.02	1859	1.51	0.02	578	0.99	0.03
LBTOT/EC	1040	1.90	0.03	2108	1.63	0.02	615	0.95	0.04
LBA/EC	1040	1.37	0.03	2108	1.38	0.02	615	0.90	0.04
LS3m/EC	1040	1.32	0.03	2108	1.32	0.02	615	0.88	0.03

LBA/ELA of ewes heterozygous and homozygous for *FecB* was 1.58 and 1.64 respectively compared to 1.00 in non-carrier ewes (Table 3). *FecB*$^{B+}$ and *FecB*^{BB} ewes had an advantage of 0.58 and 0.64 lambs at birth and of 0.52 and 0.58 lambs at the age of 3 months respectively over non-carrier ewes. This advantage declined slightly to 0.48 and 0.47 lambs at birth for *FecB*$^{B+}$ and *FecB*^{BB}

ewes respectively after considering ewes that aborted. The causes of abortions were not investigated and infectious causes cannot be ruled out. The LS3m/EC of $FecB^{B+}$ and $FecB^{BB}$ ewes was still 50% higher than non-carrier ewes. This is mainly because of the low lamb mortality which can be attributed to good mothering ability of the ewes and good management. Loss of lambs from birth to 3 months' age per ewe conceived in $FecB^{B+}$ and $FecB^{BB}$ ewes in this study was 19% and 30% respectively. In contrast, the losses between scanning and lamb marking were 64% and 89% respectively in the Booroola Merino (Walkden-Brown *et al.* 2007) due to the Australian commercial sheep rearing conditions not being conducive to multiple-born lamb survival.

The effect of *FecB* is reported to be additive on ovulation rate and varying from additive to dominant on litter size depending on the background genotype (Davis 2009), influenced by factors such as uterine capacity, perinatal survival, birth weight, level of neonatal husbandry and care. In this study, $FecB^{BB}$ ewes lost 0.38 lambs while $FecB^{B+}$ ewes lost 0.14 lambs at birth. The exact causes of these losses are not recorded in this flock but they are likely to be related to insufficient uterine capacity. The losses rendered the effect of *FecB* on LBA and LS3m partially dominant. Similarly, first and second copies added 0.62 and 0.02 lambs respectively to the litter size of the Afec-Awassi (Gootwine 2009). The meta-analysis of litter sizes of Chinese sheep (Chong *et al.* 2019) indicated an additive influence of *FecB* in some breeds and a partially dominant effect in other breeds. The litter size at birth of Indian Avishaan sheep (comprising of Garole, Malpura and Patanwadi breeds) carrying 0, 1 and 2 copies of *FecB* was 1.04, 1.70 and 1.93 respectively, also indicating partial dominance and slightly higher than the litter sizes in this study (Sharma *et al.* 2022).

CONCLUSIONS

$FecB^{B+}$ and $FecB^{BB}$ NARI Suwarna and Composite ewes selected for higher lamb survival weaned 50% more lambs than non-carrier ewes, indicating that $FecB^{BB}$ and $FecB^{B+}$ ewes perform similarly. Continued wider dissemination of *FecB* should be through $FecB^{BB}$ rams to increase heterozygosity rather than homozygosity. The introduction of *FecB* appears to be an effective way of sustainable intensification of sheep rearing on the Deccan plateau. The effect of *FecB* in other Indian sheep breeds where it is being introgressed needs to be evaluated.

ACKNOWLEDGEMENTS

The financial assistance of the Australian Centre for International Agricultural Research to this breeding program (1998-2007), that of the German Israel Fund for Rural Development (2003-05) and of the Department of Biotechnology, Government of India (2009-12) for the breeding program and a molecular biology laboratory is gratefully acknowledged.

REFERENCES

Chong Y., Liu G., and Jiang X. (2019) *Anim. Reprod. Sci.* 210: 106175.

Davis G.H., Montgomery G.W., Allison A.J., Kelly R.W. and Bray A.R. (1982) *NZ. J. Ag. Res.* 25: 525.

Davis G.H. (2009) ACIAR Proc. 133: 22.

Gilmour A.R. (2023) Echidna 1.75 Mixed Model Software for use in biological science. NSW Australia. <www.echidnamms.org>

GOI (2019). 'Breed-wise report of livestock and poultry'. Government of India, New Delhi.

Gootwine E. (2009) ACIAR Proc. 133: 119.

Nimbkar C. (2005) Ph.D. Thesis, University of New England.

Sharma, R.C., Gowane, G.R., Kumar, R., Kumar, A., Misra S.S., and Mallick, P.K. (2022) *Small Rumin. Res.* 212: 106720.

Walkden-Brown S.W., Wolfenden D.H., Charles R.J. and Maddox J.F. (2007) *Proc. Assoc. Advmt. Anim. Breed. Genet.* 17: 426.

Proc. Assoc. Advmt. Anim. Breed. Genet. 25: 142 - 145

USING GENOMIC AND PHENOTYPIC DATA TO CHARACTERISE THE GENETIC STRUCTURE OF BRAHMAN CATTLE POPULATIONS IN SOUTHERN AFRICA

W.M.S.P. Weerasinghe and B.J. Crook

Agricultural Business Research Institute, UNE, Armidale, NSW, 2350 Australia

SUMMARY

Knowledge of the genetic structure of cattle breed populations is an important consideration in genetic evaluations. This study used both genomic and phenotypic data to characterise the structure of the Brahman cattle population distributed over three countries in Southern Africa. Principal component analysis based on the genomic relationship matrix demonstrated two sub-populations, with the first principal component explaining 32% of the variation. Subsequent review of both groups showed differences in coat colour as the main source of differentiation, this being red coloured Brahmans and those that were white or grey in colour. Unsupervised analysis using ADMIXTURE with two populations revealed a unique signal in the red Brahman. Variance components and heritability estimates for 200-day weight were similar in the red, white and grey populations and the genetic correlation between the red and white types was 0.88. However, genetic correlations involving the grey type were considerably lower (0.25 with red, 0.58 with white) reflecting the limited comparisons of the grey type with either the white or red type in the same herds and contemporary groups.

INTRODUCTION

Development of the American Brahman commenced in the late 19th century with the importation of several *Bos indicus* breeds from India, followed by subsequent imports of *Bos indicus* types (such as Nellore, Guzerat, Gir and Indu-Brasil) from India and Brazil, and some infusion of local *Bos taurus* genetics (Utsunomiya *et al.* 2019). Live animal exports to Southern Africa commenced in the 1950s, with American genetics increasingly utilised via semen and embryos, and more recently from Australia and Brazil. Combined with the trade in Brahman genetics between countries in Southern Africa, this has led to the Brahman breed contributing significantly to commercial beef production in that region. The Brahman Cattle Breeders' Society of South Africa have utilised the BREEDPLAN genetic evaluation service provided by the Agricultural Business Research Institute (ABRI) since 2002, this being extended to include the Brahman Cattle Breeders' Society of Namibia in 2004 and the Brahman Breeders Society of Zimbabwe in 2021. Pedigree and performance data are combined for evaluation, with over 711,000 animals represented in a multi-trait analysis of phenotypes associated with birth (gestation length and birth weight), post-birth growth (weaning, yearling, final and mature cow weight), fertility (scrotal size and female days-to-calving), ultra-sound scan traits and net feed intake results. The genotyping of seedstock Brahman cattle represents a more recent development in Southern Africa, with a goal towards incorporating genomic data in the genetic evaluation. The objective of this study was to describe the genetic structure of the Brahman cattle population distributed across South Africa, Namibia and Zimbabwe and to investigate data structure relative to the genomic structure of the population including genetic analysis of 200-day weight.

MATERIALS AND METHODS

Genomic analysis. Genotypes were available on Brahman populations in South Africa (n=1,204), Namibia (n=749) and Zimbabwe (n=73), with SNP densities of 54K (n=1,434) and 140K (n=592). The markers located on autosomal chromosomes were considered. Quality control (QC) of genomic data was conducted using PLINK software (Chang *et al.* 2015). Individual SNPs were removed at a minor allele frequency of <0.01, a call rate <90% and a deviation from Hardy–Weinberg equilibrium of $p<1E^{-6}$, with individual genotypes being excluded if the call rate for all loci was <85%. This resulted in a dataset of 1,746 individuals. Genotypes were imputed to the highest density represented using FImpute v3 (Sargolzaei *et al.* 2014), giving 86,110 SNPs for the genetic studies. Principal component analysis (PCA) was carried out on the genomic relationship matrix (VanRaden 2008) to investigate the population structure and genomic variability within the Brahman population. The PCA highlighted two sub-groups, with review of each suggesting recorded coat colour as the main point of differentiation: between red coloured cattle (RE, n=256) distributed across all 3 countries and those recorded as white (WH, n=299) in Namibia or grey (GR, n=786) in South Africa. An unsupervised model-based clustering approach using ADMIXTURE 1.3 (Alexander *et al.* 2009) was used to explore the population structure and infer genomic admixture levels in the RE, WH and GR clusters. A cluster of animals of unrecorded colour (n=405), many from Namibia, was also included. The expected number of subgroups (*K*) was varied from 2 to 4.

Genetic analysis. For animals recorded as RE, WH or GR, their 200-day weights (200D) were extracted from the November 2022 BREEDPLAN evaluation for Southern African Brahman. Phenotypes were pre-adjusted for age at weighing and age of dam as outlined by Graser *et al.* (2005). Contemporary group was defined as herd of origin, sex, year of birth, birth number (single vs twin), birth type (natural vs ET), breeder-defined management group and weigh date. Extracted records were pruned to remove single-animal contemporary groups and those comprising ET calves. The final data set contained 138,764 records for 200D representing RE (34,103), WH (24,486) and GR (80,175) animals. Weight records for RE, WH and GR animals were defined as different traits in a multivariate analysis including additive genetic, maternal genetic (uncorrelated), maternal permanent environment and residual components within trait and a direct genetic correlation only between traits. Contemporary group was fitted as a fixed effect. Six generations of pedigree were included, giving 215,947 animals in the analysis. A genotype file and associated map file were included in the analysis, with 51% of genotyped animals having a 200D record. The GIBBSF90 program in the BLUPF90 family of software (Misztal *et al.* 2018) was used, with 50,000 rounds, a burn-in of 5,000 and every 20th round stored. (Co)variance components were obtained from posterior means using the POSTGIBBSF90 program and a burn-in of 20,000.

RESULTS AND DISCUSSION

The first and second principal components of the PCA explained 31.7% and 5.3% of total variation in the genomic data, with PC3-5 accounting for an additional 3.7%, 2.4% and 2.2% respectively. PC1 reflects a clear stratification in the genotyped population, with RE animals separated as a distinct genomic sub-type compared to WH and GR animals (Figure 1). WH and GR animals show considerable overlap, suggesting they are colour variants of more closely related types. The admixture proportions from the unsupervised analyses with *K*=2, 3 and 4 are shown in Figure 2. Based on the simplest model of *K*=2, the differentiation of RE as a distinct sub-type in the PCA reflects a significant difference in breed composition compared to WH and GR. Subsequent analyses with *K*=3 and *K*=4 were informative yet did not add considerably beyond the simplest model suggesting the predominant breed content of RE is a minor component of the WH and GR. With *K*=4, however, a breed fraction of larger representation in the WH than in the GR was evident. Based on the estimated breed allele frequencies, (i) a low frequency of mis-recorded colour codes seems evident and (ii) the unknown cohort appears to represent all 3 colour types with WH as the primary colour. One limitation of unsupervised model-based clustering is that breed allele frequencies are not explicitly specified, meaning that estimated breed allele frequencies may be biased by familiar relationships among the sample (Gobena *et al.* 2018). The genomic diversity evident in the Brahman population of Southern Africa is reflective of the heterogeneous ancestry of the breed. Although the Brahman breed societies of Southern Africa record all coat colour types among their registered cattle populations, as do their American and Australian counterparts, the results of this study describe the red Brahman as a genomically distinct sub-type within the breed.

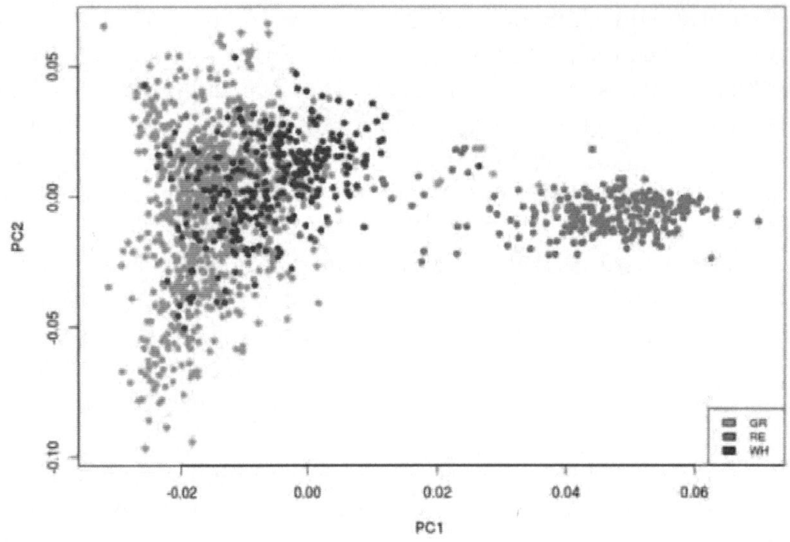

Figure 1. Plot of PC1 vs PC2 for Grey (GR), Red (RE) and White (WH) coloured Brahman

The summary statistics and variance component estimates for 200D in the RE, WH and GR are given in Table 1. The direct and maternal genetic heritability estimates are similar across the 3 types and the ratio of maternal permanent environment to phenotypic variance was 0.08 in each instance. These heritability estimates fall within the range of estimates reported for Brahman cattle in Brazil (de Oliveira Bessa *et al.* 2021), Australia (Davis 1993) and South Africa (Pico *et al.* 2004).

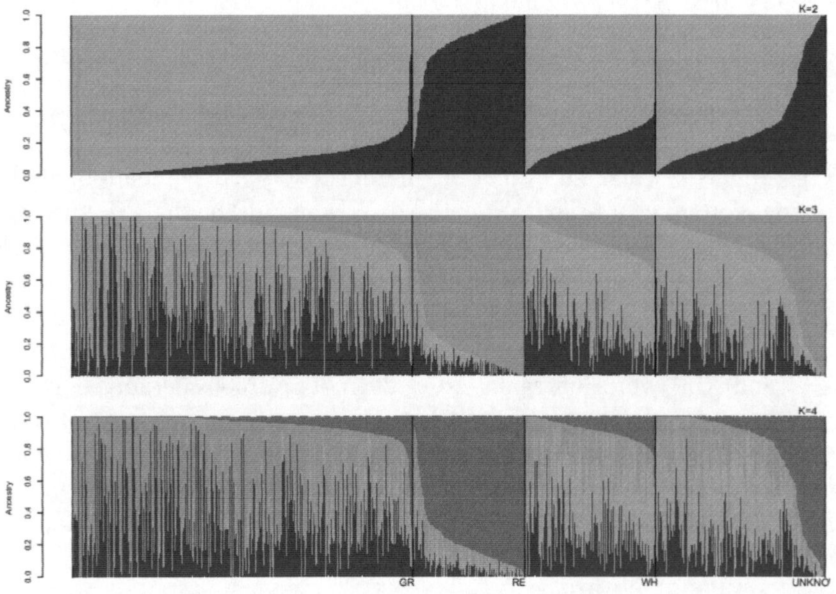

Figure 2. Estimation of admixture proportions of Grey (GR), Red (RE), White (WH) and unknown coat colour, unsupervised with K=2 (top) to K=4

These results suggest a similar mode of genetic expression for 200D in each colour group without the need for type-specific variance components. In the current study however, the genetic correlation for 200D between coat colour types was inconsistent: 0.88 for RE-WH, 0.25 for RE-GR and 0.58 for WH-GR, with high posterior density intervals (95%) of 0.85-0.90, 0.17-0.36 and 0.50-0.65 respectively. A plausible explanation is the lack of comparative data involving grey animals. Only 19 of the 18,441 contemporary groups recorded for 200D contained all 3 colour types, accounting for 320 animals in total. Of those groups comprising 2 colour types (44,003 animals in 3,013 groups), the predominantly contained red and white animals only. Most 200D records (71%) were in contemporary groups representing a single coat colour. Lower genetic correlations involving grey animals reflected limited linkage in the data available for estimation of covariance components.

Table 1. Performance statistics (mean and standard deviation, SD), variance components and heritability estimates for 200-day weight, according to coat colour. Additive genetic variance (V_A), total phenotypic variance (V_P), direct heritability (h^2_D) and maternal genetic heritability (h^2_M). All units in kilograms

Colour	Mean	SD	V_A	V_P	h^2_D	h^2_M
Red	199.8	37.7	85.72±7.01	451.37±4.55	0.19±0.02	0.08±0.01
White	203.4	34.6	75.57±5.50	412.74±4.55	0.18±0.02	0.08±0.01
Grey	206.1	36.4	80.69±5.72	480.93±3.26	0.17±0.02	0.07±0.01

CONCLUSION

Results of the genomic analysis indicate the red Brahman as a distinct sub-type within the wider Brahman population of Southern Africa, though genetic analysis suggests all 3 colour types show a similar genetic expression of 200-day weight. Data structure does, however, indicate limited linkage between the grey Brahman and the other colour types, reflecting a preference of grey Brahman breeders for grey cattle only. Genetic evaluation of the breed in Southern Africa will benefit from increasing representation of all 3 colour types in the reference population.

ACKNOWLEDGEMENTS

The authors acknowledge the Brahman breed societies of South Africa, Namibia and Zimbabwe and their members in providing access the data used in this study.

REFERENCES

Alexander D.H., Novembre J. and Lange K. (2009) *Genome Res.* 19: 1655.

Chang C.C., Chow C.C., Tellier L.C., Vattikuti S., Purcell S.M. and Lee J.J. (2015) *GigaSci.* 4: 13742.

Davis G.P. (1993) *Aust. J. Agric. Res.* 44: 179.

de Oliveira Bessa A.F., Duarte I.N.H., Rola L.D., Bernardes P.A., Neto S.G., Lôbo R.B., Munari D.P. and Buzansas M.E. (2021) *Theriogen.* 173: 261.

Gobena M., Elzo M.A. and Mateescu R.G. (2018) *Front. in Genet.* 9:9 0.

Graser H., Tier B., Johnston D., and Barwick S.A. (2005) *Anim. Prod. Sci.* 45: 913.

Misztal I., Tsuruta S., Lourenco D.A.L., Masuda Y., Aguilar I., Legarra A. and Vitezica Z. (2018) Manual for BLUPF90 family of programs, University of Georgia.

Pico, B.A., Neser, F.W.C. and Van Wyk, J.B. (2004) *Sth. Afri. J. Anim. Sci.* 34: 44.

Sargolzaei M. Chesnais J.P. and Schenkel F.S. (2014) *BMC Genom.* 15: 1.

Utsunomiya, Y.T., Milanesi, M., Fortes, M.R.S., Porto-Neto, L.R., Utsunomiya, A.T.H., Silva, M.V.G.B., Garcia, J.F. and Ajmone-Marsan, P. (2019) *Anim. Genet.* 50:5 57.

VanRaden, P. M. (2008) *J. Dairy Sci.* 91: 4414.

APPLICATION OF AN EMPIRICAL APPROACH FOR PREDICTING ACCURACY FOR GENOMIC EVALUATIONS

K.L. Moore, P.M. Gurman and D.J. Johnston

Animal Genetics Breeding Unit*, University of New England, Armidale, NSW, 2351 Australia

SUMMARY

Including genomics in genetic evaluations can effectively increase selection response, especially for hard to measure, sex limited, and late in life traits. Modelling the increase in accuracy is useful when designing reference data projects and when breeders choose animals to genotype. Theoretical equations exist to predict the EBV accuracy of un-phenotyped animals. However, there are anecdotal reports that the accuracy obtained in practice was often lower than theoretical predictions. This paper validated an empirical approach to predicting accuracy in Australian Brahman data for nine traits. The empirical approach required the accuracy of reference and target animals from a standard pedigree BLUP genetic evaluation and the accuracy of reference animals from a GBLUP genetic evaluation. Using this information, a series of equations were applied to obtain the predicted GBLUP accuracy for target animals. Forward cross-validation showed that the empirical predicted GBLUP was comparable to the actual GBLUP accuracy observed for target animals (accuracy differed between 0.9% and 3.6%). In contrast, theoretical predictions differed from the observed GBLUP accuracy between 5.2% and 21.8%. For smaller (<4,000) reference populations, the theoretical accuracy was closer to the observed GBLUP accuracy, with differences ranging from 5.2% to 11.6%. The theoretical accuracy was overestimated by between 20.7% and 21.8% for larger reference populations. Empirical estimates of the effective number of chromosome segments (M_e) were between 2.0 and 3.9 times that of theoretical M_e, with the greatest difference being for the traits with larger reference sizes. This suggests that the theoretical M_e is the reason for overestimated theoretical accuracy predictions.

INTRODUCTION

Selection response is linear with increasing EBV accuracy, and genomic selection can be an effective way of increasing accuracy, especially for hard or expensive to measure traits, late in life, and sex-limited traits. For genomic selection to be effective, reference data with genotyped and phenotyped animals are required, and generally, the larger the reference size, the greater the accuracy (Goddard and Hayes 2009). Constructing reference data to underpin genomic selection can be expensive, especially for traits not commonly recorded by the industry. Therefore, predicting EBV accuracy is useful for designing reference data projects. Accuracy predictions are also useful for breeders deciding which animals to genotype and the value they can expect from their investment. There have been several theoretical predictions formulated to predict EBV accuracy of un-phenotyped animals given different population parameters (Daetwyler *et al.* (2008), Goddard and Hayes (2009), Goddard *et al.* (2011)). However, there have been anecdotal reports that accuracy from national genetic evaluations was often lower than the theoretical predictions. Dekkers *et al.* (2021) proposed an empirical approach for predicting EBV accuracy. This method bases predictions on the accuracy of reference and target animals from pedigree BLUP and GBLUP genetic evaluations. This study aimed to apply Dekkers' empirical approach using an Australian Brahman beef cattle dataset and validate the prediction accuracies for nine traits using forward cross-

* A joint venture of NSW Department of Primary Industries and University of New England

validation.

MATERIALS AND METHODS

Full details of Dekkers' empirical method for estimating the effective number of chromosome segments (M_e) and predicted EBV accuracy are in Dekkers *et al.* (2021). In brief, this approach requires two genetic evaluations to be undertaken. The first is a BLUP evaluation with full pedigree (including target animals) and phenotypes of reference animals. The second was a GBLUP analysis using the phenotypes and genotypes of reference animals. The average BLUP and GBLUP accuracy for reference animals and average BLUP accuracy for target animals, along with population parameters (i.e. reference size, heritability and genome size) were used in a series of equations that iteratively updated M_e until estimates were stable, and predicted GBLUP accuracy for target animals. M_e was estimated with the equation below, where N was the number of reference animals, q_D^2 the proportion of genetic variance captured by the genotypes (initially $q_D^2 = 1$ but was recalculated each iteration using $q_D^2 = \frac{m}{m+M_e}$ where m = number of markers), h^2 the trait heritability and θ_{Dr} the Fisher information statistic of the reference animals. Dekkers' predicted GBLUP accuracy of target animals (r_{Gt}; equation below) was calculated based on the average accuracy of target animals from the BLUP analysis (r_{At}) and the contribution of G above that of A for target animals (r_{Dt}). For target animals, r_{Dt} was a function of the contribution of G above that of A for reference animals (calculated from average BLUP and GBLUP accuracy) and the number of generations between reference and target animals.

$$M_e = \frac{N q_D^2 h^2}{\theta_{Dr}} \qquad r_{Gt} = \sqrt{\frac{r_{At}^2 + r_{Dt}^2 - 2r_{At}^2 r_{Dt}^2}{1 - r_{At}^2 r_{Dt}^2}}$$

Pedigree, pre-adjusted phenotypes and genotypes were obtained from the Brahman BREEDPLAN genetic evaluation. Genotypes were from different commercially available SNP chips, and after imputation and QA as part of the BREEDPLAN evaluation, 67,327 SNPs were available for analysis. Nine traits were considered; four hard to measure traits (shear force, lactation anoestrus interval, percent normal sperm, age of puberty) and five that were widely recorded (ultrasound scanned EMA, scrotal size, 200, 400 and 600-day live weight) in seedstock herds. All traits were recorded following BREEDPLAN protocols.

Forward cross-validation was used to validate Dekkers' empirical method. Reference (genotyped and phenotyped) animals were split based on year of birth, with the earliest animals remaining reference animals and more recent animals considered target animals with phenotypes and genotypes assumed unknown. The birth year that defined reference and target groups varied for each trait, such that approximately 70% of the data was the reference and the remaining 30% target animals. A five-generation pedigree was built for reference and target animals, and three analyses were performed; 1. BLUP evaluation with reference phenotypes and five-generation pedigree, 2. GBLUP evaluation with reference phenotypes and genotypes, and 3. GBLUP evaluation with reference phenotypes and the genotypes of both reference and target animals. Analysis 1 and 2 were used to apply Dekkers' equations to obtain predicted GBLUP accuracy of target animals (r_{Gt}) and population M_e. While analysis 3 was undertaken to get the observed GBLUP accuracy for target animals, which was then compared with Dekkers' predictions. The same set of genetic parameters and models were used for each analysis. For all analyses, WOMBAT was used and exact accuracy based on the models and data obtained (Meyer 2007). Theoretical accuracy was calculated using Daetwyler *et al.* (2008), where $M_e = (2N_eLk)/ln(N_eL)$ from Goddard *et al.* (2011) and compared with Dekkers' prediction and the observed GBLUP accuracy. To theoretically derive M_e, the effective population size of the breed was estimated using RelaX2 (Stranden, 2014) software and was estimated to be 141.6 animals. The size of the chromosomes (L) was 1.017M (Snelling et al. 2007) with 29 autosomal chromosomes (k) represented on the SNP chips.

RESULTS AND DISCUSSION

Table 1 records the number of reference and target animals, assumed trait heritability and average BLUP and GBLUP accuracy (empirical analyses 1 and 2). The number of reference animals ranged between 982 (shear force) and 11,541 (200-day live weight). Average accuracy from the BLUP analysis ranged from 0.47 (shear force) to 0.77 (age at puberty) for reference animals and between 0.19 (percent normal sperm) and 0.39 (600-day live weight) for target animals. BLUP EBVs of target animals were based on pedigree relationships to the phenotyped reference animals. Reference animals had BLUP accuracies between 0.22 (ultrasound EMA) and 0.42 (age at puberty) higher than target animals. An additional but smaller increase in accuracy was observed for reference animals when genotypes were included in a GBLUP analysis; increases in accuracy ranged between 0.02 (lactation anoestrus interval) and 0.11 (200-day live weight).

Table 1. Number of reference and target animals, assumed heritability and average accuracy from BLUP and GBLUP analysis of Brahman reference (REF) and target (TAR) animals

Trait	Number of animals REF	TAR	h^2	Average accuracy BLUP REF	GBLUP REF	BLUP TAR
Shear force (kg)	982	511	0.26	0.47	0.50	0.21
Lactation anoestrus interval (days)	1,048	470	0.40	0.68	0.70	0.30
Percent normal sperm (%)	1,366	583	0.25	0.52	0.55	0.19
Age of puberty (day)	1,670	806	0.57	0.77	0.80	0.35
Heifer ultrasound scanned EMA (cm^2)	2,565	1,393	0.21	0.52	0.57	0.30
Scrotal size (cm)	4,351	1,988	0.48	0.67	0.73	0.32
600-day live weight (kg)	7,805	3,673	0.51	0.70	0.78	0.39
400-day live weight (kg)	8,730	4,832	0.41	0.67	0.75	0.37
200-day live weight (kg)	11,541	4,415	0.25	0.59	0.70	0.36

Table 2. The estimated effective number of chromosome segments (M_e) and predicted accuracy from Dekkers' empirical approach (Prediction), the GBLUP accuracy from forward cross-validation (observed) and the Daetwyler theoretical prediction (Theoretical)

Trait	$\mathbf{M_e}$	Accuracy of target animals Prediction	Observed	Theoretical[1]
Shear force (kg)	4,500.64	0.29	0.28	0.36
Lactation anoestrus interval (days)	3,425.78	0.42	0.40	0.45
Percent normal sperm (%)	4,252.34	0.32	0.30	0.41
Age of puberty (day)	3,997.45	0.52	0.49	0.60
Ultrasound scanned EMA (cm^2)	4,640.22	0.41	0.40	0.49
Scrotal size (cm)	5,740.04	0.56	0.53	0.74
600-day live weight (kg)	6,550.21	0.63	0.61	0.84
400-day live weight (kg)	6,359.78	0.66	0.63	0.83
200-day live weight (kg)	6,227.21	0.60	0.58	0.80

[1] theoretical prediction based on Daetwyler *et al.* (2008) method where $M_e = 1{,}680.23$ ($N_e = 141.6$)

The predicted accuracy from Dekkers' empirical (Prediction) and Daetwyler's theoretical (Theoretical) method are shown in Table 2, along with the observed GBLUP accuracy (Observed) of target animals. The difference between Dekkers' empirical and Daetwyler's theoretical accuracy was smaller (0.03 to 0.09) with smaller reference sizes, and Dekkers' empirical prediction was lower than Daetwyler's theoretical prediction. However, for traits with more than 4,000 reference animals, the difference between Dekkers' empirical and Daetwyler's theoretical predictions was much larger

(0.12 to 0.15). The observed GBLUP accuracy of target animals (analysis 3) showed that Dekkers' empirical predictions were closer to the observed accuracy than Daetwyler's theoretical accuracy. The observed accuracy was slightly lower (0.01 to 0.04) than Dekkers' empirical predictions. The comparison with Daetwyler's theoretical accuracy showed larger differences. For traits with fewer than 4,000 animals in the reference, theoretical accuracies were between 0.05 and 0.12 higher than the observed accuracy. The differences for traits with larger reference sizes ranged between 0.21 and 0.22. These differences can be explained by the theoretical M_e term being underestimated. Table 2 shows the empirically estimated M_e with estimates varying for each trait; for all traits empirical M_e was much larger than theoretical M_e. Empirical M_e increased with increasing reference size, suggesting a greater diversity of DNA represented in larger references. For traits with smaller references, empirical M_e was 2.0 to 2.8 times larger than theoretical M_e, and for traits with larger reference sizes, empirical M_e was 3.4 to 3.9 times larger. The theoretical M_e was a function of the effective population size and was constant across all traits.

These results demonstrate that Dekkers' empirical approach effectively predicted EBV accuracy, especially for larger reference sizes where theoretical methods overestimate accuracy. It was observed (results not shown) that spurious results occurred for the empirical method when the reference size was small (less than ~1,000 animals). However, with small reference sizes, genomic selection will have limited benefits over pedigree-based selection. The empirical method is only suitable once reference datasets with more than 1,000 animals exist, which limits its application for project design or breeds not yet undertaking genomic selection. It may be possible to use estimates from other breeds and traits to predict accuracy in these situations, but further work is needed to confirm this. One advantage of Dekkers' empirical method is the ability to make predictions for different subsets of target animals. This validation study obtained the BLUP accuracy for target animals from a pedigree BLUP analysis. However, an alternative may be to use an assumed accuracy for target animals. Therefore, predictions can be made for a range of scenarios, including "cleanskins" where no pedigree or phenotypes are available (i.e. BLUP accuracy=0), animals that are not phenotyped but have phenotyped relatives and already phenotyped animals (i.e. BLUP accuracy will be higher than for un-phenotyped animals). In contrast, current theoretical predictions apply to one scenario, assuming that the target animals are un-phenotyped but have pedigree recorded and do not consider other scenarios.

CONCLUSIONS

Predicting the accuracy that can be achieved from genomic selection is desirable. This paper demonstrated that an empirical approach for accuracy prediction was effective and provided better predictions than existing theoretical approaches. However, the method does rely on reference datasets being available.

ACKNOWLEDGEMENTS

The authors acknowledge funding from MLA (L.GEN.2007). We acknowledge Australian Brahman Breeders' Association and their members for access to their phenotypes and genotypes.

REFERENCES

Daetwyler H.D., Villanueva B., and Woolliams J.A. (2008) *PLoS One* 3: e3395

Dekkers J.C.M., Su H., and Cheng J. (2021) *Genet. Sel. Evol.* 53: 1.

Goddard M.E., and Hayes B.J. (2009) *Nat. Review. Genet.* 10: 381.

Goddard M.E., Hayes B.J., and Meuwissen T.H.E. (2011) *J. Anim. Breed. Genet.* 128: 409.

Meyer K. (2007). *J. of Zhejiang Uni. Sci. B* 8: 815.

Snelling *et al.*, (2007) *Gen. Biol.* 8: R165.

Stranden I. (2014) RelaX2 program for pedigree analysis, User's guide for version 1.65.

GENOMIC PREDICTION USING IMPUTED WHOLE-GENOME SEQUENCE IN AUSTRALIAN ANGUS CATTLE

N. Kamprasert[1], H. Aliloo[1], J. van der Werf[1], C. Duff[2] and S. Clark[1]

[1] University of New England, Armidale, NSW, 2351 Australia
[2]Angus Australia, Armidale, NSW, 2350 Australia

SUMMARY

Using whole-genome sequence data in genomic prediction is expected to improve the predictive ability since the whole genome sequence may contain causal variants. This study aimed to compare the accuracy of genomic prediction with three densities of genotypes, 50k, high- density and whole-genome sequence. The genomic prediction was performed to estimate breeding values for selected growth and carcass traits in Australian Angus beef cattle. Genotype imputation was conducted to retrieve genotypes at high-density and whole-genome sequence level. The dataset was split into testing and reference group to compare the accuracy of breeding values obtained from different genotype densities and for animals with different degrees of relatedness to the reference. The prediction accuracies were similar across three different genotype densities for the traits studied. We found no substantial improvement in genomic prediction accuracy using the whole-genome sequence data in this study.

INTRODUCTION

Genome-based evaluations, commonly known as genomic prediction, have become a standard approach for estimating livestock breeding values. Genomic prediction can improve the rate of response to selection by shortening generation interval and gaining more accuracy in predicting breeding value, especially for young animals and difficult-to-measure traits. The accuracy of the genomic prediction depends on two major factors; the number of DNA-tested animals recorded for the objective trait and the number of DNA markers used in genotyping. Current genomic evaluations use standardised genotyping arrays ranging from 10k to 700k in density, with 50k being the most common platform (Goddard *et al.* 2011). The advent of next-generation sequencing technologies has made it possible to obtain whole-genome sequence data at a reasonable price and such data could be used in routine genetic evaluations. Moreover, genotype imputation is a common practice to obtain whole-genome sequence with a reliable accuracy, for animals genotyped with lower densities.

Whole-genomic sequence is expected to improve the accuracy of genomic prediction since it should include actual causal variants in the data instead of depending on the association between the QTLs and markers (Meuwissen *et al.* 2016). The objective of the present study was to examine the benefit of the sequence data for genomic prediction in Australian Angus beef cattle. Different genetic marker densities, including medium-density 50k, high-density 700k and whole genome sequence were used to examine the potential improvement in prediction ability when increasing the marker density for 3 economically important traits in Australian Angus cattle.

MATERIALS AND METHODS

Animal and data. Data was obtained from the Angus Australia database. The dataset analysed was for animals born between 2013 and 2022. Animals were measured for yearling weight (400dWT), final weight (600dWT) and carcass intramuscular fat (CIMF) (Table 1.). Contemporary groups (CG) were formed according to BREEDPLAN procedures (Graser *et al.* 2005) by concatenating herd, year of birth, sex, birth type, management group defined by breeders and measurement date. The CGs were subdivided by age at measurement with slices of 45 days for the growth traits and slices of 60 days for CIMF. Genotypes for animals were also received from Angus

Australia. Medium-density genotype data (50k) were from the previous study by Aliloo and Clark (2021). A total of 1,076 animals were genotyped with 700k genotype array (HD). Genotype data contained only bi-allelic SNPs located on the autosome.

Genotype imputation. To obtain the whole-genome sequence, genotype imputation was performed. Whole-genome sequence data from 440 Angus bulls from the 1000 bull genome project (Hayes and Daetwyler 2019) were used as a reference for the imputation. The 50k genotype samples were imputed to the whole-genome sequence (WGS) level with a stepwise genotype imputation, from 50k to HD, then to WGS. The genotype imputation was performed with Minimac4 (Das *et al.* 2016) and Eagle (Loh *et al.* 2016) was used for pre-phasing with default parameters. The imputation reference panel was a combination of samples with HD and reduced genotypes from the WGS. The imputation accuracy relied on Miminac4 internal quality metric (Rsq). Post-imputation quality control was applied to the imputed genotypes. Quality control filtered out those SNPs with Miminac4 Rsq < 0.30 and minor allele frequency (MAF) < 0.05. This resulted in 44,827, 522,192, and 7,899,466 SNPs for 50k, HD and WGS, respectively, in the final genotype dataset.

Table 1. Descriptive statistics for growth traits and a carcass trait

	N	Mean	Min.	Max.	SD.	age	age mean
400-day weight, kg	56,058	398.46	235.00	622.00	67.82	301 to 500	400.29
600-day weight, kg	23,705	521.84	339.00	814.00	97.89	501 to 700	574.67
Carcass Intramuscular Fat, %	4,074	9.76	3.00	20.50	3.65	504 to 990	722.24

Statistical Analysis. Possible systematic effects were tested for their significance in the model. The effects tested were CG and a linear and quadratic covariate of age at measurement and dam age. The effects with p-value <0.05 was kept in the final model. Due to a large difference in sample size between growth and carcass traits, they were examined differently. To assess prediction accuracy, a 10-fold cross-validation was conducted using the whole dataset for CIMF. While, for the growth traits, the analysis imitated a forward prediction by splitting animals into a reference and a testing group based on their year of birth. The last two years of the data was used as the testing set and other samples were included in the reference group. Individuals in the testing set were grouped according to the level of their relatedness with the reference set by a relationship value, which extracted from a genomic relationship matrix (GRM). Then, samples in each subgroup were randomly assigned into 10 groups for cross-validation. A univariate animal model using the full dataset with 50k genotype density was used to generate phenotypes corrected for all estimated fixed-effect coefficients. GRMs with three different genotype densities were constructed based on Yang *et al.* (2011) using GCTA software. The top-30 relationship values were extracted from off-diagonal elements of the GRM using 50k and then averaged (Clark *et al.* 2012). Observed phenotypes of the testing samples were masked and genomic estimated breeding values were obtained from analyses based on 50k, HD and WGS genotypes. The genomic prediction was performed using the GBLUP approach with a univariate animal model using MTG2 (Lee and Van der Werf 2016). The accuracy of genomic prediction was calculated as the Pearson correlation coefficient between the corrected phenotypes and GEBVs of the testing group divided by the square root of the trait heritability obtained from a 50k-based analysis. The accuracies with the standard error were expressed as an average value from the cross-validation. The accuracy of genomic predictions was compared between three densities of genotypes, and was reported from the testing group and the subgroups according to the degree of relatedness.

RESULTS AND DISCUSSION

The accuracies of genomic prediction for the studied traits are presented in Table 2. For both growth traits, the prediction accuracies were similar for the three genotype densities. Although there was no significant difference, the HD density had the highest accuracy with values of 0.683 (0.017) and 0.630 (0.016) for 400dWT and 600dWT, respectively. The lowest accuracy for both traits was from the WGS, given 0.675 (0.016) for 400dWT and 0.621 (0.014) for 600dWT. The accuracy marginally increased from 50k to HD, then slightly decreased from HD to WGS. Similarly, there was no difference in the prediction accuracies for CIMF. The highest accuracy was 0.643 (0.027) retrieved from HD but there was not significantly different in a comparison. Our results agreed with previous studies showing that using WGS did not significantly improve the accuracy of genomic prediction (Raymond *et al.* 2018; Bedhane *et al.* 2021).

Table 2. Prediction accuracy[1,2] with three different genotype densities by testing group and by relatedness subgroups, and trait heritability

	n	50k	HD	WGS
400-day weight, kg				
testing group	17,942	0.677 (0.016)	0.683 (0.017)	0.675 (0.016)
medium-related	10,230	0.656 (0.020)	0.659 (0.021)	0.650 (0.022)
high-related	7,712	0.711 (0.019)	0.721 (0.020)	0.714 (0.020)
h^2		0.246 (0.007)		
600-day weight, kg				
testing group	5,117	0.627 (0.014)	0.630 (0.016)	0.621 (0.014)
medium-related	3,259	0.611 (0.013)	0.615 (0.012)	0.608 (0.011)
high-related	1,858	0.659 (0.032)	0.660 (0.036)	0.648 (0.035)
h^2		0.338 (0.001)		
Carcass Intramuscular Fat, %				
testing group		0.639 (0.024)	0.643 (0.027)	0.637 (0.027)
h^2		0.464 (0.027)		

[1] Prediction accuracy with standard error was obtained from the 10-fold cross-validation.
[2] There was no significant difference in a comparison (*p*-value <0.01).

Prediction accuracy by relatedness group. Different number of top relationship values were tested to define strength of relatedness between testing samples and reference set. The top-30 average was found to clearly split testing set into two groups (Figure 1). Then, the testing set was subdivided into two groups, which were medium- and high-related groups, and 0.25 was the threshold point. There were 17,942 and 5,117 animals in the testing group for 400dWT and 600dWT, respectively. For CIMF, a 10-fold cross-validation with the whole dataset was performed.

As expected, the high-related group obtained more accurate predictions compared to the medium-related group (Table 1). The accuracy of relatedness subgroups fluctuated with only a slight change with different genotype density. However, difference in the accuracy was not significant between the genotype densities. The highest accuracy for both the medium- and the high-related group were from HD, and the lowest accuracy was from the WGS. The accuracy by subgroups was similar to the testing group where the accuracy steadily declined as the genotype density increased. There were a small difference and no clear pattern in the prediction accuracy when increasing the genotype density. Lastly, accuracy of genomic prediction is involved by several factors, for instance, trait heritability, size of the reference and relatedness between selection samples and the reference.

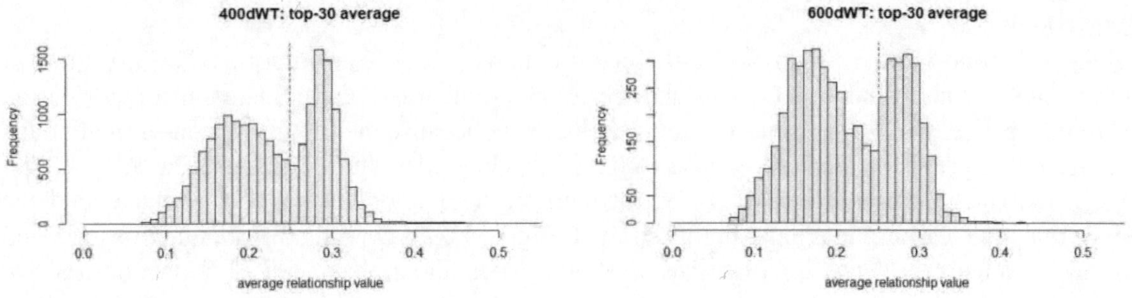

Figure 1. Relationship values for testing animals from 50k genotype by traits with the threshold point

CONCLUSION

This study has investigated the benefit of whole-genome sequence for predicting breeding values for the selected growth and carcass traits in Angus cattle. Although the highest prediction accuracies were retrieved when using the high-density genotype, the difference was not significant compared to the 50k-based prediction. For the traits studied, there was no clear evidence of increased prediction accuracy with denser genotypes, such high-density array and whole-genome sequence.

REFERENCES

Aliloo H. and Clark S.A. (2021) *Animal Production Science* 61: 1958.

Bedhane M., van der Werf J.H.J., de las Heras-Saldana S., Lim D., Park B., Park M.N., Hee R.S., Clark S. and Fortes M. (2021) *Animal Production Science* 62: 21.

Clark S.A., Hickey J.M., Daetwyler H.D. and van der Werf J.H.J. (2012) *Genetics Selection Evolution* 44: 4.

Das S., Forer L., Schönherr S., Sidore C., Locke A.E., Kwong A., Vrieze S.I., Chew E.Y., Levy S., McGue M., Schlessinger D., Stambolian D., Loh P.-R., Iacono W.G., Swaroop A., Scott L.J., Cucca F., Kronenberg F., Boehnke M., Abecasis G.R. and Fuchsberger C. (2016) *Nature Genetics* 48: 1284.

Goddard M.E., Hayes B.J. and Meuwissen T.H.E. (2011) *J. Anim. Breed. Genet.* 128: 409.

Graser H.-U., Tier B., Johnston D.J. and Barwick S.A. (2005) *Aust. J. Exp. Agric.* 45: 913.

Hayes B.J., Daetwyler H.D. (2019) *Annual Review of Animal Biosciences* 7: 89.

Lee S.H. and van der Werf J.H.J. (2016) *Bioinformatics* 32: 1420.

Loh P.-R., Danecek P., Palamara P.F., Fuchsberger C., A Reshef Y., K Finucane H., Schoenherr S., Forer L., McCarthy S., Abecasis G.R., Durbin R., L and Price A. (2016) *Nature Genetics* 48: 1443.

Meuwissen T., Hayes B., Goddard M. (2016) *Animal Frontiers* 6: 6.

Raymond B., Bouwman A.C., Schrooten C., Houwing-Duistermaat J. and Veerkamp R.F. (2018) *Genetics Selection Evolution* 50: 1.

Yang J., Lee S.H., Goddard M.E., Visscher P.M. (2011) *Am. J. Hum. Genet.* 88: 76.

Proc. Assoc. Advmt. Anim. Breed. Genet. 25: 154 - 157

COMPARING GENOMIC PREDICTION ACCURACIES FOR COMMERCIAL COWS' REPRODUCTIVE PERFORMANCE USING GA2CAT AND TWO MACHINE-LEARNING METHODS

Y. Li[1], S. Hu[1], L. Porto-Neto[1], R. McCulloch[1], S. McWilliam[1], J. McDonald[2], C. Smith[2], P. Alexandre[1], S. Lehnert[1] and A. Reverter[1]

[1]CSIRO Agriculture & Food, St Lucia, QLD, 4067 Australia
[2]MDH Pty Ltd, Cloncurry, QLD, 4824 Australia

SUMMARY

Heifers' second joining pregnancy and lactation status (PLS) is an important fertility trait for commercial cattle herds in North Queensland. Genomic prediction of a candidate bull's contribution to its female progeny's PLS presents a technical challenge because the trait has a non-ordinal multi-class nature. We previously developed a new algorithm, Genomic Attributions to a Categorical Trait (GA2CAT) to tackle the problem. However, the merit of the method has not been evaluated against those of machine learning methods. In this study, using two commercial cow populations (795 and 340 cows respectively) with high-density SNP genotypes and imbalanced PLS phenotypes, we compared the classification performance of the new method GA2CAT with two machine learning approaches (Random Forests (RF) and Support Vector Machines (SVM)). The results from a five-fold cross-validation scheme indicate that the classification accuracy of GA2CAT was greatly impacted by the coding system of PLS categories. For highly imbalanced non-ordinal multiclass datasets, using the average overall accuracy value for evaluating the classification performance of the GA2CAT and ML methods was misleading and Matthews correlation coefficient values should be applied.

INTRODUCTION

Female reproductive traits directly impact the profitability of commercial beef herds. Among many reproductive traits, fertility-related ones are the most important. In dairy and beef cattle, they are measured by a range of continuous (e.g. age of puberty, days at first calving), binary (e.g. pregnancy status) or count traits (e.g. number of inseminations) (Toghiani *et al.* 2017). However, in Australian northern commercial cattle herds, following natural syndicate joining, heifers are usually mustered and grouped based on the result of their 2nd joining pregnancy and lactation status (PLS). Females can be assigned to six PLS categories: 1. DNP = Dry and Not Pregnant; 2.WNP = Wet and Not Pregnant; 3. DEP = Dry and Early Pregnant; 4. DMP = Dry and Mid Pregnant; 5. DLP = Dry and Late Pregnant; 6. WEP = Wet and Early Pregnant (Reverter *et al.*2016). This non-ordinal multi-class phenotype presents a technical challenge when trying to rank potential sires based on their genomic relationships with phenotyped heifers. To address this issue, we have developed a new method called Genomic Attributions to a Categorical Trait (GA2CAT) to predict an individual sire's contribution to its future daughters' performance (Li *et al.* 2022). However, the performance of GA2CAT has not been benchmarked against other methods commonly used for analysing non-ordinal multi-class traits, such as the machine learning (ML) based Random Forests (RF) and Support Vector Machines (SVM). Therefore, we conducted the study to compare genomic prediction accuracies of GA2CAT and two ML methods.

MATERIALS AND METHODS

Datasets. Two datasets containing 1,135 tropical Brahman cows, 795 from the 2020 season (referred to as Cows_795) and 340 from the 2021 season (Cows_340), from a north Queensland commercial property were used for the study. All animals with PLS records were individually genotyped for 54,791 SNPs (Neogen Australasia GGP TropBeef 50K chip) which were then imputed to high density using 700K genotypes of 861 legacy BeefCRC Brahman cattle as the reference genome. Table 1 summarises the composition of the phenotype records in both populations, illustrating unevenly distributed multi-class categories.

Phenotypic data recoding. For comparison purposes, three different phenotype recording systems for PLS records were investigated (Table 1). These include: a) treating PLS as a binary trait (2PLS, Non-pregnant "1" vs pregnant "2"); b) as a four-category trait (4PLS, Dry and Non-Pregnant "1", Wet and Non-Pregnant "2", Dry and Pregnant "3", and Wet and Pregnant "4"); and c) as a six-category trait (6PLS, see Table 1 for details).

Table 1. Composition of 2nd Joining Pregnancy and Lactation Status (PLS) records of two Brahman cow populations (795 and 340 cows respectively) and three phenotype recording systems

		Cow population		Phenotype recoding system		
PLS	Code	Cows_795	Cows_340	2PLS*	4PLS*	6PLS*
Dry and Non-Pregnant	DNP	124	61	1	1	1
Wet and Non-Pregnant	WNP	358	109	1	2	2
Dry and Early Pregnant	DEP	77	109	2	3	3
Dry and Mid Pregnant	DMP	70	45	2	3	4
Dry and Late Pregnant	DLP	86	6	2	3	5
Wet and Early Pregnant	WEP	80	10	2	4	6
Total		795	340			

*2PLS: binary categories, 4PLS: four categories; 6PLS: 6 categories

Statistical methods. Three analytical methods were used for evaluating classification accuracy, including GA2CAT (Li *et al.* 2022), RF (Berriman 2001) and SVM (James *et al.* 2013). In brief, the GA2CAT algorithm applies a standard genomic relationship matrix derived from the method of VanRaden (2008) between the reference and testing populations to predict the likely contributions of an individual animal in the testing population to individual classes of a categorical trait. For PLS, a GA2CAT value of a given animal for a given PLS category is defined as the animal's average genomic relationship with other animals having that PLS category divided by its average genomic relationship across all animals. RF is based on ensemble learning of a large number of decision trees deriving from random sampling of various subsets (both SNPs and animals) of a given dataset. It takes the average of decision trees (with replacement) to improve the predicted accuracy of the dataset. The final output (variable importance value) of RF is based on the majority votes of predictions. SVM applies different kernel functions (linear or non-linear) to identify a hyperplane that maximizes the separation of the data points to their potential classes (binary or multi-classes). While a genomic relationship matrix was used for deriving the GA2CAT values, both RF and SVM directly applied SNP genotypes for the analyses.

A 5-fold cross-validation scheme was used for evaluating the classification performance of each method. Each cow population was randomly divided into 5 equal-size groups and each group (68 in Cows_340 or 159 animals in Cows_795) was in turn used as the validation set. Overall accuracy ((true positive + true negative)/(true positive + true negative + false positive + false negative)) was used for evaluating the prediction performance. The final results were based on the average prediction accuracy of five validation groups. Given the imbalanced multiclass datasets used here, we also applied the Matthews correlation coefficient (MCC, Chicco and Jurman 2020) as a measure of the quality for multiclass classification. MCC values normally range from -1 to 1, with 1 representing a perfect prediction, 0 an average random prediction, and -1 a perfect misprediction.

Hyperparameter tuning for RF and SVM. A range of hyper-parameter values was examined for each ML method to determine the critical parameters that minimize prediction errors. These include: for RF, the size of forest trees (Ntree =100, 500), and the number of SNP markers at each sampling event (Mtry = 100, 500, 1000 and 5000); for SVM, insensitivity zone (gamma = 0.001, 1, 5, 10) and the penalty parameter (C= 0.001, 1, 10). All other parameters for each method took default values. The RF and SVM classifiers in the "scikit-learn" Python package (Pedregosa *et al.* 2011) were used for classification predictions.

RESULTS AND DISCUSSION

Comparison of classification performance of GA2CAT, RF and SVM. The overall average prediction accuracies (standard deviations in the brackets) of the three methods from a five-fold cross-validation scheme are summarised in the top part of Table 2. When changing the coding of PLS from two to four to six categories, the overall classification accuracy decreased significantly in both populations for all methods in the small population Cows_340, but to a much lesser extent in the large population Cows_795.

Table 2. Classification performance of GA2CAT, RF and SVM under different PLS coding systems in two cow populations, using a five-fold cross-validation scheme. A) The overall average classification accuracies (standard deviations in brackets); b) Matthews correlation coefficients (MCC)

	Cows_340			Cows_795		
Method	GA2CAT	RF	SVM	GA2CAT	RF	SVM
A. Overall Accuracy			Cow population			
2PLS	0.46 (0.097)	0.51 (0.073)	0.47 (0.032)	0.53 (0.027)	0.61 (0.029)	0.61 (0.033)
4PLS	0.18 (0.034)	0.43 (0.063)	0.47 (0.018)	0.24 (0.027)	0.44 (0.061)	0.45 (0.052)
6PLS	0.091 (0.024)	0.25 (0.054)	0.29 (0.034)	0.12 (0.025)	0.46 (0.052)	0.45 (0.052)
B. MCC			Cow population			
2PLS	-0.071 (0.19)	0.020 (0.143)	0.000 (0.000)	0.059 (0.049)	0.077 (0.040)	0.000 (0.000)
4PLS	-0.039 (0.029)	-0.037 (0.036)	0.000 (0.000)	0.013 (0.026)	-0.027 (0.043)	0.000 (0.000)
6PLS	-0.017 (0.036)	-0.062 (0.073)	0.000 (0.000)	-0.023 (0.036)	0.053 (0.043)	0.000 (0.000)

RF: Random Forest; SVM: Support Vector Machine.

The poor performance of the three methods under 6PLS could be due to the phenotype of PLS being a non-ordinal multi-class categorical trait. The separation of animals for three Dry and Pregnant classes, i.e. early, mid, and late pregnancy was not as clean-cut as those in the binary situation (2PLS, non-pregnant vs pregnant). For the GA2CAT, the genomic relationships between animals in these three classes in the training populations were very similar, therefore the predicted contributions of the animals in the validation populations to six categories of PLS (i.e. GA2CAT values) were very similar. As a result, it made the correct assignment of the animals in the testing populations to different categories extremely difficult. The results indicate the necessity of recoding PLS records before applying different analytical methods to achieve reliable results.

Across two cow populations, for the same coding system, e.g. 6 categories (6PLS), the two ML methods (RF and SVM) seemed to outperform the GA2CAT (see the average accuracies in Table 2). The margin was large in the population Cows_795 (0.46 (RF), 0.45 (SVM) vs 0.12 (GA2CAT). The difference between RF and SVM was little in comparison to either of them with the GA2CAT. However, when investigating further on the classes correctly classified, we found that both RF and SVM assigned all of the individuals in the validation datasets to the category of Wet and Non-Pregnant. This was the class with the largest number of phenotypic observations in Cows_795. This confirms the downside of ML methods that bias toward the majority class by over-sampling the abundant classes and under-sampling minor classes (Chicco and Jurman 2020).

When evaluating the performance of three methods by the MCC values (the lower half of Table 2), all three methods had the MCC values either zero (SVM) or close to zero. These suggest that: a) the phenotype PLS is a low heritability trait, as all three methods followed a random prediction behavior (MCC values ~ 0.00). In addition, the accuracy values for the GA2CAT fitted the random sampling expected prediction accuracies of 0.5 (PLS2), 0.34 (PLS4) and 0.25 (PLS6); b) there was no significant classification performance difference among the GA2CAT, RF and SVM.

CONCLUSION

The results from a five-fold cross-validation scheme indicate that different coding systems of PLS categories greatly impacted the classification outcome of the GA2CAT. For highly imbalanced non-ordinal multiclass datasets, using the average overall accuracy value for evaluating the classification performance of the GA2CAT and ML methods was misleading and MCC values should be applied. A GA2CAT value is the weighted average of genomic relationships between reference and validation populations for a particular category, it reflects better the heritable nature of a phenotypic trait.

ACKNOWLEDGEMENT

The authors wish to acknowledge Meat and Livestock Australia (MLA) for their financial support to the project P.PSH.1211 "Validation of pooled DNA gEBV for Brahman commercial cow fertility".

REFERENCES

Breiman L. (2001) *Mach. Learn.* 45: 5.

Chicco D.,and Jurman G. (2020) *BMC genomics.* 21: 1.

James G., Witten D., Hastie T. and Tobshirani (2013) 'An Introduction to Statistical Learning'. Springer, Heidelberg, Germany.

Li Y., Lehnert S.A., Porto-Neto L., McCulloch R., McWilliam S., Alexandre P, McDonald J., Smith C. and Reverter A. (2022) Proceedings of 12th World Congress on Genetics Applied to Livestock Production. Rotterdam. The Netherlands. 3-8 July 2022.

Pedregosa F., Varoquaux G., Gramfort A., Michel V., Thirion B., Grisel O., Blondel M., Prettenhofer P., Weiss R., Dubourg V., Vanderplas J., Passos A., Cournapeau D., Brucher M., Perrot M. and Duchesnay É. (2011). *J. Mach Learn Res* 12: 2825.

Reverter A., Porto-Neto L.R., Fortes M.R., McCulloch R., Lyons R.E., Moore S., Nicol D., Henshall J. and Lehnert S.A (2016) *J Anim Sci.* 94: 4096.

VanRaden P.M (2008). *J. Dairy Sci.* 91: 4414.

Proc. Assoc. Advmt. Anim. Breed. Genet. 25: 158 - 161

ASSESSING THE VALUE OF METAFOUNDERS FOR GENOMIC PREDICTION IN AUSTRALIAN SIMMENTAL BEEF CATTLE

D.J.A. Santos, N.K. Connors, P.M. Gurman, M.H. Ferdosi, S.P. Miller and A.A. Swan

Animal Genetics Breeding Unit*, University of New England, Armidale, NSW 2351Australia.

SUMMARY

The "metafounders" framework is used to augment relationship matrixes to accommodate genetic structure in founder populations, and can be estimated from genotypes, making it useful to align pedigree and genomic relationships in single-step genomic analyses. This paper aimed to assess the value of metafounders in the genomic evaluation of beef traits in Australian Simmental cattle, and in particular the possibility of collapsing genetic groupings based on metafounder similarity. Estimated breeding values from metafounder models with different groupings had similar predictive ability across 12 beef traits, while models with higher weighting on genomic relative to pedigree information tended to perform better.

INTRODUCTION

Metafounders (MF) are pseudo-individuals included in the pedigree that allow accounting for genetic heterozygosity and relationships within and between base populations, considering unknown ancestral populations (Legarra *et al.* 2015). The MF approach may be advantageous because it derives compatibility between genomic (G) and pedigree (A) relationship matrices by modifying A to align with G (Garcia-Baccino *et al.* 2017). Currently, the BREEDPLAN genetic evaluation for Australian Simmental uses 25 genetic groups, defined based on the country of origin, breed, and year of birth of animals with unknown parentage. The influence of all these genetic groupings and structures in the pedigree of Australian Simmental need to be considered in single-step genetic evaluations. This study aimed to assess the utility of MF in the genomic evaluation of beef traits in Australian Simmental, considering the predictive ability with different MF assignment strategies in the pedigree.

MATERIALS AND METHODS

Data. The genomic data consisted of 8,245 genotyped animals with 59,678 SNPs. Traits analysed included eight live ultrasound scan body composition traits, eye muscle area, intramuscular fat, P8 fat, and rib fat in bulls and heifers (BEA, BIM, BP8, BRF, HEA, HIM, HP8 and HRF), and four body weight traits, birth (BWD), weaning (WWD), yearling (YWD), and final weight (FWD). Numbers of genotyped and pedigree-only animals recorded for each trait are shown in Table 1.

MF procedures. Metafounders were included in single-step models using an adapted inverse relationship matrix defined as $\mathbf{H}^{\Gamma-1} = \mathbf{A}^{\Gamma-1} + \begin{bmatrix} 0 & 0 \\ 0 & \mathbf{G}^{-1} - \mathbf{A}_{22}^{\Gamma-1} \end{bmatrix}$, where $\mathbf{A}^{\Gamma}$ is the pedigree relationship matrix augmented by the "gamma" matrix modelling within and across base population relationships, $\mathbf{A}_{22}^{\Gamma}$ is the sub-matrix of $\mathbf{A}^{\Gamma}$ for genotyped animals, and $\mathbf{\Gamma}$ is the gamma matrix (Legarra *et al.* 2015). The matrix G was obtained as $\lambda\mathbf{G}_m + (1-\boldsymbol{\lambda})\mathbf{A}_{22}^{\Gamma}$, where $\mathbf{G}_m$is the genomic relationship matrix as calculated via VanRaden (2008), and λ is the weighting factor between genomic and pedigree relationship matrices, set as either 0.5 or 1. For λ=1 a small positive value was added to the diagonal of $\mathbf{G}_m$ to ensure it was invertible.

As described above, the genetic groups used to define MF groups have been defined based on country of origin, breed, and year of birth of animals with unknown parents. In addition to 12

* A joint venture of NSW Department of Primary Industries and the University of New England

Simmental groups, there were 4 substantial groups of Angus origin, with minor contributions from Hereford, European, Indicus, and unrecognised breeds.

Estimation of **Γ** was performed via generalised least squares (Garcia-Baccino *et al.* 2017) using the WOMBAT software package (Meyer 2007). The correlations between MF were calculated from the **Γ** estimated, and the MF were grouped using multivariate clustering techniques.

Genetic evaluation and prediction design. Prior to cross-validation, the variance components for each trait were estimated using all data available with the WOMBAT program (Meyer 2007), and these parameters used for BLUP analyses. Variance component estimation and EBV predictions using ssGBLUP with MF were performed using a single-trait animal model with contemporary groups as a fixed effect, direct genetic effects fitted as random for all traits, and maternal genetic effects fitted for BWD and WWD only (uncorrelated with direct genetic effects). Phenotypes were pre-adjusted for fixed effects apart from contemporary group.

The performance of analyses with different MF groupings was compared across traits using a k-fold cross-validation approach with k=5. For the k-folds analyses, animals with phenotypic and genotypic data were randomly split into five parts. EBVs were calculated 5 times for each trait, omitting the phenotypes of animals in each validation set such that their EBVs were then predicted from genomic and pedigree relationships ("part" EBVs). Then, the accuracy, stability and dispersion of the predictions were assessed. Accuracy was calculated as the correlation between part EBV and phenotypes of validation animals for all traits except for the two maternally influenced traits (BWD and WWD) for which the LR method was used (Legarra and Reverter, 2018). Stability was calculated as the correlation between part and full EBVs for the validation animals, and bias as the regression of part EBVs on phenotypes. Results for each statistic were averaged across folds.

RESULTS AND DISCUSSION

Gamma matrix and MF clustering. The matrix **Γ** was estimated for 25 MFs (MF25) and the correlations between MFs grouped by similarity are shown in Figure 1. The diagonal "self-relationship" elements of **Γ** ranged from 0.29 to 0.82 with an average of 0.47 (the possible range in values is 0 to 2 with higher values indicating higher inbreeding). The average for Simmental groups was 0.43 and for Angus was 0.56. Higher values tended to be for smaller groups which by default have less diversity. Corresponding ancestral correlations were typically >0.8 within the Simmental and Angus groups, and approximately 0.2 to 0.6 between other groups (Figure 1).

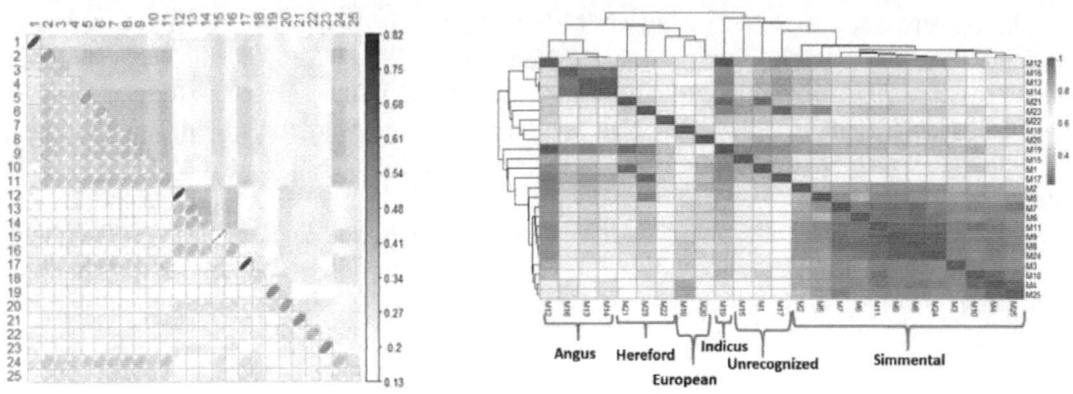

Figure 1. Metafounder clustering results: top left = Gamma matrix (***Γ***) estimated for 25 genetic groups, top right = Gamma correlation matrix with clustering and dendrogram of the genetic groups

Through a k-means algorithm, the MF were collapsed progressively into 15, 14 and 12 clusters, and new **Γ** matrices estimated. In all cases the 3 most similar groups of Angus origin were collapsed, while 12 Simmental groups were collapsed into 4 groups in MF15, 3 groups in MF14, and 1 group in MF12.

Genetic Parameters. Heritability estimates from the MF25 models are shown in Table 1. Estimates for MF12, 14, and 15 were very similar to MF25 and are therefore not shown. These results are similar to the heritabilities assumed in the BREEDPLAN analysis for the breed, although generally marginally higher. According to Legarra *et al.* (2015), genetic variance estimates obtained from MF models should not be interpreted as a genetic variance within the population but as a parameter of the statistical model used for the analysis. Heritability estimates tended to be higher for models with $\lambda = 0.5$.

Table 1.Number of genotyped (Geno) and pedigree only animals (Ped) with records for each trait, and heritability estimates for MF25 models with **λ=1 or 0.5**

Trait	Geno	Ped	**Heritability (λ=1)**	**Heritability (λ=0.5)**
BEA	1,800	21,017	0.32	0.33
BIM	1,680	11,339	0.28	0.28
BP8	1,796	20,986	0.37	0.42
BRF	1,795	20,889	0.28	0.31
HEA	483	15,787	0.35	0.36
HIM	482	9,417	0.42	0.42
HP8	479	15,759	0.56	0.57
HRF	476	15,746	0.47	0.48
BWD	3,068	111,262	0.40	0.40
WWD	2,786	115,209	0.27	0.40
YWD	2,842	118,646	0.42	0.42
FWD	1,647	64,860	0.44	0.45

Cross-validation. Accuracies across traits for MF models with $\lambda = 0.5$ and 1 are shown in Figure 2. There was no effect on accuracy for analyses with different MF groupings, but an increase in accuracy was observed with $\lambda = 1$ for body weight traits. This trend was not observed for body composition traits. Stability of part versus full EBVs (Figure 3) was also higher for models with $\lambda = 1$, but again there was no difference between MF groupings. Results for dispersion (not shown) were similar across models, and sufficiently close to the expected value of 1 across traits. These results suggest reasonable prediction accuracy can be obtained using MF models, with some evidence of higher accuracy with higher λ values. However, there was no advantage in aggregating groups based on similarity.

Before implementation, additional studies should be performed to compare these MF analyses with traditional genetic groups models, and to investigate the accuracy of estimating MF relationships for groups with low numbers of genotypes.

CONCLUSIONS

Although patterns of similarity between metafounder groups were evident, generally reflecting breed of origin, there was little apparent benefit in collapsing groups. Alternatively, simplification of groups may be possible if desired, providing the performance differences between groups to be collapsed are minimal.

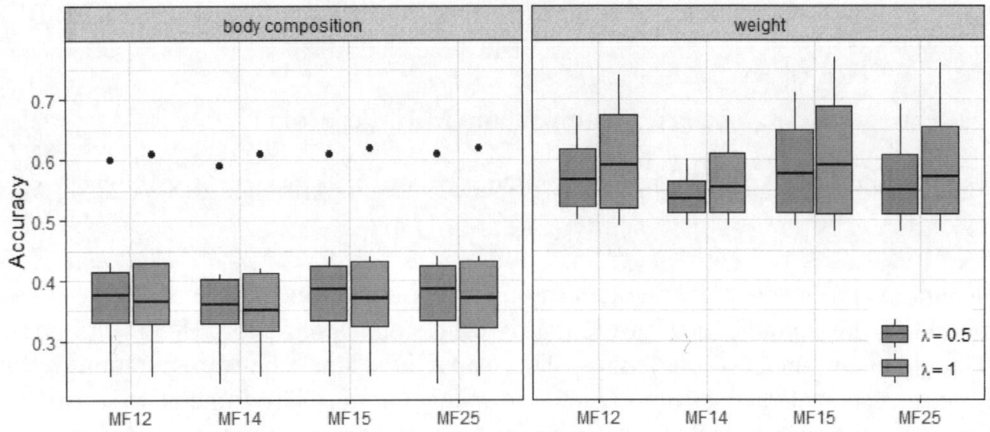

Figure 2. distribution of cross-validation accuracy across traits for MF models with **λ=1** or **0.5**

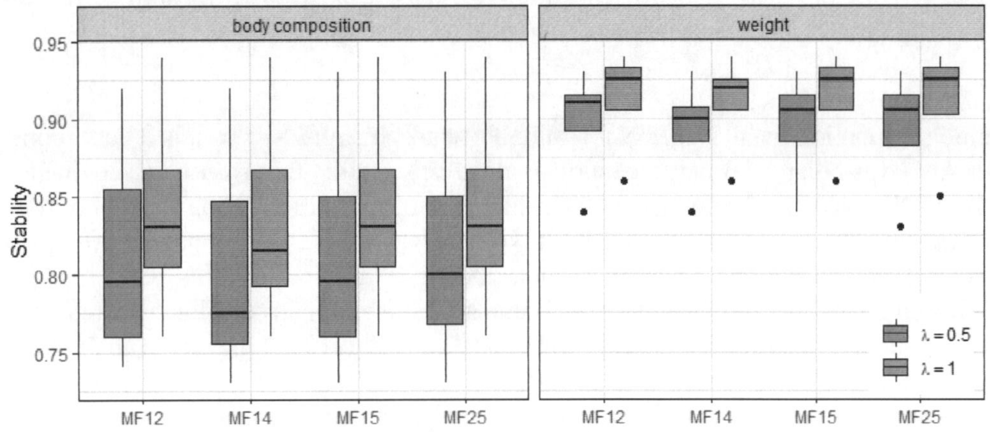

Figure 3. distribution of cross-validation stability of EBVs across traits for MF models with **λ=1 or 0.5**

ACKNOWLEDGEMENTS

This project was supported by Meat and Livestock Australia L.GEN.2204. The authors thank Simmental Australia for providing access to their data and the Agricultural Business Research Institute for providing the data extracts.

REFERENCES

Garcia-Baccino C. A., Legarra A., Christensen O. F., Misztal I., Pocrnic I., Vitezica Z. G. and Cantet R. J. C. (2017) *Genet. Sel. Evol.* 49: 34.

Legarra A., Christensen O. F., Vitezica Z. G., Aguilar I. and Misztal I. (2015) *Genetics* **200**: 455.

Legarra A. and Reverter A. (2018) *Genet. Sel. Evol.* **50**: 53.

Meyer K. (2007) *J. of Zhejiang Uni.* SCIENCE B 8: 815.

VanRaden P.M. (2008) *J. Dairy Sci.* 91: 4414.

IMPUTATION ACCURACY FOR MISSING ALLELES IN CROSSBRED BEEF CATTLE

P.K. Wahinya and M.H. Ferdosi

Animal Genetics Breeding Unit*, University of New England, Armidale, NSW, 2351, Australia

SUMMARY

Establishing a consistent set of markers through the imputation of various marker densities is crucial in building the genomic relationship matrix for genomic prediction. However, imputation in crossbred populations presents challenges. This paper investigated the imputation accuracy of purebred and crossbred Brangus, Simmental and Wagyu cattle. The imputation was carried out independently within each breed using a population-based approach. A reference population of 3,000 randomly selected purebred and crossbred animals with medium-density markers was used in each population to impute the target population with low-density markers (10,000 markers), and this process was repeated five times. On average, imputation accuracies higher than 0.9 were estimated for all three populations. However, the accuracy decreased as the relationship to the reference population within each breed decreased.

INTRODUCTION

Genomic information from Single Nucleotide Polymorphism (SNP) panels is incorporated into BREEDPLAN (Australian beef cattle genetic evaluation system; (Johnston *et al.* 2018)) to predict estimated breeding values. One of the main challenges of including genomic information is that individual animals are genotyped on different SNP panels. Often, different panels contain different SNP densities depending on the size of the panel. Therefore, imputation of un-genotyped SNPs is a standard procedure before building the genomic relationship matrix. This ensures a common consensus SNP panel from different density SNP panels enabling genotypes to be analysed together. The imputation accuracy of the missing SNPs is essential because it influences downstream analyses, such as genome-wide association studies and genomic prediction. Several factors like the relatedness between individuals in the reference and target populations, marker density, reference genome assembly and population structure influence the imputation accuracy (Ferdosi *et al.* 2021a). BREEDPLAN genetic evaluations are currently performed within breeds, however, a substantial proportion of the animals registered within breeds are crossbred at different levels. Currently, crossbred animals with low relationship (less than 80% relationship to the genomic population) to the reference population are not included in BREEDPLAN, partly because imputation in crossbred populations is challenging. Thus, this study aimed to investigate the imputation accuracy in Wagyu, Brangus and Simmental purebred cattle and commercial crossbred animals within each breed.

MATERIALS AND METHODS

Genotypes. Genotypes of 12,058 Wagyu, 8,103 Brangus and 4,778 Simmental cattle including their crosses, were extracted from BREEDPLAN data using quality controls within the BREEDPLAN genomic pipeline (Connors *et al.* 2017). Within the genomic pipeline, an estimated relationship to the reference population approximates how similar animals are to a known population, which can be inferred as breed proportion, and will be referred to as breed proportion for the remainder of this study (Boerner and Wittenburg 2018). The reference population refers to the deviation of individual allele frequencies from the mean allele frequencies in the entire population, which is considered pure based on pedigree. The total number of SNPs considered after

---

* A joint venture of NSW Department of Primary Industries and the University of New England

quality control were, 31,608, 33,302 and 33,126 for Wagyu, Brangus and Simmental breeds, respectively. Individuals and SNPs were removed if they had more than 20% of SNPs missing, while SNPs were only considered if their minor allele frequency was greater than or equal to 5%. Only autosomal SNPs were used for the analysis.

Reference and target populations. Imputation was performed independently within each breed 5 times. In each analysis, 3,000 animals were randomly selected and used as the reference while the remaining formed the target population with 10,000 randomly selected markers (the remaining SNPs were masked to missing genotypes). This is an optimal strategy to improve imputation accuracy (Ferdosi *et al.* 2021a).

Imputation. FImpute v3 software (Sargolzaei *et al.* 2014) with default parameter settings was used to impute the genotypes that were masked as missing. The imputation was population-based, and the pedigree was not utilized. To investigate the imputation accuracy, genotypes of the animals in the target population were imputed back within each breed and were compared with the true genotypes.

Imputation accuracy. The average Pearson's correlation between imputed and true SNPs was estimated to determine the imputation accuracy.

RESULTS AND DISCUSSION

Figure 1 shows the Pearson correlation coefficients between the true and imputed genotypes for Brangus, Simmental and Wagyu cattle. A relatively high imputation accuracy was observed. The average imputation accuracies were 0.96 (0.59 – 1), 0.97 (0.72 – 1) and 0.93 (0.59 – 0.99) for Brangus, Simmental and Wagyu, respectively. Overall the imputation accuracies were higher for animals with a breed proportion greater than 80% within each breed. The imputation accuracy decreased for most animals as their breed proportions decreased. Lower imputation accuracies were estimated for Wagyu crosses (shown as outliers in Figure 1) compared to Brangus and Simmental, especially for animals with less than 50% breed proportion. These results were expected and similar to what is reported elsewhere (Ventura *et al.* 2014; Aliloo and Clark 2021). Incorporating animals with a more substantial relationship to the target population enhances the detection of extended haplotypes which can improve missing SNP imputations. The reference populations in this study were a mixture of pure and crossbred commercial animals within each breed, selected at random. This strategy is important to introduce haplotypes from the other breeds present in the crossbreds. However, all the breeds involved in the crosses were not included in the reference set. The inclusion of these breeds is likely to improve the imputation accuracy, again due to enhanced detection of extended haplotypes.

The differences in the imputation accuracies between the three breeds could be attributed to differences in their population structure, effective population size and the number of crosses. The Simmental population had the least number of commercial crosses while Wagyu had the highest number of commercial crosses with multiple breed combinations. Overall, these results show that randomly missing SNPs in pure breed and crosses of Brangus and Simmental were imputed accurately. Further strategies need to be explored to improve the imputation accuracy for crosses, particularly for Wagyu, where accuracies were largely lower than 0.9 for individuals with less than 50% breed proportion. Wagyu displays a low effective population size and low haplotype diversity (Ferdosi *et al.* 2021b), which could contribute to the low imputation accuracy for animals with low Wagyu content when the parent breeds are missing in the reference population.

This study utilized a population-based imputation method and excluded pedigree information to determine the lowest possible imputation accuracy, since all individuals do not have a pedigree. Combining population and pedigree imputation methods could potentially increase accuracy, but such gains are expected to be minimal with 3,000 individuals in the reference population and a medium-density target population (more than 10k SNPs). Nevertheless, incorporating purebred individuals from which crossbreds were derived may enhance the imputation accuracy for crossbreds.

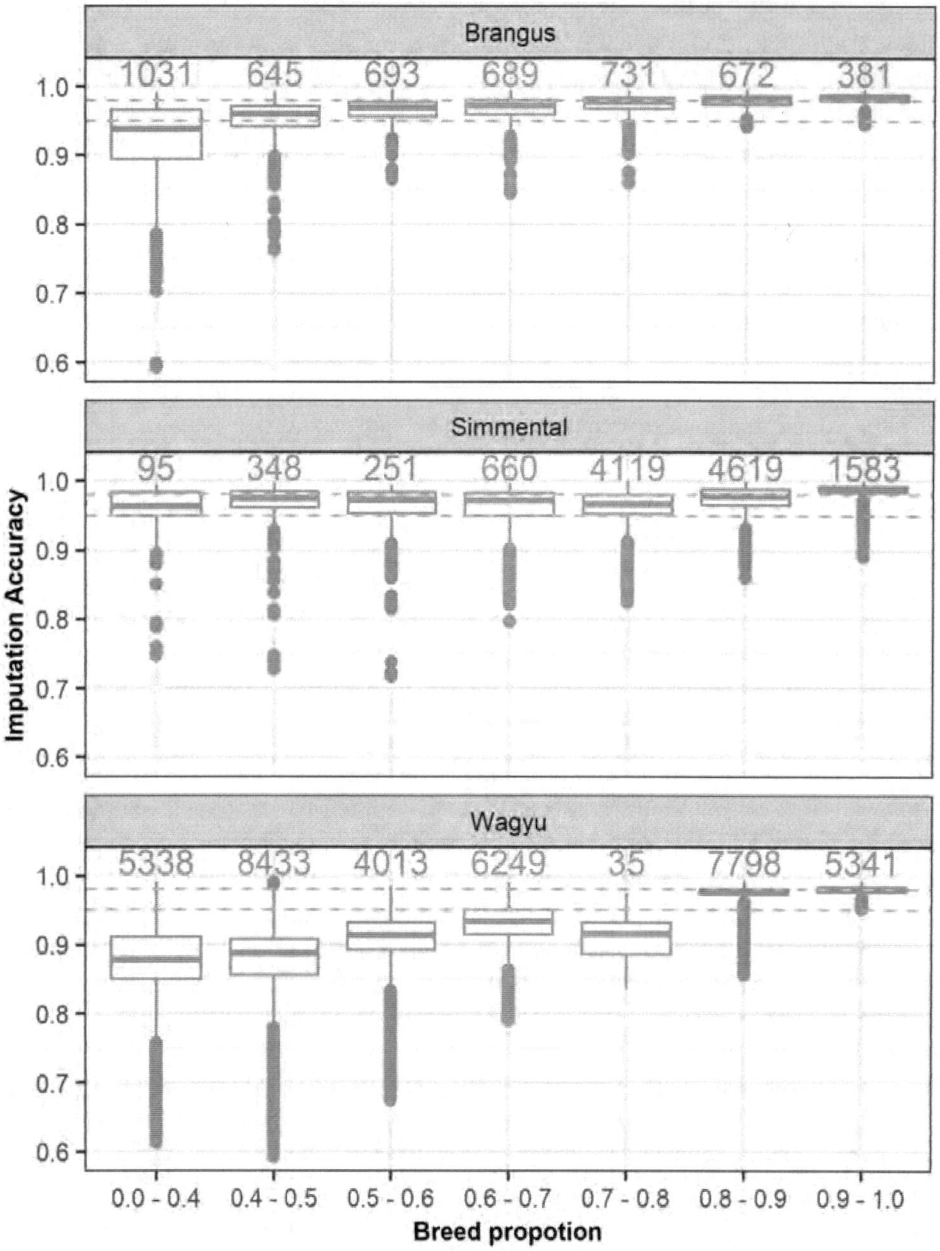

Figure 1. Boxplots showing imputation accuracy across pure-bred and crossbred Brangus, Simmental and Wagyu cattle. The number of animals analysed as target population for different breed proportion ranges combined for the five analyses are shown above each boxplot. Dotted lines represent 0.98 and 0.95 imputation accuracies

CONCLUSION

In this study we investigate the imputation accuracy for missing SNP markers in pure-bred and crossbreed Brangus, Simmental and Wagyu cattle. The results show that the imputation accuracy was on average higher for animals that are more closely related to the reference population.

However, there are differences in the imputation accuracy between the populations, and further studies should evaluate strategies to improve the imputation accuracies for individuals with lower relationships to the reference population when some of the parent breeds are missing. The increase in imputation accuracy would result in a more precise determination of relationships among individuals and an improvement in genomic prediction accuracy, particularly for crossbred individuals.

ACKNOWLEDGEMENTS

This research was partially funded by Meat and Livestock Australia project L.GEN.1704. The authors acknowledge the Agricultural Business Research Institute for the provision of and access to BREEDPLAN data, the Australian Wagyu Association, Simmental Australia and Brangus cattle association, and their breeder members who participate in BREEDPLAN.

REFERENCES

Aliloo H. and Clark S.A. (2021) *Anim. Prod. Sci.* 61: 1958.

Boerner V. and Wittenburg D. (2018) *Front. Genet.* 9:

Connors N., Cook J., Girard C., Tier B., Gore K., Johnston D. and Ferdosi M. (2017) *Proc. Assoc Advmt Anim Breed Genet.* 22: 317.

Ferdosi M., Connors N. andKhansefid M. (2021a) *Proc. Assoc. Advmt. Anim. Breed. Genet.* 24: 114.

Ferdosi M., Connors N. and Khansefid M. (2021b) *Proc. Assoc. Advmt. Anim. Breed. Genet.* 24: 118.

Johnston D.J., Ferdosi M., Connors N., Boerner V., Cook J., Girard C., Swan A. and Tier B. (2018) *Proc. World Congr. Genet. Appl. Livest. Prod.* 2018: 269.

Sargolzaei M., Chesnais J.P. and Schenkel F.S. (2014) *BMC Genomics* 15: 478.

Ventura R.V., Lu D., Schenkel F.S., Wang Z., Li C. and Miller S.P. (2014) *J Anim. Sci.* 92: 1433.

ESTIMATION OF SNP EFFECTS IN LACAUNE DAIRY SHEEP DEPENDING ON THE REFERENCE POPULATION COMPOSITION

M. Wicki[1,2], J. Raoul[1,2] and A. Legarra[1,3]

[1]INRAE, INP, UMR 1388 GenPhySE, F-31326 Castanet-Tolosan, France
[2]Institut de l'Elevage, Catstanet-Tolosan 31321, France
[3]Current address: Council on Dairy Cattle Breeding, Bowie, MD 20716, USA

SUMMARY

The Lacaune dairy sheep breed split in 1972 into two subpopulations with no exchange of genetic material but a single genetic evaluation and same selection objectives. Previous work has shown that this led to the creation of two disconnected but genetically close subpopulations. Previous work also demonstrated that the currently performed combined genomic evaluation of both subpopulations is slightly advantageous, in terms of accuracy, as opposed to within-subpopulation genomic evaluations. This paper focuses on the study of the estimated SNPs effects related to the three training populations: composed of one, the other or both subpopulations. The estimated SNP effects are strongly correlated across years within the training population. When subpopulations are predicted separately, there is low correlation between estimated SNP effects, but when they are predicted jointly, there is a strong correlation of the joint estimate with subpopulation estimates. The regression of "early" (only based on genomic information) on "late" (including progeny information) SNP predictions is lower than one for one of the subpopulations but not for the other, and close to one for the joint prediction. This shows some bias in this particular subpopulation whose origin is not understood.

INTRODUCTION

Selection in French Lacaune dairy sheep started in the 70's with Genomic selection starting in 2015. Each year, young AI rams are selected, among genotyped prospective rams, based on their Genomic Estimated Breeding Values (GEBVs) and used to inseminate females. The accuracy of Milk Yield BV of young genotyped rams (AI candidates) increased from 0.32 to 0.47 (*i.e.* a relative increase of 47%), when transitioned from pedigree-based to genomic based selection (Baloche *et al.* 2014). However, it is of interest to understand if this genomic accuracy can be enhanced further by increasing the size and optimizing the setting up of the reference population.

In 1972, the structure of genetic improvement split, with each flock participating in the AI programs of only one of two existing ram AI studs (breeding companies BC) 1 or 2), exclusively, i.e. a flock only sends rams and receives semen to and from the chosen BC. This created in fact two different subpopulations (1 and 2), subpopulations which do not exchange as breeders rarely exchange sheep and the flux of males and semen is handled by the BC within their participant flocks. Moreover, flocks respect the initial assignation of flocks to BC. Thus, for the last 5 decades, flocks have been contributing rams to a single BC and receiving semen from a single BC. In the following, we will use the wording "subpopulation" to indicate the set of animals belonging to flocks attached to each BC.

A first study (Wicki *et al.* 2023) revealed a low genetic differentiation between the two subpopulations observable, on the one hand, by a low Fst value (0.02), and on the other hand by the results of a Principal Component Analysis (PCA) of the genomic relationship matrix. Indeed, this PCA shows two distinct groups corresponding to each BC, separated on the second component. However, the percentage of variance explained (1.6%) implies that most variation is within-subpopulation, not across. Pedigree analyses showed a low and constant average pedigree relatedness between BC which confirms the very low genetic exchanges between companies.

Finally, Wicki *et al.* (2023) observed a small gain in GEBVs accuracy from the evaluations with training populations of a single BC to the evaluation based on combined reference population.

In this paper, we focus on the study of estimated SNPs effects obtained from genomic evaluations based on reference populations using one company (BC1), the other (BC2) or both of them together (T). We compare SNP effects across years, and across the three possible reference populations.

MATERIALS AND METHODS

This study used all the pedigree, genotypes (50K Illumina chip OvineSNP50) and phenotypic data obtained from regular performance recording of Milk Yield from 1972 to 2021 available in Lacaune dairy sheep (Table 1). The correlation between allele frequencies of each subpopulation is 0.905.

Table 1. Number of animals in the pedigree, number of records and animals in records and number of genotyped animals

Population	Animals in the pedigree	Animals with unknown parent(s) (%)	Number of records	Animals with records	Animals genotyped
BC 1	1,087,161	11.5%	2,968,758	908,116	16,792
BC 2	1,060,862	13.5%	3,041,612	874,329	12,225
T (BC1+2)	1,974,901	10.8%	6,010,370	1,782,445	29,017

Genomic prediction based on different reference populations. We performed genomic evaluations according to several scenarios in which the subpopulations were studied together or separately (Table1). In two scenarios, only the reference population of one subpopulation (BC1 and BC2) was included in the prediction model. In the scenario Together (T), information of both subpopulations was included.

For all the genetic evaluations we used an animal model ssGBLUP with metafounders as detailed in Wicki *et al.* (2023) using blup90iod2 (Tsuruta *et al.* 2001). We used postGSf90 to compute SNPs effects (Tsuruta *et al.* 2001; Aguilar *et al.* 2010), *i.e.* SNP effects are backsolved from GEBVs of genotyped individuals.

Validation. The scenarios were compared using the LR method (Legarra and Reverter 2018) but applied to SNP effects. We defined as "whole" the evaluation including all the phenotypes available until 2021. We compared the SNP effects estimated from this evaluation with SNP effects estimated from "partial" evaluations in which the phenotypes were truncated, i.e. phenotypes after a cut-off date were deleted, with cut-off dates ranging from 2015 to 2019. The correlation shows stability of SNP effects whereas the regression of SNP estimates on "whole" on SNP estimates on "partial" is expected to have a value of 1 for unbiased predictions.

RESULTS AND DISCUSSION

We observe very high correlations of estimated SNPs effects (Figure 1) across years within each reference population (above 0.77, 0.87 and 0.77 respectively for reference subpopulation 1, subpopulation 2 and both), which is reassuring in regards to the correctness of the model and the stability of the genomic predictions, especially for the combined reference population. The correlation is slightly higher for subpopulation 2 across years although we don't have an explanation. The low correlations between subpopulations 1 and 2 (below 0.28) are on line with previous studies investigating combined genomic evaluations where differences in SNPs effects are observed according to the reference population design. Indeed, in our previous study (Wicki *et al.* 2023) we observed that "indirect" genomic predictions using SNP estimates from one subpopulation to obtain

GEBVs in the other subpopulation had very low accuracy of 0.10 on average. In addition, these results show that, when analysing both subpopulations together, the model forces the SNP effects to be "portable" across breeds, whereas the analysis of populations alone does not impose this. The correlation between "Together" with each subpopulation is lower than 1 and lower than correlations within each subpopulation, yet the "Together" evaluation increases accuracy of GEBVs (Wicki *et al.* 2023) from 0.56 to 0.60 for subpopulation 1 and from 0.45 to 0.55 for subpopulation 2 on average (ratios of accuracies). We believe that the increase in accuracy from separate subpopulation analyses comes from the increase in the reference population size.

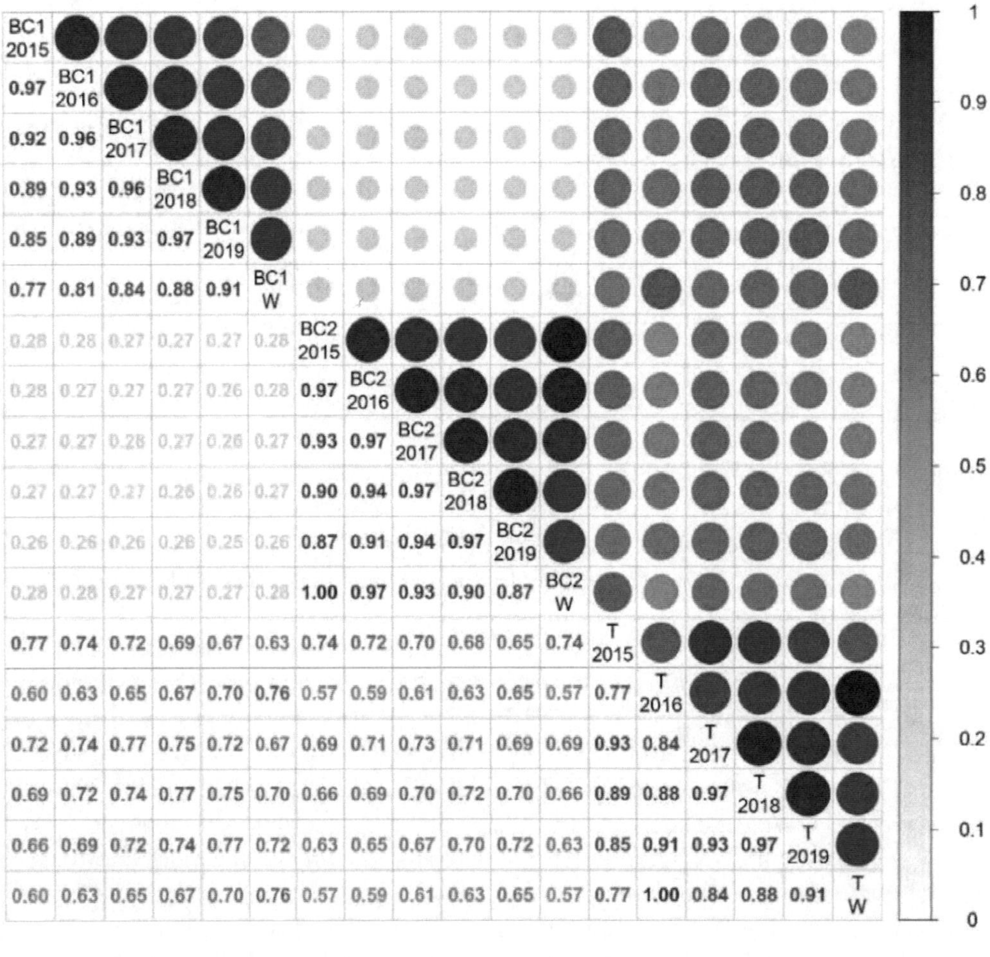

Figure 1. Correlation of estimated SNPs effects between all the studied reference populations ("BC1" = reference population based on subpopulation 1 only, "BC2" = reference population based on subpopulation 2 only, "T" = reference population based on both subpopulations, "W" = evaluation including all phenotypic information until 2021, "2015" to "2019" = evaluation with phenotypic information truncated after year 2015 to 2019)

We expected regression slopes close to 1 between SNPs effects whole and partial in each reference population. Similarly, we expected slopes slightly different from 1 between reference populations BC1 and T, BC2 and T; but far from 1 between BC1 and BC2. We indeed observed low slopes (below 0.31) when estimated SNP effects from one subpopulation were regressed on estimates from the other subpopulation. Within training population BC1, the slope increases over cohorts from 0.58 to 0.83, whereas within training population BC2 the slopes are very close to 1.

This would suggest some bias in BC1 but not in BC2 – the reasons for that are unknown. Slopes between single and combined populations are also variable across cohorts and BC but not too far from 1. Technically, they don't need to be 1 because the "partial" Together contains information that it is not in the "whole" subpopulation.

Table 2. Slopes of regression between estimated SNPs effects "whole" on "partial"

		Cohort				
Partial	Whole	2015	2016	2017	2018	2019
BC1	BC1	0.58	0.64	0.71	0.77	0.83
	BC2	0.25	0.28	0.30	0.30	0.31
	T	0.36	0.39	0.43	0.47	0.50
BC2	BC1	0.22	0.22	0.23	0.23	0.24
	BC2	1.00	1.01	1.02	1.01	1.00
	T	0.35	0.38	0.41	0.43	0.46
T	BC1	0.62	0.96	0.73	0.78	0.84
	BC2	0.94	0.92	0.96	0.95	0.94
	T	0.60	1.00	0.72	0.78	0.84

CONCLUSIONS

Although the evaluations within each subpopulation alone or combined lead to very similar results, this study showed that the estimation of SNP effects was different depending on whether each of the two Lacaune subpopulations was considered separately or together. However, the estimation of SNP effects across subpopulations were too different to be portable, leading to very poor-quality cross-subpopulations evaluations.

ACKNOWLEDGEMENTS

This study was funded by Institut de l'Elevage, INRAE and Apis-gène. We gratefully acknowledge the funding. Furthermore, we want to thank Jean-Michel Astruc for the preparation and the helpful information provided about the database. We also are grateful to the Genotoul Bioinformatics Platform Toulouse Midi-Pyrenees (Bioinfo Genotoul) for providing storage and computing resources and to the organizations responsible for the Lacaune breeding program (Upra Lacaune, Ovitest, Confederation de Roquefort) for providing access to the database.

REFERENCES

Aguilar I., Misztal I., Johnson D.L., Legarra A., Tsuruta S. and Lawlor T.J. (2010) *J. Dairy Sci.* 93: 743.

Baloche G., Legarra A., Sallé G., Larroque H., Astruc JM, Robert-Granié C. and Barillet F. (2014) *J.Dairy Sci.* 97: 1107.

Legarra A. and Reverter A. (2018) *Genet. Sel. Evol.* 50: 53.

Tsuruta S, Misztal I. and Strandén I. (2001) *J. Anim. Sci.* 79: 1166.

Wicki M., Raoul J. and Legarra A. (2023) *J. Dairy Sci.* In revision.

GENETIC EVALUATION OF LONGEVITY USING PRODUCTIVE LIFE AND SURVIVAL SCORES IN AUSTRALIAN HOLSTEIN CATTLE: PREDICTION OF EARLY SURVIVAL

M. Haile-Mariam[1,2], M. Khansefid[1,2], M. Axford[1,2,3], M.E. Goddard[1,4], and Jennie E. Pryce[1,2]

[1]Agriculture Victoria Research, AgriBio, Centre for AgriBioscience, Bundoora, Victoria, 3083, Australia
[2]School of Applied Systems Biology, La Trobe University, Bundoora, Victoria, 3083 Australia
[3]DataGene Ltd, 5 Ring Road, Bundoora, Vic. 3083, Australia.
[4]Faculty of Veterinary & Agricultural Science, University of Melbourne, Parkville, Vic. 3010, Australia.

SUMMARY

The accuracy of genetic and genomic evaluation for longevity is low due to the delay in the availability of culling data and low heritability (h^2). The h^2 and accuracy of genetic prediction for longevity is also influenced by trait definition and differences in methodologies used for estimating breeding values (EBV). This study was designed to compare the reliability, stability, and predictive ability of the current genetic evaluation of Australian dairy cattle for longevity which uses a survival score of 1 or 0 to an alternative measure that considers total months of in milk until 120 months of age (10 years). For this study, data from cows that completed their herd life (i.e., cows born before 2009, reference data) was used to assess differences in the ability of the two approaches for predicting early survival (i.e., survival to the maximum of the fourth lactation) for bulls of whose daughters were born in 2009 to 2014 (validation data). The h^2 of longevity, when the survival rate was analysed, was lower (0.04) than months in milk (0.08). However, the reliability of bull EBVs was about 10% higher for survival rate compared to months in milk. Moreover, EBV of bulls were more stable (i.e., higher correlation between EBVs) when longevity is defined as survival rate than as months in milk. Defining longevity as a survival rate provides better prediction accuracy for unobserved records than when defined as months in milk for bulls with at least 25 or more daughters in both reference and validation data. Overall reliability and stability of EBVs and prediction of unobserved phenotype is better when survival rate is used for genetic analyses than months in milk.

INTRODUCTION

Improving the longevity of cows increases profitability, animal welfare and reduces the environmental footprints of dairy production. The accuracy of genetic and genomic evaluation of longevity is low due to the delay in the availability of culling data and low heritability (h^2). The accuracy of genetic prediction for longevity is also influenced by the definition of the trait and differences in methodologies used for estimating breeding values (EBV) (Forabosco *et al.* 2009). Literature estimates show that longevity of cows when defined as productive life is more heritable (VanRaden and Klaaskate 1993; Settar and Weller 1999) than as survival rate (Madgwick and Goddard 1989) over the lifetime of the cow. However, quantifying differences between the two definitions of longevity in terms of predictive ability is important for increasing genetic progress and acceptance of the genetic evaluation results by end-users. This study was designed to compare the current genetic evaluation for longevity of dairy cattle in Australia which uses survival score 1 or 0 (Madgwick and Goddard 1989) to alternative measures such as productive life by estimating h^2, reliability, stability, and predictive ability of EBVs.

MATERIALS AND METHODS

Data of about 1.75 million Holstein-Friesian cows that were born between 1990 and 2014 from 9,742 dairy herds with valid cow termination dates were extracted from DataGene database for this study. From these data of 1.41 million cows that were born between 1990 and 2008 and that had opportunity to complete their herd life were used as a reference set. These data were used to estimate h^2 and breeding values (BV) for survival rate and total months of productive life (PL). Based on the termination dates we defined longevity in two ways at two time points in the life of the cows. The survival of cows from the first to subsequent lactations until 120 months of age (Surv120) or until the end of 7th parity (Surv7P) were coded as 1 for survived and 0 for culled. The same data was also used to defined productive life (PL) by adding the months in milk of cows in each lactation until 120 months of age (PL120) or until the 7th parity (PL7P). In all cases even if some cows were milked after 120 months of age or after 7th parity their survival rate or PL data was cut at the end of 120 months of age or 7th parity. The fixed effects that were included in the model were determined based on preliminary analyses. For survival rate (Surv120 and Surv7P), the fixed effects fitted were month and year of termination of lactations or of cows, the interaction between parity and age at calving, inbreeding of the cow (F) and month of calving and herd-year-season of calving. For PL, the fixed effects were age at 1st calving, F, the last month of calving and herd-year-season of birth. In all cases, the model included regressions on age at calving (linear and quadratic) and F (linear). When estimating h^2 and calculating reliability a sire model with numerator relationship matrix (NRM) based on sires and their ancestors was used. An animal model that considered all relationships up to 19 generations was used to calculate EBVs. ASReml was used for all data analyses (Gilmour *et al.* 2021).

Table 1. Description of the data with mean and standard deviation (SD) of the traits defined

Definition of longevity	No. records	Mean (SD) no. parities	Mean (SD) trait
Months in milk until parity 7, months	1,411,026[A]	3.69 (2.02)	39.38 (22.68)
Months in milk until 120 months of age, months	1,411,026[A]	3.65 (2.00)	38.80(21.84)
Survival until parity 7, %	5,204,131[B]	3.69 (2.02)	81.94 (38.47)
Survival until 120 months of age, %	5,153,548[B]	3.65 (2.00)	82.33 (38.14)

[A] Number of cows; [B] Includes number of repeated records for cows with more than parity.

To compare genetic evaluation based on the two definitions of longevity, reliability of EBVs were calculated from the prediction error variance. For comparing stability, data of cows that were born before 2009 was split randomly into two. Then stability was calculated by correlating EBVs for sires estimated from two group of herds (even or odd herds) based on the two definitions of longevity. To assess the predictive ability of the two definitions of longevity, data of cows born after 2008 (validation data) were used to calculate corrected phenotype and was correlated with EBV estimated based on the data of cows born before 2009 (reference set). Bulls that had only parent average (PA), at least 25 or at least 50 daughters in the reference data and at least 25 or 50 daughters in the validation data were used to compare predictive ability.

RESULTS AND DISCUSSION

Table 1 show that the mean months in milk is the same (39 months) when productive life is cut at 120 months of age or at the 7th parity. Similarly in the case of survival rates, the mean survival rate (82%) is the same (Table 1), although including lactation until the end of the 7th parity increased the total number of records by about 1%. When calculating months in milk as a measure of longevity cows are sometime evaluated based on shorter age limit such as 84 months of age (VanRaden and Klaaskate 1993; Settar and Weller 1999). In Australia most cows stay longer in the herd than in several other countries (Schuster *et al.* 2020) so editing productive life at 120 months of age or 7th parity could be more appropriate. The h^2 of longevity when survival rate was used as the trait was lower (0.035-0.036) than when defined as months in milk (0.073-0.076). The higher h^2 for months in milk compared to survival rate observed in this study agrees with estimates in literature (Forabosco *et al.* 2009). Extending the lifetime of cows by 12 months, for both survival scores or months in milk until 132 months of age increased the h^2 to 0.04 or 0.08, respectively. However, the phenotypic correlation and correlation between EBV for bulls when survival rate or months in milk was set at 120 and 132 months was effectively 1 suggesting little benefit for delaying the completion of the data beyond 10 years.

Table 2. Mean EBV, standard deviation (SD), reliability (Rel) for bulls with at least 25 and 50 daughters for survival rate or months in milk up to the 7th parity and 120 months of age

Definition of longevity	25 daughters (7,035 sires)		50 daughters (3,224 sires)	
	Mean (SD)	Rel	Mean (SD)	Rel
Months in milk until parity 7	-0.49 (3.40)	0.57	-0.34 (3.59)	0.68
Months in milk until 120 months	-0.67 (3.20)	0.56	-0.53 (3.39)	0.67
Survival until parity 7	-1.29 (4.78)	0.68	-1.40 (4.98)	0.78
Survival until 120 months	-1.38 (4.72)	0.67	-1.47 (4.94)	0.77

Table 2 shows mean EBV with SD and reliability when longevity is defined in different ways. Reliability of EBVs for bulls were consistently higher (~ 10%) when longevity is defined as survival score than months in milk (Table 2). However, the difference in reliability varied from 0.07 for bulls with less than 3 records per daughter to 0.16 for bulls with more than 4 records per daughter. Completing survival or months in milk at end of 7th parity is only slightly more reliable than at 120 months of age. This is possibly due to the slightly higher h^2 and increased number of records when end of the 7th parity was used for editing data. The higher reliability of EBVs based on survival rate in the current study agrees with VanRaden (2003) who demonstrated that reliability of EBVs were higher when repeated survival rates were analysed instead of a single measure that represents the whole lifetime information of cows. The correlation between EBVs when PL is defined as months in milk until 120 months of age and until the end of parity 7 was effectively 1. Similarly, EBVs from survival rate had a correlation of 1.0 when the data ended at the 7th parity and at 120 months of age. As a result, further analyses that assessed stability and predictive ability were based on survival rates or months in milk up to 120 months of age. The correlation between EBVs from survival rate and months in milk for bulls with at least 5 daughters was only 0.72 and increased to 0.76 and 0.80 in bulls with at least 25 and 50 daughters, respectively, suggesting that the two definitions of longevity will rank bulls differently. However, the main reason for the below 1.0 correlation between the EBVs could be due to differences in modelling of fixed effects (VanRaden 2003) and the overall lower reliability of the EBVs of bulls due to the low h^2 of trait. In the current data with an increase in the number of daughters per sire the correlation between EBVs for survival rate and months in milk increased. For bulls with at least 200 or more daughters the correlation between the EBVs was 0.87.

By defining longevity as survival rate more stable EBVs (a correlation of 0.63 between EBVs from odd herds and even herds for bulls with at least 50 daughters in whole data) were obtained than months in milk (a correlation of 0.58). On the other hand, for bulls with at least 25 daughters defining longevity as months in milk up to 120 months of age produced more stable EBVs (a correlation of 0.55) compared to defining longevity as survival rate (a correlation of 0.49). For bulls with 10 or less daughters defining longevity as months in milk provides a slightly more stable EBV (a correlation of 0.05) than defining longevity as survival rate (a correlation of 0.02). However, it is worth mentioning that stability of EBVs is a useful measure of quality of EBVs only if the EBVs have a reasonably high level of accuracy. For bulls with 100 or more daughters the correlations between EBVs from odd and even herds (0.76) were higher when survival rate was used for genetic evaluation than months in milk (0.68) suggesting that the benefit of analysing survival rates on accuracy increases with increase in the number of progenies of the bulls.

Prediction of early survival of cows (i.e., survival to a maximum of the fourth lactation) whose data were excluded was consistently higher when survival rate was a measure of longevity than months in milk. For bulls with at least 25 and 50 daughters in the data of cows that were born before 2009 (i.e., reference) and that at least had 25 daughters in the data of cows that were born after 2008 (validation) the correlation between bull EBVs and corrected bull phenotypes was 0.47 (25 daughters) and 0.48 (50 daughters) when longevity was defined as survival scored compared to 0.39 (daughters) and 0.44 (daughters), respectively, when months in milk was analysed. For bulls with at least 50 daughters in the reference and validation data the correlation between EBV and corrected phenotype was 0.57 for survival score and lower at 0.52 for months in milk. For bulls with PA only in the reference set and at least 50 daughters in the validation set the correlation between PA and corrected phenotype was lower at 0.24 for survival rate and even lower at 0.17 for months in milk.

Higher reliability, stability, and predictive ability for EBVs for bulls are useful criteria when choosing trait definitions for genetic evaluation. It is worth noting that the increased benefit in terms of reliability when survival rates are used could be peculiar to Australian conditions where the number of repeated records is higher for cows that stay in the herd longer than conditions where most cows are culled after a few lactations. The most appropriate definition could also depend on extent of recording of termination data. In Australia, about 50% of the herds that participate in herd recording do not record termination data and survival status for cows in these herds is scored based on re-calving pattern in subsequent years which delays the availability of data (Madgwick and Goddard 1989) and also milk testing is less frequent making it difficult to determine the survival status of cows at any particular time. Accurate and complete recording of termination data will improve timelines and reliability of EBVs.

CONCLUSIONS

This current study showed that EBV of bulls are more reliable and stable when longevity is defined as survival rate rather than months in milk, although the h^2 was lower. Prediction ability for unobserved data of cows was also higher when survival rate is used compared to months in milk. Overall accuracy of genetic evaluation for longevity is higher when it is defined as survival score than as months in milk which means there is no justification to change the trait used for genetic evaluation of longevity in the current production environment.

ACKNOWLEDGEMENTS

This study was undertaken as part of the DairyBio research program, which is jointly funded by Agriculture Victoria, Dairy Australia sand the Gardiner Foundation (Melbourne, Australia). Data used in this study was collected by Australian dairy farmers and maintained by DataGene.

REFERENCES

Forabosco F., Jakobsen J. H. and Fikse W. F. (2009) *J. Dairy Sci.* 92: 2338.

Gilmour A. R., Gogel B. J., Cullis B. R. and Thompson R. (2015) ASReml User Guide Release 4.0. VSN International Ltd., Hemel Hempstead, UK.

Madgwick P. A. and Goddard M. E. (1989). *J. Dairy Sci.* 72: 2624.

Settar P. and Weller J. I. (1999). *J. Dairy Sci.* 82: 2170.

Schuster J. C., Barkema H. W., De Vries A., Kelton D. F. and Orsel K. (2020) *J. Dairy Sci.* 103: 11008.

VanRaden P. M. and Klaaskate E. J. H. (1993). *J. Dairy Sci.* 76: 2758.

VanRaden P. M. (2003) Proc. Interbull Tech. Workshop, Beltsville, MD, Mar 2-3, *Interbull Bull.* 30: 43.

HERITABILITY AND REPEATABILITY OF METHANE EMISSIONS AND FEED INTAKE IN YOUNG DAIRY SIRES

R.C. Handcock[1,2], D.J. Garrick[1,2], D.P. Garrick[1,2], R.J. Spelman[3], P. van Elzakker[4], P. Beatson[4], G. Worth[3] and L.R. McNaughton[3]

[1]The Helical Company, Rotorua, Bay of Plenty, New Zealand
[2]A.L. Rae Centre for Genetics and Breeding, Massey University, New Zealand
[3]Livestock Improvement Corporation, Hamilton, Waikato, New Zealand
[4]CRV, Hamilton, Waikato, New Zealand

SUMMARY

Methane is a by-product of digestion in ruminants and is an important greenhouse gas linked to global warming. There is considerable interest in reducing methane emissions from ruminants. One method is through selection of sires with low methane emissions in their offspring. This research aimed to characterise genetic variation in methane emissions and dry matter intake (DMI) in growing dairy bulls identified from genomic prediction as genomic sires or participants in progeny testing schemes. Estimates of heritability and repeatability for methane emissions and DMI were low to moderate. Genetic and phenotypic correlations between methane emissions and DMI were positive, suggesting selection for reduced methane emissions would be associated with reduced DMI. After accounting for genetic variation in DMI, there was 77% of the genetic variation in methane emissions remaining, indicating there is scope to reduce daily methane emissions while maintaining DMI.

INTRODUCTION

Methane gas is generated as a by-product of digestion in ruminants and is one of the important greenhouse gases that have been linked with global warming (Herrero *et al.* 2016). There are global research efforts to reduce methane emissions from ruminants, especially dairy cattle (Herrero *et al.* 2016), one such method is through selection. Given that genetic progress is permanent and cumulative, it is an attractive option to reduce methane emissions. Methane emissions and dry matter intake (DMI) in ruminants have been reported to be positively phenotypically correlated (Breider *et al.* 2018; Herd *et al.* 2014), and genetically correlated (Manzanilla-Pech *et al.* 2021). Breeding dairy cattle for lower methane emissions may have an unfavourable impact on production and profitability through a correlated reduction in DMI. Measuring either methane or DMI is expensive and logistically challenging, especially in lactating dairy cattle, but unlike milk production traits can be measured in sire candidates.

The long-term goal of this project is to breed dairy cattle who produce less methane per kg DMI. This phase of the research aimed to characterise genetic variation in methane emissions and DMI in the dairy population using growing dairy bulls identified from genomic selection as genomic sires or participants in progeny testing schemes. Additionally, a genetic residual methane emissions trait was calculated using a restricted selection index (Kennedy *et al.* 1993) to assess the proportion of genetic variance in methane emissions that was genetically independent of DMI.

MATERIALS AND METHODS

Approval for animal experiments was granted by the AgResearch Animal Ethics Committee (#15176 and 15533). Experiments were conducted by LIC between February and June of each year (2021 and 2022) and by CRV between July and December 2021. Both farms were located in Waikato, New Zealand. Bulls were housed for 35 days, with the first 7-days as an acclimation period. The number of bulls per pen ranged from 5 to 12. Lucerne hay cubes (Multicube Stockfeeds,

Yarrawong, Vic) were fed in the week prior to the bulls entering the pens and then fed *ad lib.* in Hokofarm RIC2Discover Feed Intake (Hokofarm Group, Emmeloord, AX) bins, allowing individual animal intakes to be recorded. Methane emissions were measured using Greenfeed systems (GF; C-Lock Inc, Rapid City, SD). Animals were allowed to visit a GF device up to 6 times in each 24-hour period. CRV Youngstock Blend (SealesWinslow Ltd, Morrinsville, Waikato) pellets were fed as the bait feed in the GF, with animals allowed up to 24 drops of around 40g/drop in each 24-hour period.

Data handling. To obtain one daily methane "yield equivalent" value per bull, all visits within a day were summed to obtain the total methane and total time in seconds a bull was measured for methane. The calculation was as follows:

$$daily\ methane = \frac{\sum methane\ per\ visit}{\sum time\ (sec)\ at\ GF} * sec\ in\ a\ day$$

Daily DMI was calculated as the sum of the daily lucerne intake measured from the feed intake bins plus the concentrate intakes measured by the GF device.

After daily methane emissions and DMI data from the acclimation week were discarded, there were 13,109 daily methane emissions phenotypes and 12,687 daily DMI phenotypes from 486 bulls. Breeds represented were Jersey, Holstein, Friesian and crossbreeds of varying degrees.

Genotypes. The bulls included in the study were genotyped on a variety of SNP bovine panels. After applying a minor allele frequency filter of 0.1, there were 6,383 autosomal SNPs in common across panels for the 486 bulls.

Statistical analysis. A bivariate repeatability model was fitted using a Bayesian Monte Carlo Markov chain (MCMC) approach with the Julia for Whole-genome Analysis Software (JWAS v1.1.1) package (Cheng *et al.* 2018) run in a Julia computing environment (julialang.org). Inference was based on MCMC chains of 200,000 samples, retaining every 10th sample, after a burn-in of 25,000 samples which had been discarded.

The model equations were:

$$methane = CG + Year + BullPerm + M\alpha + e$$
$$DMI = CG + Year + BullPerm + M\alpha + e$$

where $methane$ and DMI are the daily measurements on methane emissions (n=13,109 records on 474 bulls) and DMI (n=12,687 records on 485 bulls), respectively; CG was the fixed class effect of contemporary group of location-group-pen-day the bull was measured with; Year was the season-year the bull was measured in (2021 or 2022); BullPerm is the random permanent effect of bull assumed to be independently and identically normally distributed with variance σ_c^2; M is a matrix whose columns are additive dosage covariates for all of the 6,410 autosomal loci with effects independently and identically normally distributed with variance σ_a^2; α is a vector of the additive effects at each locus; and e is the residual effects independently and identically normally distributed with variance σ_e^2.

The 95% credibility intervals were calculated by taking the 97.5th percentiles of the MCMC samples (or functions of them that construct heritability or correlation samples from each sample of variance parameters) as the upper bounds and the 2.5th percentiles as the lower bounds.

Prior values for genetic, permanent environment and residual variances and covariances were based on dairy cattle literature (Berry and Crowley 2013; Breider *et al.* 2019; van Breukelen *et al.* 2022; Difford *et al.* 2020; Lassen *et al.* 2012; López-Paredes *et al.* 2020; Pickering *et al.* 2015).

RESULTS AND DISCUSSION

Estimates of heritability for daily methane emissions and daily DMI were low (Table 1). For methane emissions, this is on the lower end of estimates for heritability in dairy cattle of 0.12 to 0.45

(Breider *et al.* 2018, 2019; van Breukelen *et al.* 2022; Difford *et al.* 2020; López-Paredes *et al.* 2020). A study using sniffers reported that when recorded as a mean daily value the heritability of methane emissions was 0.13, similar to the 0.10 reported in the current study. Whereas heritability was higher at 0.32 if the records were averaged over a week due to a large decline in the residual variance when averaging records (van Breukelen *et al.* 2022). In young beef cattle the heritability of methane was reported as 0.27 based on 2 consecutive 24-hour periods in respiration chambers (Donoghue *et al.* 2016).

Similar to methane emissions, the estimate for the heritability of DMI was at the low end of the range previously reported in dairy cattle of 0.11 to 0.43 (Berry and Crowley 2013; Difford *et al.* 2020; Pickering *et al.* 2015).

Table 1. Phenotypic mean and standard deviation (SD) of daily methane emissions (g/d) and dry matter intake (DMI; kg/d) and posterior means (95% credibility intervals) for the genetic variance, heritability, repeatability and genetic and phenotypic correlations

Trait	Methane		DMI
Mean ± SD	229.7 ± 52.7		10.3 ± 3.0
Genetic variance	177 (101, 272)		0.39 (0.20, 0.65)
Heritability	0.10 (0.06, 0.15)		0.09 (0.05, 0.15)
Repeatability	0.31 (0.28, 0.34)		0.37 (0.33, 0.40)
Correlations			
Genetic		0.47 (0.15, 0.71)	
Phenotypic		0.28 (0.24, 0.31)	

Estimates of repeatability for both methane and DMI were moderate (Table 1) but lower than that reported elsewhere in dairy cattle (Breider *et al.* 2019; Difford *et al.* 2020; López-Paredes *et al.* 2020), with the exception of Lassen *et al.* (2012) and van Breukelen *et al.* (2022) who reported repeatabilities of 0.3 to 0.34 for daily methane emissions.

The phenotypic correlation between methane emissions and DMI was 0.28 and was in the range of 0.01 to 0.49 based on reports in lactating dairy cattle by others (Breider *et al.* 2018; Difford *et al.* 2018, 2020; Manzanilla-Pech *et al.* 2021). To our knowledge there are no literature estimates of the correlations between methane emissions and DMI in young growing dairy bulls. Studies in young beef cattle have reported phenotypic correlations between methane emissions and DMI of 0.65 (Herd *et al.* 2014) and 0.71 (Donoghue *et al.* 2016), higher than that reported in the current study. Both of those studies in beef cattle were conducted over 2 consecutive 24-hour periods using respiration chambers.

Based on studies in growing beef cattle, a positive genetic correlation between methane and DMI was expected in growing dairy bulls (Donoghue *et al.* 2016). The posterior mean of the genetic correlation between methane emissions and DMI in the current study was 0.47 with a 95% credibility interval of 0.15 to 0.71, similar to 0.42 reported by Manzanilla-Pech *et al.* (2021), based on 2,990 lactating dairy cattle and 0.42 reported by Richardson *et al.* (2021) based on 379 lactating dairy cattle. Difford *et al.* (2020) reported the genetic correlations between methane and DMI based on two populations of Holstein-Friesian cattle in either Denmark or The Netherlands. For the Danish population, they reported a genetic correlation of 0.6 but for the Dutch population the found the correlation to be -0.09 between the two traits (Difford *et al.* 2020). The authors postulated that the difference in estimates between the two populations studied may have been due to the influence of stage of lactation and differences in diet composition between the two countries (Difford *et al.* 2020).

After accounting for genetic variation in DMI by using partial genetic regression (Kennedy *et al.* 1993), the genetic variance of methane reduced from 177 $(g/d)^2$ to 136 $(g/d)^2$, which is 77% of the genetic variation in methane remaining. The resulting heritability of methane after accounting for variation in DMI was 0.08 and the genetic correlation with DMI was zero, indicating there is scope to reduce daily methane emissions while maintaining DMI.

CONCLUSIONS

There is genetic variation in methane emissions and in DMI for growing dairy bulls. Additionally, there are positive genetic and phenotypic correlations between methane emissions and DMI, suggesting that selection to decrease methane emissions would lead to reduced DMI. Nevertheless, there is opportunity to reduce daily methane emissions from dairy bulls, whilst maintaining DMI due to a considerable proportion of the genetic variance in methane emissions being independent of DMI.

ACKNOWLEDGEMENTS

This work would not have been possible without Garry Waghorn, Hieu Trinh, Peter Cross and the LIC and CRV bull farm staff and trial technicians. This work was funded by the New Zealand Agricultural Greenhouse Gas Research Centre.

REFERENCES

Berry D.P. and Crowley, J.J. (2013) *J. Anim. Sci.* 91: 1594.

Breider I.S., Wall E. and Garnsworthy P.C. (2019) *J. Dairy Sci.* 102: 7277.

Breider I.S., Wall E., Garnsworthy P.C. and Pryce J.E. (2018) *Proc. WCGALP* 11: 134.

van Breukelen A.E., Aldridge M.A., Veerkamp R.F. and de Haas, Y. (2022) *J. Dairy Sci.* 105: 4256.

Cheng H., Fernando R.L. and Garrick D.J. (2018) *Proc. WCGALP* 11: 859.

Difford G.F., Olijhoek D.W., Hellwing A.L.F., Lund P., Bjerring M.A., de Haas Y., Lassen J. and Løvendahl P. (2018) *Acta Agr Scand A-An* 68: 25.

Difford G.F., Løvendahl P., Veerkamp R.F., Bovenhuis H., Visker M.H.P.W., Lassen J. and de Haas Y. (2020) *J. Dairy Sci.* 103: 2442.

Donoghue K.A., Bird-Gardiner T., Arthur P.F., Herd R.M. and Hegarty R.F. (2016) *J. Anim. Sci.* 94: 1438.

Herd R.M., Arthur P.F., Donoghue K.A., Bird S.H., Bird-Gardiner T. and Hegarty R.S. (2014) *J. Anim. Sci.* 92: 5267.

Herrero M., Henderson B., Havlík P., Thornton P.K., Conant R.T., Smith P., Wirsenius S., Hristov A.N., Gerber P., Gill M., Butterbach-Bahl K., Valin H., Garnett T. and Stehfest E. (2016) *Nat. Clim. Chang.* 6: 452.

Kennedy B.W., van der Werf J.H.J. and Meuwissen T.H.E. (1993) *J. Anim. Sci.* 71: 3239.

Lassen J., Løvendahl P. and Madsen J. (2012) *J. Dairy Sci.* 95: 890.

López-Paredes J., Goiri I., Atxaerandio R., García-Rodríguez A., Ugarte E., Jiménez-Montero J.A., Alenda R. and González-Recio O. (2020) *J. Dairy Sci.* 103: 7199.

Manzanilla-Pech C.I.V., Løvendahl P., Mansan Gordo D., Difford G.F., Pryce J.E., Schenkel F., Wegmann S., Miglior F., Chud T.C., Moate P.J., Williams S.R.O., Richardson C.M., Stothard, P. and Lassen J. (2021) *J. Dairy Sci.* 104: 8983.

Pickering N.K., Chagunda M.G.G., Banos G., Mrode R., McEwan J.C. and Wall E. (2015) *J. Anim. Sci.* 92: 11.

Richardson C.M., Nguyen T.T.T., Abdelsayed M., Moate P.J., Williams S.R.O., Chud T.C.S., Schenkel F.S., Goddard M.E., van den Berg I., Cocks B.G., Marett L.C., Wales W.J. and Pryce J.E. (2021) *J. Dairy Sci.* 104: 539.

TOWARDS SELECTING FOR LOWER METHANE SHEEP

P. T Fitzgerald[1], E.H. Clayton[2], A.J. Donaldson[3], D.J. Brown[4], V.H. Oddy[3] and J.H.J van der Werf [1]

[1]Animal Science, University of New England, Armidale, NSW, 2351 Australia
[2] NSW Department of Primary Industries, Wagga Wagga Ag Institute, Wagga Wagga, NSW, 2650 Australia
[3]Beef Centre, NSW DPI, Armidale, NSW, 2351 Australia
[4] Animal Genetics and Breeding Unit*, University of New England, Armidale, NSW, 2351 Australia

SUMMARY

The aim of this project is to enable Australian sheep breeders to select for reduced enteric methane emission, allowing industry to achieve a permanent and cumulative 4.2% reduction (0.8 $MtCO_2e$) in methane emissions from sheep by 2030 and 15% reduction (2.6 $MtCO_2e$) by 2040. A mobile field test using portable accumulation chambers for measuring methane emissions on 10,000 sheep across research and breeder flocks is being rolled out. Five thousand sheep will have feed intake, rumen microbiome and volatile fatty acids profiles recorded to better understand and improve CH_4 emission predictions. Combined with their genotype information, this data will allow genomic prediction of breeding values on selection candidates. Work to date has demonstrated that the protocol for methane measurement is robust and the preliminary data gathered has shown that there is sufficient variation in methane production among animals to enable selection for reduced methane production. Different technologies used to measure emissions data are highly correlated.

INTRODUCTION

Meat & Livestock Australia (MLA) has set a Carbon Neutral Target for 2030 and has initiated a number of programs to reduce the carbon footprint from the Australian livestock industry, including an Emissions Avoidance Program (EAP). The EAP aims to provide various strategies to mitigate methane, including feed additives, new forages and selection for low methane livestock. A program for selecting for more methane efficient sheep was initiated in collaboration between the University of New England, the NSW Department of Primary Industries and MLA. The aim of the project is to collect a large number of phenotypes for methane production and feed intake and to use that data to 1) estimate genetic parameters, including genetic correlations of these traits with production and reproduction traits, and 2) to predict genomically informed breeding values of young rams and ewes in order to select for these traits in studs and commercial enterprises.

Previous work in New Zealand and Australia has shown that there is variation between sheep in how much methane they produce, and this variation is heritable (Pickering *et al.* 2015), i.e. it is possible to change the average methane output via selection. Further modelling work shows that selection for methane alongside selection for other traits can simultaneously improve methane and production efficiency (Robinson and Oddy 2016; Rowe *et al.* 2019). Methane efficiency in dairy cattle has been shown to be positively correlated with feed efficiency, measured as residual feed intake (Manzanilla-Pech *et al.* 2022), but results in sheep have showed a less clear relationship (Muir *et al.* 2020). Furthermore, recent studies have shown that additional information about the methane phenotype of sheep can be obtained from VFA and rumen microbiota (Rowe *et al.* 2019; Ross *et al.* 2020). Based on reliable genetic parameters selection index theory can help clarify the best strategy for selection on productivity as well as feed- and methane efficiency.

---

* A joint venture of NSW Department of Primary Industries and the University of New England

Various trait definitions have been proposed for selection and for assessing the objective of a breeding program for methane (Johnson *et al.* 2022). We propose to simply measure methane output per head, and feed intake, and optimise a breeding objective via bioeconomic modelling, where minimizing the methane production per unit of product is likely to be an overall objective.

Sustainable genetic improvement in methane efficiency of the national sheep flock will require ongoing trait measurement and a data pipeline to turn this information into selection criteria for breeders. Developing a reference population with methane measured animals will allow genomic prediction of breeding value as well as provide more accurate estimation of genetic parameters, including correlations with other traits needed to accurately optimize selection across all economically relevant traits. This requires the development and validation of estimated breeding values and routine measurement of the trait phenotype in a model akin to the MLA resource flock. Thus, the project aims to collect methane and feed intake data on 10,000 and 5,000 animals over 5 years. Lambs from the MLA Resource Flock will be measured as well as ewes in breeders' flocks. The study commenced in 2022. This paper reports on the measurement protocols and some early results of lambs and ewe measurements.

MATERIALS AND METHODS

Methane measurement. Methods and protocols for practical use of portable chambers to measure methane production by sheep are well established and have been validated (Robinson *et al.* 2020). The animals to be measured are held on feed and water until a known time prior to measurement, generally 1 hour. Measurement of gases takes place over a period of 40-60 minutes per animal with methane measurement, oxygen consumption and carbon dioxide output all recorded.

Animals are weighed as close to measurement as possible and a sample of rumen fluid is obtained by stomach tube, after measurement of gas exchange. Rumen fluid is stored for subsequent analysis of volatile fatty acid (VFA) composition and rumen microbial composition.

For this preliminary study, 500 lambs from the MLA Resource Flock cohort were measured at the Kirby Research Station in Armidale (NSW). The Resource Flock design and the proportion of animals from the various sheep breed types is described by van der Werf *et al.* (2010). Animals were born in October 2021 and measured in early April 2022 at approximately 6 months of age. Their average body weights were 29.2 (SD = 6.3) kg. Sheep were removed from the paddock at the beginning of each day and kept in a holding yard with access to feed and water. There were up to four runs each day with 12 animals in each run housed in individual methane accumulation chambers (48 animals in total per day). Lambs were taken off feed 1 hr prior to entering the chambers and gas measurements were taken at 25 and 50 minutes after entering the chamber.

Approximately 500 adult (2016 and 2017 drop) ewes from the Merino Lifetime Productivity (MLP) project were measured at "The Vale", Temora from 31 October -5 November 2022. Their average body weights were 67.6 (SD = 10.3) kg. A total of 84 sheep were removed from the paddock at the beginning of each day and kept in a holding yard with access to feed and water. Each day there were seven runs of twelve animals (84 animals in total each day). Ewes were taken off feed 1 hr prior to entering the chambers and gas measurement on ewes were taken at 20 and 40 minutes after entering the chamber.

For both groups of animals, measurements taken included methane (CH_4, using a micro-Flame Ionisation Detector), CO_2 and O_2 (both using a FoxBox-Pro). Data obtained from these validated instruments were compared with data collected using an Eagle2 Gas Monitor (RKI Instruments).

Feed Intake. Information on methane phenotypes as well as feed intake on about 1,000 animals per year will be collected. The "Masteryard" feed intake system designed and implemented by Crown Agriculture will be installed at Kirby Research Station. The system consists of 50 feed intake units with Calan gates and EID readers, each capable of measuring approximately 6 animals at a time. Animals will be tested for a defined test period (likely 7 weeks total) and all feeding events as well as body weight will be recorded, allowing estimation of daily feed intake as well as feeding behaviour. The diet will be as close as possible to a pasture, i.e. with a high proportion of roughage, while lambs will have a higher energy diet to allow maximum growth.

Implementation. All phenotypic and genotypic data collected in this project will be submitted to the Sheep Genetics database. We are also collaborating with other groups in Australia that collect such data to ensure consistency in measurement protocols and experimental designs that include links sires between data sets.

Collaboration. We are collaborating with research partners in Western Australia at the Department of Primary Industries and Regional Development DPIRD and Murdoch University, with the aim of creating a larger genomic reference population. We are also collaborating internationally (with partners in New Zealand, Uruguay and Ireland) in order to develop more accurate estimates of genetic parameters.

Quantitative genetic and genomic analysis of methane and feed intake phenotypes will be undertaken, initially to understand sources of variation, to determine the best model for analysis and to estimate variance components and genetic parameters. Microbiome data will be analysed and association studies with methane phenotypes undertaken.

After preliminary data analysis in the first year, trait definitions and a model for genetic evaluation will be proposed. This will then be tested through the OVIS software for single step analysis and predictions will be validated and further developed. This includes data to generate estimated breeding values for methane emission as well as feed intake.

The pipeline also involves development of best methods for genomic prediction, including research on genetic markers that might have a large effect on the traits for which genomic prediction is tested. Such marker identities will be made available to genotyping service providers such that they can add them to their genotyping arrays.

Existing data on genotypes and full genome sequence will be used to impute genotypes on animals with new phenotypes. Collaboration with AGBU and Sheep Genetics will develop selection indices and help deliver this information to breeders and let them achieve the desired genetic change. The modelling work will result in clear messages to breeders on what can be achieved with selection on methane and/or feed efficiency ASBVs and how they can best implement that information in their breeding strategies.

RESULTS AND DISCUSSION

Means and standard deviations for body weight and emission traits for Kirby lambs and Temora ewes as measured by the FoxBox-Pro and EAGLE-2 are shown in Table 1 and 2 respectively.

Table 1. Means and standard deviation of lamb and ewe emission traits at Kirby and Temora

Lambs	Measure	FID/FoxBox		Ewes	Measure	FID/FoxBox	
Time	(mL/min)	Mean	SD	Time	(mL/min)	Mean	SD
25 mins	CH_4	8.45	4.88	20 mins	CH_4	6.24	7.00
	CO_2	241.4	71.4		CO_2	498.1	153.9
	O_2	-513.5	103.3		O_2	-1152.0	180.6
50 mins	CH_4	10.56	4.51	40 mins	CH_4	8.96	7.88
	CO_2	225.0	52.3		CO_2	409.9	108.6
	O_2	-358.7	68.1		O_2	-787.0	122.6

Methane output (mL/min) was significantly positively correlated with body weight for Kirby lambs (0.575, Figure 1a) but not ewes at Temora (0.11, Figure 1b). The same measurements on different machines were generally well correlated. Correlations between the FID/FoxBox-Pro and the Eagle for CH_4, CO_2 and O_2 were 0.951, 0.956 and 0.948, respectively for lambs at Kirby and 0.977, 0.970 and 0.905, respectively for ewes at Temora. Although the correlations were high, the slope of the regression was not one, which was more pronounced for CH_4, hence there is still a small scaling difference between measurements with the instruments.

Table 2. The correlation (unadjusted) between CH_4 and O_2 and CO_2 in Kirby lambs

Lambs	CH_4	O_2	CO_2
CH_4	-		
O_2	-0.682	-	
CO_2	0.749	-0.922	-

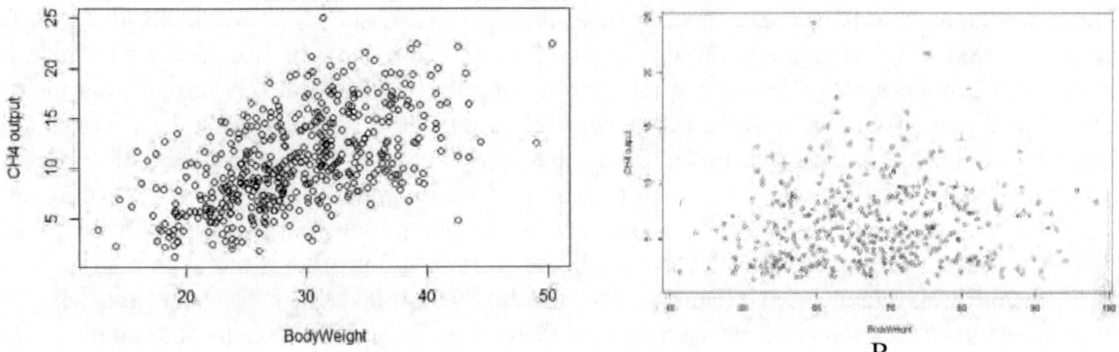

A B

Figure 1a and 1b the relationship between Methane output (ml/min) and body weight in a) lambs at Kirby; and b) mature ewes at Temora

CONCLUSIONS

The protocol for methane collection appears robust and the preliminary data gathered to date has shown that there is enough variation in methane production among animals which should enable the sheep industry to select for lower methane production without negatively impacting other traits.

ACKNOWLEDGEMENTS

Meat and Livestock Australia contributed financial support for this work.

REFERENCES

Johnson P.L., Hickey S., Knowler K., Wing J., Bryson B., Hall M., Jonker A., Janssen P.H., Dodds K.G., McEwan J.C. and Rowe S.J. (2022) *Front. Genet.* 13: 1.

Manzanilla-Pech C.I.V., Stephansen R.B., Difford G.F., Løvendahl P. and Lassen J. (2022) *Front Genet.* 13: 885932.

Muir S.K., Linden N., Kennedy A., Knight M.I., Paganoni B., Kearney G., Thompson A.N. and Behrendt R. (2020) *Small Ruminant Res.* 192: 106241.

Pickering N.K., Oddy V.H., Basarab J., Cammack K., Hayes B., Hegarty R.S., Lassen J., McEwan J.C., Miller S., Pinares-Patiño C.S. and de Haas Y. (2015) *Animal.* 9: 1431.

Robinson D.L. and Oddy, V.H. (2016) *J. Anim. Sci.* 94: 3624.

Rowe S. J., Hickey S.M., Jonker A., Hess M.K., Janssen P. and Johnson P.L. (2019) *Proc. Assoc. Advmt. Anim. Breed. Genet.* 23: 306.

van der Werf J.H.J., Kinghorn B.P. and Banks R.G. (2010) *Anim. Prod. Sci.* 50: 998.

Wahinya P.K., Oddy V.H., Dominik S., Brown D.J., Macleay C.A., Paganoni B., Thompson A.N., Donaldson A.J., Austin K., Cameron M. and van der Werf J.H.J. (2022) *Anim. Prod. Sci*, 62: 818.

Proc. Assoc. Advmt. Anim. Breed. Genet. 25: 182 - 185

USING RUMEN MICROBIAL PREDICTORS FOR GENOMIC PREDICTION OF FEED EFFICIENCY

T.P. Bilton[1], P.L. Johnson[1], F. Booker[1], H.H. Henry[1], M.K. Hess[1,2], R. Jordan[1], S.M. Hickey[3], H. Baird[1], N. Amyes[3], J.C. McEwan[1] and S.J. Rowe[1]

[1]AgResearch, Invermay Agricultural Centre, Mosgiel, New Zealand
[2]University of Nebraska-Lincoln, Institute of Agriculture and Natural Resources, Lincoln, NE 68583, USA
[3]AgResearch, Ruakura Agricultural Centre, Hamilton, New Zealand

SUMMARY

Obtaining phenotypic measures of feed efficiency requires measuring intake levels and growth rates over a period of approximately 8 weeks (2 weeks of adaption and 6 of measurement), which is expensive and low-throughput. Rumen microbial community (RMC) profiles have shown to be associated with feed efficiency traits in ruminants and so may be a suitable proxy. Using a dataset of 1298 animals across 4 genetically linked flocks that were measured through a feed intake facility (FIF), we predicted feed efficiency from RMC profiles and obtained higher prediction accuracies compared to host genomic prediction. The genetic and phenotypic correlations between feed efficiency traits measured from the FIF and predicted from RMC profiles were estimated as 0.64 and 0.33 for mid-trial intake and 0.47 and 0.30 for residual feed intake (RFI). These results suggest RMC profiles have the potential to be used as a proxy for feed efficiency traits in ruminants.

INTRODUCTION

Feed efficiency relates to the amount of feed an animal consumes to produce a fixed amount of product. There are many economic and environmental benefits from breeding for more feed efficient animals, such as reduced feed costs and reduced greenhouse gas emissions per unit of product. Feed efficiency traits are likely to play an important role in future breeding programs as competition for land resources intensifies and targets for greenhouse gas emissions are introduced. Various traits have been proposed to quantify feed efficiency, but all generally require measuring intake levels over an extended period. Although specialised facilities have been developed to measure intake, they are expensive to operate and only a limited number of animals can be measured at a given time. A potential proxy for feed efficiency in ruminants is the rumen microbial community (RMC) profile, as the fermentation process in the rumen, responsible for breaking down feed to produce volatile fatty acids that provide the majority of energy to ruminants, is driven by the microorganisms in the RMC. Previous studies have found associations between the rumen microbiome and feed efficiency in cattle (Li *et al.* 2019) and sheep (Hess *et al.* 2022). RMC profiles have previously been shown by Bilton *et al.* (2022) to be a viable proxy for methane traits. In this study, we extend this work to investigate the feasibility of RMC profiles as a proxy for feed efficiency.

MATERIALS AND METHODS

Experimental animals and protocols applied in this study were approved by the AgResearch Grasslands (Palmerston North, NZ) AgResearch Ruakura (Hamilton, NZ) Animal Ethics committees (approvals 13563, 13892, 14221, 15047 and 15386).

Animals & phenotypes. Data from 4 genetically linked performance-recorded sheep flocks were obtained and consisted of 1298 ewe lambs that were born between 2014 and 2020 (Table 1). Feed efficiency traits were measured using a sheep Feed Intake Facility (FIF) based near AgResearch's Invermay campus, Mosgiel, New Zealand. The lambs were measured at approximately 9 months of age in cohorts of approximately 200 animals across 42 days after a 14-

day introductory period and were feeding on alfalfa pellets from automated feeders. The cohorts for the animals born in 2014 and 2015 were also separated into five pens of equal size. A full description of the experiment and data collection is given in Johnson *et al.* (2022). Feed efficiency traits that were calculated were mid-trial intake (MidIntake), mid-trial metabolic liveweight (MidLWT), and residual feed intake (RFI) was computed as described in Johnson *et al.* (2022). The mid-trial traits were obtained as predictions at day 21 of the measurement period from a linear model of the measured trait values. Additional animal information and measurements were downloaded from the Sheep Improvement Limited database (Newman *et al.* 2000). The animals used in this study are a subset of the animals used by Bilton *et al.* (2022).

Table 1. Sample numbers by flock and year of birth

Flock	Dataset	Year of Birth					Total
		2014	2015	2016	2019	2020	
1	Training	87	145	154			386
2	Training	103	141	140			384
3	Training		95	93			188
4	Validation				158	182	340
Total		190	381	387	158	182	1298

Rumen microbial sampling & profiles. Rumen samples were collected from all animals via stomach intubation after the animals had been in the FIF for at least 4 weeks (2-week introductory period and 2 weeks of measurements). The protocol described in Bilton *et al.* (2022) was used to preserve, process and sequence the samples. The freeze-dried method (Kittelmann *et al.* 2014) was used for all samples except for the born 2020 samples from flock 4, which were preserved using the TNx2 solution (Budel *et al.* 2022). Sequencing was performed using a restricted enzyme-reduced representation sequencing approach (Hess *et al.* 2020) using *Pst*I and run across multiple flowcells on an Illumina HiSeq2500 or NovaSeq6000. The reference-free pipeline developed by Hess *et al.* (2020) was used to generate a count matrix of tags (unique raw sequences trimmed to 65 bp) from which a microbial relationship matrix (MRM) was computed.

Animal genotyping. To investigate prediction of feed efficiency traits from host genomics and comparing to the RMC profiles, a subset of the genomic relationship matrix (GRM) computed in Bilton *et al.* (2022) for animals included in this study was used. This GRM was computed in KGD (Dodds *et al.* 2015) using VanRaden method 1 with non-missing SNPs for each matrix entry and assuming missing data is at random. Animals were genotyped on a variety of nested SNP arrays. SNPs with a call rate of 70% were retained, resulting in 14,923 SNPs in the combined dataset.

Statistical models. Data was split into (a) a training set consisting of the 958 ewes from flocks 1 to 3, and (b) a validation set consisting of the 340 animals from flock 4. Univariate mixed models fitted to the training data were of the form:

$$y_{ijkl} = \mu + cg_j + aod_k + brr_l + a_i + e_{ijkl} \quad (1)$$

$$y_{ijkl} = \mu + cg_j + aod_k + brr_l + m_i + e_{ijkl} \quad (2)$$

where μ is the overall mean, cg_j is the jth contemporary group based on the interaction of flock, birth year, cohort and pen, aod_k is the effect of the kth age of dam (2, 3, 4+), brr_l is the effect of the lth birth/rear rank group (1/1+, 2/2, 2+/1, 3/2, 3+/3+), y_{ijkl} denotes the feed efficiency trait (MidIntake, MidLWT, RFI), $m_i \sim N(0,\sigma_m^2\text{M})$, $a_i \sim N(0,\sigma_g^2\text{G})$, $e_{ijkl} \sim N(0,\sigma_e^2\text{I})$, M denotes the MRM, G denotes the GRM and I is the identity matrix. We refer to the microbial values, m_i, as the "RMC feed efficiency trait" since it provides an estimate of the feed efficiency trait y_{ijkl} (MidIntake, MidLWT, RFI) from the RMC profiles. Predictions of the microbial values ($\widehat{m}_i$) and the animals direct genomic breeding values ($\hat{a}_i$) were made for the animals in flock 4 (validation set). Prediction accuracies were computed as the correlation between $\widehat{m}_i$ or $\hat{a}_i$ and the adjusted phenotype (y_i^*) defined as the residuals from the linear model:

$$y_{ijkl} = \mu + cg_j + aod_k + brr_l + e_{ijkl} \quad (3)$$

fitted using both the training and validation sets. The microbiability (the proportion of variance of feed efficiency trait explain the RMC profiles) was computed as $\hat{\sigma}_m^2/(\hat{\sigma}_m^2 + \hat{\sigma}_e^2)$, and the heritability was computed as $\hat{\sigma}_g^2/(\hat{\sigma}_g^2 + \hat{\sigma}_e^2)$ using all 1298 animals from both the training and validation sets.

To assess the heritability and genetic correlation of the FIF and RMC feed efficiency traits for the validation animals, a bivariate model of the form:

$$\widehat{m}_i = \mu_1 + a_{1i} + e_{1i}$$
$$y_i^* = \mu_2 + a_{2i} + e_{2i}$$

was fitted using only the animals from the validation set, where μ_1, μ_2 are the overall means, $a_{1i} \sim N(0,\sigma_{1g}^2\mathrm{G})$, $a_{2i} \sim N(0,\sigma_{2g}^2\mathrm{G})$, $e_{1i} \sim N(0,\sigma_{1e}^2\mathrm{I})$, and $e_{2i} \sim N(0,\sigma_{2e}^2\mathrm{I})$. All models were fitted in ASREML v4.2 (Gilmour *et al.* 2015). The estimated heritability was computed as $\hat{\sigma}_{1g}^2/(\hat{\sigma}_{1g}^2 + \hat{\sigma}_{1e}^2)$ for the RMC traits and $\hat{\sigma}_{2g}^2/(\hat{\sigma}_{2g}^2 + \hat{\sigma}_{2e}^2)$ for the FIF traits.

RESULTS AND DISCUSSION

Prediction accuracies of the feed efficiency traits for each birth year of flock 4 and overall from RMC profiles and host genomics is given in Table 2. The RMC profiles yielded higher accuracies for the individual cohorts for all three traits compared to host genomics with accuracies ranging between 21% and 42%. These accuracies were similar to those observed for methane traits predicted form RMC profiles in sheep (Bilton *et al.* 2022). The microbiability estimates, which ranged between 0.41 and 0.68, were also larger than the corresponding heritability estimates for all traits when computed using both the training and validation animals.

Table 2. Prediction accuracies for feed efficiency traits predicted from RMC profiles (M) and host genomics (G) for the animals in flock 4

Trait	Model	Equation	Accuracy			Microbiability	Heritability
			b19	b20	b19 & b20	(All; n=1298)	(All; n=1298)
MidIntake	M	1	0.410	0.257	0.316	0.68 ± 0.06	
	G	2	0.096	0.145	0.123		0.34 ± 0.06
MidLWT	M	1	0.312	0.210	0.244	0.41 ± 0.08	
	G	2	0.230	0.101	0.163		0.39 ± 0.06
RFI	M	1	0.417	0.220	0.313	0.54 ± 0.07	
	G	2	-0.047	0.058	0.007		0.32 ± 0.05

Table 3 reports the genetic parameter estimates from the bivariate analysis using the validation animals from flock 4. Heritability estimates for feed efficiency from FIF were slightly higher than previous reported (Johnson *et al.* 2022) and roughly double the heritability estimates of the equivalent RMC feed efficiency trait. The genetic correlation between the FIF and RMC feed efficiency traits were moderate at 0.64 (MidIntake) and 0.46 (RFI), while the phenotypic correlations were lower at around 0.32 (MidIntake) and 0.30 (RFI). These results are very similar to those reported by Bilton *et al.* (2022) for methane traits, except that the genetic correlations are lower for the feed efficiency traits. A bivariate analysis for MidLWT trait was also performed but the heritability estimate for the RMC trait was close to zero and so the results are not reported here. Nevertheless, these results suggest there is potential for using RMC profiles as a proxy for feed efficiency traits, although the small number of animals used in this study means follow-up studies are needed to confirm these results.

Table 3. Heritability, genetic and phenotypic correlations and phenotypic variances from a bivariate analysis of feed efficiency measured from the FIF and predicted from RMC profiles using flock 4 validation animals

Parameter	MidIntake		RFI	
	FIF	RMC	FIF	RMC
Heritability	0.44 ± 0.16	0.15 ± 0.11	0.45 ± 0.14	0.26 ± 0.13
Phenotypic variance	76785 ± 6610	7772 ± 606	19166 ± 1625	1094 ± 87
Genetic correlation	0.64 ± 0.30		0.46 ± 0.26	
Phenotypic correlation	0.33 ± 0.05		0.30 ± 0.05	

CONCLUSION

Our results provide evidence that microbial predictors are a suitable proxy for feed efficiency. As determining feed efficiency in ruminants via direct phenotypic measures is difficult and expensive, RMC profiles provide opportunities for ranking animals based on their feed efficiency for application in breeding programs.

ACKNOWLEDGEMENTS

Financial support was provided via the "Mapping the New Zealand ruminotype landscape" and AgResearch's Strategic Science Investment Funding (SSIF) from the Ministry of Business, Innovation and Employment (MBIE). We thank Beef and Lamb Genetics and all breeders involved for access to their animals, samples and data.

REFERENCES

Bilton T.P., Jordan R., Hickey S.M., Amyes N., Johnson P.L., Henry, H., Baird H., van Stijn T., Veenvliet B., Bain W., Pile G., Watson T., Sandoval E., Peers-Adams J., Bryson B., Hess A.S., Dodds, K.G., McEwan J.C. and Rowe S.J. (2022). *Proc. World Congr. Genet. Appl. Livest. Prod.* 12: 168.

Budel J.C.C, Hess M.K., Bilton T.P., Henry H., Dodds, K.G., Janssen P.H., McEwan J.C. and Rowe S.J. (2022) *Anim. Microbiome* 4: 39.

Dodds K.G., McEwan J.C., Brauning R., Anderson R.M., van Stijn T.C., Kristjánsson, T. and Clarke, S.M. (2015) *BMC Genomics* 16: 1047.

Hess M.K., Rowe S.J., Van Stijn T.C., Henry H.M., Hickey S.M., Brauning R., McCulloch A.F., Hess A.S., Kirk M.R., Kumar S., Pinares-Patiño C., Kittelmann S., Wood G.R., Janssen P.H. and McEwan J.C. (2020) *PLoS One* 15: e0219882.

Hess M., Zetouni L., Hess A.S., Budel J., Dodds K.G., Henry H.M., Brauning R., McCulloch, A.F., Hickey S.M., Johnson P.L, Elmes S., Wing J., Bryson B., Knowler K., Hyndman D., Baird H., McRae, K.M., Jonker A., Janssen P.H., McEwan J.C. and Rowe S.J. (2022) *Research Square.* https://doi.org/10.21203/rs.3.rs-2290213/v1

Gilmour A.R., Gogel B.J, Cullis B.R. and Thompson R. (2015) ASReml User Guide Release 4.2 structural specification. VSN International Ltd, Hemel Hempsted, UK.

Johnson P.L., Hickey S., Knowler, K., Wing J., Bryson B., Hall M., Jonker A., Janssen P.H., Dodds K.G., McEwan J.C. and Rowe S.J. (2022). *Front. Genet.* 13: 911639.

Kittelmann S., Pinares-Patiño C.S., Seedorf H., Kirk M.R., Ganesh S., McEwan J.C. and Janssen P.H. (2014) *PLoS One* 9: e103171.

Li F., Li C., Chen Y., Liu J., Zhang C., Irving B., Fitzsimmons C., Plastow G. and Guan L.L. (2019) *Microbiome* 7: 92.

Newman S.-A., Dodds K.G., Clarke J.N., Garrick D.J. and McEwan J.C. (2000) *Proc. N. Z. Soc. Anim. Prod.* 60: 195.

Proc. Assoc. Advmt. Anim. Breed. Genet. 25: 186 - 189

IMPLEMENTING GENETIC TOOLS TO REDUCE METHANE EMISSIONS IN NZ SHEEP FLOCKS

J.A. Archer[1], M.D. Aspin[1], D.E. Brier[1], S-J. H. Powdrell[1], D. Campbell[1], C.L. Winkworth-Lawrence, S.J. Rowe[2], J.C. McEwan[2] and M. Lee

[1]Beef + Lamb NZ, Wellington, New Zealand.
[2]AgResearch Ltd, Mosgiel, New Zealand.

SUMMARY

Genetics is a tool which can reduce methane emissions, and a concerted effort is required to make this technology available to ram breeders and commercial sheep producers. As a technically demanding trait to measure there is a need for a combination of cheaper and more scalable proxy measures coupled with a genomic selection strategy. Roll out will require some technical challenges within the evaluation to be addressed, particularly around genomic evaluations in multi-breed populations when phenotype records are limited. Incentives to address methane emissions using genetic selection will require different approaches to many other traits as methane is an externality to sheep production business and not directly observable. Consequently, approaches to indirectly quantify genetic reduction in methane emissions, and to incorporate these reductions into assessments of farm level methane emissions will be required to incentivise uptake.

INTRODUCTION

Methane emissions from ruminant livestock form a significant proportion of greenhouse gas emissions (GHG) in New Zealand, and so research to identify ways to reduce GHG without significant reductions in productivity has been undertaken in New Zealand and elsewhere for approximately two decades. Genetic selection of sheep to reduce methane output has been shown to be feasible, and to generate real reductions in methane with few antagonisms apparent with production and health traits (Rowe *et al.* 2019). The challenge is to implement this selection opportunity on an industry-wide scale to achieve verified reductions in the national GHG inventory. Some of the opportunities, approaches and challenges to achieving this are outlined.

CHALLENGES IN PHENOTYPING, GENOTYPING AND EVALUATIONS

Methane is, at least with current technology, a "hard to measure" trait requiring both significant expertise and capital equipment for phenotype collection. This limits the scale at which phenotyping can be applied, and a strategy (breeding scheme design) utilising genomics is the obvious mechanism to maximise the benefits from the phenotypes collected to the wider breeding population.

Phenotyping. There are multiple ways of measuring methane emissions, but many are not scalable or portable, and so not well suited to on-farm measurement of methane in industry breeding flocks. The method that has been chosen and developed with industry in New Zealand is to use Portable Accumulation Chambers (PAC) mounted on a trailer that is shifted between properties (Goopy *et al.* 2011, Jonker *et al.* 2018). In New Zealand this method as currently configured has capacity to measure 12 animals concurrently for a 30-minute period. The protocol requires animals to be fully fed prior to measurement and it is feasible to measure approximately 168 animals over 1.5 days per property. It has been successfully used for the past 10 years, and the total number of animals phenotyped per year has been more than 7,000 up to 2016 in research and resource flocks and 2,328, 3,029, and 5,574 from 2020 to 2022 respectively and the capacity is effectively fully utilised currently. A second trailer will effectively double the capacity for PAC phenotypes to be collected, and the potential will exist to collect approximately 10,000 phenotypes per year in the future. Notwithstanding the success of the PAC phenotype approach, it is logistically challenging, requires specialised technical expertise, has limited capacity, and costs NZ$40 per animal.

The challenges around PAC measurements mean that a proxy measure would have significant benefits. Ideally the proxy measure would be cheaper, able to be scaled to many more animals and utilise more widely available expertise while maintaining a high genetic correlation with the target trait. Currently the use of a metagenomics approach based on a rumen sample looks promising (Bilton *et al.* 2022). It requires expertise to sample animals while maintaining welfare requirements, but this is widely available in the veterinary and animal technician community (and readily trainable). The throughput of animals is determined mainly by the time taken to sample (approximately 2 minutes per animal) and laboratory capacity, and so can be scaled to a significantly greater number of animals than the PAC measurement. The genetic correlation with the PAC measurement is 0.76±0.14 (Bilton *et al.* 2002) so is sufficiently high as to be a useful indirect criterion. The estimated cost is currently around $50/sample, which will likely limit implementation unless technology can drive cost down (or alternative ways of funding the proxy phenotype are developed). Rumen microbiome samples have not yet been used in routine genetic evaluations.

A related approach might be to undertake a similar rumen microbial meta-biome analysis based on buccal swab samples (Kittlemann *et al.* 2015). This could further reduce the welfare cost to the animal (via less invasive sampling procedure relative to rumen fluid sampling) and make it feasible as an on-farm test without requiring highly trained technicians for sample collection. The cheaper and easier sampling could encourage further scale-up for phenotype measurement. However, currently there are several technical challenges implementing the preferred rumen metagenomics approach on buccal samples and this is under active development.

Genotyping. The challenge around genotyping is firstly to ensure that the methane phenotypes are matched with genotypes so that they contribute to methane genomic predictions, and then secondly to enhance the wider adoption of genotyping. The latter is not a methane-specific challenge, although demand for predictions on methane could conceivably assist with genomic technology uptake. Currently approximately 55,000 sheep are genotyped annually from a total of approximately 343,000 new animal identities loaded onto the national sheep genetic evaluation (nProve) database (averages per year born over 2020-2022 period). To date, all lambs with methane phenotypes collected are also genotyped, as industry subsidies for methane phenotype collection are conditional on the animal also having a genotype available in the analysis.

Evaluations. Currently research genomic breeding values (gBVs) for methane are produced in a single-step genomic BLUP (ssGBLUP) analysis, with the trait defined as absolute emissions. To make gBVs available widely across industry, key evaluation challenges will need solutions.

Firstly, a strategy will be required to enable gBV calculations across as many maternal breeds of sheep (and then Terminal breeds as the next priority) as possible. At present the NZ Genetic Evaluation (NZGE) for maternal worth uses genotypes for Romney, Perendale, Coopworth, Texel, and composites containing significant proportions of these breeds (being the most widely used maternal breeds in the NZ sheep flock), while genotypes from other breeds are excluded. A terminal evaluation is planned which will use genotypes from Suffolk, Texel and some other meat-oriented breeds. However, the NZ sheep flock is increasing in breed diversity, including moving towards fine-wool composites or shedding breeds. A strategy relying on genomics to select for methane across the wide range of breeds will require a solution which allows genotypes from multiple breeds to be used. This will depend on sufficient animals in these other breeds having both methane phenotypes and genotypes, and an evaluation solution which allows these genotypes to be incorporated and used to predict phenotypes for animals with genotypes only. This may require additional custom evaluations using genotypes from specific sets of breeds not currently included in the standard runs. In the long-term it would be desirable to include all breeds into a single analysis, but this may be more aspirational than realistic with available analysis techniques.

The ratio of phenotypes to genotypes is important to achieve stable evaluations using ssGBLUP analyses with current software. Methane will have a relatively low number of phenotypes relative to the number of animals currently genotyped, and within the current evaluation pipeline this leads to challenges in producing stable analyses with good convergence properties. Genotyping is likely to increase relative to methane phenotypes. Consequently, a different approach to evaluating methane and other sparsely recorded traits is required. One option to explore restricting the genotypes going into the ssGBLUP evaluation to animals from flocks actively recording methane phenotypes, and using a SNP co-efficients based approach to calculate gBVs for methane on animals whose genotypes are not included in the ssGBLUP.

Methane EBV definition and indexes. At present the research version of a methane BV is being calculated based on the absolute value of methane in nProve. In contrast, methane per kg dry matter intake, calculated from the same information, has been used to successfully create high and low methane selection lines in an experimental setting. However, it is still an open question as to how the methane BV should be formatted and expressed for use in industry, and the overall direction of the breeding program needs to account for the desired goals, regulatory environment and likely farmer responses. Regulatory authorities either want absolute values of methane per animal or preferably methane emitted per kilogram of dry matter ingested to be compatible with IPCC reporting conventions. In practice counting sheep and their classification is currently easier than estimating the dry matter production per farm. If the overall effect of the breeding program is mainly to improve the efficiency of production relative to methane emissions (i.e.. reduce emissions intensity), absolute methane emissions will only be reduced if production is capped and efficiency gains are used decrease the total feed consumed. Alternatively, a breeding objective which focusses on reducing the methane produced per kg of dry matter consumed could result in reduced absolute methane emissions without reducing the feed base or productivity. The latter approach is more desirable as drivers for financial sustainability of farm businesses tend to be directed towards maximising quantity of feed grown and utilised.

While the ultimate outcome of a breeding program is determined by the actual selection decisions, the format of BVs and indexes used can influence the direction substantially. Different formats of BVs can result in equivalent outcomes providing the index is generated consistently with the BV format, but a format which is easily understood and best aligns to the desired selection outcome will enhance uptake and utility. Different BV format options exist, including using the absolute value, using a percentage reduction relative to contemporary group mean, or using a residual approach (analogous to residual feed intake) by adjusting methane (on either a genetic or phenotypic basis) for another trait (e.g. total respiration, feed intake or weight) to lower the correlation between methane and size or production. Factors to consider include the ability of users to understand the trait definition and intuitively assess trait relationships, availability and accuracy of the trait data used in adjustments, and the suitability of the EBVs for use in GHG calculators.

INCENTIVISING UPTAKE

As an expensive and technically demanding trait, GHG measurement will only be sustained in the long-term if a commercial model emerges where breeders can extract sufficient additional revenue (or non-financial benefits) to cover measurement costs. This requires demand for improved rams in terms of methane production from commercial farmers. For most traits where genetic improvement results in better financial performance in the business (either cost savings, improved production or market premiums) the demand for improvement occurs from within the business. However, GHG emissions are an externality to farm businesses, and are not directly measured or observable, so improvement is unlikely to occur without other interventions.

The New Zealand government has proposed to levy agricultural businesses based on their methane emissions, which effectively converts methane emissions from an externality to having a direct financial impact on farm businesses. Industry has expressed a strong preference for this to occur at the individual farm level based on a specific farm's emissions profile calculated using models and farm performance levels as inputs to an approved emissions calculator. To incentivise using genetics as a mitigation, it will be important to develop a mechanism where genetic reduction in emissions from sheep can be accounted for and reduce the emissions levy payable for farm businesses using methane improved rams. An estimate of the flock level genetic merit for methane emissions would be required as an input to the calculator.

Calculation of improvement at the national and individual flock level might use information on ram sales/purchases over time and a gene flow model. This will need to be backed by some sort of verification system to ensure information making its way into both the individual farm level calculator (for levy calculation purposes) and the national GHG inventory (to meet international reporting obligations) is accurate. Direct measurement of methane for verification is unlikely to be viable, but there is likely to be a role for genomics in the verification processes. Genomics could be used to directly estimate flock level methane emissions from co-efficient-based predictions, or could be used to estimate the genetic contribution of rams with methane EBVs to the flock (to verify claims based on gene flow models). Both approaches would utilise information held in the national sheep genetics database (nProve). Investment in information systems to implement this will be required.

With a strong government/public interest in climate mitigation, and a real need to accelerate the uptake of genetic improvement of methane emissions, there have been joint industry/government programs to offset 50 to 100% of methane recording costs. This has led to significant uptake of phenotyping, and also removes much of the "freeloader" objection to phenotyping when competitors benefit via genomics. However, once methane phenotyping becomes routine and not subsidised longer-term strategies will be required. Much of the further work described here will be incorporated into a new "Cool Sheep" program funded by Beef + Lamb NZ and the NZ government to accelerate the application of this proven GHG mitigation technology.

CONCLUSION

Implementation and uptake of selection to improve methane emissions is a priority within the New Zealand sheep industry and is supported by Government. Challenges in improving methane are similar to other difficult to measure traits, but with additional issues which are unique to methane. These unique challenges revolve around the fact that methane is an externality to the business, while the societal and governmental drive to address greenhouse gas emissions provides opportunity to take different approaches to implementing selection to reduce methane emissions.

REFERENCES

Bilton T.P., Jordan R., Hess M.K., Hickey S.M.,.Amyes N., Johnson P., Henry H., Baird H., van Stijn T., Veenvliet B., Bain W., Pile G., Watson T., Sandoval E., Peers-Adams J., Bryson B., Hess A.S., Dodds K.G., McEwan J.C. and Rowe S.J. (2022) *Proc. Wld Congr. Genet. Appl. Livest. Prod.* 12: 168.

Goopy J.P., Woodgate R., Donaldson A., Robinson D.L. and Hegarty R.S. (2011) *Anim. Feed. Sci. Technol.* 166-167: 219.

Jonker A., Hickey S.M., Rowe S.J., Janssen P.H., Shackell G.H., Elmes S., Bain W.E., Wing J., Greer G.J., Bryson B., MacLean S., Dodds K.G., Pinares-Patino C.S., Young E.A., Knowler K., Pickering N.K. and McEwan J.C. (2018) *J. Anim. Sci.* 96: 3031.

Kittelmann S., Kirk M.R., Jonker A., McCulloch A. and Janssen P.H. (2015) *Appl. Environ. Microbiol.* 81: 7470.

Rowe S.J., Hickey S.M., Jonker A., Hess M.K., Janssen P., Johnson T., Bryson B., Knowler K., Pinares-Patino C., Bain W., Elmes S., Young E., Wing J., Waller E., Pickering N., McEwan J.C. (2019) *Proc. Assoc. Advmt. Anim. Breed. Genet.* 23: 306.

A COMPARISON BETWEEN THE USE OF PEDIGREE OR GENOMIC RELATIONSHIPS TO CONTROL INBREEDING IN OPTIMUM-CONTRIBUTION SELECTION

M. Sharif-Islam[1], M. Henryon[2], J.H.J. van der Werf[3], A.C. Sørensen[2], T.T. Chu[4], B. J. Wood[5] and S. Hermesch[1]

[1]Animal Genetics Breeding Unit*, University of New England, Armidale, NSW, 2351 Australia
[2]Danish Pig Research Centre, Danish Agriculture and Food Council, Axeltorv 3, 1609 Copenhagen V, Denmark
[3]School of Environmental and Rural Science, University of New England, Armidale, NSW, 2351 Australia
[4]Center for Quantitative Genetics and Genomics, Aarhus University, Denmark
[5]School of Veterinary Science, University of Queensland, Brisbane, QLD 4072, Australia

SUMMARY

Stochastic simulation was used to test the hypothesis that optimum-contribution selection with genomic relationships using marker loci with low minor allele frequency (MAF) below a predefined threshold (referred as TGOCS) to control inbreeding maintained more genetic variation than pedigree relationships (POCS) at the same rate of true genetic gain (ΔG_{true}). Criteria to measure genetic variation were the number of segregating QTL loci (quantitative trait loci) and the average number of founder alleles per locus. Marker alleles having a MAF below 0.025 were used in forming the genomic relationships in TGOCS strategy. For centering in establishing genomic relationships, when the allele frequency of marker loci with low MAF set to 0.5 the TGOCS strategy maintained 66% fewer founder alleles than POCS and there were 30% fewer QTL segregating. This TGOCS strategy maintained 61% fewer founder alleles than GOCS and 28% fewer segregating QTL loci. When the allele frequency of marker loci with low MAF was set to observed allele frequency these figures were 8%, 2%, 5% and 2%, respectively. Using marker loci with low MAF in the TGOCS strategy was inferior to both GOCS and POCS. Both TGOCS and GOCS were affected by the same constraint that is LD (linkage disequilibrium) between markers and QTL. Therefore, POCS is a more efficient method to maintain genetic variation in the population until a better way to use genomic information in optimum-contribution selection is identified.

INTRODUCTION

Optimum-contribution selection (OCS) can use either pedigree or genomic relationships to control inbreeding. Simulation studies showed that using pedigree relationships to control inbreeding in OCS realise more true genetic gain (ΔG) than genomic relationships at the same rate of true inbreeding (ΔF), where the true inbreeding coefficient of an individual is the observed proportion of marker loci in its genome with alleles that are identical-by-descent (IBD) (Sonesson *et al.* 2012, Henryon *et al.* 2019). Using pedigree relationships to control inbreeding in OCS (referred to as POCS) realises more ΔG than using genomic relationships based on all markers (GOCS) because POCS manages expected genetic drift without restricting selection at QTL (Henryon *et al.* 2019). By contrast, GOCS penalises changes in allele frequencies at marker loci generated by genetic drift or selection. Because these marker alleles are in linkage disequilibrium with QTL alleles, it restricts changes in favourable QTL alleles. This implies that GOCS in its current form is unlikely to realise more genetic gain than POCS at the same rate of true inbreeding.

* A joint venture of NSW Department of Primary Industries and University of New England

An alternative strategy is needed for using genomic information to control inbreeding. One possible strategy is to carry out GOCS by establishing genomic relationships using only marker loci with low minor-allele frequencies (MAF) in the generation under selection. All other marker loci are excluded in this form of GOCS. Focusing on marker alleles with low MAF is of interest because it is these marker alleles that are particularly susceptible to being lost through selection or genetic drift. While most changes in allele frequencies due to drift or selection will be reversible, the extinction of a particular allele constitutes an irreversible loss of variation. Using marker loci in establishing genomic relationships, the genotypes of animals at each locus are centred around a pre-defined allele frequency; it could reduce the loss of alleles by promoting a shift in allele frequencies towards the pre-defined frequencies. Assuming allele frequency of 0.5 for all marker loci in establishing genomic relationships, more weight will be given on the heterozygotes by moving allele frequency towards 0.5 (Meuwissen *et al.* 2020). In case of using calculated allele frequency other than 0.5, rare alleles will be given more weight than common alleles (Forni *et al.* 2011). Preventing the loss of minor alleles may maintain more segregating loci that contribute to genetic variation in the population. Since markers are in LD (linkage disequilibrium) with QTL, both the number of segregating QTL and the number of founder alleles maintained in the population can be used as criteria to assess loss of genetic variation, at least in simulation studies. This reasoning leads to the hypothesis that GOCS using marker loci with MAF below a predefined threshold in the generation under selection – hereafter referred to as TGOCS- maintains more segregating QTL and founder alleles than POCS at the same ΔG.

MATERIALS AND METHODS

Procedure. Stochastic simulation was used to estimate the number of segregating QTL and founder alleles (the average number of founders that contributed alleles to each locus, averaged over all founder loci) realised in the last generation after applying TGOCS, GOCS or POCS and making comparisons at the same ΔG_{true}. TGOCS included all low frequency alleles at the marker loci when MAF at the marker was below 0.025 in the first generation under selection. Marker allele frequencies were calculated in the OCS candidates but the allele frequencies used for the centering of genotypes was set to either 0.5 (Scenario, TGOCS_0.5) or to the allele frequency found in the base population (scenario, TGOCS_base). GOCS used genomic relationships calculated from all markers having a MAF above 0. The criterion for selection was the true breeding value (TBV) of a single trait with a genetic variance of 1. Each breeding scheme was run for ten discrete generations. Each replicate was initiated by sampling a unique base population from the founder population. Selection candidates were genotyped before selection.

Breeding scheme. A total of 25 matings were made from 250 selection candidates by OCS in each generation. Animals were selected randomly in generations 1 to 3. Selection based on TBV was introduced in generations 4 to 10. Males that we selected were allocated up to 25 matings. All male candidates were considered potential parents by OCS. The top 25 females were allocated a single mating each. The 25 sire and 25 dam matings were paired randomly. Each dam produced ten offspring resulting in 25 full-sib families and 250 offspring. Offspring were assigned as males or females with a probability of 0.5.

Genetic models. The founder population was established using a Fisher-Wright inheritance model to generate LD between QTL and markers following the study of Henryon *et al.* (2019). The genome was 30 Morgan and consisted of 18 pairs of autosomal chromosomes; each chromosome was 167 cM long. A total of 7,702 QTLs and 54,218 markers were located across the genome and were all segregating in generation $t = -1$. An additive effect of every mutant allele at each QTL followed an exponential distribution. No major QTL was simulated. Markers were distinct from QTL and were used to form genomic relationships in TGOCS and GOCS. A total of 6,012 founder loci were placed evenly across the genome in the base population (generation=0). These founder loci were not used in establishing genomic relationships.

Optimum-contribution selection. POCS was carried out by maximising $\boldsymbol{U_t(c)} = \boldsymbol{c'a} - \omega\, \boldsymbol{c'A\,c}$, where c is a vector of genetic contributions to the next generation, a is a vector of TBV, ω is a penalty applied to the average estimated relationship of the next generation, and A is a pedigree relationship matrix (Henryon *et al.* 2019). The penalty, ω, was constant across generations. It was calibrated to realise approximately 1.00 ΔG_{true} in all scenarios. GOCS was carried out by replacing A with a genomic-relationship matrix, G, which was calculated as $\frac{\boldsymbol{ZZ'}}{\sqrt{2\boldsymbol{p}'(1-\boldsymbol{p})}}$, where Z $=\boldsymbol{M} - \boldsymbol{1}(2\boldsymbol{p})'$and M is a matrix of count of mutant alleles with element M_{ij}=0, 1 or 2 for each animal at each marker. Allele frequency p, was calculated using all OCS candidates in the generation under selection for GOCS_base and TGOCS_base while the p was set to 0.5 for centering in TGOCS_0.5 and GOCS_0.5.

Data analysis. The number of founder alleles, segregating QTL and number of markers below the threshold maintained in the last generation for each of the five scenarios were calculated for each replicate. ΔG_{true} was calculated as the linear regression of G_t on t, where G_t is the average TBV of animals born at generations, t=4...10 for each replicate. All results were expressed as the mean of 300 replicates.

Software. The breeding program was simulated using the software package ADAM (Pedersen *et al.* 2009) then OCS was carried out using EVA software (Berg *et al.* 2006).

RESULTS AND DISCUSSION

The results did not support the premise that TGOCS maintains more segregating QTL or founder alleles than POCS at the same rate of ΔG_{true}. Results showed that POCS maintained more QTL alleles and IBD alleles than TGOCS (Table 1). This makes POCS a robust method to use in animal breeding. Similar to GOCS, using marker information in TGOCS does not help to maintain more alleles in the population. In addition, TGOCS_0.5 maintained significantly fewer (66% and 8%) founder alleles and (30% and 2%) segregating QTL than TGOCS_base and POCS (Table 1). To the best of our knowledge, the proposed method TGOCS has not been investigated while GOCS has been investigated previously (e.g. Sonesson *et al.* 2012; Henryon *et al.* 2019). Results show that TGOCS maintained significantly fewer founder alleles and segregating QTL than GOCS (Table 1). Consequently, TGOCS was inferior to both GOCS and POCS. GOCS was also inferior to POCS in this study, which is supported by the results found in the study of Sonesson *et al.* (2012) and Henryon *et al.* (2019). Therefore, POCS remains the worthy method to maintain more QTL alleles and founder alleles in the population.

Table 1. Numbers (N) of founder, QTL or markers alleles maintained in the last generation (standard errors) realised by scenarios of alternative optimum-contribution selection (OCS) at the same rate of true genetic gain

OCS scenarios	N founder alleles	N favourable QTL alleles	N marker alleles with MAF<0.025	N marker alleles with MAF>0.025
POCS	20.19 (0.03)	2617.17 (2.01)	3221.15 (10.48)	32840.08 (17.08)
TGOCS_0.5	6.78 (0.04)	1825.11 (5.24)	2129.75 (20.82)	24039.93 (57.37)
TGOCS_base	18.54 (0.13)	2557.70 (5.72)	4580.87 (36.17)	30779.85 (106.63)
GOCS_0.5	17.25 (0.04)	2541.87 (2.12)	2581.27 (10.78)	32509.65 (17.75)
GOCS_base	19.59 (0.09)	2596.90 (2.50)	3266.68 (10.69)	32565.85 (20.69)

Abbreviations: POCS: Optimal contribution selection (OCS) based on pedigree relationships; GOCS: OCS with genomic relationships using all marker loci; TOCS: OCS with GOCS using marker loci with low minor allele frequency (MAF) below a predefined threshold (MAF<0.025). Allele frequencies were set either to 0.5 (TGOCS_0.5/GOCS_0.5) or base population allele frequency (TGOCS_base/GOCS_base).

TGOCS_base did not maintain more IBD and favourable QTL alleles than GOCS_base. The reason could be that we simulated very small populations and LD between markers and QTL. Even if we used only a subset of markers having MAF below 0.025, still there are enough markers. Therefore, TGOCS_base could not overcome LD between markers and QTL. If we would simulate more matings, we believe that TGOCS_base and GOCS_base would produce similar results. TGOCS_0.5 also could not maintain more minor alleles than GOCS by attempting to increase allele frequency towards 0.5 at markers with low MAF. A possible reason could be that only 25 matings were simulated in this study. It had less flexibility to move allele frequency of all markers towards 0.5. However, simulating more matings might not help because allele frequency towards 0.5 is suboptimal when genetic gain is concerned. So, by giving more weight to markers with low MAF, TGOCS_0.5 ultimately lost more markers which consequently lost more QTL alleles. Therefore, TGOCS_0.5 maintained fewer favourable segregating QTL alleles than TGOCS_base. It indicates that it is difficult to maintain more genetic variation by using genomic information in its current form in OCS because of LD between markers and QTL.

By contrast, POCS can manage the expected genetic drift without restricting selection at QTLs (Henryon *et al.*, 2019). Since POCS does not depend on the markers, POCS can increase the allele frequency at some favourable QTL without much affecting allele frequency at other QTLs. Thus, POCS could maintain more favourable QTL alleles in the population than TGOCS. Since QTL and founder alleles are in LD, POCS maintained more founder alleles than TGOCS at the same ΔG_{true}. So, genomic information used in TGOCS in its current form could not help maintain more QTL and founder alleles than POCS at the same ΔG_{true}. This study gives more insight into the underlying mechanisms of why use of pedigree relationships in OCS is superior to using genomic relationships in OCS to maintain genetic variation in the population. However, this study assessed genetic variation across the whole genome but controlling genetic diversity in specific regions of genome might also be of interest. Research should be conducted how genomic relationships can be used to control genetic diversity in different regions of the genomes while maintaining rate of true genetic gain in the population.

ACKNOWLEDGEMENTS

M. Sharif-Islam is supported by an Animal Genetics and Breeding Unit Scholarship at the University of New England, Australia, and a PhD top-up scholarship from the Australasian Pork Research Institute Ltd (APRIL).

REFERENCES

Henryon M., Liu H., Berg P., Su G., Nielsen H.M., Gebregiwergis G.T. and Sorensen A.C. (2019) *Genet. Sel. Evol.* 51: 39.

Sonesson A.K., Woolliams J.A. and Meuwissen T.H.E. (2012) *Genet. Sel. Evol.* 44: 27.

Pedersen L.D., Sorensen A.C., Henryon M., Ansari-Mahyari S. and Berg P (2009) *Livestock Sci.* 121: 343.

Berg P., Nielsen J. and Sorensen M.K., (2006) *WCGALP* 27-09.

Forni S., Aguilar I. and Misztal I. (2011) *Genet. Sel. Evol.* 43: 1

Meuwissen T.H.E., Sonesson A.K., Gebreyohans G. and Woollliams J.A. (2020). *Front. Genet.* 11: 880.

DETECTION OF SIGNATURES OF SELECTION IN AUSTRALIAN BEEF CATTLE

H. Aliloo[1], B.J. Walmsley[2,3], K.A. Donoghue[4] and S.A. Clark[1]

[1]School of Environmental and Rural Science, University of New England, Armidale, NSW 2351 Australia
[2] Animal Genetics Breeding Unit*, University of New England, Armidale, NSW, 2351 Australia
[3]NSW Department of Primary Industries, Livestock Industries Centre, Armidale, NSW, 2351 Australia
[4]NSW Department of Primary Industries, Agricultural Research Centre, Trangie, NSW, 2823 Australia

SUMMARY

The 50K genotypes of 2,935 animals from the 5 most common temperate beef breeds in Australia were used to identify genomic footprints of selection based on fixation index (F_{ST}). A principal component analysis on genomic relationships between all individuals showed that Angus, Hereford and Wagyu are the most genetically differentiated breeds. Therefore, 3 pairwise F_{ST} comparisons were implemented between Angus vs. Wagyu, Angus vs. Hereford and Hereford vs. Wagyu. Genome-wide comparison of patterns of the F_{ST} values revealed 14 candidate regions under selection on chromosomes 2:6, 8, 12, 13, 20, and 24. Several of the identified candidate regions in this study have been previously reported for different economically important traits in beef cattle. In addition, our identified candidate regions for signatures of selection harboured genes in several enriched annotation clusters. If validated, the results from this study can be incorporated in genomic selection of the Australian beef cattle population.

INTRODUCTION

Understanding the genetic architecture of productivity is necessary for designing efficient breeding programs. Intensive artificial selection to increase profitability in Australian beef breeds has generated distinctive patterns at specific regions of their genome, referred to as signatures of selection (SoS). The identification of SoS may help to uncover genes and biological mechanisms responsible for breed differences in the Australian beef cattle population.

A simple, yet effective, approach to identify SoS is to compare differences between breeds in allele frequencies of their genome-wide single-nucleotide polymorphisms (SNPs) based on the fixation index (F_{ST}). A high F_{ST} value indicates large differences between the breeds of interests resulted from distinctive selection pressures. Therefore, the comparison of genome-wide patterns of F_{ST} values can help to map genomic regions contributing to the phenotype differences between Australian beef cattle breeds.

The Southern Multibreed (SMB) project has generated genomic data across the 5 most common temperate beef breeds in Australia including Angus, Charolais, Hereford, Shorthorn and Wagyu (Walmsley *et al.* 2021). The aim of this study was to use the genotypes collected in the SMB project to detect SoS in temperate Australian beef breeds using the F_{ST} method.

MATERIALS AND METHODS

Data. The genotypes of 2,938 animals were obtained by Zoetis ZBU medium density 50K (Zoetis, Kalamazoo, MI). The genotype calls with a score of <0.15 were assumed as missing (Edriss *et al.* 2013). Further quality control was undertaken using PLINK 1.9 (Chang *et al.* 2015) to remove

* A joint venture of NSW Department of Primary Industries and University of New England.

SNPs and animals with a call rate lower than 90%, SNPs that were monomorphic across all animals and those located on sex chromosomes (X and Y). Finally, 47,264 SNPs and 2,935 animals including 845 Angus, 493 Charolais, 495 Hereford, 623 Shorthorn and 479 Wagyu cattle were used in this study. The mapping information for all markers was available on the basis of ARS-UCD 1.2 bovine genome assembly.

Data Analysis. To investigate the population structure of different beef breeds, a principal component (PC) analysis based on a genomic relationship matrix (GRM) constructed using GCTA 1.94.1 (Yang *et al.* 2011) was implemented. The first and second PCs were plotted to visualize the distribution and explore the relationships among different beef cattle breeds.

The F_{ST} values were calculated by comparing the allele frequencies of pairwise SNPs between the breeds that showed the highest genetic differentiation based on the first two PCs. PLINK 1.9 (Chang *et al.* 2015) was used to calculate F_{ST} values according to the Weir and Cockerham (1984) method. To reduce the noise in estimates, and to account for linkage disequilibrium between adjacent SNPs, the '*runmed*' R function was used to smooth F_{ST} values across a moving window of 75 markers within each chromosome (Haerdle and Steiger 1995). The SNPs with smoothed F_{ST} values that were greater than 3 times the standard deviation from the mean of all smoothed F_{ST} values (the suggestive threshold) were deemed as being under selection pressure. A candidate region for SoS was defined by first identifying SNPs under selection and then searching within the 500-Kbp interval downstream and upstream (1 Mbp window) of the identified SNP for SNPs that passed the suggestive threshold. The detected region (with a 500-Kbp step size) was extended until there was no SNP with an F_{ST} value greater than the suggestive thresholds within the 500-Kbp interval from the last identified SNP. The boundaries of the candidate region were determined based on the base pair positions of the last-identified SNP in each direction. To visualize the distribution of F_{ST} values across the genome, Manhattan plots were created using the qqman 0.1.4 (Turner 2014) R package. The cattle Quantitative Trait Loci (QTL) database (https://www.animalgenome.org/cgi-bin/QTLdb/BT/index) was used to compare our identified candidate regions to literature. The candidate regions were further investigated for identification of genes residing in them using the biomaRt 2.46.3 (Durinck *et al.* 2009) R package. The identified genes were compared to the whole bovine genome background using functional annotation clustering by DAVID 2021 online bioinformatics resource (Huang *et al.* 2009) to find the biological pathways that are significantly overrepresented.

RESULTS AND DISCUSSION

Figure 1 illustrates that all animals were clearly clustered within their respective breed based on the first two PCs. The PC1 explained around 7.5% of total variation in the GRM and separated Angus and Wagyu breeds from other breeds, while PC2 explained around 5.5% of variation and showed Hereford is genetically more different to Angus and Wagyu than to the other breeds. The Shorthorn and Charolais seemed to be genetically closer to each other based on both PC1 and PC2.

Based on the results from the PC analysis, F_{ST} values were calculated between Angus vs. Wagyu (AW), Angus vs. Hereford (AH) and Hereford vs. Wagyu (HW). The averages of raw F_{ST} values were 0.15, 0.11, 0.16 from the AW, AH and HW comparisons, respectively. This showed that Angus and Hereford are genetically more similar than either is to Wagyu.

The distribution of genome-wide F_{ST} values for the 3 pairwise comparisons are shown in Figure 2. In total, 14 candidate regions on *Bos taurus* autosomes (BTA) 2:6, 8, 12, 13, 20, and 24 were found (Table 1). Here, we only focus on candidate regions that overlapped with previously reported regions in the literature from beef cattle QTL and association studies.

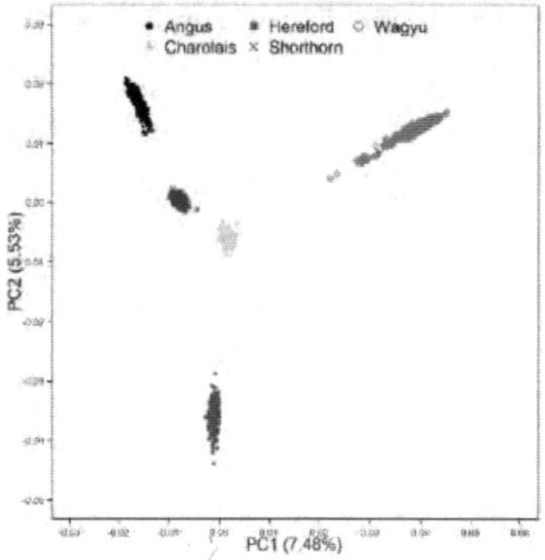

Figure 1. Plot of Principal Component 1 (PC1) vs. PC2 for 5 Australian beef cattle breeds

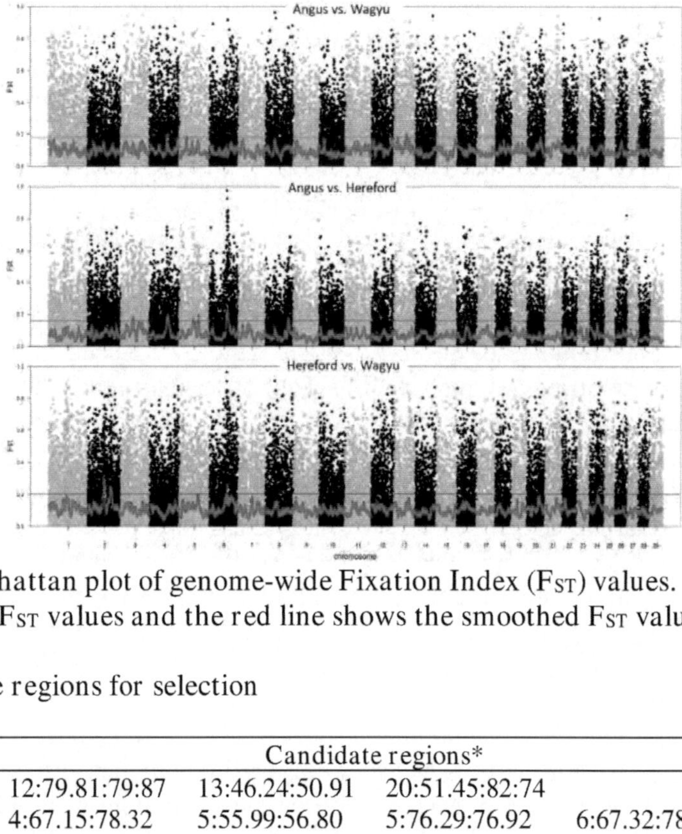

Figure 2. The Manhattan plot of genome-wide Fixation Index (F_{ST}) values. The black and grey dots show the raw F_{ST} values and the red line shows the smoothed F_{ST} values

Table 1. Candidate regions for selection

	Candidate regions*					
AW	4:69.96:74:30	12:79.81:79:87	13:46.24:50.91	20:51.45:82:74		
AH	3:49.54:54.27	4:67.15:78.32	5:55.99:56.80	5:76.29:76.92	6:67.32:78.89	8:90.69:95:74
HW	2:62.71:69:75	2:89.65:98.45	6:71.42:72.94	24:33.36:34.58		

*Chromosome:Start(Mbp):End(Mbp);

Several candidate regions found in this study have been previously reported for different economically important traits in beef cattle. One candidate region on BTA 4 overlapped between AW and AH and another candidate region on BTA 6 overlapped between AH and HW comparisons. The candidate regions on BTA 4 have been reported to contain QTLs for feed intake (Lu *et al.* 2013) and body weight (Seabury *et al.* 2017) traits in Angus and Hereford beef breeds. The candidate regions on BTA 6 intersected with regions reported to be associated with meat quality (Mateescu *et al.* 2017) and body weight (Lu *et al.* 2013) traits in Angus cattle. The candidate regions on BTA 2 from the HW comparison have been found by Snelling *et al.* (2010) to harbour variations affecting body weight in a crossbred population of different beef breeds including Hereford. A candidate region on BTA 5 between 55.99 to 56.80 Mbp from the AH comparison was also found that comprises several important genes, e.g. *INHBC*, *INHBE* and *PTGES3*, that are involved in growth and metabolism in humans. Another candidate region on BTA 8 found in this study has been associated with feed intake (Rolf *et al.* 2012) and intramuscular fat content (Bolormaa *et al.* 2011) in Angus and Hereford cattle breeds. Mateescu *et al.* (2017) performed a genome-wide association study for meat quality traits in Angus cattle and found significant associations within the candidate region on BTA 13 found here from the AW comparison.

The candidate regions in Table 1 together encompassed 40, 197 and 91 cattle genes from the AW, AH and HW comparisons, respectively. The functional annotation clustering of the identified genes resulted in 3, 25 and 15 annotation clusters from the AW, AH and HW comparison of which only 7 clusters from the AH comparison and 2 clusters from the HW comparison were significantly enriched (enrichment score $\geq$ 1.3). These enriched annotation terms are associated with some biological functions e.g. embryonic skeletal system morphogenesis and protein functional domains e.g. Insulin-like growth factor-binding proteins.

CONCLUSIONS

Genome-wide screening of F_{ST} patterns provides a straightforward method to identify genomic regions under selection. Although the results here need to be validated, several candidate regions were found that may be involved in genetic differentiation between Angus, Hereford and Wagyu cattle and could explain phenotypic differences among these breeds. The candidate regions found in this study largely overlap with previously reported regions for economically important traits in beef cattle and might be useful to be incorporated in future genomic selection of Australian beef cattle.

REFERENCES

Bolormaa S., Neto L.R., Zhang Y.D., Bunch R.J., Harrison B.E., Goddard M.E. and Barendse W. (2011) *J. Anim. Sci.* 89: 2297.

Chang C.C., Chow C.C., Tellier L.C., Vattikuti S., Purcell S.M. and Lee J.J. (2015) *GigaScience.* 4: s13742.

Durinck S., Spellman P., Birney E. and Huber W. (2009). *Nat. Protoc.* 4: 1184.

Edriss V., Guldbrandtsen B., Lund M.S. and Su G. (2013) *J. Anim. Breed. Genet.* 130:128.

Haerdle, W. and Steiger, W. (1995). *J. R. Stat. Soc. Ser. C. Appl. Stat.* 44: 258.

Huang, D.W., Sherman, B.T. and Lempicki, R.A. (2009). *Nat. Protoc.* 4: 44.

Lu D., Miller S., Sargolzaei M., Kelly M., Vander Voort G., Caldwell T., Wang Z., Plastow G. and Moore S. (2013) *J. Anim. Sci.* 91: 3612.

Mateescu R.G., Garrick D.J. and Reecy J.M. (2017) *Front. Genet.* 8: 171.

Rolf M.M., Taylor J.F Schnabel R.D McKay S.D McClure M.C Northcutt S.L., Kerley M.S. and Weaber R.L. (2012) *Anim. Genet.* 43: 367.

Seabury C.M., Oldeschulte D.L., Saatchi M., Beever J.E., Decker J.E., Halley Y.A., Bhattarai E.K., Molaei M., Freetly H.C., Hansen S.L., Yampara-Iquise H., Johnson K.A., Kerley M.S., Kim J., Loy D.D., Marques E., Neibergs H.L., Schnabel R.D., Shike D.W., Spangler M.L., Weaber R.L., Garrick D.J. and Taylor J.F. (2017) *BMC Genomics.* 18: 386.

Snelling W.M., Allan M.F., Keele J.W., Kuehn L.A., McDaneld T., Smith T.P.L., Sonstegard T.S., Thallman R.M. and Bennett G.L. (2010) *J. Anim. Sci.* 88: 837.

Turner S.D. (2014) *J. Open Source Software.* 3: 731.

Walmsley B.J., Donoghue K.A., Johnston D.J., Clark S.A., Siddell J.P., Walkom S.F., Granleese T. and Arthur PF. (2021) *Proc. Assoc. Advmt. Anim. Breed. Genet.* 24: 423.

Weir, B.S. and Cockerham, C.C. (1984). *Evolution.* 38: 1358.

Yang, J., Lee S.H., Goddard M.E. and Visscher P.M. (2011) *Am. J. Hum. Genet.* 88: 76.

Proc. Assoc. Advmt. Anim. Breed. Genet. 25: 198 - 201

DEMONSTRATING DIFFERENCES IN SURVIVAL AND FERTILITY OF HIGH AND LOW GENETIC MERIT COWS WITHIN AND ACROSS HERDS

J.E. Newton[1,2], M. Haile-Mariam[1,2] and J.E. Pryce[1,2]

[1]Agriculture Victoria, AgriBio, Centre for AgriBioscience, Bundoora, Victoria
[2]School of Applied Systems Biology, La Trobe University, Bundoora, Victoria 3083, Australia

SUMMARY

Developing value propositions that resonate with farmers can improve on-farm adoption of genetic tools. Our aim was to illustrate differences in the performance of high and low genetic merit cows in a way that was understandable to farmers and service providers. Cows (n=10,734) with lifetime performance data from 29 herds were ranked into quartiles within herd-year contemporary groups based on parent-average derived genetic merit for the Balanced Performance Index (BPI), a multi-trait index that incorporates traits contributing to farm profit. Chi-squared tests were conducted within and across herd comparing genetic merit and lifetime number of calves and presence in the herd (present =1 or absent =0) after 100 days, 12, 18 and 24 months. On average, 5.3% and 6.1% more high BPI cows (top 25%) remained in the herd after 18 and 24 months, respectively, compared to low BPI cows (bottom 25%). Over one quarter of low BPI cows only had one calf, while 55% of high BPI cows had 3 or more calves. These differences were significant ($p<0.05$) in the across-herd analyses, but few (≥20%) of the within-herd analyses. Average differences in fertility and survival breeding values of high and low BPI cows were small (<1 standard deviation). This, coupled with a small sample size for within-herd analyses limited the ability to detect differences from within-herd analyses. This study demonstrates how selection on the BPI leads to favourable responses in health and fertility, in a way that is easy for farmers to understand. Demonstrating and detecting these benefits at the individual farm level remains challenging. Studies like this one that use datasets that are representative of the information farmers are using in decision making are important to help develop meaningful case studies to support extension and engagement efforts.

INTRODUCTION

Adoption of genetic and genomic tools can be facilitated through the development of value propositions that resonate with farmers. Previous studies (i.e., Ramsbottom *et al.* (2012)) that sought to demonstrate the link between genetic potential and performance present data collated from many farms. While this approach is valuable, farmers prefer localised, region-specific examples (Nettle *et al.* 2010). The long period between investment and impact on-farm and the fact that differences in traits expressed over a lifetime (i.e., survival, fertility) cannot easily be visualised add to the complexity of demonstrating the value of genetic and genomic tools. This study has focused on detecting differences in survival and fertility as these are traits of key importance in dairy production systems. The aim of this study was to quantify differences in the performance of high and low genetic merit cows in datasets representative of the information available on-farm.

MATERIALS AND METHODS

Cow performance, pedigree and breeding value (EBV) records were extracted for 29 dairy herds for a 10-year period from the national database housed by DataGene, first described in Newton *et al.* (2017). Updated EBVs (from May 2021 genetic evaluation) were extracted from the national database to incorporate updates to individual EBVs and the Balanced Performance Index, BPI, the Australian dairy industry's national multi-trait selection index (Axford *et al.* 2021). Only cows with records for their entire productive life as well as parentage recorded were retained for analysis. After removal of cows with incomplete records, 10,734 cows remained. Average herd size was 370 cows but ranged from 122 to 1,320 cows. Lifetime number of calves, and herd persistence, defined as present (1) or absent (0), 100 days, 12, 18 or 24 months after joining the milking herd was calculated for all cows. To reduce bias from cow's own information in EBVs, parent average EBVs were derived from multi-breed models facilitating across-breed analyses. Cows were grouped into quartiles within herd-year contemporary groups based on their genetic merit for the Balanced Performance Index (BPI) (n=10,734), survival (n=10,488) and fertility EBVs (n=10,459). Quartile 1 (Q1) contained cows ranked in the bottom 25%, quartile 2 (Q2) contained cows ranked in the 26th to 50th percentiles, quartile 3 (Q3) contained cows ranked in the 51st to 75th percentiles and quartile 4 (Q4) contained cows ranked top 25%. Genetic merit was treated as a categorical variable with 4 levels. Chi-squared tests were conducted within and across herds to test if there was a relationship between genetic merit and lifetime number of calves or herd persistence (present =1 or absent =0) after 100 days, 12, 18 and 24 months.

RESULTS AND DISCUSSION

Chi-squared tests detected statistically significant associations between genetic merit and number of calves and herd persistence in across-herd analyses but few within-herd analyses. On average, 5.3% and 6.1% more high BPI cows (Q4) remained in the herd after 18 and 24 months respectively, compared to low BPI cows (Q1) (Table 1). These differences were significant at both 18 (χ^2 = 18.22, df = 3, P <0.001) and 24 (χ^2 = 22.12, df = 3, p <0.001) months in the across-herd analyses. However, for the within-herd analyses, the association between herd persistence at 18 and 24 months and the BPI was only significant in 4 herds and 3 herds, respectively. When cows were grouped on Survival EBV, significantly (P <0.001) more Q4 cows remained in the herd at 12, 18 and 24 months; 5.2%, 8.9% and 9.8% more compared to Q1 cows, respectively. Although this was larger than grouping cows on BPI, significant differences in herd persistence were only found in 5 within herd analyses. After 100 days, differences in herd persistence were ~1% across quartiles with no statistical differences found in the across or within herd analyses.

Table 1. Percentage (and number) of cows present or absent 100 days, 12, 18 and 24 months after entering milking herd where cows were grouped into quartiles based on BPI and survival breeding value (EBV); bottom 25% (Q1), 26-50% (Q2), 51%-75% (Q3), top 25% (Q4)[1]

Quartile	100 days		12 months		18 months		24 months	
	absent	present	absent	present	absent	present	absent	present
Grouping based on BPI								
Q1	3.8 (102)	96.2 (2608)	21.8 (590)	78.2 (2120)	34.1 (923)	65.9 (1787)	45.6 (1236)	54.4 (1474)
Q2	4.1 (107)	95.9 (2481)	21.1 (546)	78.9 (2042)	31.8 (822)	68.2 (1766)	43.0 (1112)	57.0 (1476)
Q3	3.6 (95)	96.4 (2562)	20.4 (542)	79.6 (2115)	30.9 (820)	69.1 (1837)	41.5 (1103)	58.5 (1554)
Q4	3.1 (86)	96.9 (2693)	19.2 (533)	80.8 (2246)	28.8 (800)	71.2 (1979)	39.5 (1098)	60.5 (1681)
Across	χ^2= 4.32, df=3, p=0.23		χ^2=6.16, df=3, p=0.10		χ^2=18.22, df=3, p<0.001		χ^2=22.12, df=3, p<0.001	
Within	0 significant herds		0 significant herds		4 significant herds		3 significant herds	
Grouping based on survival EBV								
Q1	4.3 (95)	95.7 (2140)	22.6 (506)	77.4 (1729)	35.9 (803)	64.1 (1432)	47.8 (1069)	52.2 (1166)
Q2	3.6 (86)	96.4 (2329)	21.8 (526)	78.2 (1889)	32.4 (783)	67.6 (1632)	43.7 (1056)	56.3 (1359)
Q3	3.8 (102)	96.2 (2606)	21.6 (584)	78.4 (2124)	32.0 (867)	68.0 (1841)	42.1 (1141)	57.9 (1567)
Q4	3.1 (98)	96.9 (3032)	17.4 (546)	82.6 (2584)	27.0 (846)	73.0 (2284)	38.1 (1191)	61.9 (1939)
Across	χ^2= 4.86, df=3, p=0.18		χ^2= 28.33, df=3, p<0.001		χ^2= 50.6, df=3, p<0.001		χ^2= 52.97, df=3, p<0.001	
Within	1 significant herd		1 significant herd		5 significant herds		4 significant herds	

[1]Chi-squared test of genetic merit and herd persistence reported across and within herds

Around 26% of low BPI and low fertility EBV cows only had 1 calf (Table 2). In contrast, 54.8% of high BPI cows and 56.1% of high fertility EBV cows had 3 or more calves. As genetic merit increased, the proportion of cows who had 1 or 2 calves decreased and the proportion of cows who had 3 or more calves increased. These differences were significant in the across-herd analyses (BPI $\chi^2 = 38.2$, df = 6, p = <0.001; Fertility EBV $\chi^2 = 60.4$, df = 6, p = <0.001). In the within-herd analyses these differences were significant in 1 and 6 herds when cows were grouped on BPI and fertility EBV, respectively.

Table 2. Percentage (and number) of cows who have 1, 2, or 3+ calves over their lifetime where cows were grouped into quartiles based on BPI and fertility breeding value (EBV); bottom 25% (Q1), 26-50% (Q2), 51%-75% (Q3), top 25% (Q4)

Quartile	BPI			Fertility EBV		
	1	2	3+	1	2	3+
Q1	25.8 (700)	27.1 (735)	47.0 (1275)	26.1 (614)	27.8 (654)	46.2 (1087)
Q2	24.0 (622)	25.7 (666)	50.2 (1300)	23.6 (584)	27.7 (684)	48.7 (1205)
Q3	23.1 (614)	25.9 (688)	51.0 (1355)	22.9 (598)	25.7 (671)	51.5 (1346)
Q4	20.3 (563)	24.9 (693)	54.8 (1523)	20.6 (622)	23.3 (702)	56.1 (1692)

As the BPI places substantial weight on fertility and survival (Axford *et al.* 2021) it was expected high BPI cows would be more fertile and last longer. The statistically significant relationship between the BPI and lifetime number of calves and herd persistence in the across-herd analyses undertaken in this present study support this. Although there were few significant differences in the within-herd analyses, high BPI cows had more calves in 90% of herds (26/29) and had a higher proportion of animals present after 18 and 24 months in 76% of herds (22/29) which supports the across-herd analyses. These findings also align with our earlier studies (Newton *et al.* 2017; Newton *et al.* 2018), which used a previous iteration of the BPI. Here we also found few herds (<8%) had significant differences in fertility, expressed as number of calves/cow/year or calving interval. These previous studies analysed performance at the individual cow level whereas this present study was conducted at the herd level.

We have previously focused on productive life, the total length of time an animal remained in the milking herd as a key measure of survival (i.e. Newton *et al.* 2018). While this approach was able to consistently illustrate that high BPI cows lasted longer in the herd overall, survival is multi-faceted. The probability of culling due to infertility is high in early parities (Workie *et al.* 2021) so getting heifers back in calf for their second lactation can be a challenge on-farm. The measures of herd persistence, which used the binary definition of present or absent in the herd at a series of time points, chosen in consultation with industry were designed to test this. This approach successfully showed that improved survival of high BPI cows (and high survival EBV cows) begins to be expressed during the period of getting cows back in calf for the second lactation (12-18 months). As one of the barriers to uptake of genetics is the long period between investment and impact on-farm (Axford *et al.* 2015), the ability to show that the impact of genetic selection for survival can be seen as soon as 12 months after entry to the milking herd will be particularly helpful in extension activities.

Few statistically significant differences in herd persistence or number of calves were found in within-herd analysis in this current study. While there was 2.3 standard deviation units difference in BPI between Q1 and Q4 cows, average differences in fertility and survival EBVs were small, 0.6 and 1.1 standard deviation units, respectively. This may have limited the ability to detect differences from within-herd analyses. Herds where significant differences were detected, were generally larger than the average herd size and had above average variation in fertility or survival EBVs. This suggests that a small sample size, coupled with lower within-herd and within-herd-year contemporary group variation in fertility and survival EBVs limits the ability to detect significant differences at the individual herd level. Encouraging accurate reporting of reproduction and health events on-farm and ensuring pipelines exist to facilitate transfer of data captured on-farm into centralised databases will not only improve routine genetic evaluation but also improve our ability to illustrate the impact of genetics in farm businesses. Developing case studies on the value of genetics that are 1) scientifically robust, 2) accessible to non-scientific audiences, 3) use easily accessible data and 4) that align to farmer preferences for region- (or farm-) specific case studies remain a challenge. By working collaboratively with service providers and farmers and seeking iterative feedback, it is possible to develop case studies that align with this.

CONCLUSION

This study aimed to illustrate differences in the performance of high and low BPI cows in a way that was understandable to farmers and service providers. We focused on survival and fertility measures, as it has previously been found to be particularly challenging to detect and demonstrate the impact of genetic improvement on these traits. We demonstrated how selection on the multi-trait index, BPI, leads to favourable responses in survival and fertility. Significant differences in lifetime number of calves and herd persistence after 18 and 24 months were seen in across-herd analyses. Given the lag between investment in genetics and impact is a barrier to uptake of genetics, the ability to show genetic merit impacting survival as soon as 12 months after entry to the milking herd will be particularly helpful in extension activities. Detecting differences within-herd was made difficult by small sample size and low variation within contemporary groups. Trends in the within herd analyses were in support of across-herd analyses though. Illustrating significant differences at an across-herd level and showing similar trends at within-herd level can support the development of case studies that are scientifically robust, but also met farmer need for localised, regionally specific case studies.

ACKNOWLEDGEMENTS

This study was undertaken as part of the DairyBio research program, which is jointly funded by Dairy Australia (Melbourne, Australia), the Gardiner Foundation (Melbourne, Australia), and Agriculture Victoria (Melbourne, Australia). The authors acknowledge the farmers who contributed their data to this study.

REFERENCES

Axford M., Santos B., Stachowicz K., Quinton C., Pryce J.E. and Amer P. (2021) *Anim. Prod. Sci.* 61: 1940.

Axford M., Williams P.W., Abernethy D.P., Nieuwhof G.J. (2015) *Proc. Assoc. Advmt. Anim. Breed. Genet.* 21: 229.

Nettle R., Paine M. and Penry J. (2010) *Anim. Prod. Sci.* 50: 429.

Newton J.E., Goddard M.E., Phuong H.N., Axford M.A., Ho C.K.M., Nelson N.C., Waterman C.F., Hayes B.J., and Pryce J.E. (2017) *Proc. Assoc. Advmt. Anim. Breed. Genet.* 23:19.

Newton J.E., Phuong H.N., Axford M., Hayes B.J., Oakes E.M., and Pryce J.E. (2018) *WCGALP* 11: 381.

Ramsbottom G., Cromie A.R., Horan B.and Berry D.P. (2012) *Anim.* 6: 1031.

Workie Z.W., Gibson J.P. and van der Werf J.H.J (2021) *Anim. Prod. Sci.* 61: 680.

POPULATION SCALE LONG-READ SEQUENCE DATABASES: ARE THEY USEFUL FOR ACCURATE SNP AND INDEL DISCOVERY?

I.M. MacLeod[1,2], T.V. Nguyen[1], J. Wang[1], C.J. Vander Jagt[1] and A.J. Chamberlain[1,2]

[1] Agriculture Victoria, Centre for AgriBioscience, Bundoora, VIC, 3083 Australia
[2] School of Applied Systems Biology, La Trobe University, Bundoora, VIC, 3083 Australia

SUMMARY

Several animal industries, including cattle, have built population scale whole-genome reference databases of genetic variants, SNPs and small INDELs, that have been discovered using short-read sequencing. These databases have proved invaluable: enabling development of genetic tools to breed healthier and more productive animals. However, while accurate and cost effective, short-read sequencing is not well suited to the discovery of larger genetic variants called structural variants (defined as > 50 base pairs in length). Thus, there is interest in creating population scale long-read databases for structural variant discovery and downstream applications. Ideally, for cost efficiencies, these would also contribute to the sequence database of SNPs and INDELs and enable imputation of all variants. Therefore, we explored the effect of long-read coverage on accuracy of SNP and INDEL discovery compared to a truth set from short-read sequence. The results show that at all read depths, recall and precision of SNP was considerably higher than for INDEL. At ≥ 10X read depth, SNP recall was 0.95 and reached 0.99 at 50X cover. The precision for SNPs and particularly INDELs suggested that the long-read variant calls included a relatively high, but likely overestimated proportion of false positives. We conclude that SNP and INDEL discovery in long-read data is useful, particularly if extensive 'truth' variant sets exist that could help remove false positives.

INTRODUCTION

Several animal industries, including cattle, have built population scale whole-genome reference databases of small genetic variants (SNPs, and INDELs < 50 base pairs) that have been discovered using short-read sequencing (Daetwyler *et al.* 2017). These databases have proved invaluable for the detection of recessive deleterious mutations, for sequence imputation and enabling the development of genetic tools to breed healthier and more productive animals. However, while short-read sequencing is highly cost effective and accurate for SNP and INDEL discovery, it is not well suited to the discovery of larger genetic variants (> 50bp in length) called structural variants (SVs). Instead, long-read sequencing is much better suited to genome-wide SV discovery. Limited research in livestock, and experience from human genetics research suggests that SVs may often have large effects on both mendelian and quantitative traits (reviewed by Nguyen *et al.* 2023a).

Until recently, two major deterrents for long-read sequencing have been the higher cost and lower per base accuracy, where the latter resulted in low quality SNP and INDEL calls compared to short-read sequencing. However, two key competitors in the field of long-read sequencing, Pacific Biosciences (PacBio) and Oxford Nanopore Technologies (ONT), have made significant improvements in both per base accuracy and cost. Thus, there is now considerable interest in exploring the SV landscape at a population scale in cattle (Chamberlain *et al.* 2023) and potentially other livestock. For livestock studies, it is critical to consider how to reduce costs per individual without unduly compromising on the accuracy of variant discovery. The sequencing read depth is a key factor regulating cost, and Nguyen *et al.* (2023b) have used ONT long-read sequencing to explore the impact of read depth on the accuracy of SV discovery. Additionally, to maximise the cost effectiveness of long-read sequencing and to enable SV imputation, it is desirable to use these same sequences to develop new, or expand existing, whole-genome SNP and INDEL databases. Therefore, the aim of this paper was to explore the accuracy of SNP and INDEL discovery in long-read sequencing at a range of read depths. Additionally, the paper considers the impact of incomplete discovery of these variants for population scale studies or smaller scale studies of recessive deleterious mutations.

MATERIALS AND METHODS

Three Holstein animals were each sequenced at approximately 50X coverage using an ONT PromethION sequencer, with flow cell 9.4.1 and ligation kit LSK110. To achieve maximum accuracy, the bases were re-called using Guppy v6.1.7 with the 'super high accuracy' setting (SUP). The output FASTQ files were trimmed using Filtlong (default settings: https://github.com/rrwick/Filtlong). Filtered reads were mapped to the ARS-UCD 1.2 reference genome (Rosen *et al.* 2020) using Minimap2 (Li 2018). Clair3 software was used to call SNPs and INDELs in individual sequences (default settings: Zheng *et al* 2022) and for comparison, Longshot software was also used to call SNPs (default settings: Edge and Bansal 2019).

Next, mapped reads at 50X coverage for each individual were subsampled using Sambamba (default settings: Tarasov *et al* 2015) to 3X, 5X, 10X, 15X and 20X coverage and the data at each read depth was processed as for the 50X coverage to re-call SNPs and INDELs. For each of the three animals, three chromosomes were chosen as technical replicates (chromosome 1, 19 and 25) to investigate the accuracy of SNP and INDEL discovery at each of these read depths. The same three animals had also been sequenced using short-read Illumina technology at approximately 12X, 15X & 18X read depth and were previously processed in Run8 of the 1000 Bull Genomes Project according to project guidelines (Daetwyler *et al.* 2017) with GATK joint variant calling according to GATK best practices (DePristo *et al.* 2011). The SNPs and INDELs discovered in the short-read data of the three animals were used as the gold standard 'truth set' of variants for comparison with the SNPs and INDELs discovered in the long-read sequencing for the same animals. To ensure a high quality truth set, we retained only biallelic variants with minor allele count of > 3, GATK Variant Quality Score Recalibration Tranche < 99.0, and indel < 50bp.

Hap.py software (https://github.com/Illumina/hap.py) was used to compare the variant truth set with the SNPs and INDELs discovered in the long-read sequencing that passed default software filters at each read depth ('query sets'). The following three sets of variants were identified from this comparison: 1) true-positive variants/genotypes (TP) that match in truth and query variant sets, 2) false-negative variants (FN) missed in the query set but present in the truth set, and 3) false-positive variants (FP) that have mismatching genotypes or alternate alleles in query versus truth set. The summary statistics calculated were; Recall = TP/(TP+FN) and Precision = TP/(TP+FP).

RESULTS AND DISCUSSION

The results were calculated for the combined truth variant sets across the three animals and three chromosomes, resulting in comparisons for a total of 1,894,775 SNPs and 158,338 INDELs at each read depth. As expected, accurate discovery of both SNPs and INDELs in long-read sequence was affected by read depth: declining more rapidly once read depth fell below 10X coverage, compared to higher read depths of 15X, 20X and 50X. The "recall" statistic (Figure 1A) indicates the proportion of variants that were discovered in the long-read data that were also in the truth set ("true positives": TP). There was excellent recall of SNPs from the long-read sequencing at 10X to 50X read coverage using Clair3 software, plateauing at between 0.95 to 0.99 (i.e. only 1 to 5% of SNPs in the truth set were not detected in the long-read sequence). Even at 5X coverage, Clair3 only missed 14% of SNPs. Longshot software showed much lower SNP recall, particularly at lower read coverage and even at 50X read depth 17% of SNPs were missed. This was expected because Longshot implements a less sophisticated variant calling approach (pileup only) compared to Clair3 which combines both pileup and full alignment in a deep learning-based variant calling algorithm (Zheng *et al.* 2022). Furthermore, Longshot is recommended for use with at least 30X read depth and it cannot be used to call INDELs. The precision of SNP discovery was very similar for both Clair3 and Longshot (Figure 1B) and suggested that the proportion of false positives among all SNPs discovered in long-read sequences was between 12 to 28%. The precision was lower than for high quality human data reported to be 0.99 at 20X coverage (Zheng *et al.* 2022). However, there are several reasons why we would expect our precision to be lower: (1) our strict filtering of variants to create the 'truth set' from the short-read data would likely result in a proportion of real SNPs and INDELs being excluded so if found in the long-read data they appear to be false positives, (2) the human field has put tremendous effort into creating high quality truth sets through the "Genome in a Bottle Consortium" with higher short-read depth (35X) (e.g. Olson *et al.* 2022) while our lower coverage short-read data likely missed some real variants, and (3) Clair3 software algorithms were trained on human data with difficult to map regions excluded. Thus, our less accurate truth set compared to the human field will inflate the estimated false positive rate and this biases downwards our estimate of precision. There is clearly a need for high accuracy truth sets in cattle for improved benchmarking.

The recall and precision for INDELs using Clair3 was much lower than for SNPs, for example, recall ranged from 0.27 at 3X to 0.89 at 50X read depth (Figure 1a). Additionally, the recall rate kept improving with increased coverage compared to the plateau observed for SNP at around 15X coverage. As mentioned above, there is likely to be some downward bias in the estimate of precision. However, even in more accurate human data the precision for INDELs at 20X coverage was lower than for SNPs at around 0.87. INDEL calls in long-read data are known to be more error prone than for short-read sequence, particularly in homopolymer regions (consecutive repeat bases) where sequencing difficulty creates false positives (Amarasinghe *et al.* 2020; Delahaye and Nicolas 2021).

The high recall rates for SNPs suggests that long-read data of at least 10X coverage is likely to be of considerable value in augmenting or developing whole-genome SNP databases at population scale. This would be convenient because the study by Nguyen *et al.* (2023b) also suggested that read depth of ≥ 10X is preferable for population scale structural variant discovery. Furthermore, if the false negative rate for SNP in long-read data is around 10% or less and is largely sporadic (i.e., there is a different set of SNPs missing in each animal) this should enable highly accurate imputation of the missing SNPs where there are reasonable sized sequence databases. We examined the distribution of missing variants in our animals at 10X read depth (Chromosome 1) and found that only 4% of missing SNPs overlapped between each pair of animals on average. However, the overlap of the missing INDEL sets was much higher than for SNPs, averaging 16% between pairs of animals at 10X coverage. Therefore, if these INDELs are missed in most or all individuals and given the higher overall missing rate of INDELs compared to SNPs, then accurate imputation would require an existing reference population with accurately genotyped INDELs. If SNPs and SVs are accurately genotyped in long-read data then it will be possible to impute SVs into large populations of cattle with SNP panel genotypes using a reference population with long-read sequences.

Although the results suggest relatively high false positive rates, if there are existing short-read databases of variants (such as the 1000 Bull Genomes project: Hayes and Daetwyler 2019) then these could be used as a filter/training set to help remove false positive SNPs and INDELs from long-read data. In the case where research may be undertaken to discover a mendelian mutation of large effect in a small cohort of animals, Nguyen *et al.* (2023b) recommend long-read sequencing at ≥ 20X coverage for high accuracy discovery of a causal SVs in the data. Thus, if the mendelian mutation might equally be a SNP or INDEL, and no short-read sequence was available on the same animals, then sequencing (≥ 20X) of parent-offspring trios would be necessary to filter putative false positive variants (particularly INDELs) that do not show mendelian inheritance (although this would remove *de novo* mutations). Although INDELs constitute around 10% of all variants in Run8, they are important. For example, in Run8 of the 1000 Bull Genomes project Variant Effect Predictor software (VEP: McLaren *et al.* 2016) annotated 0.28% of INDEL, versus only 0.01% of SNP, to have a high impact on a protein (i.e. loss of function, truncation and/or triggering nonsense mediated decay). A caveat of our study is that the ONT flow cell 9.4.1 used here for long-read sequencing is now superseded by a newer flow cell that should increase accuracy. Nonetheless, our results provide a useful benchmark, with the expectation that a range of advances will result in improved accuracy.

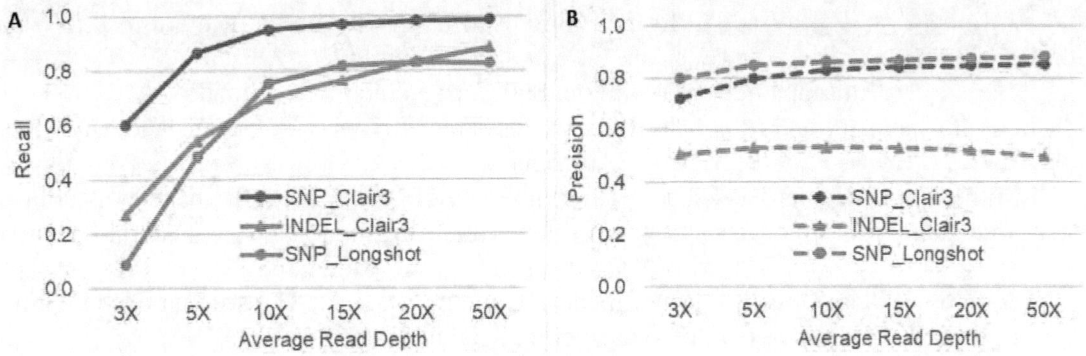

Figure 1. Recall (A) and precision (B) for SNP and INDEL discovery in long-read sequence of different read depths, using Clair3 (SNP and INDEL) and Longshot software (SNP only)

CONCLUSIONS

This study shows that with the use of existing truth sets of SNPs and INDELs, we can curate useful SNP and INDEL databases from long-read sequences. While there are some limitations particularly for small INDEL discovery in long-read sequence, it is likely that this will continue to improve with modifications in hardware, chemistry and variant calling algorithms. Also, there is a need to further develop truth sets in cattle of sequence variants for future benchmarking studies.

ACKNOWLEDGEMENTS

The authors acknowledge financial support from DairyBio, a joint venture between Agriculture Victoria, Dairy Australia and the Gardiner Foundation. We acknowledge partners in the 1000 Bull Genomes Project for access to the data and thank Prof. Paul Stothard for providing VEP annotation.

REFERENCES

Amarasinghe S.L., Su S., Dong X. *et al.* (2020) *Genome Biol.* 21: 30.

Chamberlain A.J., Nguyen N.V., Wang J. and MacLeod I.M. (2023) *Proc. Assoc. Advmt. Anim. Breed. Genet.* 25: *These proceedings.*

Daetwyler H.D., Brauning R., Chamberlain A.J., *et al.* (2017) *Proc. Assoc. Advmt. Anim. Breed. Genet.* 22: 201.

Delahaye C. and Nicolas J. (2021) *PLOS ONE* 16(10): e0257521.

DePristo M.A., Banks E., Poplin R., *et al.* (2011) *Nat. Genet.* 43: 491.

Edge P. and Bansal V. (2019) *Nat. Comms.* 10: 1.

Hayes B.J. and Daetwyler H.D. (2019) *Annual Review of Animal Biosciences* 7: 89.

Li H. (2018). *Bioinformatics*. 34: 3094.

McLaren W., Gil L., Hunt S.E., *et al.* (2016) *Genome Biology* 17: 122.

Nguyen T.V., Vander Jagt C.J., Wang, J. *et al.* (2023a) *Genet. Sel. Evol.* 55: 9.

Nguyen T.V., Wang J., Chamberlain A.J. and MacLeod I.M. (2023b) *Proc. Assoc. Advmt. Anim. Breed. Genet.* 25: *These proceedings.*

Olson N.D., Wagner J., McDaniel J. *et al.* (2022) *Cell Genom.* 2: 100129.

Rosen, B.D., Bickhart, D.M., Schnabel, R.D., *et al.* (2020) GigaScience, 9: 1.

Tarasov A, Vilella AJ, Cuppen E, Nijman IJ and Prins P. (2015). *Bioinformatics*. 31: 2032.

Proc. Assoc. Advmt. Anim. Breed. Genet. 25: 206 - 209

EPIGENETIC REPRESSION OF GENES ASSOCIATED WITH RIBEYE AREA OF NELORE CATTLE

J. Afonso[1], M. R. S. Fortes[2], W. J. Shim[2,3], T. F. Cardoso[1], J. J. Bruscadin[4], A. O. de Lima[5], W. J. S. Diniz[6], B. Silva-Vignato[7], W. L. A. Tan[2], A. S. M. Cesar[8], M. Boden[2], G. B. Mourão[7], A. Zerlotini[9], L. L. Coutinho[7] and L. C. de Almeida Regitano[1]

[1]Embrapa Pecuária Sudeste, São Carlos, São Paulo, Brazil.
[2]School of Chemistry and Molecular Biosciences, The University of Queensland, Brisbane, Australia.
[3]Institute for Molecular Bioscience, The University of Queensland, Brisbane, Australia
[4]Post-graduation Program of Evolutionary Genetics and Molecular Biology, Federal University of São Carlos, São Carlos, São Paulo, Brazil.
[5]Division of Medical Genetics, Department of Genome Sciences, Department of Medicine, University of Washington, Seatle, WA, USA.
[6]Department of Animal Sciences, Auburn university, Auburn, AL, USA.
[7]Department of Animal Science (ESALQ), University of São Paulo, Piracicaba, Brazil.
[8]Faculty of Animal Science and Food Engineering (FZEA), University of São Paulo, Pirassununga, Brazil.
[9]Bioinformatic Multi-User Laboratory, Embrapa Informática Agropecuária, Campinas, São Paulo, Brazil.

SUMMARY

Understanding the epigenetic repression role in regulating genes involved with the ribeye area (REA) of bovine muscle can help us to predict this trait in the future. Here, we identified genes putatively regulating REA in Nelore cattle and divergently epigenetically repressed between contrasting sample groups. For that, we applied the TRIAGE method with a Rank Product analysis using bovine muscle expression data on high versus low REA groups. Further, we identified over-represented pathways and biological processes linked to candidate genes, searching for their regulatory direction. This result advances the knowledge about how epigenetic regulation may impact production traits in Nelore cattle.

INTRODUCTION

The ribeye area (REA) of the bovine *Longissimus dorsi* muscle is used as an indirect measure of carcass composition (Miar *et al.* 2013). The complete regulation of this trait is not known. As such, identifying candidate genes modulating REA is important. Additionally, delineating the mechanisms underlying the modulation of candidate genes would lead to a better understanding of this complex trait. Based on that, we aimed to identify genes regulating REA and that are also being putatively epigenetically repressed in one of the contrasting sample groups for REA. The lack of data on epigenetic repression mechanisms linked to bovine muscle tissue is a limitation. However, our approach can predict genes discordantly activated by epigenetic repression mechanisms considering only expression data. This methodology, named TRIAGE, consists of a repressive tendency score calculated for human genes. We applied this score to the expression value for each gene, in each sample, to calculate a bovine discordant score that can predict genes being affected by repressive mechanisms in each sample (Shim *et al.* 2020). TRIAGE was then expanded using a Rank Product analysis (Afonso *et al.* 2023) to allow us to compare discordant scores between the REA contrasting groups.

MATERIALS AND METHODS

Samples, phenotypes and expression. The genetic estimated breeding values (GEBV) for Ribeye area (REA) from contrasting Nelore steers groups and their *Longissimus thoracis* muscle's expression data from an RNA-Seq experiment were previously described by Silva-Vignato *et al.* (2017). In short, we used the RNA-Seq data of 12 Nelore steers muscle samples representing contrasting GEBV groups for REA. These 12 animals were selected out of 385 samples from a research population from Embrapa (Brazilian Agricultural Research Corporation, São Paulo, SP, Brazil), representing the Brazilian breeding lineages in 2009.

DRGs identification. We implemented a combination of the TRIAGE method (Shim *et al.* 2020) with the RankProd R package (Hong *et al.* 2006) using the expression data to identify putatively epigenetic repressed genes affecting the REA trait, called herein discordantly regulated genes (DRGs). The TRIAGE method is based on the inverse relationship between H3K27me3 histone modification and human gene expression and can be extrapolated to any species (Shim *et al.* 2020). The outputs of this analysis are ranks of genes per sample regarding their discordant score (DS). These DS represent the discordance between the expected expression and the real one based on the chance of this gene being epigenetically repressed. These DS were compared between the contrasting groups with the RankProd R package to identify the DRGs.

Putative relationship between DRGs and REA. In search of the link between the DRGs and REA, we used the PCIT algorithm (Reverter *et al.* 2008) and the Cytoscape software (Shannon *et al.* 2003) to construct a correlation network. The correlation analysis with PCIT was made with all the expression data and REA GEBV for all 12 samples. The correlated pairs containing at least one DRG or the REA GEBV were considered for the network analysis. The genes significantly correlated with each DRGs were used in separate functional annotation analysis with the STRING software (Pertea *et al.* 2015) to retrieve GO terms and metabolic pathways from known protein-protein interaction, considering the product proteins of the DRGs. Subsequently, different sources of information were used to characterize genes present in the network: 1) enriched terms from our functional annotation analysis; 2) previously published differentially expressed genes (Silva-Vignato *et al.* 2017); 3) bovine transcription factors (de Souza *et al.* 2018); 4) bovine known miRNAs. Thus, we identified putative regulatory processes by the functional annotation analysis and other known regulatory (miRNA or TF) or REA related genes (DEG or correlated to REA), depending on their attributes and their correlation with REA or a DRG.

RESULTS AND DISCUSSION

DRGs for REA. We identified six DRGs for REA. The DRGs are the candidate regulators for REA that are also putatively being affected by a repressive epigenetic mechanism. They were differentially ranked between contrasting groups by our choose method because they have a high tendency to be repressed in several tissues but presented an expression between contrasting groups for REA. This is an indicator of epigenetically repression. One DRG was significant in the comparison considering High REA x Low REA and five DRGs were significant in the comparison considering Low REA x High REA ($pfp < 0.01$). The difference in expression between both contrasting groups shows that the only DRG for the comparison High REA x Low REA (*CDH22*) presented higher expression in the Low REA group. Based on the methodology assumption, this can be interpreted as an indication of epigenetic repression of this DRG in the High REA group. The same is valid on the contrary for the other five DRGs, being DRGs for the comparison Low REA x High REA and presenting a higher expression in the High REA group. Figure 1 shows the DRGs, the percentage of false positive (pfp) indicating its significance in the analysis (A) and their expression differences between the groups (B). DRGs can affect the trait in the study by regulating biological processes, while being epigenetically repressed by H3K27me3 or other epigenetic repressive mechanisms (Afonso *et al.* 2023), proposing a new layer of understanding regarding the biological regulation linked to REA.

Putative relationship between DRGs and REA. Figure 2 presents the correlation network considering the significant correlations containing at least one DRG or REA, and its attributes pointing to regulatory functions (miRNAs, TFs and DRGs) and its known relationship with REA (previously published Differentially expressed genes, DEGs, for REA, Silva-Vignato *et al.* 2017). No DRG for REA was previously published as DEG for REA (Silva-Vignato *et al.* 2017).

A)

DRGs	high_up_pfp	low_up_pfp
CDH22	**4.11E-04**	9.62E-01
LBX1	1.22E+00	**2.01E-03**
ZIC4	1.19E+00	**6.18E-03**
CNTFR	9.53E-01	**7.94E-03**
COL2A1	9.12E-01	**9.17E-03**
EN1	9.27E-01	**9.43E-03**

B)

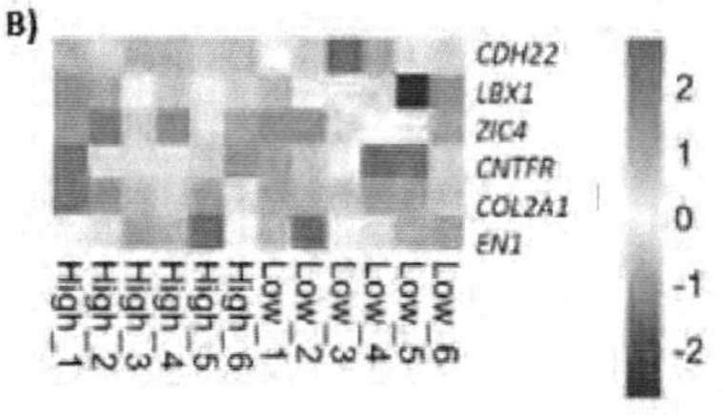

Figure 1. Discordantly regulated genes (DRGs) for Ribeye area (REA) in Nelore

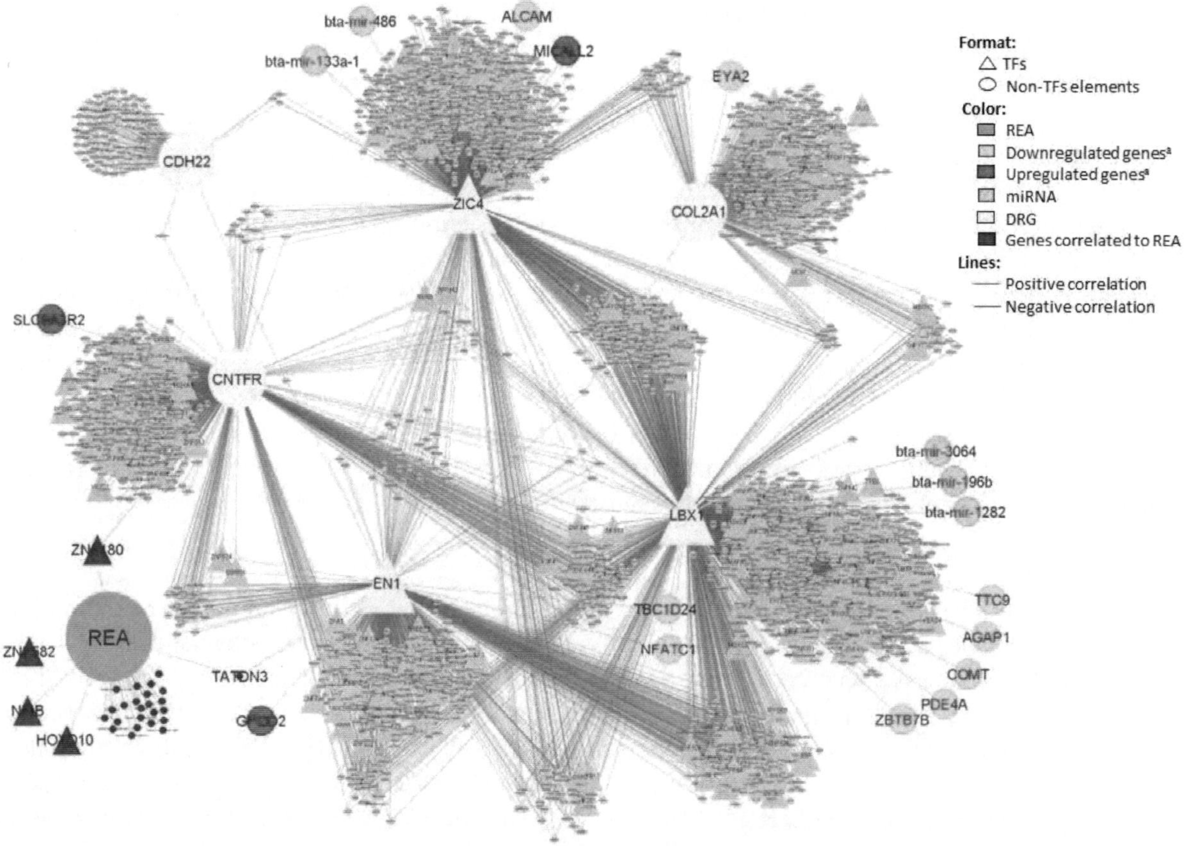

Figure 2. Correlation network focused on the first neighbours of Discordantly Regulated Genes (DRGs) and Ribeye Area (REA) in Nelore [a]Genes downregulated or upregulated in the Low REA group when compared to the High REA group.

All five DRGs for the comparison Low REA x High REA are correlated to at least one DEG for REA, showing its previously known link to REA. Three of the six DRGs are TFs *(ZIC4, LBX1* and *EN1*), and two of these TFs are correlated to miRNAs (*ZIC4* and *LBX1*), indicating its regulatory nature. Considering our network (Figure 2), the expression of the DRGs is not directly correlated to the REA GEBV, but there are two genes directly correlated to REA that are correlated to two different DRGs (the TF *ZNF180* correlated to the DRG *CNTFR* and *TATDN3* correlated to the DRG and TF *EN1*), which are also candidate regulators to the REA, with all the DRGs.

The DRG correlated to more DEGs for REA (Silva-Vignato *et al.* 2017) is *LBX1*, a TF also correlated to genes enriched for the two pathways enriched for the DEGs related to REA (Silva-Vignato *et al.* 2017): MAPK signalling and endocytosis pathways. Considering all the results from the functional annotation analysis, we noted that the *ZIC4* and *CDH22* DRGs were mainly correlated to genes enriched for pathways and processes related to immunity and metabolism. The *CNTFR* gene was involved with protein and DNA regulation, *EN1* to histone modification, protein transport and chromatin regulations, *LBX1* to protein, transcription, DNA-template, growth and cell death regulations, and *COL2A1* to an extracellular matrix organization, synthesis and degradation and protein digestion and absorption. All these pathways and processes can be related to muscle growth, organization, degradation and fat deposition, which are key biological process to REA (Silva-Vignato *et al.* 2017).

CONCLUSIONS

Our approach helped us to point to candidate regulatory genes for REA, also being putatively epigenetically regulated. Also, we identified the possible pathways and biological processes being regulated by each DRG and other candidate regulatory genes underlying REA.

ACKNOWLEDGEMENTS

We wanted to thank FAPESP (2019/04089-2, 2019/25639-0) and CNPq for financing this research, the staff of Embrapa Pecuária Sudeste (São Carlos, São Paulo, Brazil) for the animal caring and all the researchers involved in this project from The University of Queensland (Brisbane, Queensland, Australia) for all the collaboration. W.J.S. Diniz was financially supported by the Alabama Agricultural Experiment Station – Hatch program of the National Institute of Food and Agriculture, U.S. Department of Agriculture. L. L. Coutinho and L. C. de A. Regitano were granted CNPq productivity fellowships.

REFERENCES

Afonso J., ShimW.J., Bodén M., Fortes M.R.S., Diniz W.J.S., de Lima A.O., Rocha M.I.P., Cardoso T.F., Bruscadin J.J., Gromboni C.F., Nogueira A.R.A., Mourao G.B., Zerlotini A., Coutinho L.L. and Regitano L.C.A. (2023) *Biochem. Biophys. Rep.* 33: 101420.

de Souza M.M., Zerlotini A., Geistlinger L., Tizioto P.C., Taylor J.F., Rocha M.I.P., Diniz W.J.S., Coutinho L.L. and Regitano L.C.A. (2018) *Sci. Rep.* 8: 13747.

Hong F., Breitling R., McEntee C.W., Wittner B.S., Nemhauser J.L. and Chory J. (2006) *Bionformatics* 22: 2825.

Miar Y., Plastow G.S., Bruce H.L., Moore S.S., Durunna O.N., Nkrumah J.D. and Wang Z. (2013) *Can. J. Anim. Sci.* 94: 273.

Pertea M., Pertea G.M., Antonescu C.M., Chang T.C., Mendell J.T. and Salzberg S.L. (2014) *Nat. Biotechnol.* 33: 290.

Reverter A. and Chan E.K.F. (2008) *Bioinformatics* 24: 2491.

Shannon P., Markiel A., Ozier O., Baliga N.S., Wang J.T., Ramage D., Amin N., Schwikoski B. and Ideker, T. (2003) *Genome Res.* 13: 2498.

Shim W.J., Sinniah E., Xu J., Vitrinel B., Alexanian M., Andreoletti G., Shen S., Sun Y., Balderson B., Boix C., Peng G., Jing N., Wang Y., Kellis M., Tam P.P.L., Smith A., Piper M., Christiaen L., Nguyen Q., Bodén M. and Palpant N.J. (2020) *Cell Syst.* 11: 625.

Silva-Vignato B., Coutinho L.L., Cesar A.S.M., Poleti M.D., Regitano L.C.A. and Balieiro J.C.C. (2017) *BMC Genomics.* 18: 506.

GENOMIC PREDICTION OF CONSUMER SATISFACTION TRAITS OF AUSTRALIAN BEEF

A.M. Lynn[1], P. McGilchrist[1], H. Aliloo[1], R. Polkinghorne[1,2], M. Forutan[3], B.J. Hayes[3] and S.A. Clarke[1]

[1]School of Environmental and Rural Science, University of New England, Armidale, NSW, 2350 Australia
[2]Birkenwood International, 45 Church Street, Hawthorn, Victoria, 3122 Australia
[3]Queensland Alliance for Agriculture and Food Innovation, The University of Queensland, St Lucia, Queensland 4067, Australia

SUMMARY

Consumer satisfaction has become a key focus for beef producers as eating quality traits such as tenderness and flavour dictate purchasing choices and, ultimately, the price consumers are willing to pay. Due to the difficulty in measuring eating quality traits and the inability to predict those traits prior to slaughter, beef producers opt to select for correlated traits and indirectly select for eating quality. Genotyping of animals offers the opportunity for the selection of cattle with superior eating quality directly for both breeding and market allocation. The aim of this study was to determine the accuracy of genomic prediction along with heritabilities for eating quality traits; tenderness, juiciness, flavour and overall liking as well as the overarching consumer satisfaction trait known as MQ4 in a 10-fold cross validation. Phenotypes from 1,701 cattle recorded in eating quality trials held across Australia were collected for the 5 eating quality traits. Those same cattle were genotyped using varying Illumina SNP arrays between 50k and 100k density and then imputed up to high density 700k using a reference set of 4,506 cattle representing most breeds and crossbreds composites of the Australian beef herds. A linear mixed model was used with cohort, days aged, carcase weight, principal components 1-4 and heterozygosity fit in the model. Heritabilities ranged from 0.21 to 0.32 between juiciness and tenderness respectively, while tenderness and MQ4 had the highest accuracy of 0.27 from the cross validation and juiciness and flavour having the lowest accuracies of 0.23. While accuracies observed in this study were low, moderate heritabilities indicate selection for eating quality traits is feasible.

INTRODUCTION

Beef eating quality has been identified as the leading factor in Australian consumer purchasing habits (Bonny *et al.* 2018). This has led to an increased emphasis on the selection for eating quality traits such as tenderness, juiciness, flavour and overall liking in beef herds for both domestic and export beef herds (Watson *et al.* 2008a). Consumer derived eating quality traits are expensive to test so large-scale measurement is not viable. Processors currently rely on the Meat Standards Australia (MSA) model to predict consumer satisfaction based on objective carcase measurements such as intramuscular fat (IMF), ossification (physiological maturity), paying producers based on meeting phenotype thresholds. This results in producers having to rely on selection of related traits such as IMF content which is known to increase tenderness through muscle fibre dispersion as well as impacting flavour due to modified fatty acid profiles. Genomic analysis offers opportunities to select for eating quality traits prior to slaughter with the possibility of the implementation of genomic estimated breeding values (GEBVs) in commercial Australian beef herds. The aim of this study is to examine the heritability and accuracy of genomic prediction for beef eating quality traits within the diverse Australian cattle herd using Genomic Best Linear Unbiased Prediction (GBLUP)

MATERIALS AND METHODS

Phenotypes. The Striploin muscle (*longissimus lumborum*) was collected and consumed by Australian consumers from 1,701 genotyped Australian cattle between 1997 and 2019. The animals were from 65 cohorts and encompass a diverse breed profile to represent the Australian beef herds. Breeds represented in this research covered tropical *Bos indicus* (261 Brahman), *Bos taurus* (285 Angus, 274 Hereford, 38 Shorthorn), composite breeds (100 Belmont Red and 83 Santa Gertrudis), dairy *Bos taurus* (72 Holstein, 23 Jersey), 121 crossbred cattle and 444 cattle with unidentified breed profiles. The study used steers (n=1319), heifers (n=345) and bulls (n=37) however breed and sex were found to be completely confounded with cohort. Carcase weight ranged from 50.6kg to 576kg, averaging 261.4kg. Steak samples were grilled to protocol as described in Watson *et al.* (2008). Steak samples from each animal were consumed by ten consumers for tenderness, juiciness, flavour and overall liking on a sliding bar scale from poor to excellent. Scores were clipped by removing the top and bottom two scores with the remaining six averaged. Consumers were given seven samples during the sitting with the first sample (link) being removed from analysis (Watson *et al.* 2008b). The four eating quality traits are then used to calculate a singular satisfaction score known as Meat Quality 4 (MQ4) which is based of weightings to reflect Australian consumers preferences; MQ4 = 0.3 x tenderness + 0.1 x juiciness + 0.3 x flavour + 0.3 x overall liking (Thompson *et al.* 2010). Animal phenotypic data included cohort, days aged (post slaughter proteolysis period), and carcase weight.

Genotypes. The genotypes for the 1,701 animals were obtained using five different Single Nucleotide Polymorphism (SNP) chips (Illumina BovineSNP50 Genotyping Beadchip v1, v2, GeneSeek Genomic Profiler (GGP) Bovine 50K, GGP Bovine 100K and the TropBeef chip). SNP densities ranged between 50k and 100k with the TropBeef SNP chip having approximately 19k overlap with those used for *bos taurus*. Cleaning of genotypes removed any SNP with missing rates >0.1, minor allele frequencies (MAF) <0.01 and those departing from Hardy-Weinberg equilibrium at $p < 1x10^8$. All genotypes were imputed to high density 709,068 SNP with findhap4 (VanRaden *et al.*, 2013) utilising a reference set of 4,506 individuals of which were originally genotyped with Illumina HD array. This reference set spans most breeds, composites and crossbreds in Australia and was adequately suited for the imputation of this dataset. The first four principal components from a genomic relationship matrix (GRM) based on GCTA (Yang *et al.*, 2011) explained around 25% of the genetic variance and were used to represent the breed proportion effect in the model. The proportion of heterozygous loci for animals were calculated from the imputed genotypes to be used in the model.

Analysis. A univariate mixed linear model based on GBLUP approach was performed using airemlf90 from the BLUPf90 family of programs (Aguilar *et al.*, 2018) for each of the five eating quality traits to obtain estimates of fixed effects along with heritability of the trait:

$$\boldsymbol{y} = \boldsymbol{Xb} + \boldsymbol{Zu} + \boldsymbol{e} \quad (1)$$

Where y is the phenotype, b is the estimated fixed effect of group and effect of covariates; days aged, carcase weight, principal components 1-4 and heterozygosity, u is the vector estimated genomic breeding values (GEBV) of animals, e is the residual term. X and Z are incidence matrices relating to observations to effects fitted in the model. It was assumed that $v(u) = G\sigma^2$ where G was the genomic relationship matrix based on VanRaden (2008) and σ^2 is the additive genetic variance.

Animals were randomly assigned into ten groups of equal size. A 10-fold cross validation was performed by removing the phenotypes of each fold allowing the information from the approximately 1,530 animals to estimate breeding value for the remaining 170 animals with deleted phenotypes. Correlations between the EBV and adjusted phenotypes for each group were calculated and averaged across the ten folds to calculate accuracies based on the trait heritability.

RESULTS AND DISCUSSION

Estimated heritabilities were moderate for all five traits (0.21-0.32; Table 1) with tenderness having the highest observed heritability. Heritabilities were similar to other studies for both Australian and international herds suggesting the potential for selection for eating quality traits directly in commercial herds (O'Connor *et al.* 1997; Johnston *et al.* 2003). Johnston et al. (2003) reported heritabilities for temperate *Bos Taurus* and tropical *Bos indicus* separately and found that tropically adapted cattle had higher heritabilities (0.2 – 0.32) across all traits while temperate breeds had markedly lower heritabilities (0.05 – 0.15) for all traits. The current study reports heritabilities from commercial herds in Australia which resembles mixed breed profiles of temperate, tropical and composite animals. Re-estimation of heritabilities of eating quality traits for different breeds may benefit the industry in the future as many commercial herds still have singular classes of cattle to reflect the market and climate.

Accuracies were low with tenderness and MQ4 having the highest accuracy of 0.27, while juiciness and flavour had the lowest accuracies of 0.23. This dispersal of accuracies is reflected in Forutan *et al.* (2023) who used the same data set to examine four separate strategies for prediction through a BayesR model. Observed accuracies were between 0.2 and 0.5 for tenderness with Strategies 2,3 and 4 outperforming the GBLUP model used in this study with juiciness repeatedly having the lowest accuracies of >0.3 (Forutan *et al.* 2023). Miller *et al.* (2014) used GBLUP method for the prediction of breeding values for mechanical tenderness (shear force) in a Canadian beef herd consisting of *Bos Taurus* breeds (predominantly European breeds) and found correlations of 0.1 to 0.5 between GEBV's and adjusted phenotypes but they observed a lower heritability of 0.19 for shear force tenderness. However, due to the correlation between mechanical tenderness and panel tenderness being approximately -0.72 it is expected that studies utilising shear force as a phenotype would differ in heritability estimates than that of panel derived tenderness (Destefanis *et al*, 2008). There is valid argument as to utilising shear force over consumer tenderness scores due to most consumers being able to only differentiate changes of around 1kg of force rather than the minute increments detectible by machine. Zwambang *et al.* (2013) examined the heritability of beef tenderness (shear force) at differing aging points and found that the heritability of beef tenderness reduced from 0.19 to 0.05 when comparing the same beef at 7 and 21 days aged suggesting that genetic variance is reduced by longer days aging. However, the study examined only *Bos Taurus* breeds (predominantly European breeds) and did not need to consider the declined aging potential of *Bos indicus* breeds due to their altered enzyme production. The increased heritabilities in this study may be owed to the diverse breed profile of the data set.

Table 1. Means (± SD), heritabilities (± SE) and accuracy of GBLUP prediction of phenotype (± SE) for tenderness, juiciness, flavour, overall liking and MQ4

Trait	Mean	h^2	Accuracy
Tender	57.35 ± 16.5	0.317 ± 0.07	0.27 ± 0.04
Juiciness	57.72 ± 14.18	0.213 ± 0.07	0.23 ± 0.03
Flavour	59.11 ± 12.24	0.268 ± 0.07	0.23 ± 0.03
Overall liking	58.34 ± 14.19	0.272 ± 0.07	0.25 ± 0.04
MQ4	57.81 ± 13.76	0.301 ± 0.07	0.27 ± 0.04

The current study was hindered by the confounding nature of cohort, where both sex and identified breed were completely confounded by the group ID. Other difficulties identified in this research was the lack of uniformity of phenotypes when utilising a large number of datasets where the effects being observed differ. This was evident when age (days) was available for a proportion of the cattle in this study but not for others. While this could have been addressed in the same way that the MSA model utilises ossification as an indication of physiological age or maturity, it was decided that only effects that can be measured or predetermined prior to slaughter be used. Carcase weight as an effect in this study could be interpreted in multiple ways as the effect of size or maturity due to the large range in recorded weight. For simplicity however, it was used as an indication of size alone however further manipulation on the way carcase weight could be fitted will be examined in further research. Although breed was confounded with cohort, there were a large proportion of animals unidentified, or misidentified when examining a plot of the first two principal components. Principal components were fitted to rectify the lack of breed information for a proportion of the dataset by also giving an indication of breed proportion. Even though it is likely that breed would still be confounded with group due to the nature of these projects not assessing breed effect, a reliable identification of breed or cross for all animals would have been of value in assessment.

CONCLUSION

Economically important traits such as tenderness and consumer satisfaction can be predicted and selected for through GBLUP models in diverse beef herds. However, improvements to the model and data structure with increased consistency of phenotype records, reduced data collection periods and a controlled breed profile may strengthen the low accuracies observed in this study. Genomic prediction of eating quality traits is a financially viable option for both commercial and seed stock breeding herds.

REFERENCES

Aguilar I., Tsuruta S., Masuda Y., Lourenco D., Legarra A. and I. Misztal. (2018) *Proc. World Congress on Genetics Applied to Livestock Production* p 11-16.

Bonny S.P., O'Reilly R. A., Pethick D. W., Gardner G. E., Hocquette J.-F. and Pannier L. (2018) *J. Integr. Agric.* 17:1641.

Casas E., White S., Wheeler T., Shackelford S., Koohmaraie M., Riley D., Chase Jr C., Johnson D. and Smith T. (2006) *J. Anim. Sci.* 84: 520,

Destefanis G., Brugiapaglia A., Barge M.T. and Dal Molin E. (2008) *Meat Sci.* 78: 153.

Forutan M., Lynn A., Aliloo H., Clark S.A., McGilchrist P., Polkinghorne R. and Hayes B.J. (2023) *Front. Genet.* 14

Johnston D., Reverter A., Ferguson D., Thompson J.M. and Burrow H. (2003) *Aust. J. Agric. Res.* 54: 135.

Miller S., Lu D., Vander Voort G. and Mandell I. (2014) *10th World Congress of Genetics Applied to Livestock Production*

O'connor S., Tatum J., Wulf D., Green R. and Smith G. (1997) *J. Anim. Sci.* 75: 1822.

Thompson J., Polkinghorne R., Gee A., Motiang D., Strydom P., Mashau M., Ng'ambi J., Kock R. D. and Burrow H. (2010) *ACIAR Technical Reports Series* 72.

VanRaden P.M., Null D.J., Sargolzaei M., Wiggans G.R., Tooker M.E., Cole J.B., Sonstegard T.S., Connor E.E., Winters M., van Kaam J.B., Valentini A., Van Doormaal B.J., Faust M.A. and Doak G.A. (2013) *J. Dairy Sci.* 96: 668.

Watson R., Gee A., Polkinghorne R. and Porter M. (2008a) *Aust. J. Exp. Agric.* 48: 1360.

Watson R., Polkinghorne R. and Thompson J.M. (2008b) *Aust. J. Exp. Agric.* 48: 1368.

Yang J., Lee S.H., Goddard M.E. and Visscher P.M. (2011) *Am. J. Hum. Genet.* 88: 76.

Zwambag A., Kelly M., Schenkel F., Mandell I., Wilton J. and Miller S. (2013). *Can. J. Anim. Sci.* 93: 307.

Proc. Assoc. Advmt. Anim. Breed. Genet. 25: 214 - 217

RELATIONSHIPS OF SIRE BREEDING VALUES FOR MERINO PRODUCTION TRAITS WITH EATING QUALITY OF LAMB

S.I. Mortimer[1], B.W.B. Holman[2], S.M. Fowler[3], T.I.R.C. Alvarenga[1], D.L. Hopkins[4], K.L. Egerton-Warburton[5], J.L. Smith[6], B.C. Hine[6] and A.A. Swan[7]

[1]NSW Department of Primary Industries, Armidale, NSW, 2351 Australia
[2]NSW Department of Primary Industries, Wagga Wagga, NSW, 2650 Australia
[3]NSW Department of Primary Industries, Cowra, NSW, 2794 Australia
[4]Canberra, ACT, 2903 Australia
[5]NSW Department of Primary Industries, Orange, NSW, 2800 Australia
[6]CSIRO, Agriculture and Food, F.D. McMaster Laboratory, Armidale, NSW, 2350 Australia
[7]Animal Genetics and Breeding Unit*, University of New England, NSW, 2351 Australia

SUMMARY

Regression coefficients were estimated of sensory and objective eating quality (EQ) traits on sire Australian Sheep Breeding Values (ASBVs) for a range of Merino production traits to identify if genetic relationships were likely to exist between these traits. The sire ASBVs were not associated with either overall liking scores of loin, knuckle and topside cuts, or intramuscular fat and shear force of the loin. This preliminary study has shown that it is likely that selection on sire ASBVs to improve Merino production traits would yield negligible responses in EQ traits.

INTRODUCTION

For the current MERINOSELECT indexes where the breeding objective includes improvement of carcass traits (Dual Purpose+ and Dohne+), it is predicted that small unfavourable responses in eating quality (EQ) traits would occur (A.A. Swan, personal communication). With around 30% of Merino breeding ewes being mated for crossbred lamb production (MLA and AWI 2021), considering EQ traits in these Merino breeding objectives is warranted. Like the lamb EQ indexes for Terminal sires (Swan *et al.* 2015), refinement of these indexes would contribute to ensuring that lamb produced by Merino dual purpose production systems are of acceptable quality, when eaten by consumers. For those Merino ewes mated to Terminal sires to produce crossbred lambs, it would be prudent to know if the MERINOSELECT objectives used to generate those ewes are consistent with the EQ objectives of the LAMBPLAN Terminal sire indexes. Based on low to negligible genetic correlations, Mortimer *et al.* (2017) had concluded that Merino breeding programs emphasising wool production would have little or no effect on the objectively measured EQ traits of intramuscular fat and shear force. The genetic relationships of wool production traits with sensory scores for EQ traits have not yet been reported.

The diversity of the sires selected to generate progeny of Australian Wool Innovation's Merino Lifetime Productivity (MLP) project (Ramsay *et al.* 2019) and the availability of data from consumer testing of sensory EQ traits of meat samples from MLP wether carcasses provide a means to detect if genetic relationships exist between EQ traits and sire Australian Sheep Breeding Values (ASBVs) for production traits. This preliminary study estimated relationships between ASBVs for a range of MERINOSELECT breeding objective traits and EQ traits, sensory and objective, assessed on 3 cuts of Merino lamb sampled from carcasses produced at 2 MLP sites.

---

* A joint venture of NSW Department of Primary Industries and the University of New England

MATERIALS AND METHODS

The design of the MLP project (Ramsay *et al.* 2019) and the pre-slaughter procedures (Mortimer *et al.* 2021) that produced the carcasses for this study have been described elsewhere. Sensory EQ data were recorded on loin, knuckle and topside samples aged for 5 days taken from carcasses of 2018-born F1 wethers at the Macquarie (fine/medium ewe base) and New England (ultrafine ewe base) MLP sites. Sample collection and preparation, cooking procedures and sensory testing protocols applied to the grilled samples and tasted by panels of untrained consumers have been described by Pannier *et al.* (2014). Briefly, for each consumer tasting session (57 sessions), 10 sub-samples were prepared from each meat sample following grilling under standardised conditions and provided to 10 consumers. The EQ traits were assessed by the consumers using a 0-100 scale (100 being most preferred) and included tenderness, juiciness, liking of flavour and overall liking of loin, topside and knuckle cuts, respectively. The EQ record for each sample was then based on the mean of the 10 consumer responses. For this study, the overall liking scores for the loin (llike), knuckle (klike) and topside (tlike) samples only were analysed; 408, 409 and 403 records respectively were available from the Macquarie carcasses, while 152, 156 and 157 records were available from the New England carcasses. Objective EQ data were recorded on samples taken from the other loin of each carcass. Intramuscular fat (imf, %) was measured using procedures described by Hopkins *et al.* (2014), while shear force (sf5, N) was tested as described by Hopkins and Thompson (2001).

The ASBVs were available for 14 (extracted from MERINOSELECT analyses 21 September 2017) and 12 (extracted from MERINOSELECT analyses 21 January 2018) of the sires used to generate progeny at the Macquarie and New England sites (Table 1), respectively. The ASBVs for the production traits included: yearling (ycfw) and adult (acfw) clean fleece weight (%); yearling (yfd) and adult (afd) fibre diameter (micron); yearling (yfdcv) and adult (afdcv) coefficient of variation of fibre diameter (%); yearling (yss) and adult (ass) staple strength (N/ktex); yearling (ysl) and adult (asl) staple length (mm); yearling (ywt) and adult (awt) live weight (kg); yearling ultrasound fat depth (yfat, mm); and yearling ultrasound eye muscle depth (yemd, mm).

Table 1. Summary statistics for eating quality and ASBV (minimum and maximum in brackets) traits for Macquarie and New England samples

	Macquarie samples		New England samples	
Eating quality trait				
	Mean (SD)	Range	Mean (SD)	Range
llike	68.7 (8.20)	42.3 - 88.8	69.5 (8.16)	49.0 - 87.6
klike	65.0 (7.10)	40.4 - 84.2	65.6 (6.97)	43.3 - 88.0
tlike	53.7 (9.08)	26.6 - 73.8	52.2 (8.99)	28.1 - 74.3
imf	4.7 (1.39)	2.2 - 10.8	4.5 (1.34)	2.2 - 8.2
sf5	24.5 (5.28)	14.1 - 41.4	24.1 (4.91)	13.4 - 41.4
Australian Sheep Breeding Value				
	Yearling	Adult	Yearling	Adult
cfw	22.98 (10.35, 41.63)	17.95 (5.35, 37.67)	13.37 (-34.96, 30.95)	8.29 (-34.94, 21.23)
fd	-1.21 (-2.71, 0.05)	-1.13 (-2.79, 0.05)	-2.29 (-4.19, -0.78)	-2.55 (-4.78, -1.02)
fdcv	-0.14 (-1.95, 1.65)	-0.08 (-1.73, 1.52)	-0.69 (-2.17, 1.63)	-0.60 (-2.04, 1.36)
ss	-0.80 (-5.9, 5.35)	-0.88 (-5.96, 3.27)	-1.06 (-5.58, 2.57)	-1.33 (-6.46, 2.45)
sl	5.96 (0.62, 13.31)	5.75 (-1.4, 12.54)	3.82 (-12.62, 17.31)	2.96 (-18, 14.8)
wt	6.33 (1.94, 13.55)	5.20 (0.66, 12.42)	3.36 (-5.81, 6.62)	2.04 (-6.50, 5.59)
fat	-0.03 (-1.06, 2.15)	-	-0.08 (-0.96, 1.70)	-
emd	0.17 (-1.72, 2.39)	-	-0.11 (-1.30, 2.72)	-

Separate analyses for each site's data were performed to estimate the regression coefficients of each EQ trait on sire ASBV for each production trait using ASReml (Gilmour *et al.* 2021). The model fitted to the data included a fixed effect of contemporary group (accounting for management and slaughter group effects) and a random effect of sire. Although fixed effects of birth type, rearing type, dam age and their interactions were also tested, these effects were found to be not significant and were excluded from the model.

RESULTS AND DISCUSSION

Average scores tended to be similar for overall liking for each of the 3 cuts across the sites (Table 1). The average scores for overall liking of topside samples were lower than scores for loin samples from both sites, with differences of 10 units for Macquarie samples and 17 units for the New England samples. Average overall liking scores for knuckle samples were 4 units lower than the average scores for loin samples at both sites.

For both data sets, significant ($P < 0.05$) regression coefficients were not detected for any of the EQ traits with the sire ASBVs for wool traits (Table 2). Nonetheless across the 3 cuts from the

Table 2. Regression coefficients for eating quality traits of Macquarie and New England samples on sire ASBVs

	llike	klike	tlike	imf	sf5
Macquarie samples					
ycfw	0.09 ± 0.09	0.13 ± 0.08	0.12 ± 0.10	-0.01 ± 0.02	-0.05 ± 0.74
acfw	0.07 ± 0.08	0.05 ± 0.08	0.09 ± 0.09	0.00 ± 0.02	-0.02 ± 0.07
yfd	-0.17 ± 0.95	0.34 ± 0.94	0.10 ± 1.08	0.10 ± 0.21	-0.56 ± 0.77
afd	-0.10 ± 0.92	-0.07 ± 0.92	0.01 ± 1.05	0.10 ± 0.20	-0.38 ± 0.76
yfdcv	0.62 ± 0.65	0.02 ± 0.67	0.49 ± 0.76	-0.09 ± 0.15	0.03 ± 0.56
afdcv	0.79 ± 0.75	0.30 ± 0.78	0.89 ± 0.87	-0.09 ± 0.17	-0.09 ± 0.65
yss	-0.21 ± 0.21	0.17 ± 0.21	0.00 ± 0.25	-0.01 ± 0.05	-0.05 ± 0.18
ass	-0.35 ± 0.27	0.13 ± 0.28	-0.19 ± 0.32	0.01 ± 0.06	-0.07 ± 0.24
ysl	-0.09 ± 0.18	0.00 ± 0.18	-0.05 ± 0.21	0.03 ± 0.04	-0.03 ± 0.15
asl	-0.02 ± 0.20	-0.07 ± 0.2	0.00 ± 0.23	0.04 ± 0.04	-0.04 ± 0.16
ywt	-0.02 ± 0.22	-0.03 ± 0.22	-0.17 ± 0.24	-0.02 ± 0.05	0.20 ± 0.17
awt	0.02 ± 0.21	-0.11 ± 0.21	-0.16 ± 0.23	-0.02 ± 0.05	0.21 ± 0.16
yfat	-0.40 ± 0.77	0.10 ± 0.78	-0.39 ± 0.89	-0.08 ± 0.17	0.28 ± 0.65
yemd	-0.48 ± 0.56	-0.46 ± 0.56	-0.71 ± 0.63	-0.17 ± 0.12	0.47 ± 0.46
New England samples					
ycfw	0.01 ± 0.05	-0.03 ± 0.04	-0.02 ± 0.05	-0.01 ± 0.01	0.01 ± 0.04
acfw	0.00 ± 0.05	-0.03 ± 0.05	-0.01 ± 0.06	-0.01 ± 0.02	0.00 ± 0.04
yfd	0.84 ± 0.64	0.18 ± 0.62	0.76 ± 0.71	0.00 ± 0.22	-0.21 ± 0.50
afd	0.63 ± 0.50	0.10 ± 0.48	0.65 ± 0.55	0.01 ± 0.17	-0.20 ± 0.39
yfdcv	-0.57 ± 0.53	-0.25 ± 0.51	-0.07 ± 0.26	-0.15 ± 0.18	0.10 ± 0.42
afdcv	-0.67 ± 0.60	-0.22 ± 0.58	0.41 ± 0.68	-0.15 ± 0.20	0.14 ± 0.47
yss	0.46 ± 0.29	0.16 ± 0.28	0.13 ± 0.34	0.06 ± 0.10	-0.06 ± 0.23
ass	0.44 ± 0.28	0.17 ± 0.27	0.20 ± 0.33	0.04 ± 0.09	-0.10 ± 0.22
ysl	0.09 ± 0.09	-0.06 ± 0.08	-0.07 ± 0.10	-0.02 ± 0.03	0.08 ± 0.06
asl	0.09 ± 0.09	-0.07 ± 0.08	-0.07 ± 0.10	-0.02 ± 0.03	0.07 ± 0.07
ywt	0.15 ± 0.23	-0.12 ± 0.21	-0.07 ± 0.26	0.00 ± 0.08	0.09 ± 0.17
awt	0.05 ± 0.22	-0.14 ± 0.20	0.07 ± 0.24	0.01 ± 0.07	0.06 ± 0.16
yfat	0.83 ± 0.92	0.76 ± 0.84	-1.07 ± 0.96	0.29 ± 0.29	-0.07 ± 0.69
yemd	0.98 ± 0.60	0.53 ± 0.58	-0.88 ± 0.65	0.10 ± 0.21	0.17 ± 0.47

Macquarie carcasses, there may be a possibility that improving sire ASBVs for clean fleece weight and fibre diameter variability could lead to slight favourable and unfavourable responses, respectively, in overall liking scores. For the New England cuts, improving sire ASBVs for fibre diameter may lead to slight unfavourable responses, while improving sire ASBVs for fibre diameter variability and staple strength may yield slight favourable responses. In the case of the objective EQ traits (imf, sf5), the lack of associations was consistent with the negligible to low genetic correlations of imf and sf5 with wool production traits reported by Mortimer *et al.* (2017), which were generally less than 0.20 in size.

No significant regression coefficients were detected for any of the EQ traits with the sire ASBVs for the live weight and ultrasound traits (Table 2). The effect of increasing sire ASBVs for yfat and yemd on overall liking scores, though, may vary between data sources: slightly unfavourable effects on the scores on cuts from the Macquarie carcasses versus slightly favourable effects on scores of loin and knuckle cuts and slight unfavourable effects on scores of topside cuts from New England carcases. For imf and sf5, negligible genetic correlations have been estimated for these objective EQ traits with ywt, awt, yfat and yemd (Mortimer *et al.* 2018).

CONCLUSION

This preliminary study suggests that selection on sire ASBVs to improve Merino production traits would yield negligible responses in sensory and objective EQ traits. Estimation of genetic correlations among the traits will verify if at most weak genetic associations do exist between EQ and wool production traits. Based on a combination of data from the Macquarie and New England flocks and data from other resource flocks that have assessed eating quality of Merino lamb cuts, analyses are underway to estimate the accurate genetic parameters needed to include an EQ breeding value in MERINOSELECT indexes.

ACKNOWLEDGEMENTS

We acknowledge the co-investment by Meat & Livestock Australia in the conduct of this study (MLA L.EQT.1908 and MLA P.PSH.1032). Australian Wool Innovation Limited is the owner of pedigree and fixed effect data from the MLP Project used by this study. Staff are thanked at: NSW DPI Cowra for sample collection and measurement; NSW DPI Trangie and CSIRO Armidale for sheep management; and Fletcher International Exports for their support.

REFERENCES

Gilmour A.R., Gogel B.J., Cullis B.R., Welham S.J. and Thompson, R. (2021) 'ASReml User Guide Release 4.1 Functional Specification'. VSN International Ltd, Hemel Hempstead, UK.

Hopkins D.L., Clayton E.H., Lamb T.A., van de Ven R.J., Refshauge G., *et al.* (2014) *Meat Sci.* 98: 135.

Hopkins D.L. and Thompson J.M. (2001) *Meat Sci.* 57: 1.

MLA and AWI Wool and Sheepmeat Survey Report (2022) Accessed 14 April 2023: https://www.mla.com.au/globalassets/mla-corporate/prices--markets/documents/trends--analysis/mlw-stats/mla-and-awi-wool-and-sheepmeat-survey-june-2022-final-report-mla.pdf

Mortimer S.I., Hatcher, S., Fogarty, N.M., van der Werf, J.H.J., Brown, D.J., *et al.* (2017) *J. Anim. Sci.* 95: 4260.

Mortimer S.I., Fogarty, N.M., van der Werf, J.H.J., Brown, D.J., Swan, A.A., *et al.* (2018) *J. Anim. Sci.* 96: 3582.

Mortimer S.I., Smith J.L., Hine B.C., Fowler S.M., Holman B.W.B., *et al.* (2021) *Proc. Assoc. Advmt. Anim. Breed. Genet.* 24: 255.

Pannier L., Gardner G.E., Pearce K.L., McDonagh M., Ball A.J., *et al.* (2014) *Meat Sci.* 96: 1076.

Ramsay A.M.M., Swan A.A. and Swain, B.C. (2019) *Proc. Assoc. Advmt. Anim. Breed. Genet.* 23: 512.

Swan A., Pleasants T. and Pethick D. (2015) *Proc. Assoc. Advmt. Anim. Breed. Genet.* 21: 29.

COST EFFECTIVE DETECTION OF STRUCTURAL VARIANTS IN LONG-READ SEQUENCE – HOW DEEP IS ENOUGH?

T.V. Nguyen[1], J.Wang [1], A.J. Chamberlain[1,2] and I.M. MacLeod[1,2]

[1] Agriculture Victoria, AgriBio, Centre for AgriBiosciences, Bundoora, Victoria, 3083, Australia
[2] School of Applied Systems Biology, La Trobe University, Bundoora, VIC, 3083, Australia

SUMMARY

In recent years, the 1000 Bull Genomes Project has brought together short-read sequence for more than 6,000 cattle, playing a key role in detection of small variants and generating improved accuracy of genomic prediction for complex traits. While this continues to be an invaluable resource for SNP and small INDEL studies, it is not suited to detecting more complex structural variants (SV: variants with length > 50bp). However, SV often show large and sometimes deleterious effects on phenotypes and remain largely unexplored in livestock. Here, we use long-read sequences of two bovine parent-offspring trios to explore the optimal read depth to be cost effective whilst still maintaining a high chance of detecting SVs. This study shows that while sequencing from between 10X to 15X coverage resulted in some reduction in the SV discovery rate versus higher read depth, this may be an acceptable compromise for population scale studies to spread sequencing costs over a larger number of animals. However, if the purpose of using long-read sequencing is to discover a deleterious Mendelian mutation among a small group of known affected or carrier animals, the results here suggest that at least 20X cover would be preferable.

INTRODUCTION

Structural variants (SV) are genetic variations that involve the insertion, deletion, or rearrangement of large segments of DNA, typically affecting > 50 base pairs (Freeman *et al.* 2006). These types of variants can have significant impacts on gene function and expression, but their detection in livestock has been challenging due to limitations of short-read sequencing technology. To improve the accuracy and sensitivity of SV detection, several livestock genomics studies have deployed long-read sequencing technologies, mostly using PacBio and Oxford Nanopore Technologies (ONT) platforms. These studies have generally sequenced a relatively small sample of individuals at high read coverage, either to build reference pan-genomes or to pinpoint a deleterious SV (reviewed in Nguyen *et al.* 2023). Long-read sequencing is still relatively expensive for large population scale analyses, therefore it is critical to optimise read-depth for cost effective SV discovery. Therefore, we conduct a pilot experiment to study the effect of read-depth on discovery rate statistics of SV using two cattle parent-offspring trios.

MATERIALS AND METHODS

A flowchart of the methodology is illustrated in Figure 1. In brief, two Holstein trios (parents and offspring) were sequenced at ~60X coverage using ONT PromethION sequencer (flow cell 9.4.1 and ligation kit LSK110) following the manufacturer's recommendations. Post sequencing, the FAST5 files were re-basecalled using Guppy (v6.1.7) with the super high accuracy setting (SUP). The output FASTQ files were then trimmed using Filtlong (https://github.com/rrwick/Filtlong) with the default setting. Filtered reads were mapped to the ARS-UCD 1.2 + Btau5.0.1 Y reference genome ARS-UCD1.2 (Rosen *et al.* 2020) with additional Btau5.0.1 Y (Bellott et al. 2014) using Minimap2 aligner (Li 2018). Post sequencing and alignment, the recorded mapped coverage is estimated at 50X, so we considered this as the "baseline" read coverage. The aligned reads were used to detect SV with Sniffles2 (Sedlazeck *et al.* 2018) in individual samples and then merged using Sniffles2 joint genotyping function (default settings). Next, mapped reads at 50X coverage were scaled down using Sambamba (default settings: Tarasov *et al.* 2015) to an estimate of 3X, 5X, 10X, 15X, 20X coverage. These alignment files were re-exported to FASTQ format. The scaling down of read coverage was replicated three times for each animal at each coverage and replicate samples were then subjected to the same SV detection pipeline described above. For ease of analysis, we only considered SVs detected from autosomes (Chr 1 – 29). Finally, we deployed RTG Tools (https://github.com/RealTimeGenomics/rtg-tools) and its Mendelian plugin on merged SV calls to count the number of Mendelian consistent and inconsistent SV calls across each of the two trios (per replicate at each read cover). This plugin only counts SV that are genotyped in all individuals (i.e. excluding SV where one or more of the trio had a missing genotype).

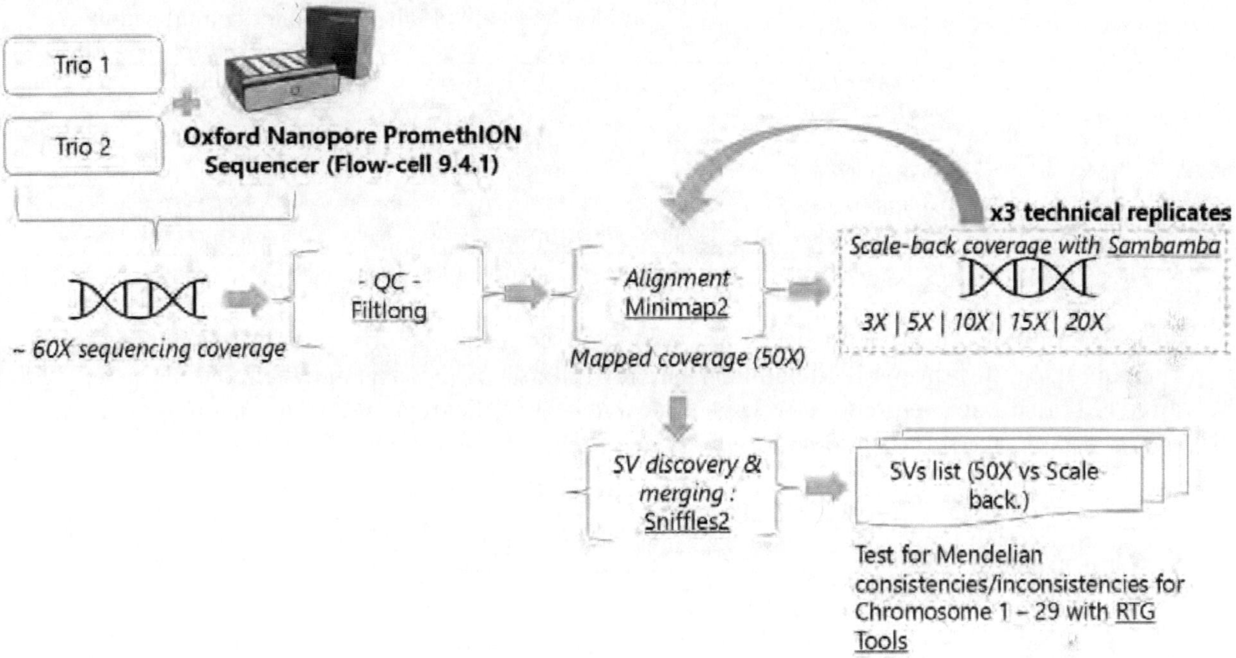

Figure 1. Schematic workflow of the experiment to detect structural variants (SV) with different read coverage. Software used is shown with underlined text

RESULTS AND DISCUSSION

Table 1 summarises the SV discovery statistics at each scaled back read depth compared to the 50X cover. For example, on average at 3X we discover approximately 20% fewer SV, 93% fewer SVs with high quality genotypes (GQ > 10), and the sporadic missing genotype rate can be up to 21%. Missing rate is an important statistic to consider because the missing rate may impact the accuracy of imputing these sporadic missing for downstream use of SV genotypes. At 15X cover the summary statistics are much closer to the 50X cover compared to the 3X or 5X. In Table 2 we summarise the observed proportion of SV that violate Mendelian consistency at each scaled back read depth. Interestingly, the rate of Mendelian inconsistency only slightly increased with lower read cover: varying from 3-11% with larger variability between replicates at lower read depth. This estimate has some bias because we can only assess SV for which no individual in the trio has a missing genotype. Given that this number is relative to the total SV discovery at each coverage, this demonstrates that even at lower read-depth, if the SV are confirmed, the majority would be accurately genotyped in all animals. This might be expected because as coverage reduces it is likely that the merged set will be those SV that are relatively the easiest to detect.

It is important to note that a small number of Mendelian inconsistencies may arise from *de novo* mutation: for SNP this is in the order of ~30 given a bovine genome size of 3 Gb and a per base per generation mutation rate of $1x10^{-8}$. It is likely that the number due only to SV would be similarly low. Therefore, here we made no attempt to differentitate *de novo* mutation from false positive SV as it would have no impact on our conclusions.

Table 1. Summary of Structural Variant (SV) discovery for different scaled back read depths, averaged across 2 trios of parent-offspring bovine (3 replicates) after join calling as well as in the original 50X read coverage. Average standard deviation between 2 trios is shown in brackets

	Number of SV called	% non-missing genotypes among trio individuals	% genotypes where of all trio individual genotypes had GQ score[1] > 10
3X	30,484.7 (40.4)	78.3 (0.1)	3.3 (0.1)
5X	37,015.5 (60.5)	93.9 (0.04)	29.4 (0.2)
10X	38,042.7 (57.8)	98.6 (0.04)	59.3 (0.1)
15X	38,292.7 (31.1)	99.2 (0.01)	73.9 (0.2)
20X	38,507.3 (8.5)	99.3 (0.02)	87.1 (0.1)
50X	38,513.5 (0)	99.3 (0)	96.1 (0)

[1] GQ is a composite mapping quality score that estimate the quality of the identified SV

Table 2. Summary of Structural Variants (SVs) observed in the offspring of two parent-offspring trios that show Mendelian consistency (Cons.) or inconsistency (Incons.) for a range of scaled back sequence coverage (average of 3 replicates) and in the original 50X coverage. Average standard deviation between the 2 trios is shown in brackets

	Number of Cons. SV	Number of Incons. SV	Rate of Incons.(%)
3X	21,585 (68.8)	2,189 (35.4)	9.2 (0.11)
5X	30,698 (29.7)	3,912 (58.3)	11.3 (0.13)
10X	34,470 (47.9)	3,001 (32.2)	8.0 (0.07)
15X	35,682 (43.4)	2,269 (26.3)	6.0 (0.06)
20X	36,616 (38.5)	1,598 (25.8)	4.2 (0.06)
50X	37,064 (0)	1,194 (0)	3.1 (0)

Our results demonstrate that the lower coverage of mapped reads increases the difficulty for the SV detection software to confidently call the genotype across multiple animals, particularly at 3X and 5X cover. This is likely partly due to the merging approach relying on there being at least one individual with good evidence of the SV and there being at least 5X cover of the SV region to call the genotype in each animal (as we are running with default settings), this perhaps explain the poor result of these two read depth coverages. Observing the missing rate, we can see that even at the highest read depth, in the merged calling of SV there are still around 1% of sporadically missing genotypes. We believe some of these missing genotypes are due to (i) complex SV that perhap require manual curation for accurate genotype calling, (ii) false positive SV that were either merged and/or joint called incorrectly. In addition, this study shows that while sequencing from between 10X to 15X coverage resulted in some reduction in the SV discovery rate versus higher read depth, this may be an acceptable compromise for population scale studies to spread sequencing costs over a larger number of animals. However if the purpose of using long-read sequencing is to discover a deleterious Mendelian mutation among a small group of known affected or carrier animals, the results here suggest that at least 20X cover would be preferable.

In this pilot study it is important to note that we deployed just one SV discovery program (Sniffles2), while there are several other programs currently available for this purpose. However, at the time of running this analysis, Sniffles2 was the software recommended for ONT long-read sequence with the best accuracy. In addition, Sniffles2 has an automated global/joint calling module that can automate calling of SV across population scale samples. Undoubtedly, results from this pilot study highlight the needs for further studies in resolving precise breakpoints and therefore, leading to more accurate genotyping of SV.

CONCLUSIONS

This study analysed the impact of sequencing read-depth on the detection of SV using two deeply sequenced bovine trios. Our results provide a means for future research to make decisions on optimising cost effective long-read sequencing cover for SV detection of either: (i) specific deleterious SV in a few individuals (iii) population scale genome-wide SV discovery or (iii) characterize an original set of SV such as for pan-genomes.

ACKNOWLEDGEMENTS

We are grateful for funding from DairyBio: a joint venture project between Agriculture Victoria (Melbourne, Australia), Dairy Australia (Melbourne, Australia) and the Gardiner Foundation (Melbourne, Australia).

REFERENCES

Bellott D.W., Hughes J.F., Skaletsky H., *et al.* (2014) *Nature* 508: 494.0.
Nguyen T.V., Christy V.D.J., Wang C.J. *et al.* (2023) *Genet. Sel. Evol.* 55: 9.
Freeman J.L., Perry G.H., Feuk L., *et al.* (2006) *Genome Res.* 16: 949,
Li H. (2018). *Bioinformatics*.34: 3094.
Rosen B.D., Bickhart D.M., Schnabel R.D., *et al.* (2020) *GigaScience* 9: 1.
Sedlazeck F.J., Rescheneder P., Smolka M., *et al.* (2018) *Nat Methods.* 15: 461.
Tarasov A, Vilella A.J., Cuppen E., Nijman I.J., Prins P. (2015) *Bioinformatics*..31: 2032.

HERITABILITY AND REPEATABILITY OF PATERNAL HAPLOTYPE RECOMBINATION RATE IN BEEF CATTLE AUTOSOMES

M.H. Ferdosi[1], S. Masoodi[2] and M. Khansefid[3]

[1] Animal Genetics Breeding Unit*, University of New England, Armidale, NSW, 2351 Australia
[2] Computer Engineering Group, University of Payam Noor, Tehran, Iran
[3] AgriBio Centre for AgriBioscience, Agriculture Victoria, Bundoora, VIC 3083, Australia

SUMMARY

Recombination and de novo mutations generate genetic diversity in a population, which is the key element for evolution and selective breeding. The variation in recombination rate across the genome and the recombination hotspots can be estimated by haplotype analysis. However, the crossing-over rate is not uniform across different individuals. In this research, we estimated the recombination rate across the autosomal chromosomes of 4 Australian beef cattle breeds. Further, we estimated variance components, heritability and repeatability of recombination rate within each breed.

INTRODUCTION

During meiosis, haplotypes exchange Deoxyribonucleic acid (DNA) strands as a result of recombination processes, which contribute to the genetic diversity of the next generation. Genetic diversity is an essential element for natural and artificial selection. The change in genetic diversity across generations mainly depends on selection, the reduction in genetic variation due to genetic drift and inbreeding, and the amount of generated variation as a result of de novo mutations and recombination events (REs). In humans, recombination rates vary by gender and on average there are 1.65 times more autosomal crossing-over events in maternal than paternal haplotypes. In addition, recombination rate is higher near centromeres in females and near telomeres in males (Kong *et al.* 2002). In male beef cattle, mutations in REC8 (Sandor *et al.* 2012), CLPX1, (Ma *et al.* 2015) and RNF212 (Kong *et al.* 2002; Sandor *et al.* 2012) genes have been reported to affect genome-wide recombination rates. Progeny of sires with high recombination rates may have higher genetic diversity at each chromosome. Hence, depending on the selection criteria, the recombination rate of paternal chromosomes can be considered in selecting superior individuals to produce the next generations.

Based on phased data generated by Beagle (Browning and Browning 2007) and DAGPHASE2 (Druet and Georges 2010), Weng *et al.* (2014) estimated the recombination rates in Angus and Limousin cattle breeds. They tried to minimise the effect of wrong phasing in their results by removing anomalies in the phased genotypes like double crossover at short intervals, more than three crossovers per chromosome, and haplotype mismatch. These factors could substantially affect the ability to identify the number of REs correctly. Ferdosi *et al.* (2016) developed a maximum likelihood algorithm to identify paternal haplotype REs. This method was an extension to hsphase (Ferdosi *et al.* 2014) to identify REs in the paternal strand of half-sib families. It is robust to genotyping errors and does not require phased genotypes to identify REs. Our aim in this study was to estimate the heritability and variation of genome-wide recombination numbers in paternal haplotypes (GRNP) of Brahman, Hereford, Santa Gertrudis, and Wagyu without phasing their genotypes.

* A joint venture of NSW Department of Primary Industries and University of New England

MATERIALS AND METHODS

Genomic Data and estimation of recombination rate. The genomic data for this study was extracted from the BREEDPLAN genomic pipeline (Connors *et al.* 2017). The BREEDPLAN genomic pipeline was developed at the Animal Genetics and Breeding Unit (AGBU) and is commercialised by the Agricultural Business Research Institute (ABRI). This pipeline performs several quality control steps and consolidates several marker densities together. For example, the individuals were removed if they failed parent verification due to Mendelian inconsistency or other issues, had less than 79% calls with GC score less than 0.6, less than 80% call rate, average GC less than 0.6 or had more than 80% homozygosity rate (for more details, please refer to (Connors *et al.* 2017). To be able to estimate the paternal chromosomal REs accurately, the sires with more than eleven genotyped progenies were used in our study (Table 1). The pedigree was also extracted from the BREEDPLAN genomic pipeline for the selected individuals up to 3 generations. The GRNP was estimated in each offspring using hsphase 2 (Ferdosi *et al.* 2016).

Table 1. Number of sires and genotyped progeny and range of half-sib family size in different beef breeds after quality control and removing half-sib families with less than 12 progenies

Breed	Number of Sires	Range of Half-sib family size (mean ± s.d.)	Number of Individuals
Brahman	789	12 to 288 (33.65 ± 26.54)	26,491
Hereford	1,125	12 to 584 (34.32 ± 37.96)	38,609
Santa Gertrudis	164	12 to 145 (34.28 ± 23.88)	5,622
Wagyu	1,760	12 to 3245 (61.23 ± 148.22)	107,763

Variance components – repeatability model. The heritability and repeatability of recombination rate for each breed were estimated using the following model: $y = Xb + Zu + Wp + e$, where X, Z and W are design matrices that relate observations to their corresponding effects, and y, b, u, p, and e are the vectors containing the number of REs of paternal autosomal chromosomes in progeny, fixed effects (mean), predicted breeding values, sire permanent environment effects (PE) and random residual terms, respectively. The variance of EBVs, PE and residual effects were assumed to be normally distributed with $u \sim N(0, A\sigma 2u)$, $p \sim N(0, I\sigma 2pe)$, and $e \sim N(0, I\sigma 2e)$, respectively, where A is the Numerator Relationship Matrix (NRM) built using pedigree and I is an identity matrix. ASReml-R was used to estimate the variance components, heritability and repeatability of GRNP (Gilmour *et al.* 2015).

RESULTS AND DISCUSSION

The estimated GRNP in four cattle breeds using hsphase 2 in Santa Gertrudis, Wagyu, Hereford, and Brahman had on average 28, 27, 25, and 25 REs in autosomes, respectively. The normal distributions of estimates and the range of GRNP were in line with the previously published articles (Chowdhury *et al.* 2009; Weng *et al.* 2014). Weng *et al.* (2014) reported GRNP of 27.4 and 26.9 for Angus and Limousin, respectively. The number of genome wide REs ranged from 0 (Brahman and Hereford) to 59 (Wagyu). Figure 1 shows the median, first quartile and third quartile of REs in each half-sib group. There was large variation in GRNP across half-sib groups, which could be partially explained genetically (Table 2).

The boxplot of the number of REs by chromosome is shown in Figure 2. The average number of REs in chromosomes 1 to 20 was higher (close to 1) than other autosomal chromosomes (close to 0). Weng *et al.* (2014) removed the individuals with more than three REs in each chromosome from their study. However, the individuals which had high GRNP in Figure 2 were not removed in our study due to the high reliability of the hsphase 2 algorithm in detecting crossing-over events.

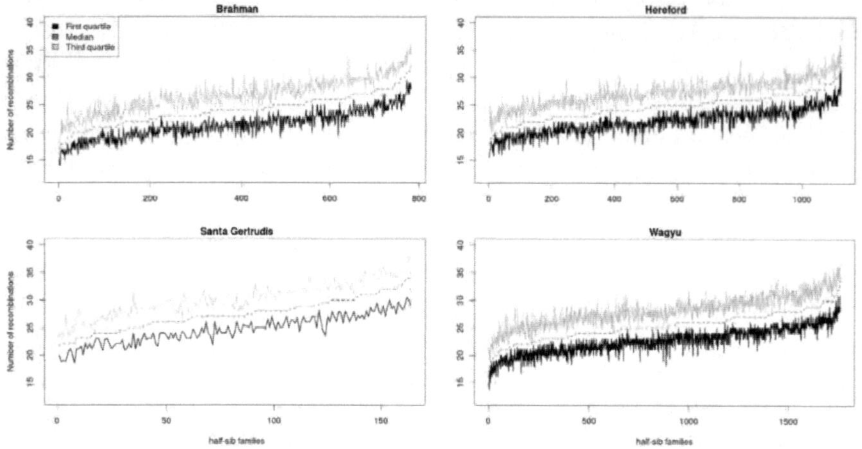

Figure 1. First quartile, median and third quartile of number of recombination events in different half-sib families sorted by median of number of genome-wide recombination numbers in paternal haplotypes

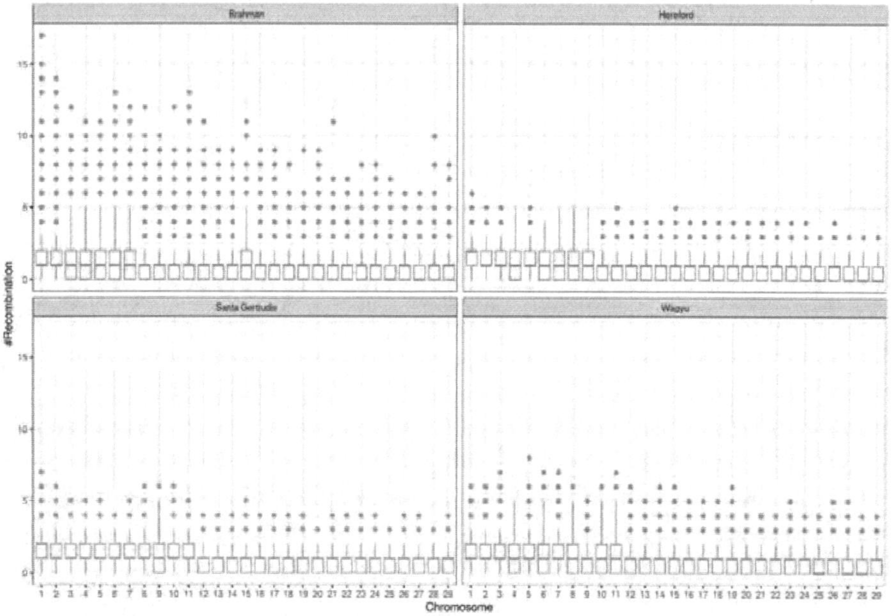

Figure 2. Boxplot of number of recombination events in 29 autosomal chromosomes in different breeds

The GRNP in some Brahman individuals was higher than our expectations. These individuals must bc investigated further to identify the possible reason behind their strangely high recombination number estimates. This issue may be caused by the *Bos Taurus* map assemblies, as this map may not be adequate for mapping SNPs in the *Bos Indicus* cattle genome. However, removing these individuals had a negligible effect on the variance component estimation.

Variance components, heritability and repeatability of the number of REs are shown in Table 2. Weng *et al.* (2014) have reported heritability of 0.26 ± 0.030 and 0.23 ± 0.042 for recombination rate in Angus and Limousin sires, respectively, which were higher than our estimates. The rate of chromosome recombination is proportional to chromosome length and also varies between individuals. However, the identification of crossing-overs can be influenced by the level of heterozygosity in the parents (Weng et al. 2014). Assuming the sire is completely homozygous, no REs can be detectable in the progeny. High homozygosity caused by low quality genotypes was not a concern in our study, as the BREEDPLAN genetic data passed the stringent quality control pipeline, and any individual with greater than 80% homozygosity was eliminated from the dataset. For example, although Australian Wagyu had very low haplotype diversity (Ferdosi *et al.* 2021), the number of detected REs in Wagyu was very similar to other breeds in our study.

Table 2. Additive genetic (σ^2_u), permanent environment (σ^2_{pe}) and **residual (σ^2_e)** variances, and the estimated heritability (h^2) ± s.e., and repeatability (r) ± s.e. of genome-wide recombination numbers in paternal autosomal chromosomes of different beef breeds

Breed	σ^2_u	σ^2_{pe}	σ^2_e	h^2	r
Brahman	1.57 ± 0.68	4.80 ± 0.67	19.10 ± 0.17	0.06 ± 0.03	0.25 ± 0.01
Hereford	3.08 ± 0.43	1.21 ± 0.31	17.90 ± 0.13	0.14 ± 0.02	0.19 ± 0.01
Santa Gertrudis	3.84 ± 2.03	3.56 ± 1.77	21.86 ± 0.42	0.13 ± 0.08	0.25 ± 0.09
Wagyu	2.97 ± 0.38	2.30 ± 0.25	20.33 ± 0.09	0.12 ± 0.02	0.21 ± 0.02

CONCLUSIONS

There was a large variation in the frequency of GRNP across individuals. The heritability of the number of REs was similar in different beef cattle breeds in our study, except Brahman, which was lower and could be a result of the *Bos Taurus* genome assembly used. A high GRNP in sires may contribute to an increase in population diversity. However, the underlying mechanisms and consequences of variation in REs in different individuals need to be investigated in future studies.

ACKNOWLEDGEMENTS

This study was supported by Meat and Livestock Australia project L.GEN.1704 and L.GEN 2204. The authors want to thank ABRI, the Australian beef societies and their members for providing the data for this study.

REFERENCES

Browning S.R.and Browning B.L. (2007) *Am. J. Hum. Genet.* 81: 1084.
Chowdhury R., Bois P.R.J., Feingold E., Sherman S.L. and Cheung V.G. (2009) *Plos Genet.* 5: 9
Connors N., Cook J., Girard C., Tier B., Gore K., Johnston D. and Ferdosi M.H. (2017) *Proc. Assoc. Advmt. Anim. Breed. Genet.* 22: 317.
Druet T. and Georges M. (2010) *Genetics* 184: 789.
Ferdosi M.H., Boerner V. and Tier B. (2016) *Proc. Inter. Conf. Quant. Genet. 5:* 150.
Ferdosi M.H., Connors N.K. and Khansefid M. (2021) *Proc. Assoc. Advmt. Anim. Breed. Genet. 24:* 118.
Ferdosi M.H., Kinghorn B.P., van der Werf J.H.J., Lee S.H. and Gondro C. (2014) *BMC Bioinformatics* 15: 172.
Gilmour A.R., Gogel B.J., Cullis B.R., Welham S. andfThompson R. (2015) *VSN International Ltd, Hemel Hempstead, HP1 1ES, UK. www.vsni.co.uk*
Kong A., Gudbjartsson D.F., Sainz J., Jonsdottir G.M., Gudjonsson S.A., Richardsson B., Sigurdardottir S., Barnard J., Hallbeck B., Masson G., Shlien A., Palsson S.T., Frigge M.L., Thorgeirsson T.E., Gulcher J.R. and Stefansson K. (2002) *Nat Genet* 31: 241.
Ma L., O'Connell J.R., VanRaden P.M., Shen B.T., Padhi A., Sun C.Y., Bickhart D.M., Cole J.B., Null D.J., Liu G.E., Da Y. and Wiggans G.R. (2015) *Plos Genet* 10: 8.
Sandor C., Li W.B., Coppieters W., Druet T., Charlier C. and Georges M. (2012) *Plos Genet* 8: 7.
Weng Z.Q., Saatchi M., Schnabel R.D., Taylor J.F. and Garrick D.J. (2014) *Genet Sel Evol* 46: 34.

CHARACTERISING HETEROZYGOSITY OF THE X CHROMOSOME IN THE AUSTRALIAN WAGYU POPULATION

D.P. Garrick[1,2,3], R.C. Handcock[1,3] and C. Teseling[4]

[1]The Helical Company Limited, Rotorua, BOP, New Zealand
[2]Theta Solutions LLC, Lacey, WA, USA
[3]A.L. Rae Centre for Genetics and Breeding, Massey University, New Zealand
[4]Australian Wagyu Association, Armidale, NSW, Australia

SUMMARY

A pedigree inbreeding coefficient is the probability that two alleles in an individual will be identical by descent which requires them to be homozygous for that locus. Homozygosity at the non-pseudo autosomal region (nPAR) of the sex chromosomes is complicated by their unique inheritance patterns. Heterozygosity at the nPAR X chromosome region is frequently used to predict the sex of genotyped animals for quality control purposes but the characteristics of the X chromosome in the Australian Wagyu population can make such sex predictions inconclusive.

INTRODUCTION

Managing the trade-offs between genetic diversity, inbreeding, and genetic gain is a known challenge in animal breeding programs. Pedigree based inbreeding coefficients provide an on-average indication of inbreeding across the entire genome but do not reflect the true genomic diversity, especially in specific genome locations that may be under more intense selection pressure. For example, non-identical twins, just like any siblings, will inherit different combinations of their parents' genomes. When their parents are related, it increases the chance of the twins inheriting more regions that are identical by descent (IBD). However, the specific IBD regions each twin inherits could differ due to the random assortment of genes, leading to unique genetic variations between the twins.

The X chromosome has a unique inheritance pattern in that mammalian males inherit only one copy of the X chromosome, from their dam. Loss of genetic diversity in the X chromosome can be exacerbated by widespread usage of a small number of sires, particularly in closed populations where a historical population bottleneck has reduced diversity. In other words, the X chromosome has a smaller effective population size than the autosomes which are not involved in determining the sex of an individual. A smaller effective population size means a faster accumulation of homozygosity for the same selection strategy.

Mammalian females inherit one X chromosome from each parent and typically inherit alleles that result in some of the loci being heterozygous, e.g., "AB". Meanwhile males who inherit one X and one Y chromosome are said to be hemizygous and will appear to exhibit homozygosity at loci that are unique to the X chromosome, known as the non-pseudo autosomal region (nPAR).

When parents are related the offspring will exhibit increased rates of homozygosity and reduced genetic diversity relative to offspring of matings from unrelated parents. Consider the extreme case where a female's father was also her maternal grand sire, i.e., her mother's father. In that case, in the absence of recombination along the X chromosome, there would have been a 50% probability that two identical X chromosomes would have been inherited.

Reduction in genetic diversity and an increase in inbreeding within a population can have negative effects on the adaptability or fitness of the population. It can also impact genomic predictions which rely on X chromosome data. The X chromosome is routinely used to predict the likely gender of a genotype sample in order to assist with sample identification and quality control for curation and animal evaluation purposes (McClure *et al.* 2018). The heterozygosity of the nPAR

X chromosome SNPs is commonly used where a threshold value (or values) of heterozygosity indicates whether the "genotype sex" of a sample can be determined as male, female, or ambiguous. Reduced diversity in the X chromosome can lead to a higher rate of ambiguous or incorrect sex predictions for genotype samples of legitimate females.

This research characterises heterozygosity of X chromosomes in Australian Wagyu cattle.

MATERIALS AND METHODS

The data utilised in the study is the Australian Wagyu Association's genotype database containing > 323,000 SNP genotype samples representing more than 3,600 different chips or manifests.

The 280 nPAR X chromosome SNPs, the 101 PAR X SNPs, and 7 Y nPAR chromosome SNPs provided by McClure *et al.* 2018 were utilised for the study. The data set was first reduced to those samples with a raw locus call rate >= 0.95. Second, samples had to have been recorded as males or females in the pedigree and whose genotype samples were predicted as male or females respectively using the Irish Cattle Breeding Federation (ICBF) Y chromosome sex prediction described by McClure *et al.* 2018. This resulted in a final set of 73,814 female and 48,818 male samples. The test checks for called Y chromosome SNPs only on samples genotyped on chips where at least 6 Y chromosome SNPs were present, i.e., samples where no Y chromosome SNPs were available on the chip were discarded. The Y chromosome test was chosen due to its greater accuracy over the X chromosome sex test according to McClure *et al.* 2018 and Garrick 2019. Males required 6 or 7 called Y chromosome SNPs to be verified and females could not have more than 1 called Y chromosome SNP. The chrX sex test considers the heterozygosity of the nPAR X SNPs where if the heterozygosity is low (<= 5%) the sample is considered male, if high (>= 15%) considered female, with moderate heterozygosity (>5% and <15%) considered ambiguous.

The complete Australian Wagyu Association pedigree was utilised to compute pedigree inbreeding coefficients for all animals.

The heterozygosity of each sample was computed as the percentage of loci with heterozygote AB calls divided by the total number of loci with called values, e.g. #AB/(#AB+#AA+#BB).

The average pedigree inbreeding and chrX heterozygosity by birth year were computed as the simple mean of those coefficients across the sex verified genotyped animals born in that year.

The pedigree inbreeding coefficient calculation, genotype database extracts, sex tests, manipulations and analysis of genotype data were all undertaken using the "helical" command-line software package (Garrick *et al.* 2023).

RESULTS AND DISCUSSION

The animals with sex verified genotypes were examined by plotting the pedigree inbreeding coefficient against the nPAR and PAR X chromosome heterozygosities for females (Figure 1), and males (Figure 2). There is no evidence of a relationship between pedigree inbreeding and heterozygosity. The Pearson correlation for chrY verified females between pedigree inbreeding and heterozgosity was -0.28 for PAR X SNPs and -0.09 for nPAR X SNPs. For males the PAR and nPAR correlations with inbreeding are -0.009 and. -0.001 respectively. Interestingly, while the -0.28 correlation for PAR X SNPs for females is considered a low correlation, the value is notably higher compared to the nPAR X SNPs. The pedigree inbreeding coefficient is limited by recorded pedigree information, and while only 3,632 of the 73,814 genotyped females were missing either sire or dam information in the pedigree, without a complete pedigree inbreeding coefficients will be underestimated for some animals.

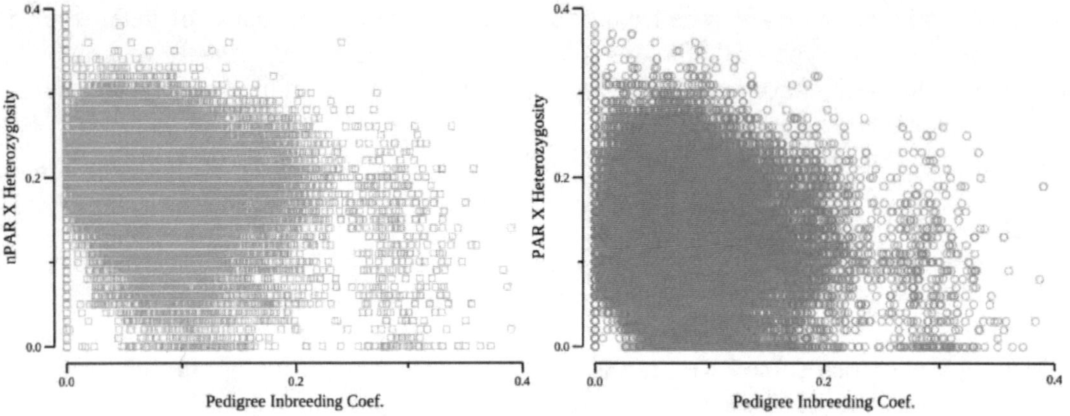

Figure 1. nPAR (left) and PAR (right) chrX heterozygosity versus pedigree inbreeding coefficients for chrY verified females

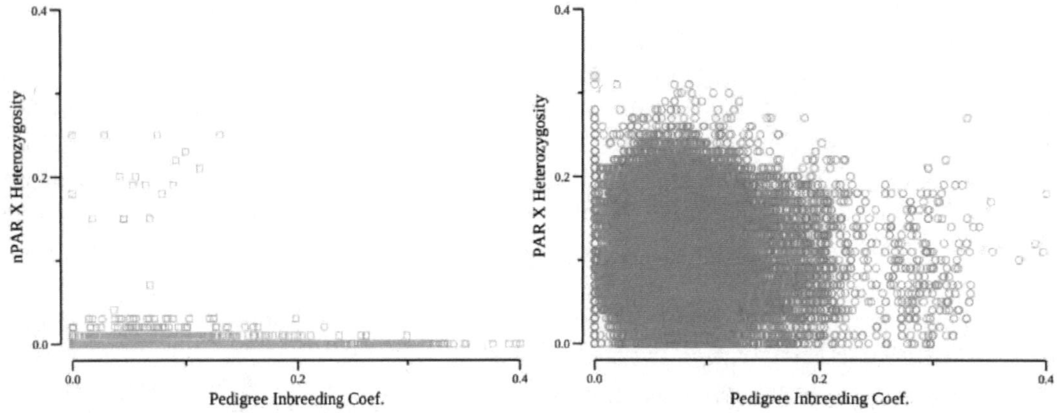

Figure 2. nPAR (left) and PAR (right) chrX heterozygosity versus pedigree inbreeding coefficients for chrY verified males

Comparing Figures 1 and 2 highlights the capability of the nPAR X chromosome SNPs to help confirm genotype samples originating from males and demonstrates no relationship between inbreeding and heterozygosity. However, using nPAR X chromosome heterozygosity alone, no clear demarcation point can be identified to accurately classify all females as some legitimate females exhibit low heterozygosity.

Consider a heifer who inherits two haplotypes on the nPAR X chromosome – one from the single copy in her sire, and one from one of the two copies in her dam. Cases of inbreeding where the X carried by her sire is unrelated to the X carried by the mother – for instance if her paternal and maternal grand sires were the same animal – then inbreeding will not be related to homozygosity. This is because her sire inherits his nPAR X chromosome from his mother. If however the heifers sire is also her maternal grand sire, then she may have inherited the same X regions from her sire and dam, and inbreeding will be related to homozygosity. A pedigree-based measure to characterise the unique inbreeding associated with the nPAR inheritance pattern may help management of nPAR X chromosome diversity.

Table 1 summarises nPAR X heterozygosity and inbreeding by year of birth for chrY verified females grouped by the chrX nPAR sex predictions of low, medium, and high heterozygosity. Approximately 1.2% of the females are not distinguishable from males according to the chrX sex prediction, while approximately 9.9% are ambiguous. The average inbreeding in the low heterozygosity group is over double that in the high heterozygosity group at 0.13 versus 0.06, compared to 0.09 for the medium heterozygosity (ambiguous) group. The average pedigree inbreeding increased from roughly 0.03 to 0.07 over the last 20 years for the high heterozygosity group, while at the same time the number of females classified into this group dropped from ~98% to 90%.

Table 1. Grouped by the X chromosome heterozygosity as per chrX sex prediction class, we calculate: the average pedigree inbreeding (F), the mean heterozygosity of chrX (h_m), and the percentage of total individuals N within a birth year that are chrY verified females

nPAR X	h <= 0.05 (predicted male sex)			0.15 > h > 0.05 (ambiguous)			h >= 0.15 (predicted female sex)			
Birth yr.	h_m	F	% of N	h_m	F	% of N	h_m	F	% of N	N
2000	0	0	0.0	0.09	0.06	2.2	0.22	0.03	97.8	45
2002	0.05	0.03	0.9	0.11	0.1	2.6	0.2	0.03	96.6	117
2004	0	0	0.0	0.12	0.14	8.5	0.2	0.05	91.5	141
2006	0.03	0.09	0.8	0.12	0.08	5.1	0.2	0.05	94.1	389
2008	0.03	0.12	1.3	0.12	0.08	8.1	0.21	0.05	90.7	1190
2010	0.04	0.14	1.0	0.11	0.1	10.0	0.21	0.06	89.0	1171
2012	0.03	0.12	0.9	0.12	0.09	7.0	0.21	0.06	92.1	1716
2014	0.02	0.15	1.3	0.11	0.09	9.4	0.21	0.06	89.2	2965
2016	0.03	0.11	1.5	0.12	0.08	10.4	0.21	0.06	88.1	5153
2018	0.03	0.13	1.3	0.11	0.09	11.2	0.21	0.06	87.5	8523
2019	0.03	0.13	1.3	0.11	0.09	11.5	0.21	0.07	87.2	9211
2020	0.03	0.13	1.3	0.11	0.1	10.0	0.21	0.07	88.8	10914
2021	0.03	0.13	1.1	0.11	0.1	9.3	0.21	0.08	89.6	12412
2022	0.03	0.12	0.9	0.12	0.09	9.4	0.21	0.07	89.7	2584

CONCLUSIONS

The nPAR X chromosome has a unique inheritance pattern which means standard pedigree inbreeding coefficients cannot accurately characterise the probability of identical inheritance by descent of two alleles. A new pedigree-based inbreeding measure could account for the fact that homozygosity in the non-pseudoautosomal (nPAR) region of the X chromosome is expected only when both parents of an offspring share a common ancestor. This allows for the inheritance of identical nPAR X chromosome segments from both the sire and dam. Reduction in the heterozygosity of the nPAR X chromosome in the Australian Wagyu population over the last 20 years creates challenges in sex prediction associated with genotype quality control.

REFERENCES

McClure M.C., McCarthy J., Flynn P., McClure J.C., Dair E., O'Connell D.K. and Kearney, J.F. (2018). Front. Genet. 9: 84.

Garrick D.P. (2019) *Proc. Assoc. Advmt. Anim. Breed. Genet.* 23: 23.

Garrick D.P., Howse, J. E. and Garrick, D.J. (2023). Retrieved from https://www.helicalco.com/

Proc. Assoc. Advmt. Anim. Breed. Genet. 25: 230 - 233

MEASURED GOATS: AN OVERVIEW OF A MEAT GOAT REFERENCE POPULATION

T. Granleese[1], S.I. Mortimer[1], T. Atkinson[4], G. Refshauge[3], T. Bird-Gardiner[4], F. Haynes[1], D.J. Brown[2], P. Alexandri[2] and S.F. Walkom[2]

[1]NSW Department of Primary Industries, Armidale NSW, 2351 Australia
[2] Animal Genetics Breeding Unit*, University of New England, Armidale, NSW, 2351 Australia
[3] NSW Department of Primary Industries, Cowra, NSW 2794 Australia
[4] NSW Department of Primary Industries, Trangie NSW2823 Australia

SUMMARY

New South Wales Department of Primary Industries' second-most western station, Condobolin Agricultural Research and Advisory Station will be host to the "Measured Goats in the Rangelands project". This five-year co-investment between New South Wales Department of Primary Industries and the Meat and Livestock Donor Company project will also work collaboratively with the Animal Genetics and Breeding Unit. The project will utilise three goat breeds – Boer, Kalahari Red and wild "Rangeland", both in purebred and crossbred forms to become a multi-breed genomic reference population. All project animals born will have a goat specific 70k SNP genomic test to identify parentage, breed composition and heterozygosity. Furthermore, performance, health, reproduction and structural traits will be recorded in large contemporary groups. The aim is to breed and measure over 8,000 animals in a self-replacing style breeding nucleus over 4 years. This project aims to provide trait and breed means, update genetic parameter estimates for meat goats, obtain heterosis estimates, and provide new links into the KIDPLAN database. The project will also generate new traits and knowledge to update the assumptions used for the KIDPLAN analysis. Finally, the project has a major adoption and extension focus to increase the uptake and adoption of KIDPLAN breeding values at a seedstock and commercial level.

INTRODUCTION

Currently little is known about the phenotypic and genetic performance of Rangeland goat (captured feral goats) and their crossbred progeny. The Australian Rangeland goat population, however, underpins the Australian goat meat industry which is an export industry valued at $235 million per annum (MLA 2020). Rangeland does will continue to be the basis of the goatmeat industry owing to their numbers relative to the limited number of pure or crossbred does. To date, the majority of goatmeat is supplied from harvest enterprises that capture goats from a semi-feral state. However, the National Goat Meat Forecasting Committee and other industry sources, as well as an industry survey (Williams and Williams 2019), are reporting a rapid and unprecedented increase in the number of producers in NSW and Queensland operating managed and semi-managed goat enterprises.

Given the limited numbers of animals contributing to genetic parameter estimations in a single breed for meat goats, producers want to know whether crossbreeding approach will achieve production gains within their herds while maintaining the rangeland goat's hardiness and suitability to semi-arid and arid environments. Well-designed Research and Development programs can capitalise on this interest to engage producers in herd improvement using genetic strategies. There is also a need to investigate the performance and genetic variation within rangeland herds to determine and demonstrate the potential for production gains through selection.

---

* A joint venture of NSW Department of Primary Industries and University of New England

This proposal has been developed through a consultation process with goatmeat producers from NSW and Queensland who operate a range of enterprises (including extensive commercial, seedstock and harvesting) typical of the industry. With the industry focused on increasing national supply, transitioning harvest enterprises to managed production systems, improving farm profit and meeting consumers preferences, there is a clear need to prioritise working with producers to establish and implement selection and crossbreeding strategies that will realise the potential within rangeland herds for relatively rapid genetic gain in production and welfare traits.

There is considerable potential for genetic improvement in Rangeland herds, although the benefits from research in genetics and genomics are yet to be realised (Kijas 2012; Aldridge and Pitchford 2018). Kijas (2012) reported that the Rangeland population was one of the most genetically diverse in domesticated species in the world. One of his main conclusions was "if selection pressure was applied to almost any trait, the population would quickly respond and exhibit strong genetic gain.". Both Aldridge and Pitchford (2018) and Williams and Williams (2019) suggest that performance recording and understanding genetic parameters of Rangeland goats would help increase breeding goat enterprise's profitability by better understanding the genetic capability of Rangeland goat genetics and hence make more informed selection decisions.

Currently a small population of 400-600 Boer goats (KIDPLAN database) are performance recorded annually where KIDPLAN breeding values are calculated by Sheep Genetics (MLA). This pipeline needs to be used more as the benefits of genetic selection to improve performance, reproductive, health and survival traits has been well documented across all domestic livestock species farmed around the world. Given the potential for rapid gains due to high diversity, Rangeland goat production systems can rapidly increase performance and hence on-farm profitability.

This project aims to collect a dataset of up to 9,000 animals that will be genome tested using a 70k goat specific SNP chip while also recorded for phenotypes. The data collected will also be submitted to the KIDPLAN database. The dataset will include 3 breeds (Boer, Rangeland and Kalahari) so breed-by-breed comparisons can be made for an eventual multi-breed evaluation. Furthermore, the project has a comprehensive adoption and extension arm to help deliver a larger uptake of breeding values at a seedstock and commercial breeder level. Finally, the project will be available to overlay/sister projects eliciting considerable interest has been made if this project proceeds. This includes but not limited to providing links between current KIDPLAN participants, a potential KIDPLAN analysis upgrade and "Going Ahead with Goats" project led by Local Land Services and meat science overlays.

MATERIALS AND METHODS

Site. Condobolin Agricultural Research and Advisory Station will host the project in central-west New South Wales, Australia. The site is a New South Wales Government research station. Condobolin has a semi-arid climate with median annual rainfall of 386mm with a pasture growing profile between Autumn and Spring and is well-suited for goat research.

Time. The project will be run over 5 years starting in 2024. The project will require four years to achieve five kidding events and grow progeny out for performance recording. The final twelve months will be used for data analysis and adoption activities.

Breeds and numbers. Up to 1,000 breeding does will be joined 5 times in a self-replacing style breeding nucleus. The initial 1,000 breeding does will as near as possible equally representing the following breeds.

1) Rangeland goat population that have been sourced from multiple wild-harvest depots
2) High Boer content
3) High Kalahari Red content

Three breeds of bucks will be mated via artificial insemination and then back up syndicate mated in a self-replacing style reference population over the 4 years. Both purebred and crossbred (including reciprocal crosses) matings will be undertaken. Link sires to the KIDPLAN database will be used so all data can be used in future KIDPLAN analyses.

Does born in the project will be first eligible for mating at eight months of age. Female progeny will only be culled from the project for one of the following reasons:

1) Animal ethics issue where a physical and health affliction affects the animal's state of health
2) If a doe is scanned empty after two consecutive mating events
3) If a doe is unsuccessful in raising kids at two consecutive events

Traits recorded including genomics. Traits measured on project progeny are defined in Table 1. All parents and progeny in the project will be genotyped on Neogen's 70k goat specific panel via a tissue sample unit. This will allow pedigree to be assigned. Furthermore, genomic testing of each project animal will facilitate genomic based analysis including genetic parameter estimation, genome wide association studies, and estimation of breed composition and heterosis.

Table 1. List of traits to be measured on experimental animals. Wt: weight

Growth	Birth & Reproduction	Carcase & EQ	Hard to Measure	Others
Weaning Wt	Conception	Fat Depth	Faecal egg count[1]	Temperament
Post-weaning Wt	Litter size	Eye muscle depth		Structure
Yearling Wt	Kids weaned	Condition score		Body condition score
Adult Wt	Udder score			Horn Score
				Coat colour

[1]Where enough phenotypic variation is available

Sire selection. Sires will be used via artificial insemination and natural mating. Artificial insemination sires will be nominated by industry for a fee like the Merino Central Test Sire Evaluation (Swan *et al.* 1998) systems. Nominated bucks will be accepted according to their influence in industry and/or pedigree supplied and/or genetic diversity and/or a breeder seeking a genetic links to the KIDPLAN database to begin submitting performance records. Eligibility for inclusion in artificial insemination will require bucks who are currently in KIDPLAN and/or sires or sons of sires who have contributed significantly to the goatmeat population. This will provide sire linkage between databases. In addition, naturally mated back-up bucks will be purchased by NSW DPI from key industry seedstock herds according to selection criterion similar to that adopted for the selection of AI sires. Sire purchasing will also strive to capture as much of the goat meat population's genetic diversity. Given there is little pedigree recorded in meat goats, the project accepts it will not capture all the diversity of the goat meat population. However, it will be a foundation block to build on over time, like the beginnings of the Sheep CRC Information Nucleus Flock (van der Werf *et al.* 2010). As the project progresses, more sophisticated approaches to sire selection can be used such as optimisation of current and future contributions using optimal contribution methods (Wray and Goddard 1994).

Base dam selection. Base dams will be sourced from commercial breeders who can provide as high a content of each breed (Rangeland, Kalahari Red, Boer) as possible. The project will aim to source from no more than three properties each. If a depth of pedigree can be obtained (e.g. sires of the sourced females are genotyped to provide sire pedigree), this would be advantageous. Prior to the project beginning, dams will be given time to kid any potential existing pregnancies. This is necessary given many goat producers have little control of bucks (owned or feral entering properties) mating does.

Statistical analysis. Univariate animal models for single and/or repeated measures will be fitted using linear mixed model equations to do genetic parameter estimation of each trait. Genomic relationships will then be added to further refine estimations where breed proportions and heterosis can be accounted for.

Once univariate models are finalised, multi-trait analysis will be undertaken to better understand genetic relationships among traits.

Furthermore, genomic information will facilitate genome wide association (GWAS) analysis to investigate genetic markers that have large phenotypic effects such as horns, coat colour, muscling and potentially ovulation rate.

A feasibility study will also be conducted to examine whether data from this reference population will be suitable to become part of the KIDPLAN database. If successful, the KIDPLAN database would be expected to grow by 80% through this project, thereby providing key genetic links for existing and new studs and breeds into the KIDPLAN analysis. It will be key to providing a foundation to improve and update the KIDPLAN analysis plus adding new traits and industry relevant indexes.

CONCLUSION

The Measured Goats in the Rangelands project will form a valuable resource population for the growing goat meat industry. The project will introduce new breeds Kalahari Red and wild "Rangeland" to the KIDPLAN database as well as providing a well-structured genomic reference At the time of writing this paper, the project is at the stage of sourcing females and sires.

ACKNOWLEDGEMENTS

The Measured Goats in the Rangelands project is an initiative co-funded by NSW Department of Primary Industries and Meat and Livestock Australia Donor Company. The authors give thanks to the efforts of the Condobolin Agricultural Research and Advisory Station manager and their staff, the skilled team of technicians and scientists. Further, invaluable support and advice has been given by the project's steering committee and well as initial support from The Goat Industry Council of Australia (GICA) and the goat industry research, development and adoption committee (GIRDAC).

REFERENCES

Aldridge M., and Pitchford W. (2018) *B.STU.0295 MLA Final Report*

Kijas J. (2012) *B.GOA.0072 MLA Final Report*

Swan A.A, Coelli K., Casey A.E., and Atkins K.D. (1998) *Central Test Sire Evaluation Results*, 5, 32.

van der Werf J.H.J., Kinghorn B.P. and Banks, R.G. (2010) *Anim. Prod. Sci.* 50: 998.

Williams N. and Williams S. (2019) *B. GOA. 1902 MLA Final Report*

Wray N.R. and Goddard, M.E. (1994) *Gen. Sel. Evol.* 26: 431.

SELECTING FOR MORE METHANE EFFICIENT SHEEP

J.H.J. van der Werf

School of Environmental & Rural Science, University of New England, Armidale, NSW, 2531 Australia

SUMMARY

A bioeconomic model was developed to predict outcomes of scenarios to select for methane efficiency in Merino sheep. The model determines economic values for trait improvement according to different breeding objectives. A selection index approach is used to predict response to selection, assuming knowledge or assumptions about genetic parameters, economic values of breeding objective traits and phenotypic measurement information used to select animals, including information from reference populations for genomic selection. Breeding objectives were based on profit per DSE, reduction in overall flock methane output and methane production per kg lamb produced (defined here as methane efficiency). Results showed that methane production to produce a certain amount of lambs is affected not only by methane production per head, but also by reproductive rate. Methane output per kg lamb produced can be decreased by 3.5% per annum, with the effect of improving production and reproduction efficiency being stronger than the effect of reducing methane production per head.

INTRODUCTION

In the need to reduce the production of enteric methane by ruminants breeding programs can be used to select for sheep that produce less methane. However, reducing methane emission per head might not be the most optimal strategy, as methane production is correlated to feed intake and productivity traits. Bio-economic modeling of sheep production systems can be used to determine breeding objectives and economic values of traits in multiple trait selection indices. An overall breeding goal is required and this can optimize profit per unit of production, e.g. profit per ha or profit per product. Equally we can minimize methane output, either per head or per kg product. How these various breeding objectives compare can be explored by deriving the index weights for the various traits in the breeding objective and predicting selection response when applying these weights. The purpose of the paper is to compare various selection strategies in sheep breeding programs with special emphasis on reducing overall methane output in sheep production systems.

MATERIALS AND METHODS

Increasing reproduction rate and other output traits for a fixed number of breeding ewes results in more lambs per breeding ewe and more feed requirement and more output overall. Therefore, it is relevant to calculate profit and output for a fixed amount of feed resource input, i.e. profit per dry sheep equivalent (DSE). The methane production of the flock was calculated for a fixed number of breeding ewes, in which case an increased reproductive rate increases the number of lambs produced by the flock, which increases the overall methane yield. However, the methane yield per kg lamb carcass produced could be lower, as fewer ewes are needed to produce the same number of lambs. Therefore, methane yield, assessed as the amount of CO_2 Eqvt (kg) produced per kg of lamb carcass is a relevant measure of measuring methane efficiency.

A production model was used to calculate profit and outputs based on the average phenotypic value of breeding objective traits. A model was based on a Merino flock with 100 breeding ewes, focusing on ten key traits that define profitability. Note that the flock size of 100 is just for convenience and actual size is not relevant as outputs are scalable per breeding ewe. All assumptions in the production model are initial 'ballpark' guesses with the aim to compare different breeding objectives and their associated selection response. The average methane production (MP) per mature ewe per day in the base situation was 24 g/day which was 8.76 kg/year. The MP of a slaughter lamb was assumed to be 90% of that of an adult ewe. The methane yield per breeding ewe, including the slaughter and young replacement ewes associated with each breeding ewe is then 15.35 kg/per breeding ewe per year which is 42.98 CO_2 Eqvt tonne per year for a flock with 100 breeding ewes. The MEMJ requirement for different periods in the life of lambs, replacement ewes and mature ewes were derived from Thompson *et al.* (1985), with a price of feed ($/MJME) between $0.01 and $0.03. It was assumed that 1 kg feed contains on average 10 MJME, with some variation in feed quality leading to differences in MJME cost for different ages. The price for lamb carcass was $6.50/kg whereas the wool price was $10/kg. The carbon price was assumed to be $40 per tonne CO_2 Eqvt. Variation in fleece weight and fibre diameter did not contribute to differences in methane production and neither was any variation in carcase fat and carcass eye muscle depth assumed to be related to feed requirement or methane production. It is important to note that a bioeconomic model for the purpose of deriving economic weights for genetic improvement requires partial derivatives of the profit, or any other objective, with respect to the trait means. This means that to calculate the economic value of one trait, only one trait at a time is changed, assuming that other traits are constant. This might seem counter-intuitive, e.g. for mature body weight, as typically one would expect more feed intake if animals become larger. However, these relationships are captured by the correlations applied in the selection index model that determines optimal trait responses for a given set of economic values and a given amount of information measured to select animals. Note that only changes in the mean of reproduction traits affect the methane production in the flock as for the same number of breeding ewes, more lambs per ewe will lead to more methane production. The ten breeding objective traits modelled are given in Table 2 along with their means in the base situation.

Economic values were calculated as partial derivative of a breeding objective criterion with respect to trait means. Breeding objective criteria were i): optimising profit per ewe for a fixed feed resource, i.e. per dry sheep equivalent (DSE) (BrObj1) ii) total methane production of a 100-ewe flock (BrObj2) and iii) the amount of methane (kg CO_2 Eqvt) produced per kg of lamb (BrObj3). Table 1 gives the results for economic weights of some objective traits for these three breeding objectives. The weights in BrObj2 suggest that to reduce methane, one should select against more reproductive ewes as fewer lambs per breeding ewe produce less methane per breeding ewe overall. The third breeding objective is more relevant where kg CH_4 per kg lamb carcase is minimized. This objective results in positive weights for slaughter weight and reproduction traits and a negative weight for methane production per ewe.

Table 1. Economic weight (standardized) for four different breeding objectives

	Profit per head for fixed DSE (BrObj1)	Methane yield per ewe (BrObj2)	kg methane/ kg carcass (BrObj3)
Slaughter Weight (9 mo)	$3.03	0.00	3.03
Fertility (pregnancy rate)	$146.03	-89.00	173.01
Lambing rate (lambs weaned/lambing)	$72.81	-44.50	85.48
Mature ewe Weight kg	$0.75	0.00	0.00
Daily DM Feed Intake (kg/day)	-$34.98	0.00	0.00
Methane production (g per ewe/day)	-$0.89	-6.29	-6.29

Selection response was calculated based on selection index theory. Genetic and phenotypic parameters were taken from Brown and Swan (2015) with parameters related to methane yield and feed intake largely based on Robinson *et al.* 2016. The genetic correlation between feed intake and methane production (MP) was assumed to be 0.8 and genetic correlations of MP with mature weight, fertility and lambing rate were 0.6, 0.10 and 0.10, respectively. Selection of breeding animals was optimised across age classes with information used for selection typical for traits routinely measured on-farm and MP and feed intake traits predicted from a reference population via genomic testing, assuming a reliability (accuracy-squared) of 0.10.

RESULTS AND DISCUSSION

Selection response per annum are given for breeding objective traits in Table 2. The results of genetic change in the MP per ewe, the MP per 100 ewe flock and MP per kg lamb are also given in Table 2. Results show that a breeding objective that maximizes profit per head for a fixed amount of DSE increases the methane output per breeding ewe because ewes have more lambs and there is a correlated response to selection for larger ewes and lambs. However, the methane yield per kg lamb carcase produced is about 2.7% lower than without selection.

Selecting only for lower MP per ewe in BrObj2 leads to lower methane production per ewe, but all traits respond negatively, such that profit decreases by $5.52 per annum, rather than increasing by $6.65 per annum. Note that BrObj2 only uses (negative) weights for reproduction traits and methane yield (Table 1), but these traits are correlated to weight traits. BrObj3 leads to the largest reduction in methane yield per kg lamb carcase produced. Under this scenario, the methane yield per breeding ewe increases because of an increase in mature size and because each breeding ewe produces more lambs, but the increase is about half of that when selection for profit (BrObj1). The increased productivity under BrObj3 means that every year of genetic improvement gives a 3.5 percent reduction in methane for the same amount of lamb meat. The profit increase due to genetic improvement is about 17% lower than under BrObj1. This indicates that the carbon price of $40 per ton CO_2 Eqvt is not having a large effect on the selection response.

The predicted response under BrObj1 varies little between no price on carbon and a carbon price of $400, with the response of MP changing most, from 0.22 to 0.14 g/day. A carbon price of ~$900 would be required for zero response in MP under BrObj1. A very high carbon price would result in a similar response as BrObj2. In a scenario under BrObj3 where MP and feed intake are not measured in a reference population, the increase in MP would be 35% higher compared to results in Table 2, whereas the increase in feed intake would be 10% higher. This would also allow a slightly higher (~3%) response in production and reproduction trait responses, and the overall effect on methane output per kg lamb produced would be small. However, the effect on profit increase per annum would be 14% lower, mainly due to the higher cost of MP. Therefore, measuring MP and feed intake has a limited effect on methane/kg lamb, but it allows more improvement in productivity and reproduction traits while limiting and increase in methane output per kg lamb produced .

Knowledge of genetic parameters of feed intake and MP in sheep is still limited and a current MLA-EPA project aims to collect a lot more data on these traits. It is also unclear how methane production changes between lamb and ewe stages and whether the genetic correlation between MP measured in these different stages is close to 1. Therefore, results in this paper are preliminary. However, they already give a clear picture of the various perspectives from which methane efficiency breeding objectives can be based on. The paper has not considered functions of traits such as residual methane production (methane production adjusted for body weight and production traits) or methane yield (methane production per kg feed intake). Whereas breeding objectives can be defined as productivity ratios, it is not useful to define objective traits as function of traits. Especially ratios of traits have undesirable properties in selection index schemes as they tend to be less normally distributed and could give rise to non-linearity in the breeding objective.

Previous studies such as Robinson *et al.* (2016) and Gebbels (2022) have also shown that optimal breeding strategies do not aim to reduce methane production per ewe, as this tends to result in lower feed intake, lower growth rates and lower reproductive performance, hence overall reducing efficiency of lamb production. Therefore, sheep production systems that aim for less overall methane output should aim for genetic improvement resulting in increased productivity and reproductive performance. Including feed intake and MP in the breeding strategy is now possible due to genomic selection and the creation of reference populations. Such a strategy allows for maintaining selection for increase productivity while limiting the increase in methane production. In that aspect, this is akin to selection strategies to improve feed efficiency, where optimal genetic improvement will not result in lower feed intake per animal, but rather in improved productivity while limiting the correlated increase in feed intake.

Table 2. Breeding objective traits, their means before selection, their annual change with three different breeding objectives, and the effect on methane efficiency parameters

Breeding Objective Trait (units)	Current mean	Breeding Objective Profit / DSE	 Reduce CH_4/ ewe	 kg CO_2 Eqvt/ kg lamb
		Annual change (trait units)		
Slaughter Weight (9 mo)	47.27	0.97	-0.79	0.77
Carcase Eye Muscle Depth (mm)	28.00	0.31	-0.17	0.12
Carcase Fat Depth (mm)	7.00	0.10	-0.05	0.00
Fleece Weight (kg)	4.00	0.01	-0.01	-0.01
Fibre Diameter (micron)	18.00	-0.04	0.00	0.03
Fertility (pregnancy rate)	0.75	0.01	-0.01	0.01
Lambing Rate (lambs weaned/lambing)	1.50	0.01	-0.01	0.01
Mature ewe Weight (kg)	55.00	1.27	-1.11	1.13
Daily DM Feed Intake (kg/day)	1.20	0.02	-0.02	0.01
Methane Production (g per ewe/day)	24.00	0.22	-0.28	0.10
Change in profit (p.a.)		\$6.65	-\$5.52	\$5.47
Change in CH_4 output /ewe (% of mean)		0.92%	-1.16%	0.43%
100 ewe flock CO_2 Eqvt tonne/yr	53.33	101.8%	98.6%	100.8%
kg CO_2 Eqvt per kg lamb produced	28.04	97.3%	102.5%	96.5%

CONCLUSION

Breeding strategies to reduce the amount of methane produced in sheep production systems rely mainly on improving productivity and reproductive performance while measuring and selecting for methane production and feed intake allow increased productivity while limiting an increase in methane production and feed intake. The amount of methane produced per kg lamb product can be reduced via genetic improvement by about 3.5% per annum.

ACKNOWLEDGEMENTS

Meat and Livestock Australia contributed financial support for this work through EAP project PSH2011. Daniel Brown and Sam Walkom contributed through useful discussion

REFERENCES

Gebbels J.N., Kragt M.E., Thomas D.T. and Vercoe P.E. (2022) *Animal* 16:100490.
Robinson D.L. and Oddy V.H. (2016) *J. Anim. Sci.* 94: 3624.
Brown D.J. and Swan A.A. (2015) *Anim. Prod. Sci.* 56: 1449.
Thompson J.M., Parks J.T. and Perry D. (1985) *Anim. Prod.* 40: 55.

Proc. Assoc. Advmt. Anim. Breed. Genet. 25: 238 - 241

INVESTIGATING THE GENETIC CAUSE OF WRY FACE IN AUSTRALIAN JERSEY CATTLE

C.J. Vander Jagt[1], C.M. Reich[1], I.M. MacLeod[1,2], M.E. Goddard[1,3], B.J. Hayes[4], T.T.T. Nguyen[5], G.J. Nieuwhof[5] and A.J. Chamberlain[1,2]

[1]Agriculture Victoria, Centre for AgriBiosciences, Bundoora, VIC, 3083, Australia
[2]School of Applied Systems Biology, La Trobe University, Bundoora, VIC, 3083 Australia
[3]Australia Faculty of Veterinary & Agricultural Science, The University of Melbourne, Parkville, VIC, 3052, Australia
[4]Queensland Alliance for Agriculture and Food Innovation, University of Queensland, St Lucia, QLD, 4072, Australia
[5]DataGene Ltd, Centre for AgriBiosciences, Bundoora, VIC, 3083, Australia

SUMMARY

Wry Face (WF) is a mammalian condition resulting in facial asymmetry. It is most obvious in long-faced species (e.g. horses and cattle) and ranges in severity. A mild hereditary form of WF is seen at a low frequency in the Australian Jersey cattle population. To investigate the underlying genetics and mode of inheritance of WF, a pilot study was performed. Four WF Australian Jersey cows and one unaffected half-sibling were whole genome sequenced (WGS) and included in Run 9 of the 1000 Bull Genomes Project (1kbulls). A subset of genetic variants found in the WF cows compared to the unaffected half-sibling were in or near genes associated with disorders involving facial deformities. This study is being expanded to validate these results and increase the power to detect more potential WF causal variants. Identifying WF causal variants and including them in routine DNA testing may allow farmers to avoid high risk matings that could result in WF offspring.

INTRODUCTION

WF is a mammalian condition causing facial asymmetry. WF is typically congenital, resulting in maxilla deviation and sometimes involves the mandible. It is most obvious in long-faced species such as horses and cattle (Abdelhakiem and Elrashidy 2017). The condition ranges in severity from a slight < 5° lateral deviation, only impacting aesthetics, to severe > 60° lateral deviation impacting breathing and feeding (Aiello and Moses 2016). Individuals with severe WF often do not survive to adulthood. A mild hereditary form of WF is seen at low frequency in the Australian Jersey cattle population, but at high frequency within some herds (mode of inheritance is unclear). This form of WF does not appear to impact quality of life or production of affected cattle.

Despite the incidence of WF in cattle, there have been few studies examining the inheritance mode or underlying genetics. Here we present a small study examining WGS from four WF Australian Jersey cows (two different herds) and one unaffected half-sibling cow. We demonstrate that we can identify candidate genetic variants and genomic regions which may underlie WF. We suggest that it would be advantageous to expand this study to validate these results and increase the power to determine the inheritance mode and detect the most likely candidate causal variants.

MATERIALS AND METHODS

Tail hair samples were obtained from four Australian Jersey cows that had been visually assessed as having WF by conformation classifiers and one unaffected half-sibling Australian Jersey cow (two different herds). DNA was extracted using the DNeasy Blood and Tissues Kit (Qiagen) and WGS libraries prepared with the NEBNext Ultra II DNA Library Prep Kit. Libraries were sequenced in a 150 cycle paired-end run on a NovaSeq6000 (Illumina). Raw sequence reads were processed according to the 1kbulls guidelines and submitted for inclusion in 1kbulls-Run9 (reviewed in Hayes and Daetwyler 2019).

Jersey cattle genotypes in 1kbulls-Run9 were obtained (179 individuals) and filtered for variants that were homozygous for the reference allele in the unaffected half-sibling and either heterozygous or homozygous for the alternate allele in all four WF cows. Assuming there may be more WF individuals in 1kbulls unknown to us (at low frequency), further filtering was applied so that the alternate allele would be seen in ≤10 individuals (~5% of individuals). Remaining variants were run through a haplotype detector (custom in-house program) and annotated using Ensembl Variant Effect Predictor (VEP) (McLaren *et al.* 2016). *Bos taurus* genes identified by VEP as being associated with these variants were included in over-representation analysis of GO Biological Processes (BPs) and KEGG pathways using DAVID (Huang *et al.* 2009). MalaCards (Rappaport *et al.* 2013) was used to determine if these genes had an association with human diseases affecting craniofacial skeletogenesis and/or other disorders involving maxillofacial dysmorphism.

To investigate the WF inheritance mode, pedigree information for the four WF cows was provided by DataGene Limited and interrogated to identify common ancestors (manually and using custom scripts). The pedigree was visualised using the R package visPedigree (Luan 2018).

RESULTS AND DISCUSSION

Pedigree analysis revealed all five cows could be traced to two common ancestors: "Secret Signal Observer" (paternal line) and "Soldierboy Boomer Sooner of CJF" (combination of maternal and paternal ancestry) (Figure 1). However, we cannot confirm that either bull had WF as there are no phenotypic records and only side-profile photographs available. In an early study by Ewing (1957) examining frequency (21.5%) and inheritance mode of "twisted face" (presumably WF) in a North American Jersey cattle herd, it was concluded that "twisted face" was most likely a simple recessive trait. Our pedigree cannot definitively corroborate this conclusion as there is an insufficient number of animals and inadequate phenotypic information. While it appears not to be a dominant trait, we cannot rule out a reduced (or "incomplete") penetrance inheritance mode. Also, while Secret Signal Observer from our pedigree is an American Jersey bull, he was born in 1953, and if WF was at high levels in Ewing's 1957 study (21.5%), WF is highly unlikely to have arisen initially from him.

Filtering 1kbulls-Run9 Jersey cattle genotypes found 16,771 variants homozygous for the reference allele in the unaffected half-sibling and heterozygous or homozygous for the alternate allele in all four WF cows (and at <5% of the total 179 Jersey cattle). We observed these variants tended to cluster in regions, most likely the result of high linkage disequilibrium as Australian Jersey cattle are particularly inbred (Scott *et al.* 2021). Interrogating these variants with VEP identified 84 variants within 66 genes as having either a low, moderate or high impact. Most variants were intronic (49%) or intergenic (39%) with no impact assigned. Of the coding variants, 53% were synonymous, 42% were missense, 2% were nonsense, and 2% were frameshift mutations.

The only high impact variant within an annotated gene was a C/A SNP at Chr6:13249536 which creates a stop codon in the *AP1AR* gene (Table 1). AP1AR is involved in negative regulation of receptor recycling and vesicle targeting between the trans-Golgi network and endosomes (Stelzer *et al.* 2016). While there is no known link between *AP1AR* and diseases affecting craniofacial skeletogenesis or maxillofacial dysmorphism, several genes associated with variants classified as having a "moderate" impact do have known associations (Table 1), including *SCARF1*, *BANK1* and *RAB2A* (Rappaport *et al.* 2013). *SCARF1* is expressed in endothelial cells and regulates the uptake of chemically modified low density lipoproteins (Stelzer *et al.* 2016). It has been implicated in Van Den Ende-Gupta syndrome, a congenital autosomal recessive malformation syndrome that effects facial features and the skeletal system in humans (Rappaport *et al.* 2013). *BANK1* is associated in humans with both Parry-Romberg syndrome (facial hemiatrophy), a rare condition involving atrophy of facial components (including the jaw) and Potocki-Shaffer syndrome which effects craneo bones and facial appearance (Rappaport *et al.* 2013). RAB2A belongs to the Rab family, membrane-bound proteins involved in vesicular fusion and trafficking (Stelzer *et al.* 2016). *RAB2A* has been implicated in cleft palate malformation as well as Warburg Micro syndrome 1, a rare autosomal recessive syndrome effecting facial appearance (Rappaport *et al.* 2013). Interestingly, examining the region around this *RAB2A* variant revealed a large cluster of variants which were heterozygous in the WF cows and homozygous for the reference allele in the unaffected half-sibling.

Most variants were in intronic and intergenic regions (49% and 39% respectively). Regulatory elements (e.g. enhancers) are also located in these regions, therefore these variants should not be completely dismissed. Another unexplored category of variants in this study are structural variants (>50 bp long) that can be more accurately detected using long read sequencing.

Over-representation analysis of VEP genes identified several significantly implicated GO BPs and KEGG pathways (P<0.05). Of particular interest were those genes involved in "endochondral ossification" and "positive regulation of osteoblast differentiation", both essential for bone formation. The ENSBTAG00000037710 gene (unannotated) containing a "high" impact frameshift and *ZNF536* containing two missense SNP with a "moderate" impact were linked to "regulation of transcription, DNA-templated" (Table 1). Also of interest was the KEGG pathway "folate biosynthesis" involving the ENSBTAG00000016748 gene (unannotated) which has two "moderate" impact missense SNP (Table 1). Folate (vitamin B_9) is an essential nutrient long acknowledged as important for foetal growth and development and plays an important role in maintaining bone health.

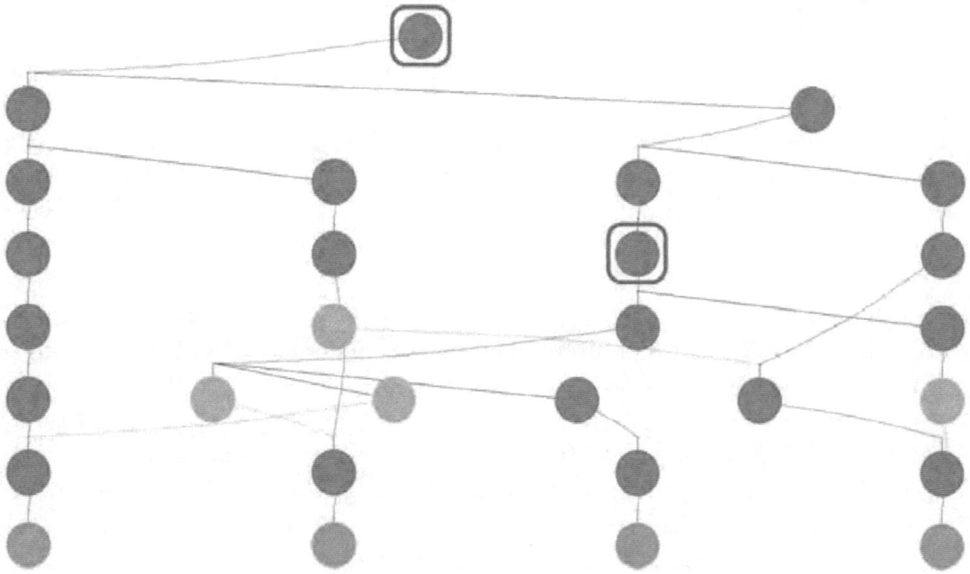

Figure 1. Simplified pedigree for wry face affected cows

Green circles represent the four affected cows, blue circles represent sires and yellow circles represent dams. Common ancestors are highlighted by red boxes. To visually simplify the pedigree, sires and dams not connecting back to common ancestors have been removed.

CONCLUSION

This small study has demonstrated that it may be possible to uncover the genetic cause for WF in the Australian Jersey population. Since multiple putative causal genes have been identified, extra sequencing, including long read sequencing, of affected and unaffected relatives is required to identify the causal mutation(s). This study is being expanded to validate these results and increase the power to detect more potential WF causal variants. Inclusion of causal variants in routine DNA testing may allow farmers to avoid high risk matings that could result in WF offspring.

Table 1. Variants classified as having high or moderate impact

Chr	Position	Variant	Consequence	Impact	Gene	Ensembl ID
6	13249536	C/A	stop codon gained	HIGH	AP1AR	ENSBTAG00000003941
12	71304228	C/T	splice donor	HIGH	-	ENSBTAG00000026070
18	57221530	G/GA	frameshift	HIGH	-	ENSBTAG00000037710
2	117389921	T/A	missense	MODERATE	DNER	ENSBTAG00000016063
2	120041472	C/G	missense	MODERATE	-	ENSBTAG00000016748
2	120041475	C/G	missense	MODERATE	-	ENSBTAG00000016748
3	60257454	C/T	missense	MODERATE	TTLL7	ENSBTAG00000003322
3	76782778	C/T	missense	MODERATE	DEPDC1	ENSBTAG00000001343
4	116914236	G/C	missense	MODERATE	PAXIP1	ENSBTAG00000017505
4	118147886	G/A	missense	MODERATE	RNF32	ENSBTAG00000020335
6	22868432	T/A	missense	MODERATE	BANK1	ENSBTAG00000015297
9	13265164	G/C	missense	MODERATE	CD109	ENSBTAG00000013222
12	72839314	C/A	missense	MODERATE	-	ENSBTAG00000023309
13	23414943	C/T	missense	MODERATE	-	ENSBTAG00000051361
13	27934712	A/G	missense	MODERATE	-	ENSBTAG00000047869
13	42905566	G/A	missense	MODERATE	-	ENSBTAG00000035572
13	43143962	C/T	missense	MODERATE	CALML5	ENSBTAG00000013854
14	8564112	C/T	missense	MODERATE	TMEM71	ENSBTAG00000017138
14	26252117	A/T	missense	MODERATE	RAB2A	ENSBTAG00000000948
14	76196018	T/C	missense	MODERATE	RMDN1	ENSBTAG00000015734
15	6405728	T/G	missense	MODERATE	BIRC3	ENSBTAG00000024918
15	81534709	T/A	missense	MODERATE	OR5B12	ENSBTAG00000049719
18	36479140	G/A	missense	MODERATE	COG8	ENSBTAG00000001665
18	38353565	C/G	missense	MODERATE	ZFHX3	ENSBTAG00000014636
18	41070602	C/T	missense	MODERATE	ZNF536	ENSBTAG00000007262
18	41071302	C/T	missense	MODERATE	ZNF536	ENSBTAG00000007262
18	62698984	C/T	missense	MODERATE	-	ENSBTAG00000049820
18	62921228	C/G	missense	MODERATE	-	ENSBTAG00000050536
18	62972057	G/C	missense	MODERATE	-	ENSBTAG00000038797
18	62972059	A/G	missense	MODERATE	-	ENSBTAG00000038797
19	22747077	G/A	missense	MODERATE	SCARF1	ENSBTAG00000011483
23	13275715	T/C	missense	MODERATE	KIF6	ENSBTAG00000027197
28	35535151	G/T	missense	MODERATE	-	ENSBTAG00000048082
28	35647259	C/T	missense	MODERATE	SFTPA1	ENSBTAG00000023032

REFERENCES

Abdelhakiem M. and Elrashidy M. (2017) *Assiut Vet. Med. J.* 63: Article 11.

Aiello S.E. and Moses M.A. (2016) 'The Merck Veterinary Manual' 11th ed. Wiley.

Ewing M.B. (1957) Masters Thesis, Kansas State College of Agriculture and Applied Science.

Hayes B.J. and Daetwyler H.D. (2019) *Annu. Rev. Anim. Biosci.* 7: 89.

Huang da W., Sherman B.T. and Lempicki R.A. (2009) *Nat. Protoc.* 4: 44.

Luan S. (2018) URL https://github.com/luansheng/visPedigree.

Rappaport N., Nativ N., Stelzer G., Twik M., Guan-Golan Y., Stein T.I., Bahir I., Belinky F., Morrey C.P., Safran M. and Lancet D. (2013) *Database (Oxford)* 2013: bat018.

Scott B.A., Haile-Mariam M., Cocks B.G. and Pryce J.E. (2021) *J. Dairy Sci.* 104: 11832. Stelzer G., Rosen N., Plaschkes I., Zimmerman S., Twik M., Fishilevich S., Stein T.I., Nudel R., Lieder I., Mazor Y., Kaplan S., Dahary D., Warshawsky D., Guan-Golan Y., Kohn A., Rappaport N., Safran M. and Lancet D. (2016) *Curr. Protoc. Bioinformatics* 54: 1.30.1-1..3.

Proc. Assoc. Advmt. Anim. Breed. Genet. 25: 242 - 245

IMPORTANCE OF INCORPORATING FEEDING RATE WHEN DEVELOPING PREDICTIONS OF FEED INTAKE

P.L. Johnson, K. Knowler, T.P. Biton and S.J. Rowe

[1]AgResearch Invermay, Puddle Alley, Mosgiel, New Zealand

SUMMARY

Obtaining individual feed intake data under pastoral grazing studies is important for work relating to feed efficiency and greenhouse gas emissions, but is nearly impossible to obtain. Accelerometer technology has been used to determine the duration of grazing events, but data from feed intake facilities suggests that between-animal variation in feeding rate makes duration alone a poor proxy for feed intake. This study explored in detail the trait of feeding rate (feed eaten/feeding duration) on data collected through a feed intake facility. Feeding rate was demonstrated to be a very consistent trait of an individual animal across their feeding events with a high heritability (0.60 ± 0.14) and considerable between-animal variation. Using feeding rate and feeding duration accurately predicted feed intake. Future accelerometer work to predict feed intake should therefore emphasise whether or not feeding rate can be accurately determined in addition to feeding duration.

INTRODUCTION

There is increasing interest in being able to accurately determine the individual feed intake for use within studies relating to feed efficiency and greenhouse gas emissions. Whilst this can be achieved in feed intake facilities through either cut and carry of feed, or the use of feed intake recorded against electronic identification tags, limited options are available when animals are grazing at pasture. One possible approach in which the feed intake of animals grazing at pasture could be estimated is through the use of accelerometer data that can be classified to describe the behaviour of the animal at any point in time. Smith *et al.* (2016) used "the head of the cow is tilted downwards and positioned near the ground. The cow is either taking bites of the pasture or searching for the pasture" to classify animals as grazing. This definition was used by Greenwood *et al.* (2017) to estimate the individual intake of animals by multiplying the length of time an animal was classified as grazing by a constant to estimate intake. However, such a model assumes that all individuals consume feed at a constant rate. In feed intake facility studies, significant between animal variation in the rate at which animals eat has been demonstrated (Durunna *et al.* 2011; Johnson *et al.* 2022). Johnson *et al.* (2022) estimated the heritability of feeding rate to be 0.29 ± 0.10. Utilising the data set described by Johnson *et al.* (2022), the question of whether feeding rate could be a useful metric, together with feeding duration, to predict feed intake is explored.

MATERIALS AND METHODS

All animal experiments were conducted to meet the guidelines of the 1999 New Zealand Animal Welfare Act and were approved by the AgResearch Animal Ethics committees. Specific approval numbers were AEC13563, AEC13892, and AEC14221.

Animals. The data used in this study is described in detail by Johnson *et al.* (2022). In brief, individual feed intake data over 42 days (after 14 days adaptation) was collected on 986 ewe lambs in a feed intake facility utilising automated feeders which captured the weight of individual feeding events and their duration through the feeders being fitted with RFID readers which recorded which animal was in the feeder during a feeding event. Five cohorts of lambs recorded over three years made up the data set, with the animals sourced from two progeny test flocks and the AgResearch methane selection line flock. The animals were fed a lucerne pellet *ad libitum*.

Data cleaning and analysis. In the study of Johnson *et al.* (2022), data cleaning was carried out at the population level. In this study, data editing was carried out at the individual animal level as follows. For each animal, intake was regressed on duration, the upper and lower bounds of the 95% confidence interval determined, and any values lying outside of the bounds were deleted and the regression step was repeated to estimate the model goodness of fit (R^2). Approximately 6% of the data was deleted. From this revised data set, an estimate of feeding rate (FR) was determined, calculated as feed eaten divided by the duration of the individual feeding event. The overall FR was calculated as the average of all FR data across the 42 days for each individual.

To determine the consistency and utility of FR, the data set was split into two equal time periods of 21 days (PER1 and PER2). The measured average daily intake in PER2 was calculated. The average FR was calculated for each time period independently. The FR value from PER1 was multiplied by daily feeding duration (FD) in PER2 to provide an estimate of intake in PER2. The derived trait data were subsequently analysed to determine their relationships.

The heritability of FR across all of the data was estimated using a model fitted as described by Johnson *et al.* (2022) including fixed effects of birth-rearing rank, age of dam, contemporary group (cohort*flock) and a covariate of birthday deviation.

RESULTS AND DISCUSSION

The potential for accelerometers to generate feed intake data has been explored but models to date have been limited to using grazing duration as a proxy for intake. Whilst the dataset used in this study is generated from a feed intake facility it allows the value of the inclusion of FR to better predict intake to be assessed. Figure 1 demonstrates that feeding event duration alone is not an accurate predictor of the intake.

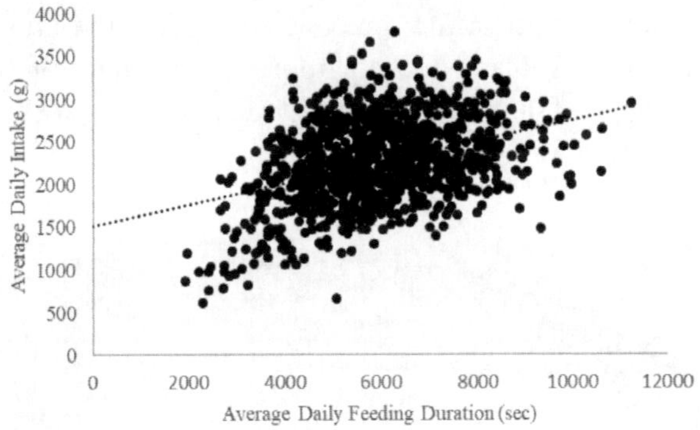

Figure 1. Average daily feeding duration for all animals plotted against their daily feed intake across full 42 day time period ($R^2 = 0.14$)

Figure 2 demonstrates between-animal differences in FR, and the consistency with which it presents for an individual animal. Contrasting between one animal which exhibited a low FR and as such for a FD of 500 seconds it only consumed 132g, compared with another animal exhibiting a high FR, consuming 327g of feed over 500 seconds. The R^2 of the associated regression models for these two animals was more than 0.86 indicating that the concept of rate is highly consistent across the 42 days of measurement for each animal. Across all animals, the average R^2 after one round of data cleaning was 0.89 with a range of 0.61 to 0.96, with the R^2 value greater than 0.80 for 96% of the animals. The average FR across all animals was 0.40 with a standard deviation of 0.11 and a coefficient of variation of 29%. Combined, these results indicate that for the majority of animals in the dataset, there is a very consistent relationship between the length of time that they are feeding and the amount of feed they are consuming within that time, but that individual animals feed at different rates such that some animals are "nibblers" with a very sow feeding rate and others are "guzzlers" with a very high feeding rate.

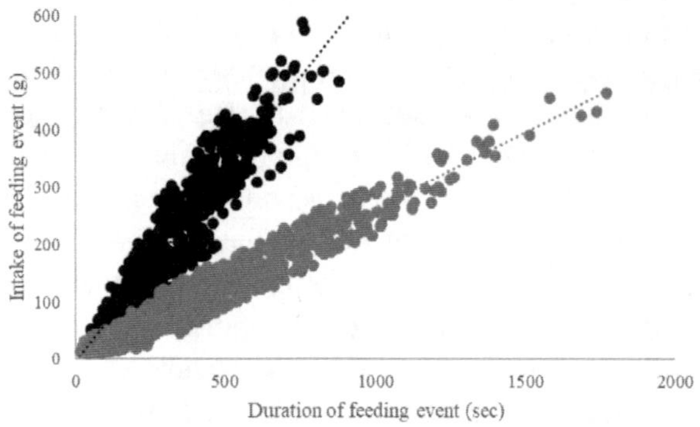

Figure 2. Examples of animals with high (black) feeding rate (model $R^2 = 0.87$) and low feeding rate (model $R^2 = 0.95$). Data points plotted are all feeding events across 42 days of individual feed intakes being measured, with outlier data points removed (beyond 95% confidence interval of original regression removed)

The next step was to explore the potential of FR to more accurately predict feed intake than FD alone. The dataset was split into two 21-day periods and FR was calculated for each period. Estimated feed intake in PER2 was calculated by multiplying the FR of PER1 trait by the daily FD. Figure 3 a) shows that FR for PER1 and PER2 were highly correlated. Although PER1 and PER2 were contiguous periods, it does demonstrate the consistency of the trait over 42 days. Figure 3 b) shows that by utilising the PER1 FR and FD an improved estimate of feed intake was obtained compared with using FD alone, and that FR at the individual animal level calculated on one data set was robust enough to be used with independent data.

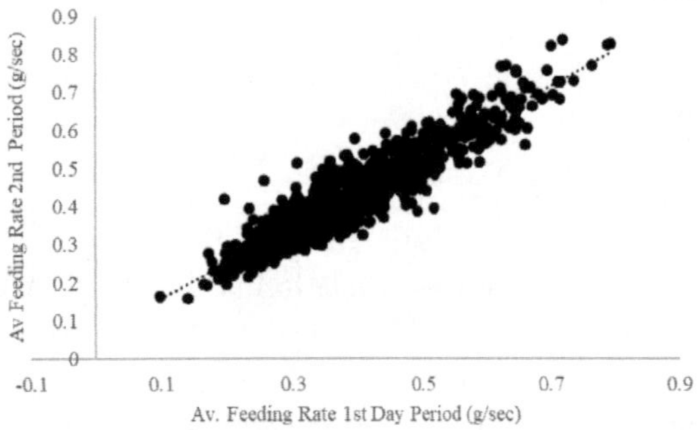

a)

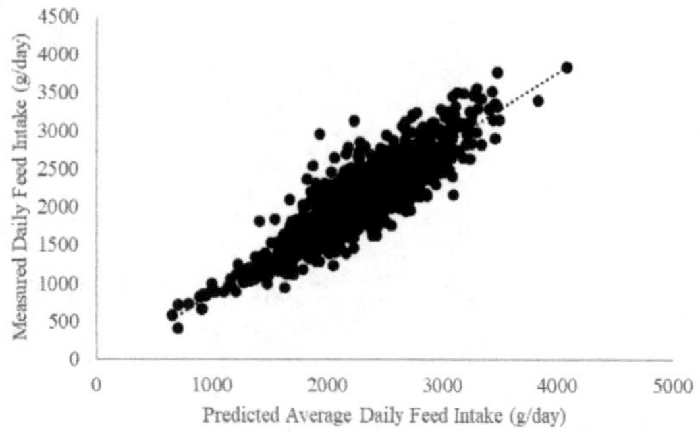

b)

Figure 3. a) Average feeding rate calculated using data from the first 21-day time period plotted against the average feeding rate calculated using data from the second 21-day time period ($R^2 = 0.87$); b) Predicted average daily feed intake for the second 21-day period (using feeding rate calculated from the first 21 day period feeding rate multiplied by the average daily duration of feeding from the second 21 day period) plotted against the average measured feed intake for the second 21 day period (R^2=0.78)

The heritability estimate for FR was 0.60 ± 0.14. This value is considerably higher than 0.29 ± 0.10 reported for the same data set in Johnson *et al.* (2022), however, in that dataset rate data was cleaned at the population level, versus the individual animal level as was carried out in this current study, highlighting that whilst some values might be within population limits, they are inconsistent and anomalous for an individual animal.

Using the two animals in Figure 1 the intake predictions using FD or FD and PER1, FR are given in Table 1. Whilst both animals were measured to have eaten nearly identical amounts of feed, their FD were over two-fold different and as such a model only considering FD resulted in very different estimates of intake for the two animals, whereas the model incorporating FD and FR improved the estimates relative to their measured intakes.

Table 1. Predicting feed intake using feeding duration with and without feeding rate (FR) data for two animals with similar total measured intakes but very different durations and one animal exhibiting a high FR (Guzzler) and the other a low FR (Nibbler). FR was calculated on two consecutive 21-day subsets of the full dataset (PER1 and PER2)

Trait/Model Description	High FR	Low FR
PER1 Feeding Rate (g/sec)	0.65	0.27
PER2 Feeding Rate (g/sec)	0.63	0.26
Daily Feeding Duration (sec/day)	4540	11235
Model: Intake=Dur (g)	2079	2907
Model: Intake=Dur* PER1 Rate (g)	2951	3033
Actual Intake (g/day)	2950	2949

CONCLUSION

This work demonstrates that feeding rate is a unique attribute of an individual and is a highly heritable trait. As such future work on accelerometer, or other, technology used to predict feed intake should place a strong emphasis on determining whether or not the rate at which an animal is feeding can be determined versus just predicting feeding duration. If the accelerometer data can predict FR, models combining FD and FR will result in improved predictions of feed intake.

ACKNOWLEDGEMENTS

The feeders used in the feed intake facility were developed and manufactured by the AgResearch Engineering Team. The feed intake measurements were funded by Beef + Lamb NZ Genetics which is funded by the Ministry for Business, Innovation and Employment (MBIE) and Beef + Lamb Zealand. Maintenance of the flocks used in the study was funding Beef + Lamb NZ Genetics, the New Zealand Agricultural Greenhouse Gas Research Centre and Pastoral Greenhouse Gas Research Consortia and AgResearch's Strategic Science Investment Funding (SSIF) from theMBIE). This work is aligned with the GrasstoGas consortia (2019EN202) programme of research.

REFERENCES

Durunna O.N., Wang Z., Basarab J.A., Okine E.K. and Moore S.S. (2011) *J. Anim. Sci.* 89: 3401.

Greenwood P.L., Paull D.R., McNally J., Kalinowski T., Ebert D., Little B., Smith D.V., Rahman A., Valencia P., Ingham A.B. and Bishop-Hurley G.J. (2017) *Crop Pasture Sci.* 68: 1091.

Johnson P.L., Hickey S., Knowler K., Wing J., Bryson B., Hall M., Jonker A., Janssen P.H., Dodds K.G., McEwan J.C. and Rowe S.J. (2022) *Front. Genet.* 13: 911639.

Smith, D., Rahman, A., Bishop-Hurley, G.J., Hills, J., Shahriar, S., Henry, D. and Rawnsley, R. (2016) *Com. Elec. Agri.* 131: 40.

REMOTE SENSOR COLLARS MEASURE AGE AT PUBERTY IN TROPICAL BEEF HEIFERS IN NORTHERN AUSTRALIA

D.J. Johnston[1], M. Dayman[2], T.P. Grant[2], K. Hubbard[2], K. Goodwin[2], A.K. Doughty[3] and J.D. Cook[1]

[1] Animal Genetics Breeding Unit*, University of New England, Armidale, NSW, 2351 Australia
[2] DAF, Brian Pastures, Gayndah, Qld, 4625 Australia
[3] Allflex, 33 Neumann Road, Capalaba, 4157 Australia

SUMMARY

A total of 173 heifers were fitted with SenseHub™ collars and recorded for their activity and rumination behaviour over a 530-day period. Daily averages and their change over time were plotted in a heat map for each heifer. Patterns in the plots were assessed by five scorers independently to determine the date of each heifer's first oestrus event. Her age was referred to as collar-determined age at puberty (CollarAP). Heifers were also regularly ultrasound ovarian scanned to determine age at first observed corpus luteum (AGECL), this was used to assess the relationship with CollarAP. The results from this study show the collars can accrue large amounts of data but interpretation was not simple and differences existed between the five scorers. However, mean CollarAP and AGECL were the same and correlations were positive and strong, ranging from 0.69-0.83 across scorers. When observations on CollarAP were pooled across scorers, the correlation with AGECL was 0.86. Further, least squares means for sires for the two traits were also highly related (R^2=0.89). This study has shown the collar data can be used to determine age at puberty in tropical beef heifers.

INTRODUCTION

Heifer age at puberty is a highly heritable trait (Johnston *et al.* 2009) and is an early in life genetic predictor of lifetime reproductive performance in tropical beef cattle (Johnston *et al.* 2014). However, measuring the trait currently requires serial ovarian scanning measures that are costly, invasive and require regular musters and handling. The SenseHub™ collars and associated algorithms and software system DataFlow™ (www.allflex.global/au/wp-content/uploads/sites/3/2021/06/SH_4_A4_Eng_March-2020_low-1.pdf) have been optimised for mature *Bos Taurus* dairy cattle for monitoring their health and reproductive status. This project investigated the off-label use of the collars in tropical beef heifers managed in an extensive northern grazing environment. The aim of the project was to establish if the collar data could be used to assign each heifer's age at puberty and compare that to age at first observed corpus luteum (CL) determined by serial ultrasound ovarian scanning.

MATERIAL AND METHODS

Animals. Heifers used in this study were the 2019 cohort of the Repronomics Project (Johnston *et al.* 2017) at the DAF managed Brian Pastures Research Facility, Queensland. Heifers were born between September and December 2018 and comprised of 62 Brahman, 49 Droughtmaster, and 62 Santa Gertrudis. Heifers were managed as a single group from birth. The SenseHub™ collars were deployed at the start of November 2019 when the average age of the heifers was 13 months and they remained on until 19th April 2021. Collars recorded activity and rumination every 20 minutes, and this was compressed into daily averages. During the recording period the heifers were rotated through a range of paddocks that were generally open grazing

* A joint venture of NSW Department of Primary Industries and the University of New England

native pastures, undulating, with some trees. Watering was mainly troughs or dams with some gullies and intermittent water courses. Receivers were positioned to gain maximum coverage of each paddock and required the use of mobile solar booster stations with an average distance of 2.7 km between stations and the largest coverage was 240 hectares. A daily activity log was maintained to record all group cattle movements (e.g. change of paddock, muster) and any known interactions with individuals that may have influenced the data from a given day.

Collar data. At the completion of the recording period the daily files were downloaded and the daily activity log file was used to remove records for all animals for a particular day (e.g. a mustering event) or the record for an individual (e.g. 3-day sickness). A total of 13 heifers lost their collars during the period. Of these three occurred early in the recording period and three stopped working and these animals and their data were deleted. The remainder had their collars replaced or the loss occurred sufficiently late in the recording that cycling had already commenced and had been detected. This yielded a final dataset for analysis of 167 heifers with collar data.

The change in the daily activity and rumination time in dairy is used to identify individual cow oestrus behaviour based on increased activity and decreased rumination. In these data the change in activity and rumination for each heifer was combined into a single metric for each day. Excel heat maps were developed to visualise the change in this daily parameter over the 530-day period for each heifer. The heat map for each heifer was examined by five novice scorers to visually assign the date of her first oestrus event. This was based on examination of the entire period and establishing if a cyclic pattern (between 16-28 days) existed, and if so, traced back until to the date of the initial occurrence. The ability to call the first cycling event varied greatly due to differing strength of signals and the amount of "noise", so each scorer allocated a confidence score (*viz.* 0=unable to call, 1=very uncertain, 2=uncertain, 3=moderately certain, 4=very certain) for the calling of each record. This whole process was done completely independently of each other and without knowledge of the ovarian scanning data.

Ovarian scanning. Ovarian scanning of the heifers also commenced in November 2019 and were scanned by two experienced ultrasonographers. Heifers were scanned approximately every four weeks, until an individual heifer was observed to have a CL on two successive scanning events. After this she was no longer scanned but was still mustered and yarded at each subsequent scanning event. In total, there were 14 scanning events over the duration of the collar experiment. Heifer age at puberty (AGECL) was computed as their age at their first observed CL. Any heifer not pubertal at the completion of the collar study were assigned a CL date as the completion date. Likewise, if a collar record determined no detectable oestrus event it was also assigned the completion date.

Analyses. Individual CollarAP was compared to AGECL from each scorer separately and was also pooled across scorers. Firstly, averaged across scorers for each heifer (MEAN) and secondly, as the average after deleting scorer records that deviated more than 100 days from the median value for each heifer (CLIP). In this process there were four animals where only records from two scorers remained and these animals had their record deleted. Sire (n=22) least squares means were computed using SAS (SAS Institute Inc., Cary, NC, USA) for the entire dataset for AGECL (n=173) and for the pooled CollarAP and plotted for those with more than three daughters recorded.

RESULTS AND DISCUSSION

The study showed the collars were effective in recording vast amounts of behavioural data over a large period from an extensive grazing system. In general, the retention of collars was good, however there were loss of data due to periodic system failure primarily related to storms and wet weather. There were also issues with some collars/individuals that failed to read effectively. Removal of whole day data also occurred for all heifers due to the frequent mustering and handling, particularly for the monthly ultrasound ovarian scanning and changing paddocks.

Table 1. Correlations between scorers for collar-determined age at puberty (above diagonal) and correlations between scores for confidence score in assessment of the collar record (below diagonal)

	Scorer 1	Scorer 2	Scorer 3	Scorer 4	Scorer 5
Scorer 1		0.57	0.67	0.75	0.75
Scorer 2	0.32		0.59	0.56	0.47
Scorer 3	0.35	0.24		0.73	0.59
Scorer 4	0.37	0.30	0.25		0.67
Scorer 5	0.63	0.42	0.36	0.38	

Table 1 presents the pairwise correlations between each of the scorers for CollarAP and correlations of their calling confidence scores. The CollarAP correlations between scorers were moderate to high ranging from 0.56 to 0.75 with scorer 2 generally lower than the others. The confidence score correlations were generally low (0.25-0.62) and reflected the naivety in this process and the overall difficulty experienced by novice assessors.

Table 2. Mean, standard deviation of the difference (std diff) and correlation (corr) for collar-determined age at puberty (CollarAP) and age at first CL (AGECL), by each scorer (1-5) and combined (MEAN and CLIP)

Variable	Scorers					Combined	
	1	2	3	4	5	MEAN	CLIP
N collar records	166	166	155	154	159	167	163
mean Confidence Score	2.8	1.7	2.6	2.5	3.1		
mean CollarAP, d	610.0	607.1	599.7	573.5	587.6	598.3	595.8
mean AGECL, d	599.1	597.5	592.5	599.5	593.0	599.8	597.5
std diff (AGECL-CollarAP)	83.9	116.5	62.7	69.4	75.3	56.7	57.5
corr (CollarAP, AGECL)	0.72	0.60	0.83	0.81	0.69	0.85	0.86

Results in Table 2 show that for each of the five scorers their mean CollarAP and AGECL were very similar, and correlations ranged between 0.60 and 0.83. Scorer 2 had the lowest mean confidence score, the lowest correlation and the largest standard deviation of the difference between the two measures. When records were pooled across scorers, the mean CollarAP and AGECL were almost identical and the high correlations (e.g. 0.86) show the collars were accurately determining age at puberty with an average standard deviation of the difference of 57 days. Removing of outlier scorer records had little effect on the results.

Biologically it might have been expected that CollarAP would be on average lower than AGECL, given the frequency of scanning. This was the case for scorers 4 and 5 and may reflect the other scorers not being as confident in assigning the first oestrus event. However, breed differences may have existed, with Brahmans averaging 7.9 days older for CollarAP compared to AGECL average across all scorers, whereas Santa Gertrudis were 11.9 days younger. This may be due to differences in the strength of the first oestrus event or issues with the operation of the collar (e.g. amount of dewlap) for the three breeds. It is also possible that in some cases the first collar-detected oestrus was removed if it coincided with a scanning event where the first CL was observed. Therefore, any subsequent first collar event would be greater than the AGECL. Also, the ovarian scan data is not 100% accurate as the frequency of scanning was only every four weeks, thus there is a chance a heifer was pubertal but had no observable CL on the day of scanning.

Sire least squares means based on daughter records pooled across scorers showed a strong relationship between CollarAP compared to AGECL. This relationship was improved using CLIP (R^2=0.89, b=0.96 d/d see Figure 1) vs MEAN (R^2=0.86, b=0.86 not plotted) and suggest that even though assigning CollarAP could be difficult, it was closely related to ultrasound determined AGECL, especially when pooled across scorers and averaged across the daughters of a sire.

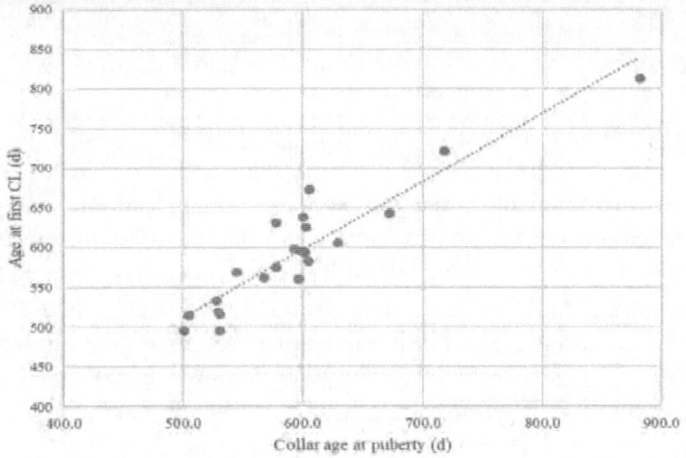

Figure 1. Sire least squares means for collar-determined age at puberty and age at first CL

CONCLUSIONS

This study has shown the continuous recording of heifer activity and rumination using collars is an alternative method of obtaining individual heifer age at puberty. However, the use of untrained scorers probably was not the best approach as each scorer developed their own method and confidence scoring. Collar assessment could be improved by having a basic tutorial, however a better approach will be to develop automated algorithms to interpret this beef application of the collars. Collar data accuracy could be improved by decreasing the amount of cattle handing, but at this stage this was not possible given the importance of ovarian data to the overall project.

Opportunities exist to use the collar data to study other aspect of reproduction such as cycle length, cycle strength and relationships with mating outcomes. This study has shown collars can determine heifer puberty, however it now requires a full cost assessment compared to serial ovarian scanning and the collection of more data to determine its utility as a trait in a genetic evaluation.

ACKNOWLEDGEMENT

This study used heifers that were part of the MLA-funded Repronomics Project (L.Gen.0174). The Staff at Brian Pastures, in particular Johnathon Conroy, are acknowledged for their assistance with the development and movement of collar infrastructure and cattle handing. Nick Corbet and Paul Williams are acknowledged for their ultrasound ovarian scanning. Assistance and contributions of Allflex are gratefully acknowledged.

REFERENCES

Johnston D.J., Barwick S.A., Corbet N.J., Fordyce G., Holroyd R.G., Williams P.J. and Burrow H.M. (2009) *Anim. Prod. Sci.* 49: 399.

Johnston D.J., Barwick S.A., Fordyce G., Holroyd R.G., Williams P.J. and Corbet N.J. (2014) *Anim. Prod. Sci.*54: 1.

Johnston D.J., Grant T.P., Schatz T.J., Burns B.M., Fordyce G. and Lyons R.E. (2017) *Proc. Assoc. Advmt. Anim. Breed. Genet.* 22: 385.

GENETIC DIVERSITY OF DOMESTIC GOATS FROM CENTRAL LAOS

S.V. Le[1], S. de las Heras-Saldana[2], P. Alexandri[2], L. Olmo[1], S.W. Walkden-Brown[1] and J.H.J. van der Werf[1]

[1]School of Rural & Environmental Science, University of New England, Armidale, NSW, 2351 Australia
[2]Animal Genetics Breeding Unit*, University of New England, 2351, Armidale, NSW, 2351 Australia

SUMMARY
Maintaining genetic diversity and variation in livestock populations is critical for allowing natural and artificial selection genetic improvement well as for avoiding inbreeding. The Laotian government and farmers are concerned that there has been a decrease in genetic diversity and an increase in inbreeding among native goats in their village-based smallholder system. The objective of this study was to investigate the genetic diversity of Lao native goats in a small-scale farming system in central Laos using genotype data. The results showed that there was a close genetic relationship between Lao native goats with Chinese goat breeds and a low to moderate genetic differentiation among goat populations in Central Laos ranging from 0.0112 to 0.0427. This goat population had close to zero inbreeding coefficients (-0.093 to 0.052).

INTRODUCTION
In developing countries, local livestock breeds are crucial for the welfare of rural communities producing a wide range of products and requiring low levels of management and health care. The genetic diversity of goat populations has been studied in Asia (Tarekegn *et al.* 2019; Hermes *et al.* 2020), Africa (Nandolo *et al.* 2019; Tarekegn *et al.* 2019), Europe (Oget *et al.* 2019; Danchin-Burge *et al.* 2012) and Oceania (Brito *et al.* 2017). These studies used genetic indexes such as genetic differentiation (F_{ST}) and inbreeding coefficient (F_{IS}) to describe population characteristics and history. French goat breeds are genetically very well described with an average F_{ST} and F_{IS} of 0.092 and 0.0213 respectively, between breeds and regions in France (Oget *et al.* 2019). However, the genetic diversity of goats in other Asian countries like the Lao People's Democratic Republic (Laos) has not been investigated, and detailed information about this population would benefit its management and production. In Laos, almost all goat production comes from smallholder systems with 2 to 5 goats per household (Windsor *et al.* 2018). Goat production is an essential source of income for household incomes and in recent years, Lao's goat population has significantly increased due to the high demand of goat meat to export to neighbouring Vietnam (Stür and Gray 2014). The objective of this study was to identify the genetic diversity of Lao's native goats and assess levels of inbreeding.

MATERIALS AND METHODS
Sample collection and genotyping. During the period from February to April 2022, a total of 420 ear-notch samples were collected in the Savannakhet province of Laos. These samples were obtained from 140 households situated in seven villages located in Phin, Songkhone, and Sepon districts (Figure 1). The samples were then genotyped using the goat 50K Illumina BeadChip at GenomNZ in New Zealand.

Reference population. To assess the diversity at a Asian level, a reference dataset comprising genotype information from 1132 goats, including 185 indigenous goats from Mongolia (Mukhina *et*

* A joint venture of NSW Department of Primary Industries and University of New England

al. 2022), 237 local Russian goats (Deniskova *et al.* 2021), 193 Chinese goats (Berihulay *et al.* 2019), 416 Pakistan goats, 16 Iranian goats and 85 goats from Turkey (Colli *et al.* 2018).

Quality control. A quality control assessment of genomic data was performed in PLINK v1.9 (Purcell *et al.* 2007). SNPs on the sex chromosomes, unmapped location, minor allele frequency (MAF) lower than 0.05, call rate lower than 90% and a deviation from Hardy-Weinberg equilibrium of $p < 10^{-6}$ were excluded from the analysis. Individuals with more than 10% missing SNPs were also excluded. After quality control, 419 genotyped goats with 42666 SNPs remained in the analysis. The Laos goat dataset was merged with the reference population which, after quality control included 43429 autosomal SNPs and 1546 individuals.

Genetic diversity and inbreeding analysis. A principal component analysis (PCA) was performed in PLINK and visualized with the R package "ggplot2". Pairwise genetic differentiation (F_{ST}; Weir and Cockerham 1984) and inbreeding coefficient (F_{IS}; Wright 1965) as well as expected and observed heterozygosity for each population were calculated in R using the "hierfstat" package (Goudet 2015).

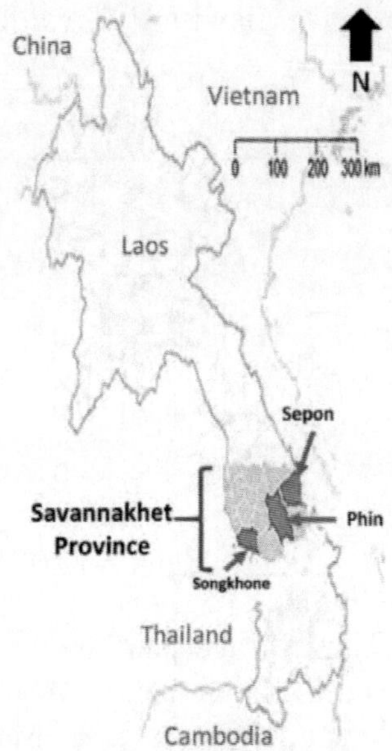

Figure 1. Geographical location of goat samples used in this study

RESULTS AND DISCUSSION

The principal component analysis of the combined Laos and reference population showed clear differentiation between indigenous Lao goats and other Asian breeds (Figure 2A). The first principal component (PC1) accounted for 35.36% of the total variation and separated Lao native goats and Chinese breeds from other Asian breeds. The PCA for Lao goats suggested that genetic structure exists within this population with goats in Sepon clearly differentiated from those in the Phin and Songkhone districts in the PC1 and PC2, which accounted for 12.64% and 8.75% of the total variation, respectively (Figure 2B).

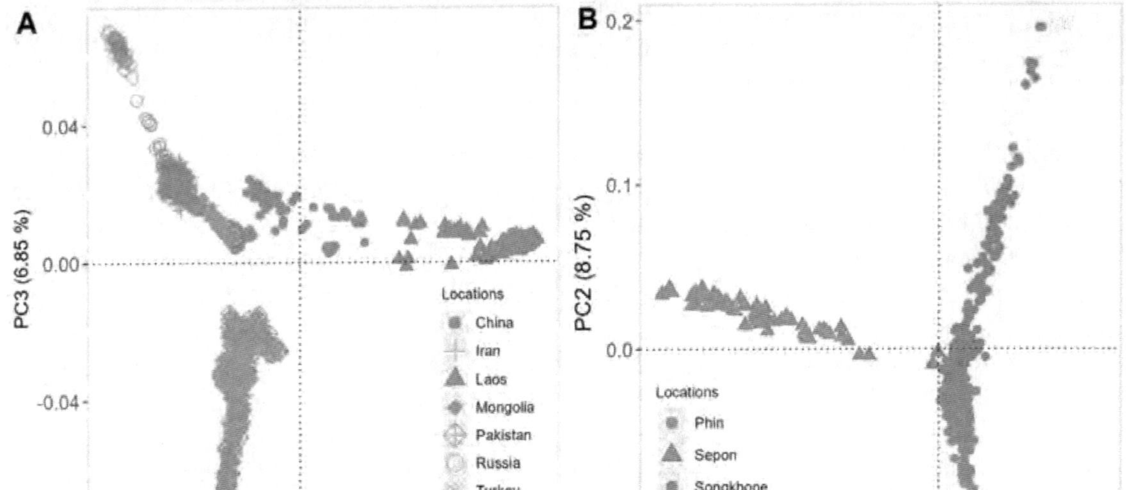

Figure 2. Principal component analysis for (A) Lao goats and Asian goat breeds, and (B) goat populations from different districts in Laos

Table 1. Pairwise FST values for goats from different Asian countries

	China	Iran	Laos	Mongolia	Pakistan	Russia
Iran	0.0408	-				
Laos	0.0481	0.0731	-			
Mongolia	0.0135	0.0327	0.0512	-		
Pakistan	0.0351	0.0433	0.0507	0.0291	-	
Russia	0.0295	0.0195	0.0614	0.0197	0.0371	-
Turkey	0.0385	0.0201	0.0703	0.0292	0.0418	0.0143

The average pairwise F_{ST} values between breeds in seven Asian countries were 0.0418, and ranged from 0.0135 to 0.0731 (Table 1), indicating a low to moderate genetic differentiation among these populations. The Mongolian and Chinese goat breeds showed the lowest level of differentiation (0.0135), while the highest level of differentiation (0.0731) was observed between Laos and Pakistan goat breeds. Results from the pairwise F_{ST} between Lao goats and other goat breeds from Asia confirmed the PCA results as there was a low genetic difference between Laos and Chinese goat breeds ($F_{ST} = 0.0481$), and there were moderate genetic differences between Lao goat breeds with other goat breeds in Asia ($F_{ST} > 0.05$).

In this study, the F_{ST} for Laos was 0.0223 and ranging from 0.0211 to 0.0815 which is slightly higher than the F_{ST} reported in Mongolia (0.009 to 0.035; Mukhina *et al.* 2022), but lower than F_{ST} in Russia (0.06 to 0.11; Deniskova *et al.* 2021) and Chinese goat population (0.02 to 0.16; Berihulay *et al.* 2019).

Table 2. Pairwise fixation index (F_{ST}) among districts in Laos

	Phin	Sepon
Sepon	0.0404	-
Songkhone	0.0112	0.0427

Genetic differentiation among goats from three districts in Central Laos revealed low genetic differentiation with greater differentiation between Sepon and Phin and Songkhone districts than between Phin and Songkhone (Table 2). This was supported by the PCA results that showed a gene flow between goats in Phin and Songkhone districts. Berihulay *et al.* (2019) described that genetic differentiation between two populations could be explained by natural geographic isolation. Phin and Songkhone were not located far from each other and they both have access to the main road, known as Route 13. The farmers in these two districts were able to easily trade goats between each other's locations in Xathamua town which is located almost halfway between the two districts and is close to the Savannakhet province centre. On the other hand, Sepon is more culturally isolated, being of a more distinct ethnic and linguistic group (Mong Kong), which may reduce the trade with other two districts. Additionally, Sepon is closer to the Vietnam border and it is easier for farmers to trade goats with Vietnamese dealers.

Overall, the observed heterozygosity values were less than the expected heterozygosity, excepted in Sepon (Table 3). The Ho and He in three Lao districts were lower than those in any Asian country. The average value of Ho and He in Asia were 0.3655 and 0.3956, respectively. The average Ho and

He in three districts in Laos were very similar, varying from 0.288 to 0.299 (Ho), and from 0.264 to 0.308 (He), respectively.

Table 3. Genetic diversity indices for goats from seven Asian countries and three districts in Laos. Ho: observed heterozygosity, He: expected heterozygosity, F_{IS}: inbreeding coefficient

	Asian country							Districts in Laos		
	China	Iran	Mongolia	Pakistan	Russia	Turkey	Laos	Phin	Sepon	Song
Ho	0.3672	0.343	0.4006	0.3379	0.409	0.408	0.292	0.299	0.288	0.2888
He	0.4023	0.419	0.4071	0.3831	0.434	0.421	0.303	0.308	0.264	0.3046
F_{IS}	0.0873	0.181	0.0158	0.118	0.056	0.03	0.037	0.028	-0.093	0.052

In general, F_{IS} were close to zero with positive F_{IS} values for Laos population (F_{IS}=0.0367), Phin (0.0283) and Songkhone (0.052), while a negative value was observed for Sepon (-0.093) indicating low outbreeding depression in this district. Similarly, low F_{IS} was found in Mongolian goat breeds, ranging from -0.013 to 0.025 (Mukhina *et al.* 2022) and it varied from -0.014 to 0.062 in the Chinese goat breeds (Berihulay *et al.* 2019). Those values were close to zero, indicating low levels of inbreeding in Lao natives. However, additional genomic analysis can assess the levels of relatedness between individuals and give insights into the history of Laos populations.

CONCLUSION

The results of Lao native goats revealed a closer genetic relationship with Chinese goat breeds than with other Asian goat breeds. Low to moderate genetic differentiation within Lao native goats was found, especially between Sepon and other locations. Inbreeding coefficients were close to zero being negative in Sepon and low but positive in Phin and Songkhone districts. These findings are not consistent with inbreeding depression being a major cause of small body size and low productivity in Lao native goats. Further analyses such as run of homozygosity are needed to identify levels of homozygosity in genomic regions.

ACKNOWLEDGEMENTS

This study was supported by ACIAR under project LS/2017/034.

REFERENCES

Berihulay H., Li Y., Liu X., *et al.* (2019) *Anim. Genet.* 50: 242.
Brito L.F., Kijas J.W., Ventura R.V., *et al.* (2017) *BMC Genomics* 18: 229.
Colli L., Milanesi M., Talenti A., *et al.* (2018) *Genet. Sel.* 50: 58.
Danchin-Burge C., Allain D., Clément V., *et al.* (2012) *Small Rumin. Res.* 108: 36.
Deniskova T.E., Dotsev A.V., Selionova M.I., *et al.* (2021) *Front. genet.* 12: 1.
Goudet J., Jombart T., Goudet, M.J. (2015) Package 'hierfstat'. *R package version* 0.04-22
Hermes T.R., Frachetti M.D., Voyakin D., *et al.* (2020) *PLoS One* 15: e0233333.
Mukhina V., Svishcheva G., Voronkova V., *et al.* (2022) *Animals* 12: 221.
Nandolo W., Mészáros G., Banda L.J., *et al.* (2019) *Front. genet.* 10: 537.
Oget C., Servin B., Palhiere I. (2019) *Anim. Genet.* 50: 54.
Purcell S., Neale B., Todd-Brown K., *et al.* (2007) *Am. J.Hum. Genet.* 81: 559.
Stür W., Gray G.D. (2014) ACIAR Monograph No. 177
Tarekegn G.M., Wouobeng P., Jaures K.S., *et al.* (2019) *PloS one* 14: e0214843.
Weir B.S., Cockerham C.C. (1984) *Evolution* 1358.
Windsor P.A., Nampanya S., Putthana V., *et al.* (2018) Vet. Parasitol. 251: 95.
Wright S. (1965) *Evolution* 395.

IMMUNE COMPETENCE AND MICRO-ENVIRONMENTAL SENSITIVITY

M.D. Madsen[1], J.H.J. van der Werf[1], A. Ingham[2], B. Hine[3], A. Reverter[2] and S. Clark[1]

[1] School of Environmental and Rural Science, University of New England, Armidale, NSW 2351, Australia
[2]Queensland Biosecurity Precinct, Agriculture & Food, CSIRO, St Lucia, QLD, 4067 Australia.
[3]F.D. McMaster Laboratory, Agriculture & Food, CSIRO, Armidale, NSW, 2350 Australia

SUMMARY

The aim of this study was to estimate the genetic relationship between immune competence and micro-environmental sensitivity (ES) of weaning weight, eye muscle area, and rib and rump fat depth. Variation in micro-environmental sensitivity among livestock leads to variability of phenotypes. The genetic correlations indicated that animals with higher immune competence tended to have lower micro-ES of weaning weight and eye muscle area, and higher micro-ES of rib and rump fat depth.

INTRODUCTION

Selecting to improve the immune competence (IC) of livestock could potentially lead to increased health and welfare of the animals, and decrease the livestock industry's reliance on antibiotics (Dominik *et al.* 2019; Hine *et al.* 2019; Hine *et al.* 2021; Reverter *et al.* 2021a; Reverter *et al.* 2021b). Furthermore, improved immunity could reduce the production loss and cost of medical intervention associated with disease incidences thus increasing profits (Hine *et al.* 2021).

The immune system is a complex system affecting many other systems in the animals, which can influence many phenotypes. The relationship between IC and live weight traits, growth and eye muscle area have been found to be unfavourable, while carcass traits and dry matter intake have a less straight forward relationship with IC (Reverter *et al.* 2021b). Aside from the direct relationship between IC and production traits, it is possible the IC affects the variability of production traits. The variability of phenotypes can vary between animals of different genetic backgrounds, in which case the genotypes exhibit micro-environmental sensitivity (micro-ES). Animals with less micro-ES are expected to respond less to disturbances in their environments and can be quantified at a genetic heterogeneity of the environmental variance (SanCristobal-Gaudy *et al.* 1998; Hill and Mulder 2010). The relationship between IC and micro-ES has not yet been reported.

The aim of this study was to investigate the relationship between IC and some production traits and between IC and the micro-ES of production traits in Australian Angus cattle.

MATERIALS AND METHODS

Data. Antibody- and cell-mediated immune response (AMIR, CMIR) were the IC component traits. The AMIR and CMIR records were provided by CSIRO and Angus Australia. The records were collected in 2012-2020 in accordance with the procedures described by Hine *et al.* (2019). The AMIR phenotypic values represent the level of antigen-specific serum IgG1 antibody in response to vaccination with Ultravac 7in1 vaccine (Zoetis) and were calculated from the square root transformed optical density values generated using an enzyme-linked immunosorbent assay and corrected for inter-plate variation. The CMIR phenotypes were calculated from the log-transformed ratio between the measured double skinfold thickness at test (intradermal vaccine injection) and control site (intradermal saline injection) (Hine et al. 2019). To account for initial double skinfold thickness, the pre-injection log-transformed ratio between the double skinfold thickness at test and control site was used as a covariate in the analysis.

Production traits consisted of weaning weight (WW), scan eye muscle area (EMA), scan rib fat depth (RIB) and scan rump fat depth (P8). The production traits were provided by Angus Australia and were part of the routine recording scheme between 2012 and 2020.

For the IC records, contemporary groups (CG) were constructed by concatenating herd, year and test cohort. For the production traits, trait specific CGs were concatenations of herd, birth year, observation date for the trait, breeder defined management group, birth type and embryo transfer status. Age slicing further subdivided CGs for WW, RIB, P8 and EMA. Age slices covered 45 days for WW and 60 days for RIB, P8 and EMA as per Graser *et al.* (2005), and slices were symmetric around the average age of the CG. Summary statistics are shown in Table 1. Two pedigrees were used for analysis, one for sire (10948 animals) and one for rearing dams (98151 animals).

Table 1. Summary statistics for the final dataset

Parameter	Statistic	WW (kg)	RIB (mm)	P8 (mm)	EMA (cm^2)	AMIR	CMIR
Records	Count	31699	83034	83314	83486	3910	3908
Phenotype	Mean	254.60	6.11	7.89	80.49	0.85	1.89
	SD	51.63	2.76	3.73	17.56	0.43	0.42
	Range	77-496	1-22	1-33	31-157	0.01-2.13	0.85-4.92

Analysis. The data was analysed using 8 two-trait models with an IC trait as one trait and a production trait as the second trait. The production traits were fitted with a double hierarchical generalised linear model (DHGLM) for estimating the micro-ES of the production traits resulting in a trivariate model. The general model was:

$$\begin{bmatrix} \mathbf{y}_{IC} \\ \mathbf{y}_{PT} \\ \mathbf{y}_{mES} \end{bmatrix} = \begin{bmatrix} \mathbf{X}_{IC} & \mathbf{0} & \mathbf{0} \\ \mathbf{0} & \mathbf{X}_{PT} & \mathbf{0} \\ \mathbf{0} & \mathbf{0} & \mathbf{X}_{mES} \end{bmatrix} \begin{bmatrix} \mathbf{b}_{IC} \\ \mathbf{b}_{PT} \\ \mathbf{b}_{mES} \end{bmatrix} + \begin{bmatrix} \mathbf{Z}_{IC} & \mathbf{0} & \mathbf{0} \\ \mathbf{0} & \mathbf{Z}_{PT} & \mathbf{0} \\ \mathbf{0} & \mathbf{0} & \mathbf{Z}_{mES} \end{bmatrix} \begin{bmatrix} \mathbf{s}_{IC} \\ \mathbf{s}_{PT} \\ \mathbf{s}_{mES} \end{bmatrix} + \begin{bmatrix} \mathbf{e}_{IC} \\ \mathbf{e}_{PT} \\ \mathbf{e}_{mES} \end{bmatrix}$$

where $\mathbf{y}_{IC}$, $\mathbf{y}_{PT}$ and $\mathbf{y}_{mES}$ were the IC trait (AMIR or CMIR), the production trait phenotype and calculated micro-ES phenotype of the production trait, respectively. $\mathbf{b}_{IC}$ contained the fixed effects of sex and CG for AMIR and the fixed effect of CG and the pre-injection covariate for CMIR, $\mathbf{b}_{PT}$ and $\mathbf{b}_{mES}$ contained the fixed effects of sex and CG and covariate of age for the production traits (and the covariate of dam age and squared dam age for WW). $\mathbf{s}_x$ and $\mathbf{e}_x$ were the fixed effects, additive genetic sire effects and residuals of trait x ($x \in (IC,\ PT, mES)$). The micro-ES phenotype was calculated as $\mathbf{y}_{mES} = \hat{\mathbf{e}}^2_{PT} / (\mathbf{1} - \mathbf{h}_{PT})$, where $\mathbf{h}_{PT}$ was the diagonal element of the part of the hat-matrix corresponding to $\mathbf{y}_{PT}$ ($\hat{\mathbf{y}}_{PT} = \mathbf{H}\mathbf{y}_{PT}$) also known as the leverage (Hoaglin and Welsch 1978). For models where WW was the production trait, the model also included maternal genetic ($\mathbf{c}$) and permanent environmental ($\mathbf{pe}$) effects.

The distribution assumptions for the random genetic sire effects were $\begin{bmatrix} \mathbf{s}_{IC} \\ \mathbf{s}_{PT} \\ \mathbf{s}_{mES} \end{bmatrix} \sim MVN\left(\begin{bmatrix} \mathbf{0} \\ \mathbf{0} \\ \mathbf{0} \end{bmatrix}, \begin{bmatrix} \sigma^2_{s_{IC}} & \sigma_{s_{IC}s_{PT}} & \sigma_{s_{IC}s_{mES}} \\ & \sigma^2_{s_{PT}} & \sigma_{s_m s_{mES}} \\ & & \sigma^2_{s_{mES}} \end{bmatrix} \otimes \mathbf{A} \right)$, where $\mathbf{A}$ was the numerator relationship matrix among sires based on the sire pedigree and $\otimes$ is the Kronecker product. The distribution assumptions of the residuals were $\begin{bmatrix} \mathbf{e}_{IC} \\ \mathbf{e}_{PT} \\ \mathbf{e}_{mES} \end{bmatrix} \sim MVN\left(\begin{bmatrix} 0 \\ 0 \\ 0 \end{bmatrix}, \begin{bmatrix} \mathbf{I}\sigma^2_{e_{IC}} & 0 & 0 \\ 0 & \mathbf{W}^{-1}_{PT}\sigma^2_{\epsilon_{PT}} & 0 \\ 0 & 0 & \mathbf{W}^{-1}_{mES}\sigma^2_{\epsilon_{mES}} \end{bmatrix} \right)$, where $\mathbf{I}$ was an identity matrix of appropriate size, and $\mathbf{W}_{PT}$ and $\mathbf{W}_{mES}$ were matrices containing weights for the residual variances of the DHGLM. $\boldsymbol{W}_{PT} = diag(\widehat{y_{mES}})^{-1}$ and $\boldsymbol{W}_{mES} = diag((1 - \boldsymbol{h}_{PT})/2)$.

Post analysis corrections of variance components to obtain additive genetic and residual variances on the animal level were applied as described in Madsen *et al.* (2021). The variation and heritability of micro-ES ($h^{2^*}_{mES}$) was converted from the logarithmic to the measurement level following Mulder *et al.* (2007) and Mulder *et al.* (2009).

RESULTS AND DISCUSSION

The results showed additive genetic variance of micro-ES in all production traits (Table 2). The heritabilities were in line with the heritabilities reported for production traits in Nellore beef cattle by Neves *et al.* (2011) and Iung *et al.* (2017). The genetic coefficient of variation (GCV) of micro-ES was low to moderate, with higher values for the fat traits. The higher GCV of RIB and P8 indicate that some response to selection could be obtained.

The heritability of AMIR and CMIR were in line with those previously reported (Dominik *et al.* 2019; Hine *et al.* 2019; Reverter *et al.* 2021a; Reverter *et al.* 2021b). Likewise, the heritabilities of EMA and P8 were within previously reported values for Australian beef cattle, while the heritability of RIB was slightly higher than previously reported (Meyer *et al.* 2004; Jeyaruban *et al.* 2009). In contrast, the heritability of WW was higher than the 0.13-0.35 reported for Australian beef cattle (Meyer *et al.* 2004; Jeyaruban *et al.* 2009; Torres-Vázquez *et al.* 2018). Slightly larger heritabilities can be expected when a trait is fitted with a DHGLM as the genetic variation due to micro-ES is removed from the observed residual variance of the phenotype reducing the denominator used to calculate the heritability.

Table 2. Estimated heritabilities and genetic coefficient of variation

	AMIR	CMIR	WW	RIB	P8	EMA
$\boldsymbol{h}^2$ (%)	36.18	35.62	43.50	34.58	35.11	25.13
$\boldsymbol{h}^{2^*}_{mES}$(%)			0.03	1.22	1.42	0.35
$\boldsymbol{GCV}_{mES}$ (%)			13	24	27	11

Table 3. Genetic correlations between the production and immune traits in Angus cattle*

	AMIR			CMIR		
	$r_{AMIR,PT}$	$r_{AMIR,mES}$	$r_{PT,mES}$	$r_{CMIR,PT}$	$r_{CMIR,mES}$	$r_{PT,mES}$
WW	*-0.35*	-0.12	*0.18*	*-0.26*	-0.15	*0.18*
RIB	0.11	0.14	*0.87*	0.15	0.09	*0.87*
P8	0.06	0.00	*0.90*	0.16	0.12	*0.90*
EMA	-0.13	-0.34	*0.30*	0.04	-0.17	*0.31*

*Italic values had 95% confidence intervals not including 0

The genetic correlations between the IC traits and RIB and P8 indicated that animals with higher fatness also tended to have higher IC (Table 3). In contrast, the genetic correlations indicated that animals with higher IC had lower WW, showing that immune response may be utilising resources that would otherwise have contributed towards growth. The genetic correlations between micro-ES of production traits and IC tended to be moderately negative for WW and EMA and non-existing to lowly positive for RIB and P8. The genetic correlations involving the IC traits had large SEs and therefore only the genetic correlations between WW and either IC trait had a 95% confidence interval not including 0. The larger SEs were likely due to the small data size of the two IC traits.

The genetic correlations between the production traits and their micro-ES were strongly positive for RIB and P8 fat showing that selection to reduce fatness would have a correlated decrease in the micro-ES of fatness and vice versa.

CONCLUSIONS

All production traits showed micro-ES. The heritabilities and genetic coefficient of variance of micro-ES was higher for RIB and P8 than the other production traits. Selection to decrease micro-ES may be possible for these traits.

Results showed that mounting immune responses might direct resources away from growth.

The positive genetic correlation between the fat and IC traits indicated that animals with higher fatness also have higher ICs.

The genetic correlations between the IC traits and micro-ES of production traits showed a tendency for animals with higher genetic potential for IC to have lower micro-ES of WW and EMA and higher micro-ES of RIB and P8.

ACKNOWLEDGEMENTS

M. D. Madsen acknowledges the Livestock Genomics Scholarship provided by University of New England for funding their studies. The authors thank C. Duff and H. Aliloo for providing the dataset containing the production traits and the pedigrees, and P. Madsen for their contributions towards analysis.

REFERENCES

Dominik S., Porto Nero L., Duff C., Byrne A., Hine B., Ingham A., Colditz I. and Reverter A. (2019) *Proc. Assoc. Advmt. Anim. Breed. Genet.* 23: 402.

Graser H. U., Tier B., Johnston D. J. and Barwick S. A. (2005) *Anim. Prod. Sci.* 45: 913.

Hill W.G. and Mulder H.A. (2010) *Genet. Res.* 92:

Hine B.C., Bell A.M., Niemeyer D.D., Duff C.J., Butcher N.M., Dominik S., Ingham A.B. and Colditz I.G. (2019) *J. Anim. Sci.* 97: 4053.

Hine B.C., Bell A.M., Niemeyer D.D., Duff C.J., Butcher N.M., Dominik S., Porto-Neto L.R., Li Y., Reverter A., Ingham A.B. and Colditz I.G. (2021) *J. Anim. Sci.* 99:

Hoaglin D.C. and Welsch R.E. (1978) *Am Stat.* 32: 17.

Iung L.H.S., Neves H.H.R., Mulder H.A. and Carvalheiro R. (2017) *J. Anim. Sci..* 95: 1425.

Jeyaruban M.G., Johnston D.J. and Graser H.U. (2009) *Anim. Prod. Sci.* 49: 584.

Madsen M.D., van der Werf J.H.J, Börner V., Mulder H.A. and Clark S. (2021) *Animal* 15: 100411.

Meyer K., Johnston D.J. and Graser H.-U. (2004) *Aust. J. Agric. Res,* 55: 195.

Mulder H.A., Bijma P. and Hill W.G. (2007) *Genetics.* 175: 1895.

Mulder H.A., Hill W.G., Vereijken A. and Veerkamp R.F. (2009) *Animal.* 3: 1673.

Neves H., Carvalheiro R., Roso V. and Queiroz S. (2011) *Livest. Sci.* 142: 164.

Reverter A., Hine B.C., Porto-Neto L., Alexandre P.A., Li Y., Duff C.J., Dominik S. and Ingham A.B. (2021a) *Anim. Prod. Sci.* 61: 1919.

Reverter A., Hine B.C., Porto-Neto L., Li Y., Duff C.J., Dominik S. and Ingham A.B. (2021b) *J. Anim. Sci..* 99: 1.

SanCristobal-Gaudy M., Elsen J.-M., Bodin L. and Chevalet C. (1998) *Gen. Sel. Evol.* 30: 423.

Torres-Vázquez J.A., van der Werf J.H.J. and Clark S.A. (2018) *J. Anim. Sci.* 96: 4521.

AN ACROSS-FLOCK ANALYSIS ON FAECAL WORM EGG COUNTS IN MERINO SHEEP IN SOUTH AFRICA

S.W.P. Cloete[1,2], Z. Mpetile[1] and K. Dzama[1]

[1]Department of Animal Sciences, Stellenbosch University, Matieland, 7602 South Africa
[2]Directorate Animal Sciences, Western Cape Department of Agriculture, Elsenburg, 7607 South Africa

SUMMARY

Previous South African studies on faecal worm egg count (FWEC) in South African Merinos have been limited to analyses within flocks. This study details an across-flock-season analysis of FWEC at the Tygerhoek and Elsenburg research farms using 9080 records collected between 1995 and 2016. Two discrete environments were identified, namely an autumn lambing season at Tygerhoek, and a winter lambing season at Elsenburg. The exchange of sires across environments allowed the estimation of the sire x site variance as an indication of a genotype x environment interaction for FWEC. At 0.12 ± 0.02, FWEC was lowly heritable across environments. Additionally, variance ratios for the dam permanent environment and sire x site contributed respectively 0.03 ± 0.01 and 0.014 ± 0.006 to the observed phenotypic variation. Selection for a reduced FWEC across flocks would likely result in genetic gains, while the probability of a major reranking of sires across sites appears to be small.

INTRODUCTION

So far, research in South Africa has focused on deriving genetic parameters for faecal worm egg count (FWEC) and on correlations of FWEC with other traits of economic importance within flocks or localities. For FWEC to be considered as an indicator of resistance to round worm infection in South Africa (as advocated by Cloete *et al.* 2014) it is important to conduct analyses across flocks. Across-flock analyses allow for the estimation of genotype x environment interactions (G x E; van Wyk *et al.* 2008). Such analyses became commonplace for FWEC in other sheep producing countries such as Australia (Brown *et al.* 2016; Brown and Fogarty 2017) and New Zealand (Pickering *et al.* 2012). Sheep farmers in these countries are thus benefiting from advances brought about by using across-flock genomic breeding values for FWEC for the selection of replacements with resistance to gastro-intestinal nematodes. South Africa has been lagging with respect to these advances. This study, therefore, reports the first across-flock analysis for FWEC in South African sheep. Linkage provided by sires across flocks additionally allowed the estimation of the sire x flock/season variance as an indication of G x E as hypothesized for FWEC.

MATERIALS AND METHODS

The study combined data from Merino flocks maintained on the Tygerhoek and Elsenburg research farms. Both farms are situated in the Mediterranean region of South Africa, Tygerhoek at 34°08' S and 21°11' E and at an elevation of 425 m. Elsenburg is located at 33°51' S and 18°50' E, at an elevation of 177. Rainfall averages 425 mm at Tygerhoek and 606 mm at Elsenburg, with respectively 60% and 77% of the precipitation recorded from April to September (Cloete *et al.* 2016). The management, breeding and husbandry of both flocks are well described (Tygerhoek: Cloete *et al.* 2007; Elsenburg: Mpetile *et al.* 2015). Further information on these topics will thus be omitted. Faecal grab samples were obtained from the rectum of individual sheep and counted at an accuracy of 100 eggs per gram (epg) wet faeces at the Regional Veterinary Laboratory at Stellenbosch. Worm challenge at the respective localities was not quantified, but Cloete *et al.* (2016) suggested that a mixed challenge of *Teladorsagia* spp, *Trichostrongylus* spp and *Nematodirus* spp was more likely at Tygerhoek. A greater reliance on irrigated pastures at Elsenburg resulted in hematophagous parasites like *Haemonchus contortus* becoming more important (Cloete *et al.* 2016). Data at Tygerhoek were recorded from 1995 to 2016, except for 2004 when no data were available (Cloete *et al.* 2007). The data at Elsenburg were recorded over the same period, except for 1997 to 1996 and 2000 (Mpetile *et al.* 2015). Flock data at Tygerhoek and Elsenburg contributed respectively 6,527 hogget and 2,563 yearling records to the study. Age at recording (± s.d.) was 498 ± 38 days at Tygerhoek and 322 ± 30 days at Elsenburg.

Mpetile *et al.* (2017) reported that season had a profound effect on genetic variation of FWEC at Tygerhoek, with the heritability of FWEC using spring samples being substantially higher than for samples collected in autumn. As lambs were born in autumn at Tygerhoek and winter at Elsenburg, samples for FWEC were taken during spring at Tygerhoek and autumn at Elsenburg. This seasonal effect was confounded with location, but eight sires with progeny at both locations and having, on average, 40 ± 12 recorded offspring at Tygerhoek and 17 ± 5 recorded offspring at Elsenburg linked the data recorded on the two locations.

Given the well-established deviations from normality in FWEC data, individual records were transformed to natural logarithms after 100 was added to account for zero counts. Previous studies on the respective resource flocks also tested the cube root transformation at Tygerhoek (Cloete *et al.* 2007) and Elsenburg (Mpetile *et al.* 2015). Genetic parameters stemming from the alternative approaches did not differ and the analysts preferred the log transformation for its lower coefficient of variation. The data so derived were analysed by single-trait analyses using ASREML (Gilmour *et al.* 2015). Fixed effects fitted included contemporary group (90 levels involving year-site-season-sex combinations), age of dam (2-6+ years) and birth type (single vs. multiple). Random effects were sequentially added to the fixed-effects analysis as described in Table 1.

Likelihood Ratio tests (LRT) were used to test the significance of random effects. A random effect was considered significant when its inclusion in the model improved the log likelihood ratio using the Chi^2 distribution as a test statistic. When models had the same number of random effects, the model with the highest log likelihood was preferred. After the the most appropriate model was determined, the random effect of sire x site (encompassing 566 levels) was added to the model by fitting an identity matrix linking sire x site effects to the data (see Table 1). The LRT was then conducted additionally to assess this effect for significance. Phenotypic variance was expressed as the total of all the estimated variance components. Variance ratios were derived by dividing significant ($P < 0.05$) variance components by the phenotypic variance. The pedigree file used in all analyses contained 14832 animals, the progeny of 830 sires and 4342 dams.

RESULTS AND DISCUSSION

The raw data were leptokurtic and skewed with extreme individual variation of FWEC records ranging from 0 to 32700 epg of wet faeces and an overall mean of 1960 ± 2599. The log transformation improved the distribution of the data appreciably resulting in a normal distribution (skewness = -0.32; kurtosis = -0.58) and a coefficient of variation of 17.9% with a mean of 6.97 ± 1.25. These results were consistent with previous studies on these flocks (Cloete *et al.* 2007; Mpetile *et al.* 2015) and are not discussed. Contemporary group exerted a marked effect on the data ($P < 0.001$), while FWEC depended less on age of dam ($P = 0.57$) and birth type ($P = 0.07$).

The LRT suggested that the log likelihood improved markedly from a model consisting of only fixed effects to a model including additive genetic effects (Table 1). Compared to this model with only one random effect, the addition of maternal additive effects did not result in an improvement ($P > 0.05$). Adding dam permanent environmental (PE) effects improved the log likelihood, though. Including both maternal genetic and PE effects did not change the log likelihood when added to the latter model. Adding the sire x site variance to the model including additive and dam PE effects resulted in a further improvement in the log likelihood.

Table 1. Log likelihood ratios for the various models fitted in the across-flock analysis conducted on the Tygerhoek and Elsenburg Merino flocks (Chi² values are for the more comprehensive model compared to the simpler model with 1 less random effect)

Effect fitted	Random effects	Log likelihood value	#Chi²
Fixed effects (FE) only	0	-3708.39	NA
FE + σ^2_a (Model 1)	1	-3641.62	133.54**
FE + σ^2_a + σ^2_m (Model 2)	2	-3640.21	2.82ns
FE + σ^2_a + σ^2_{pe} (Model 3)	2	-3637.75	7.74**
FE + σ^2_a + σ^2_m + σ^2_{pe} (Model 4)	3	-3637.75	0.00ns
FE + σ^2_a + σ^2_{pe} + $\sigma^2_{sire:site}$ (Model 5)	3	-3635.17	5.16*

σ^2_a = additive variance; σ^2_m = maternal genetic variance; σ^2_{pe} = dam permanent environmental variance; $\sigma^2_{sire:site}$ = sire x site variance; #Critical values: 3.84 (P = 0.05); 6.64 (P = 0.01); * P < 0.05; ** P < 0.01; ns – not significant

The phenotypic variance components and variance ratios for additive genetic, dam genetic, dam PE and sire x site effects are presented in Table 2 for the respective models. The across flock heritability of FWEC ranged from 0.12 for Model 5 (the model of choice) to 0.16 for Model 1. Dam PE consistently contributed 0.03 to the phenotypic variation, while the sire x site variance amounted to somewhat more than 1% of the phenotypic variance. As FWEC was variable and heritable, genetic gains across flocks seems feasible although these gains may not necessarily be fast. The estimated heritability is within the ranges of 0.00 to 0.52 reported in the literature (Greeff *et al.* 1995; Safari and Fogarty 2003; Snyman 2007) and a fair reflection of previous heritability estimates within the flocks contributing data to this study (Cloete *et al.* 2007; Mpetile *et al.* 2015). The across-flock heritability of FWEC amounted to 0.16 for Australian meat sheep (Brown *et al.* 2016) and to 0.16 and 0.17 for Australian Merino yearlings and hoggets, respectively (Brown and Fogarty 2017). Maternal effects were not important in both latter studies. More comprehensive data on FWEC in the South African small stock industry is needed to allow the incorporation of this important input trait in the formal genetic evaluation scheme.

Table 2. The estimated phenotypic variance components and variance ratios for FWEC in across-flock analyses on Tygerhoek and Elsenburg Merinos for the random models fitted

Random model	σ^2_p	h²	m²	pe²	sire.site
Model 1	0.817	0.16 ± 0.02	NA	NA	NA
Model 2	0.816	0.15 ± 0.02	0.02 ± 0.01	NA	NA
Model 3	0.815	0.14 ± 0.02	NA	0.03 ± 0.01	NA
Model 4	0.815	0.14 ± 0.02	0.00 ± 0.00	0.03 ± 0.01	NA
Model 5	0.815	0.12 ± 0.02	NA	0.03 ± 0.01	0.014 ± 0.006

σ^2_p – phenotypic variance; h² – heritability; m² – dam genetic effect; pe² – dam permanent environmental effect; sire.site – sire x site variance ratio; NA – not applicable

The dam PE estimate of 0.03 is somewhat lower than comparable estimates for FWEC of around 0.05 derived previously for the Tygerhoek flock (Cloete *et al.* 2007). It may well be that the accrual of additional pedigree information as well as sire x site effects partitioned animal variances away from dam PE in the present study. Corresponding values for PE effects sourced from the literature were variable from 0.02 to 0.16 (Safari and Fogarty 2003).

Although the observed variation for sire x site/season was significant in a Mediterranean climate, it contributed less than 2% to the overall phenotypic variation. Baker *et al.* (2004) did not find a significant G x E for FWEC in Red Masaai and Dorper sheep maintained under either sub-humid or semi-arid conditions. Carrick and van der Werf (2007) reported highly variable genetic correlations between extreme quintiles for FWEC in Australian Sheep Genetics data. Some correlations involving FWEC were below 0.8, thus indicating the possibility of G x E. Since different methods were used, it is difficult to relate the present results to those of Carrick and van der Werf (2007). Both studies suggest the possibility of G x E for FWEC, although reranking among sires may be considered as small when the outcome of the present study is considered. The sire x site variance ratio in this study was on the lower end compared to previous estimates of between 2.2 and 2.5% of the variation attributed to sire x contemporary group for production traits in an across-flock analysis on South African Dohne Merinos (van Wyk *et al.* 2008). To our knowledge, there are no comparable studies exploring G x E for FWEC. An alternative approach that is worthy of exploration in future is the usage of random regression methods (Pollot and Greeff 2004; Hollema *et al.* 2018).

CONCLUSION

This study confirmed significant across-flock genetic variation in FWEC in South African sheep flocks. It therefore paves the way for further exploration of the genetic improvement of FWEC as an input trait in the local sheep industry. The derived heritability was not particularly high but backed by sufficient phenotypic variation to sustain genetic progress. Moreover, it is foreseen that further across-flock studies incorporating more flocks will provide more accurate estimations of other sources of variation, such as maternal effects and sire x flock effects.

ACKNOWLEDGEMENTS

Financial support of the Western Cape Agricultural Research Trust and Cape Wools SA is gratefully acknowledged.

REFERENCES

Baker R.L., Mugambi J.M., Audho J.O., Carles A.B. and Thorpe W. (2004) *Anim. Sci.* 79: 343.

Brown D.J. and Fogarty N.M. (2017) *Anim. Prod. Sci.* 57: 209.

Brown D.J., Swan A.A., Gill J.S., Ball A.J. and Banks R.G. (2016) *Anim. Prod. Sci.* 56: 1442.

Carrick M.J. and van der Werf J.H.J. (2007) *Proc. Assoc. Advmt. Anim. Breed. Genet.* 17: 248.

Cloete S.W.P., Mpetile Z. and Dzama K. (2016) *Small Rumin. Res.* 145: 33.

Cloete S.W.P., Olivier J.J., Du Toit E. and Dreyer F.H. (2007) *S. Afr. J. Anim. Sci.* 37: 237.

Cloete S.W.P., Olivier J.J., Sandenbergh L. and Snyman M.A. (2014) *S. Afr. J. Anim. Sci.* 44: 308.

Gilmour A.R., Gogel B.J., Cullis B.R., Welham S.J. and Thompson R. (2015) ASReml User Guide Release 4.1, VSN International Ltd, Hemel Hempstead, HP1 1ES, UK.

Greeff J.C., Karlsson L.J.E. and Harris J.F. (1995) *Proc. Aust. Assoc. Anim. Breed. Genet.* 11: 117.

Hollema B.L., Bijma P. and van der Werf J.H.J. (2018) *J. Anim. Breed. Gen.* 135: 357.

Mpetile Z., Dzama K. and Cloete S.W.P. (2017) *J. S. Afr. Vet. Assoc.* ISSN: (Online) 2224, (Print) 1019.

Mpetile Z., Cloete S.W.P., Kruger A.C.M. and Dzama K. (2015) *S. Afr. J. Anim. Sci.* 45: 510.

Pickering N.K., Dodds K.G., Blair H.T., Hickson R.E., Johnson P.L. and McEwan J.C. (2012) *J. Anim. Sci.* 90: 1411.

Pollott G. and Greeff J.C. (2004) *J. Anim. Sci.* 82: 2840.

Safari A. and Fogarty N.M. (2003). Genetic parameters for sheep production traits: estimates from the literature. Technical Bulletin vol. 49. NSW Agriculture, Orange, Australia.

Snyman M.A. (2007) *Grootfontein Agric.* 7: 29.

Van Wyk J.B., Swanepoel J.W., Cloete S.W.P., Olivier J.J. and Delport G.J. (2008) *S. Afr. J. Anim. Sci.* 38: 31.

GENETIC VARIATION IN SUSCEPTIBILITY TO FACIAL ECZEMA IN DAIRY CATTLE IN NEW ZEALAND

A. Ismael, E. Donkersloot, F. Wallace, G. Worth, S. Davis, L. McNaughton R. J. Spelman

Research and Development, Livestock Improvement Corporation, Hamilton 3240, New Zealand

SUMMARY

Facial eczema is a disease of ruminant animals, caused by a fungal toxin that grows on pastures and causes liver damage. The objective of this study was to investigate the genetic variation for susceptibility to facial eczema (FE) in dairy cattle using phenotypes collected under on-farm conditions. Weekly vat milk samples from progeny test herds were monitored for a marker of liver damage, to identify when to blood sample cows in the herd. Gamma-glutamyltransferase (GGT) enzyme concentrations in blood were used as the proxy to measure response to the fungal toxin on the animals. Log transformation (logGGT) and Box-Cox transformation (boxGGT) were applied to GGT before running the genetic analysis. The highest heritability found was for the logGGT (0.26). Genetic correlations between logGGT and production traits were all weak and positive, ranging from 0.02 to 0.12 indicating that, the trait is almost independent from production and selection for tolerance to FE could be performed without compromising milk production. The moderate heritability for logGGT indicates that 26% of the total variation of tolerance to FE among animals was attributable to genetic variance breeding values for this trait could be predicted with accuracy, enabling the identification of sires with tolerance to FE to be used in the breeding program in dairy cattle in New Zealand.

INTRODUCTION

Facial eczema (FE) is caused by the ingestion of the spores of the fungus *Psuedopithomyces chartarum* (Di Menna *et al.* 2009; Ariyawansa *et al.* 2015). The mycotoxin sporidesmin A causes damage to the liver (Smith and Towers 2002). Affected animals have reduced milk production (Mason *et al.* 2022); the worst affected animals may die or require euthanasia. Diagnosis of FE is typically via measurement of gamma glutymyltransferase (GGT) in the serum (Towers and Stratton 1978). The disease has been reported in grazing systems in Australia, South Africa, Brazil and parts of Europe (Di Menna *et al.* 2009). FE has traditionally been a problem on farms located in the North Island of New Zealand.

Genetic variation in susceptibility to sporidesmin has been demonstrated in sheep and cattle (Mcrae *et al.* 2016; Morris *et al.* 2013). In sheep, the Ramguard programme, dosing rams with sporidesmin and measuring the GGT response, is used to identify resistant sires (Amyes *et al.* 2018). However, this is not feasible for dairy sires with greater value than an individual ram. Nor acceptable to public. Previous research in cattle has demonstrated that blood sampling herds that have experienced a 'natural challenge' can be used to gather data for the estimation of genetic parameters. However, given the primarily subclinical nature of the disease, it is not easy to identify herds that have been exposed. A biomarker that can be used to screen herds that have liver damage has been identified. The aim of this work was to use the biomarker as a screening technique to identify herds where there has been a natural FE challenge and confirm that these phenotypes can be used to estimate genetic parameters.

MATERIALS AND METHODS

This work was carried out with the approval of the AgResearch Animal Ethics Committee, approval numbers 15236 and 15576. Herds were identified via screening the bulk tank milk for what, or by veterinarians volunteering herds for the study. Blood samples were collected from 9,866 animals from 34 commercial dairy herds that were naturally challenged by FE between April 2021 and May 2022. Tolerance to FE was evaluated based on gamma-glutamyltransferase (GGT) enzyme concentrations in blood as evidence of liver damage caused by sporidesmin. Genetic analysis in the present study applied to the raw GGT measurements, log-transformed GGT (logGGT) and Box-Cox transformed GGT (boxGGT) (Box and Cox 1964).

To estimate the genetic correlations between tolerance to FE and production traits, average first lactation 305-d test days yield deviations for milk, fat, and protein were extracted from the animal evaluation database after adjusting for the lactation stage included in the analysis. Descriptive statistics of each trait are summarised in Table 1.

Table 1: Mean, standard deviation, minimum, and maximum of all traits in the present study

Trait	Mean	SD	Min	Max
GGT (IU/L)	402.5	831.26	2.0	5352
logGGT (IU/L)	3.98	2.066	0.69	8.59
boxGGT	2.88	1.13	0.66	4.97
Milk (litre)	12.5	3.49	2.99	34.12
Fat (kg)	6.16	1.48	0.88	12.48
Protein (kg)	4.96	1.26	1.27	14.95

Genetic analyses were performed with AI-REML algorithm in ASReml-R v4 statistical package (Butler *et al.* 2017). A univariate animal model was performed to estimate variance components and heritability for each trait separately, whereas bivariate model was performed to estimate genetic correlations among traits. The following animal model was used for the analysis:

$$\mathbf{y} = \mathbf{Xb} + \mathbf{Za} + \mathbf{e},$$

where y is the vector of phenotypes, X is the matrix relating fixed effects to phenotypes, b is the vector of fixed effects, Z is the matrix relating phenotypes to animals, and a is the polygenic random additive genetic effect which was assumed to be normally distributed following var(a) ~N(0, Aσ_a^2), where σ_a^2 is the additive genetic variance and A is the average numerator relationship matrix (Wright 1922), and e is the vector of random residual, ~ND (0, I σ_e^2), where I is the identity matrix and σ_e^2 is the residual variance. The fixed effects in the model include cow age category, contemporary groups (herd-year-month of blood sampling), cow breed proportions (proportion of cow's breed ancestry that was Jersey, Holstein, Friesian, Ayrshire), heterosis effects (Friesian × Jersey, Jersey × Ayrshire, Jersey × Holstein) and cow's inbreeding coefficients. The genetic correlations between traits (r_a) were estimated as: $r_a = \frac{\sigma_{a_1 a_2}}{\sqrt{\sigma_{a_1}^2 \sigma_{a_2}^2}}$ where $\sigma_{a_1 a_2}$ is the additive genetic covariance among traits; and $\sigma_{a_1}^2$ and $\sigma_{a_2}^2$ are the additive genetic variances.

RESULTS AND DISCUSSION

The laboratory defined 'adequate' range for GGT is 3-47 IU/L (Gribbles Veterinary, Hamilton). Herds that were sampled had elevated GGT concentrations, averaging 402.5 IU/L across all herds sampled, indicative of liver damage. Figure 1 shows the difference of distribution between raw GGT, logGGT and boxGGT. Both logGGT and boxGGT were able to remove the skewness of the raw data so the distribution was more suitable for genetic analysis.

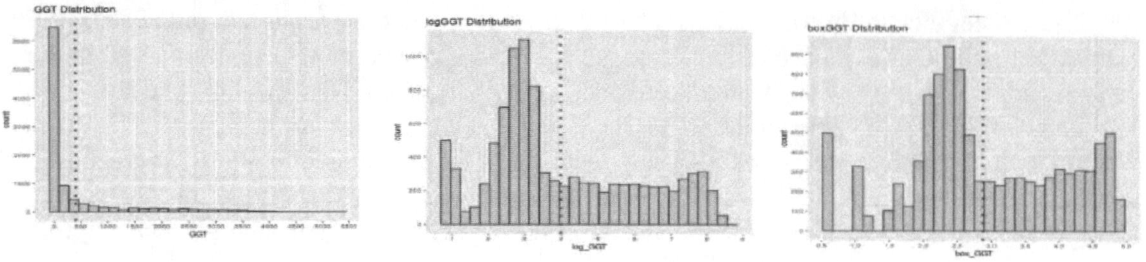

Figure 1. Distribution of raw GGT, logGGT, and boxGGT from cow blood samples with the mean represented in the dotted red line

Figure 2 shows associations between each of age (years) and breed and GGT. Average serum GGT concentrations were highest for 3-year old animals (515.6 IU/litre), and lowest for 10-year old animals (278 IU/litre) indicating that younger animals are more susceptible to FE. Holsteins and Jerseys had similar average serum GGT concentrations (approximately 48x IU/litre for each breed), but crossbred animals (Holstein x Jersey) had lower GGT (363.9 IU/litre), suggesting a possible heterosis effect on tolerance to FE.

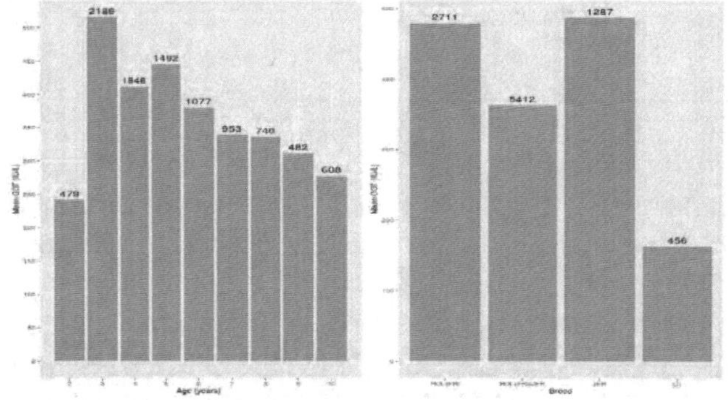

Figure 2. Associations between each of age and breed and raw GGT presented as means with the number of observations in each class annotated above the bars

Variance components and genetic parameter estimates. Variance components and heritability estimates for GGT, logGGT and boxGGT are presented in Table 2. The highest heritability estimate was for the logGGT (0.26). The raw GGT had the lowest heritability estimate (0.15). The heritability estimate for logGGT was slightly lower than previously reported heritability estimates in dairy cattle in New Zealand (Cullen *et al.* 2011; Morris *et al.* 1990; Morris *et al.* 1998), which ranged from 0.29 to 0.34. The moderate heritability for logGGT indicates that selection for tolerance to FE is possible in dairy cattle after a natural challenge from infected pasture.

Table 2. Variance components and heritability estimates with their standard errors between parentheses for raw GGT, logGGT and boxGGT

Trait	$\sigma_a{}^2$	$\sigma_e{}^2$	h^2
GGT	72026.05	408747	0.15 (0.02)
logGGT	0.76	2.21	0.26 (0.03)
boxGGT	0.24	0.70	0.25 (0.03)

Estimates of genetic correlations between production traits and logGGT are shown in Table 3. Genetic correlations between logGGT and production traits were generally weak, and positively correlated for all traits ranging from 0.02 ± 0.03 (fat) to 0.12 ± 0.03 (volume). Morris *et al.* (1990), reported opposite results where the genetic correlations between logGGT and production were negative in Jersey cattle in New Zealand. The positive genetic correlations in the current study were unfavourable when selecting for tolerance to FE, given that the goal is to reduce logGGT. However, the estimate for fat was close to zero. Furthermore, for milk, volume and protein, these weak genetic correlations indicate that logGGT is almost independent of production traits and one could select for tolerance to FE without compromising milk production.

Table 3. Estimates of genetic correlations (r_g) between logGGT and production traits, with their standard errors between parentheses

Trait	r_g
Milk volume	0.12 (0.03)
Milk fat	0.03 (0.03)
Milk protein	0.09 (0.03)

CONCLUSIONS

Bulk milk screening for the biomarker was able to identify herds with elevated GGT in individual animals. Tolerance to FE in naturally challenged dairy herds is moderately heritable and genetic gain would be expected with selection for improved tolerance to FE. The genetic correlations between tolerance to FE and production traits are weak, indicating that FE tolerance is almost independent of production and selection for sires with tolerance to FE is possible without affecting milk production.

ACKNOWLEDGEMENTS

The authors would like to thank the farmers and veterinarians that participated in this work and acknowledge the assistance of Jayden Calder in extracting the pedigree data.

REFERENCES

Amyes N. C., Johnson P. L. and Alexander M. C. (2018) *Proc. World Congress on Genetics Applied to Livestock Production.* 11: 354.

Ariyawansa H.A., Kevin D.H., Subashini C.J., Bart B., Thilini Chethana K.W., Dong Q.D., Yu C. D., *et al.* (2015) *Fungal. Divers.* 75: 27.

Box G.E.P. and Cox D.R. (1964) *J. R. Stat. Soc. B.* 26: 211.

Butler D.G., Cullis B.R., Gilmour A.R., Gogel B.J. and Thompson R. (2017). ASReml-R Reference Manual Version 4. ASReml-R Reference Manual.

Cullen N.G., Morris C.A., Hickey S.M. and Henderson H.V. (2011*) Proc. NZ. Soc. Anim. Prod.* 71: 117.

Mason W.A., Cuttance E.L., Laven R.A., Jamieson P. and Davis S R. (2022) *NZ Vet J.* 70: 40.

Mcrae K.M., Cullen N.G., Amyes N.C. and Johnson P. L. (2016) *Proc. New Zeal Soc. Anim. Prod.*76: 43.

Menna M.E.D., Smith B.L. and Mues C.O. (2009) *NZ. J. Agr. Res.* 52: 345.

Morris C.A., Burton L.J., Towers N.R., Cullen N.G., Rendel J.M. and Johnson D.L. (1998) *NZ. J. Agr. Res* .41: 347.

Morris C.A., Phua S.H., Cullen N.G. and Towers N.R. (2013) *NZ. J. Agr. Res.* 56: 156.

Morris C.A., Towers N.R., Tempero H.J., Cox N.R. and Henderson H. V. (1990) *Proc. NZ Soc. Anim. Prod.* 50: 255.

Smith B.L. and Towers N.R. (2002) *NZ. Vet J.* 50: 28.

Towers N.R. and Stratton G.C. (1978) *NZ. Vet J.* 26: 109.

Wright, Sewall. (1922) *Am. Natt.* 56: 330.

FAECAL MICROBIOTA OF ANGUS CATTLE WITH DIVERGENT IMMUNE COMPETENCE

B.N. Maslen[1], B.C. Hine[2], C. Duff[3], P.A. Alexandre[4], S.A. Clark[5], J.H.J van der Werf[5], J.D. White[6] and S.D. Pant[1]

[1]Gulbali Institute, Charles Sturt University, Boorooma Street, Wagga Wagga, NSW, 2678 Australia
[2]CSIRO Agriculture & Food, F.D. McMaster Laboratory, Chiswick, New England Highway, Armidale, NSW, 2350, Australia
[3]Angus Australia, Armidale, NSW, 2350 Australia
[4]CSIRO, Agriculture & Food, Queensland Bioscience Precinct, 306 Carmody Rd., St Lucia, Brisbane, QLD, 4067 Australia
[5]School of Environmental and Rural Science, University of New England, Armidale, NSW 2350 Australia
[6]Charles Sturt University, Wagga Wagga, NSW, 2650 Australia

SUMMARY

Microorganisms inhabiting the gut (gut microbiota) have been shown to influence immune responsiveness of the host in a variety of species. It has also been discovered that specific species of gut microbiota may contribute to immunity in multibreed cattle. In this study, faecal samples were obtained from Angus cattle that were concurrently phenotyped for cell-mediated and antibody-mediated immune responsiveness (IR) at weaning. Both IR phenotypes, and an ImmuneDex score, were calculated and used to identify high, medium and low IR cohorts (n=20/group). 16s rRNA gene sequence data was used to infer the relative abundances of different phyla in the sampled animals. A total of 6 phyla were found to significantly differ in relative abundances for at least one of the IR phenotypes. Of these, Bacteroidota, Euryarchaeota and Proteobacteria may be biologically relevant due to their relationship with gut health and disease.

INTRODUCTION

Gut microbiota play an important role in modulating host immune responses. Specifically in livestock, recent studies indicate that host immune responsiveness is linked with gut microbiota profiles in both pigs and multibreed cattle (Fan *et al.* 2021; Ramayo-Caldas *et al.* 2021). Gut microbiota have also been reported to differ between different cattle breeds (Fan *et al.* 2021), and there is evidence indicating the relative abundance of some groups of gut microorganisms may be heritable (Fan *et al.* 2021). Therefore, the aim of this study was to determine whether there were significant differences in the faecal microbiota profiles of Angus cattle cohorts with high, medium, or low immune responsiveness. This could be used to develop a better understanding of microbial profiles and specific gut microorganisms differing between IR cohorts and could further lead to the development of a variety of selection or intervention tools that makes Angus production more profitable.

MATERIALS AND METHODS

Rectal faecal samples were collected from 444 Angus weaners (6months of age) run on pasture at Charles Sturt University and Talooby farms over 2021 and 2022. Rectal faecal samples were put on dry ice immediately after collection and stored in the laboratory at -20°C until further processing.

Measurement of immune responsiveness. During rectal faecal sampling, all weaners were concurrently phenotyped for cell-mediated and antibody-mediated immune responsiveness. A measurement representing a combined cell-mediated and antibody-mediated response, known as ImmuneDEX, was also estimated as previously described (Reverter *et al.* 2021).

DNA extractions and 16s rRNA gene sequencing. Quick-DNA™ Fecal/Soil Microbe Miniprep Kit (Zymo Research, Irvine, CA) was used to extract genomic DNA from faecal samples as per manufacturer's instructions. NanoDrop 2000 Spectrophotometer (Thermo Scientific, Australia) was used to determine final yield and quality of extracted DNA. Finally, samples were subjected to paired end 16s rRNA gene sequencing at the Novogene sequencing facility in Singapore.

Statistical analysis. All IR phenotypes were calculated via linear regression after accounting for contemporary groups (based on herd ID, calf year of birth, cohort, sex, dam year of birth), age at measurement and weaning weight. These phenotypes were subsequently transformed into z-scores, and 20 animals with the highest, lowest, or z-scores closest to zero, were classified into high, low and medium IR cohorts respectively. Sequence data was used to create relative abundance graphs in R (using packages ggplot2 and ggpubr), and analysis was limited to the taxonomic level of phyla due to space limitations. Statistical analyses was performed using MANOVA in R to identify significant differences in relative abundances of the top 15 most abundant phyla between different IR cohorts.

RESULTS AND DISCUSSION

The average z-scores along with standard deviations are presented in Table 1. While there were some animals shared between high, medium and low cohorts of different IR phenotypes, a majority of animals were different. For instance, only one animal was common between the antibody-mediated and cell-mediated IR phenotypes in the high IR cohort. On the other hand, ImmuneDEX which is strongly correlated with the other two IR phenotypes, had more animals shared in common in its high cohort when compared to the high antibody-mediated IR cohort (8 animals) and the high cell-mediated IR cohort (7 animals).

Table 1. Z-scores (Mean ± Standard deviation) for high (n=20), medium (n=20) and low (n=20) IR cohorts for each of the three IR phenotypes

IR Cohorts	IR phenotypes		
	Antibody-mediated	*Cell-mediated*	*ImmuneDEX*
High	2.084 ± 0.21	2.285 ± 0.31	2.736 ± 0.49
Medium	0.008 ± 0.04	-0.006 ± 0.03	-0.003 ± 0.02
Low	-1.969 ± 0.21	-1.909 ± 0.26	-2.452 ± 0.35

The average relative abundances of different phyla, inferred based on 16s rRNA gene sequences (Figure 1), revealed Firmicutes and Bacteroidota to be the two most abundant phyla, which is consistent with scientific literature (Fan *et al.* 2021). Together these phyla account for ~ 90% of all microorganisms represented in the faecal samples. Analysis of the relative abundance data also revealed several significant differences between phyla of high, medium and low cohorts of different IR phenotypes. These differences have been presented in Table 2.

Bacteroidota was the only phylum whose relative abundance was found to significantly differ in the antibody-mediated IR phenotype. Bacteroidota have been previously reported to contribute to the development of the immune system, and to anti-inflammatory responses (Gibiino *et al.* 2018). They have also been linked to the activation of Th1 systemic immune responses, as well as stimulation of B cells (Ivanov *et al.* 2008).

The cell-mediated IR phenotype had two phyla that were found to significantly differ in terms of relative abundance, Euryarchaeota and Fusobacteriota. Existing evidence suggests Euryarchaeota are comprised of methanogenic species existing in the gut. In humans, it is possible that these methanogens have either a direct or indirect contribution to the development of gastrointestinal disorders and therefore, can adversely impact host health (Horz *et al.* 2010).

In the ImmuneDEX IR phenotype, Proteobacteria, Fusobacteriota and Acidobacteriota were found to differ significantly. Proteobacteria have previously been reported to increase in abundance in the gut of individuals with a compromised immune system and could potentially be indicative of a diseased state (Shin *et al.* 2015).

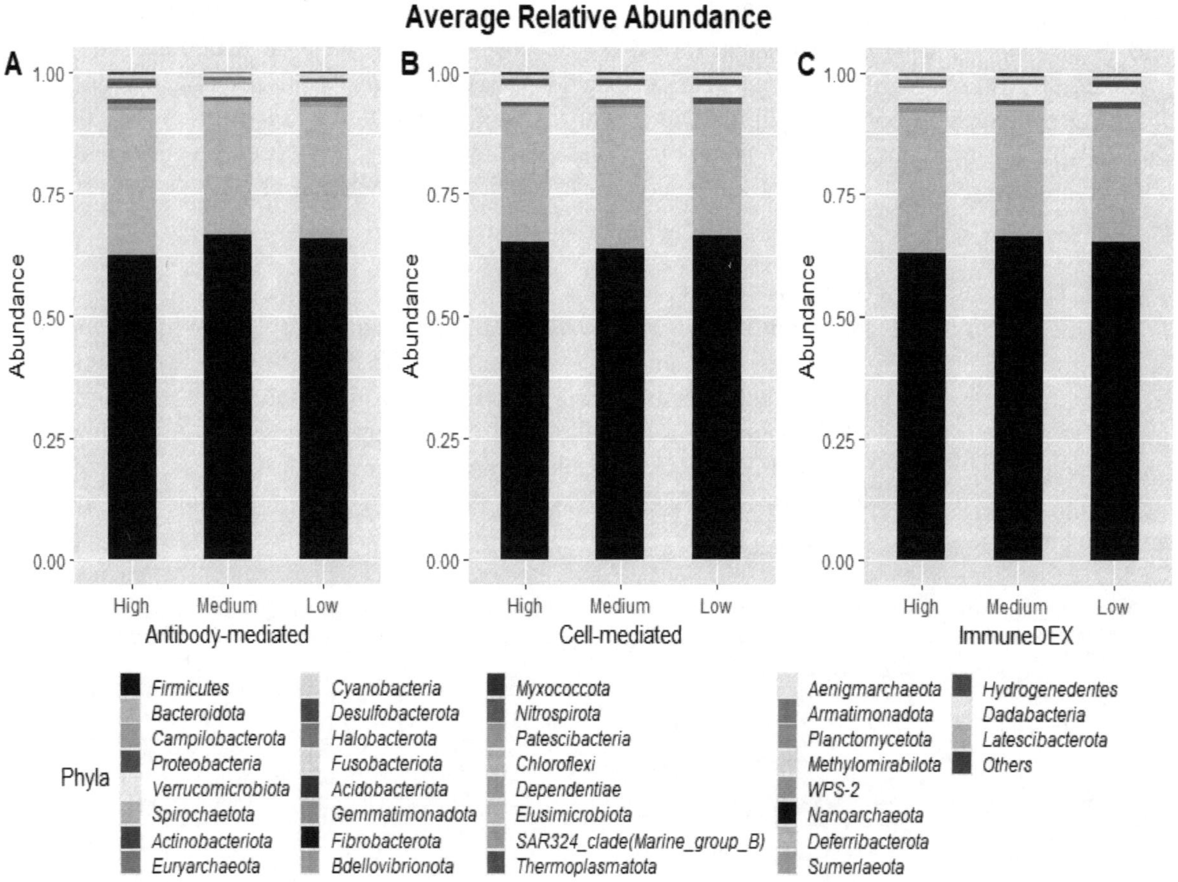

Figure 1. Phylum-level faecal microbiota assortment. Bar chart representing the average relative abundance of all bacterial ASVs taxonomically classified for A) antibody-mediated B) cell-mediated and C) ImmuneDEX, high, medium and low cohorts

Table 2. Represents the phyla that significantly differed between high, medium and low cohorts for antibody-mediated, cell-mediated and ImmuneDEX phenotypes

Antibody-mediated (P=0.05)	
Phyla	*P-value*
Bacteroidota	0.044*
Cell-mediated (P=0.05)	
Phyla	*P-value*
Euryarchaeota	0.035*
Fusobacteriota	0.040*
ImmuneDEX (P=0.05)	
Phyla	*P-value*
Proteobacteria	0.022*
Fusobacteriota	0.002*
Acidobacteriota	0.049*

Note. * Indicates a significant value (P=0.05)

CONCLUSION

Overall, 6 phyla out of the top 15 phyla, in terms of relative abundances, were found to significantly differ between the high, medium and low cohorts of the three IR phenotypes. The presence of Bacteroidota in the antibody-mediated phenotype, Euryarchaeota in the cell-mediated phenotype and of Proteobacteria in the ImmuneDEX phenotype, could be biologically relevant and warrants further in-depth investigation including investigating the heritability of these abundances and how much variation is explained by the host. Characterising microbiome-based signatures of different IR states could help identify at-risk animals and afford opportunities for early intervention that could improve animal health, welfare and productivity.

ACKNOWLEDGEMENTS

The authors are grateful for the assistance extended by J. Stephens, D. Niemeyer, B. Allworth, the staff at CSIRO and staff at Angus Australia. Angus Australia and the Australian Research Training Scholarship contributed financial support for this work.

REFERENCES

Fan P., Nelson C.D., Driver J.D., Elzo M.A., Peñagaricano F. and Jeong K.C. (2021) *ISME J.* 1.

Gibiino G., Lopetuso L.R., Scaldaferri F., Rizzatti G., Binda C. and Gasbarrini A. (2018) *Digest Liver Dis.* 50: 635.

Horz H.-P. and Conrads G. (2010) *Archaea.* 2010.

Ivanov I.I., de Llanos Frutos R., Manel N., Yoshinaga K., Rifkin D.B., Sartor R.B., Finlay B.B. and Littman D.R. (2008) *Cell Host Microbe.* 4: 337.

Ramayo-Caldas Y., Zingaretti L.M., Pérez-Pascual D., Alexandre P.A., Reverter A., Dalmau A.,Quintanilla R. and Ballester M. (2021) *Animal Microbiome.* 3: 11.

Reverter A., Hine B.C., Porto-Neto L., Li Y., Duff C.J., Dominik S. and Ingham A.B. (2021) *J Anim.Sci.* 99: 384.

Shin N.-R., Whon T.W. and Bae J.-W. (2015) *Trends Biotechnol.* 33: 496.

MULTI-TRAIT GENOME WIDE ASSOCIATION META-ANALYSIS OF BODY WEIGHT, CARCASE COMPOSITION AND EATING QUALITY TRAITS IN AUSTRALIAN SHEEP

N. Moghaddar[1], A.A. Swan[2] and J.H.J. van der Werf[1]

[1] School of Environmental & Rural Science, University of New England, Armidale, NSW 2351, Australia
[2] Animal Genetics Breeding Unit*, University of New England, Armidale, NSW, 2351 Australia

SUMMARY

The objective of this study was to perform a multi-trait meta-analysis of summary statistics of a single-trait genome-wide association study (GWAS) on 10 body weight, carcase composition and eating quality traits in Australian sheep. Meta-analysis was performed based on an approximate chi-squared test according to estimated SNP effects and their associated standard errors obtained in a single-trait GWAS. Single-trait association testing was based on single-marker regression analysis in linear mixed models using imputed whole genome sequence data and between 2,707 and 135,022 adjusted phenotypes across the traits studied. Meta-analysis showed higher power of QTL detection compared to single trait GWAS, it confirmed the highly significant QTL regions in single-trait GWAS and revealed numerous pleiotropic QTLs on chromosomes 1, 3, 6, 8, 11, 16 and 18, affecting two or more traits. In total meta-analysis showed 4,823 SNPs in strong association with at least one trait ($-Log\ P \geq 6.0$) but did not show any new highly significant QTL regions across the traits.

INTRODUCTION

Identification and estimation of the genetic parameters of Quantitative Trait Loci (QTLs) is valuable for understanding the biology of traits and is useful to accelerate the rate of genetic gain of economically important traits in plant and livestock breeding programs. Literature shows higher genomic evaluation accuracy and hence faster genetic progress by prioritizing and weighting genetic variants with larger effect in genomic prediction statistical models (e.g. MacLeod *et al.* 2017). Moghaddar *et al.* (2019) showed improvement in prediction accuracy of weight and eating quality traits in a combined dataset from multiple research and commercial sheep populations using information about polymorphisms affecting the genetic variation of the traits.

Identification of QTLs in polygenic traits has been broadly based on single-trait GWAS. However, when multiple correlated phenotypes are available, a joint analysis of multiple traits enabled via meta-analysis, could increase the statistical power of detecting genetic associations. This could be more important for traits with smaller numbers of observations which show weaker associations with the genetic variants in single-trait GWAS (Fang & Pausch, 2019). The objective of this study was to perform a multi-trait GWAS meta-analysis using summary statistics of single-trait GWAS which was performed on two growth trait, four carcase trait and four eating quality traits using imputed whole genome sequence data.

MATERIALS AND METHODS

Studied population and phenotypes. Phenotypic records of two body weight, four carcase composition and four eating quality traits were derived from the Australian national Sheep Genetics database (https://www.sheepgenetics.org.au/). Table 1 shows the names of the traits studied, phenotypic summary statistics and heritability estimates derived from phenotypes and pedigree information. The phenotypes belonged to a multi-breed/admixed sheep population from both

* A joint venture of NSW Department of Primary Industries and the University of New England

research (Information Nucleus Flocks and MLA resource flock) and industry flocks (Sheep Genetics). More than 30 breeds were represented in the data set which was constructed by merging data from three separate Sheep Genetics evaluations, for maternal breeds, terminal sire breeds and Merinos. The dominant breeds represented in the data either as purebreds or crossbred/admixed with other breeds were Border Leicester, Coopworth and maternal composites, White Suffolk, Poll Dorset, Suffolk, Texel, and Merino. The definition and measurement of the traits between research and industry flocks and within maternal breeds, terminal sire breeds and Merinos were based on the same standard.

Table 1. Trait abbreviation and definition, phenotypic summary statistics and heritability estimates for studied traits

Trait	N of Records*	Average	sd	h^2(se)**
PWT (post weaning weight, kg)	92,586	45.44	12.65	0.30 (0.001)
CWT (carcass weight, kg)	20,831	21.889	3.29	0.31 (0.001)
CCFAT (carcass scanned fat, mm)	20,281	3.84	1.96	0.27 (0.001)
CEMD (carcass scanned eye muscle depth, mm)	20,393	31.32	4.23	0.31 (0.001)
DRESS (dressing percentage,%)	15,977	44.34	2.86	0.30 (0.001)
IMF (intra muscular fat, %)	20,320	4.47	1.15	0.55 (0.001)
LMY (lean meat yield, kg)	2,707	56.58	9.69	0.48 (0.02)
PCF (post weaning scanned fat, mm)	51,319	2.84	0.79	0.26 (0.001)
PEMD (post weaning EMD, mm)	51,597	27.15	3.85	0.31 (0.001)
SF5 (shear force at day5 aging, Newton)	20,474	33.45	13.57	0.32 (0.001)

*: number of records with both phenotypes and genotypes, sd: standard deviation, **: heritability (standard error)

Genotypes. The whole genome sequence data on 26 Ovine autosomes which were imputed from a mixture of different low, medium and high-density SNP genotypes were used in this study. In the imputation pipeline, research and industry data with low and medium density SNP genotypes (12k, 15k and 50k) were imputed to a common 60k genotype based on a large reference set. In the next step the 60k genotypes were imputed to high-density genotypes (500k) using a multi-breed reference set of 2,266 animals. Animals with high-density genotypes were then imputed to whole genome sequence using 726 multi-breed animals as a reference set. Genotype phasing and imputation was performed in Beagle 5.3 (Browning *et al.* 2021). The final set of sequence data was comprised of 31,154,249 genetic variants after applying quality control on genotypes and removing genetic variants with low imputation accuracy based on a significant threshold level suggested by software ($r < 0.63$).

Statistical analysis. Phenotypes used in single-trait association studies were obtained as outputs of the multi-trait industry evaluation analyses for Merino, maternal breeds, and terminal breeds respectively run by AGBU (Animal Genetics Breeding Unit). These analyses use phenotypes corrected for known environmental effects such as age and birth/rearing status, then fit a multi-trait mixed model with contemporary group fitted as a fixed effect, and genetic groups, direct and maternal genetic effects, maternal permanent environment, and sire by flock-year fitted as random. The three analyses were run with genetic effects fitted with a pedigree relationship matrix, and pre-corrected phenotypes derived as the sum of the estimated breeding values for direct genetic effects and residual values ($\boldsymbol{y}^* = \widehat{\boldsymbol{a}} + \widehat{\boldsymbol{e}}$).

Single-trait GWAS was performed by single SNP regression analysis in a linear mixed model using Gemma V0.96 software (Zhou *et al.* 2014) and based on the following equation: $y^* = Xb + Zu + e$. In this equation y^* refers to the pre-corrected phenotypes explained above, b includes a fixed effect modelling the mean of each of the three analyses described above and the SNP effect at each marker, *u* refers to the random additive genetic effect of the animal fitted by genomic relationship matrix (G), and *e* is the residual effect. G was calculated according to Yang *et al.* 2011 using 50k genotypes and *X* and *Z* are incidence matrix which relate fixed and random effects to phenotypes.

Meta-analysis was performed based on multi-trait approximate chi-squared test for each SNP, distributed according to chi-square distribution and estimated as $\chi^2_{df,n} = t'V^{-1}t$ (Bolormaa *et al.* 2014). In this equation *n* is number of traits, *t* is the signed *t*-value derived from single trait GWAS SNP effect and its standard error across all 10 traits ($t=b/se(b)$), and V^{-1} is the inverse of correlation matrix derived from SNP effect (signed t-values between traits). An arbitrary p-value of equal or less than 1.0×10^{-6} was considered as the SNP significance threshold level. Significant SNPs were pruned for high LD in Plink (Purcell *et al.*2007) according to window size of 5000 SNPs, sliding window of 200 SNPs and LD≥0.95).

RESULTS AND DISCUSSION

Figure 1 shows the results of the multi-trait meta-analysis as a plot of SNP p-values versus chromosomal position. The significant regions in Figure 1 are those which are significant for at least one trait. Compared to p-values derived in a single-trait GWAS and meta-analysis, meta-analysis showed higher power of QTL detection which was in line with results of comparison of single-trait GWAS and a meta-analysis in cattle (Bolormaa et al. 2014). Meta-analysis confirmed the highly significant regions in single trait GWAS, and furthermore showed stronger association for some regions that had weaker association with phenotype. However, there were regions around significant thresholds with weak association with phenotypes in single-trait GWAS which were not significant in the meta-analysis, particularly for traits with smaller number of phenotypes. Numerous highly significant pleiotropic QTL region were found across the studied traits, including regions on chromosome 1 (CCFAT, PCF, IMF and PEMD), chromosome 2 (CWT, CWT, CCFAT and IMF), chromosome 3 (affecting both weight traits), chromosome 6 (weights and EQ traits except SF5 and CEMD), chromosome 8 (PWT, PCF and PEMD), chromosome 11 (PWT, CWT, PCF and PEMD), chromosome 16 (weights and PCF traits) and chromosome 18 (PWT, IMF, CEMD and SF5). Meta-analysis did not reveal new significant QTL regions across these traits compared to single-trait GWAS on sequence data. In total, meta-analysis discovered 4,823 SNPs that were in significant association with at least one trait ($-log\ p\geq6.0$) after pruning for high LD.

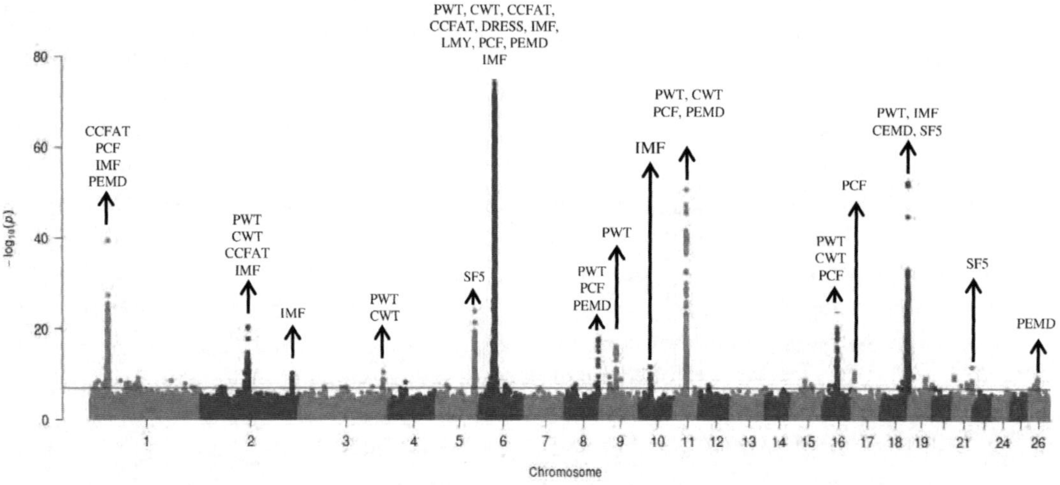

Figure 1. Manhattan plot of results of multi-trait GWAS meta-analysis of weight, carcase composition and eating quality traits

Comparison of genetic correlations between traits based on signed t-values derived from p-values, and the genetic correlation obtained based on pedigree information showed a similar correlation direction. However, the strength of correlation coefficients was different in some cases, most particularly for traits with smaller number of records. LMY had the smallest number of phenotypes and genotypes in the meta-analysis which showed the most different correlation estimated in meta-analysis in comparison to the genetic correlation estimated from pedigree.

Meta-analysis uses information in summary statistics of the results of single-trait GWAS on genetically related traits to improve the power of QTL detection, and together with single-trait GWAS is useful to confirm pleiotropic QTL effects. In this study meta-analysis showed notably stronger evidence of QTL affecting the traits with moderate to high genetic correlations. Meta-analysis was also useful to flag those regions which were close to the significance threshold in single-trait GWAS. Meta-analysis did not show new highly significant regions. This could be related to a high resolution in the single-trait GWAS due to strong linkage disequilibrium provided by using whole genome sequence data.

CONCLUSIONS

Multi-trait meta-analysis of weight and eating quality traits using SNP effects derived from single-trait GWAS on imputed whole genome sequence data showed higher power of detecting genetic variants in significant association with phenotypes compared to single-trait GWAS. Meta-analysis was also useful to flag those genetic markers which were close to the significance threshold in single-trait. GWAS Results of meta multi-trait analysis and single-trait GWAS revealed numerous pleiotropic QTL regions affecting two or more traits in this study. In total, meta-analysis showed 4,206 genetic variants in significant association with at least one trait (-log p>7.0), however, it did not show new highly significant QTL compared to results of single-trait GWAS on whole genome sequence data.

ACKNOWLEDGEMENTS

This research was partly funded by Meat Livestock Australia project L.GEN.1815. The authors thank Sheep Genetics and contributing breeders for access to the data used in this study.

REFERENCES

Bolormaa S., Pryce J.E., Reverter A., Zhang Y., *et al.* (2014) *PLoS Genet.*

Browning B.L., Tian X., Zhou Y. and Browning S.R. (2021) *Am. J. Hum. Genet.* 108: 1880.

Fang Zih-Hua. and Pausch H. (2019) *BMC Genomics* 20: 695.

Gilmour A.R., Gogel B.J., Cullis B.R., Welham S., *et al.* (2015) *Hemel Hempstead: VSN International Ltd.*

MacLeod I.M., Bowman P.J., Vander Jagt C.J., Haile-Mariam M., *et al* (2017) *BMC Genomics* 17: 144.

Moghaddar N., Khansefid M., Van der Werf J.H.J., Bolormaa S., *et al.* (2019) *Genet. Sel. Evol.* 51: 72.

Purcell S., Neale B., Todd-Brown K., Thomas L., *et al.* (2007) *Am. J. Hum. Genet* 81.

Zhou X and Stephens M., (2012) *Nature Genetics* 44: 821,

Yang J., Benyamin B., McEvoy B.P., *et al.* (2010) *Nature Genetics* 42: 565.

DISCOVERING THE MISSING VARIATION IN THE BOVINE GENOME; A LONG READ SEQUENCING PILOT STUDY INTO THE STRUCTURAL VARIATION IN TWO DAIRY BREEDS.

A. Chamberlain[1,2], T. Nguyen[1], J. Wang[1] and I. Macleod[1,2]

[1] Agriculture Victoria, Centre for AgriBioscience, Bundoora, VIC, 3083 Australia
[2] School of Applied Systems Biology, La Trobe University, Bundoora, VIC, 3083 Australia

SUMMARY

Structural variation has been posited to contribute equal or greater diversity at the nucleotide level than any other form of genetic variation. Short read sequencing technologies are limited in their ability to characterise structural variants (SVs), however long read sequencing, which is now cost effective, poses as a solution to this problem. The Bovine Long Read Consortium (BovineLRC) aims to use long read sequencing technologies to sequence cattle at population scale to characterise the structural variation of the bovine genome for downstream applications. This pilot study sequenced 41 animals from two breeds in an effort to understand how much SV variability exists within and across breeds. A total of 76,572 SVs were detected across all samples, one third of which were segregating in only one breed. Insertions and deletions tended to be smaller and duplications larger. Insertions and deletions more often segregated across both breeds, while inversions were more often breed specific. Few duplications were detected but they tended to be slightly more likely to be breed specific. The results highlight that it would be beneficial to have a dataset with large numbers of animals and breeds to understand the structural variation that exists and explore the impact of SVs on traits of interest.

INTRODUCTION

The 1,000 bull genomes project has had a massive impact on cattle genomics worldwide cataloguing single nucleotide polymorphisms (SNPs) and small insertions and deletions (INDELs) in more than 6,000 cattle genomes (Daetwyler *et al.* 2014; Hayes, Daetwyler 2019). However, limitations of short read sequencing technologies mean that SVs are not easy to characterise. SVs can be large INDELs (>50 basepairs), inversions, translocations, copy number variations or segmental duplications and studies in human estimate that SVs together occupy a proportion of the genome that is equal to or greater than that of SNPs and small INDELs (Feuk *et al.* 2006; Ho *et al.* 2020) and contribute greater diversity at the nucleotide level than any other form of genetic variation (Chaisson *et al.* 2019). Multiple studies in cattle have demonstrated that SVs impact classic mendelian traits, quantitative traits and gene expression (Kadri *et al.* 2014; Rothammer *et al.* 2014; Lee *et al.* 2021).

Long read sequencing, such as nanopore sequencing from Oxford Nanopore Technologies (ONT) and single molecule real time sequencing from Pacific Biosciences (PacBio), have recently become cost-effective. Both claim costs of <$1,000US per genome at 30x coverage and have the advantage of being able to sequence across large SVs and therefore better characterise them compared to short read technology (Chaisson *et al.* 2019).

To date genome wide SV detection in cattle at population scale has largely used short read sequence data (Boussaha *et al.* 2015; Mesbah-Uddin *et al.* 2017; Mielczarek *et al.* 2018; Hu *et al.* 2020; Mei *et al.* 2020; Chen *et al.* 2021; Upadhyay *et al.* 2021) or limited long read sequence data (Low *et al.* 2020; Crysnanto *et al.* 2021) or a combination of the two (Couldrey *et al.* 2017). Like the Human Genome Structural Variation Consortium (Chaisson *et al.* 2019) the Bovine Long Read Consortium (BovineLRC) (Nguyen *et al.* 2023) aims to use long read sequencing technologies to sequence cattle at population scale to characterise the structural variation of the bovine genome.

Such a reference dataset will empower imputation of SVs into larger populations to examine their impact on quantitative traits as well as better resolve segmental duplication regions with copy number variants, understand the evolution of SVs and identify deleterious causal variants.

As a pilot study we have sequenced 41 animals from two breeds with ONT in an effort to understand how much variability exists within and across breeds in SV.

MATERIALS AND METHODS

DNA sequencing. 19 Holstein and 22 Jersey animals were selected, avoiding full and half sib relationships to maximise diversity. High molecular weight DNA was extracted from semen, liver tissue or whole blood using Gentra Puregene kit (Qiagen). Sequencing libraries were prepared using ligation sequencing kit v9 or v10 (ONT) according to manufacturer's instructions and sequenced on R9.4.1 flowcells on a MinION or PromethION (ONT). Super high accuracy basecalling was undertaken with Guppy v6.1.7 and reads with q-score greater than 7 retained for analysis.

Data analysis. Reads were quality trimmed using FiltLong (https://github.com/rrwick/Filtlong accessed December 2022) with default settings and samples with short reads (6 Holstein and 6 Jersey, 150 cycle paired reads) polished. Filtered reads were then mapped to ARS-UCD1.2 (Rosen *et al.* 2020) with additional Btau5.0.1 Y (Bellott *et al.* 2014) using Minimap2 (Li 2018). Sniffles2 (Sedlazeck *et al.* 2018) was used to detect SVs for each sample and subsequently merge SVs from multiple individuals and re-genotype. SVs larger than 3Mb or with a genotype quality score less than 20 were excluded.

RESULTS AND DISCUSSION

A mean of 26x and 20x read coverage was achieved with mean read length N50 of 30kb and 26kb for Holstein and Jersey samples respectively. On average 20,770 deletions, 19,620 insertions, 234 inversions and 38 duplications were detected for each Holstein and 19,815, 18,458, 177 and 39 respectively for each Jersey. After merging and filtering data from all animals a total 76,572 SVs were detected. This is more than studies using short read data with similar sample numbers (Boussaha *et al.* 2015; Couldrey *et al.* 2017; Mesbah-Uddin *et al.* 2017; Mielczarek *et al.* 2018; Hu *et al.* 2020; Upadhyay *et al.* 2021) and similar to small studies using long read data in a pangenome approach (Crysnanto *et al.* 2021) but less than the largest pangenome approach with short read data and almost 900 samples (Zhou *et al.* 2022) which detected greater than 3.6 million SVs. 14,526 SVs were segregating in Holstein only and 11,264 only in Jersey (Figure 1A). 50,782 (66%) were detected in both breeds, therefore one third of all SVs were breed specific.

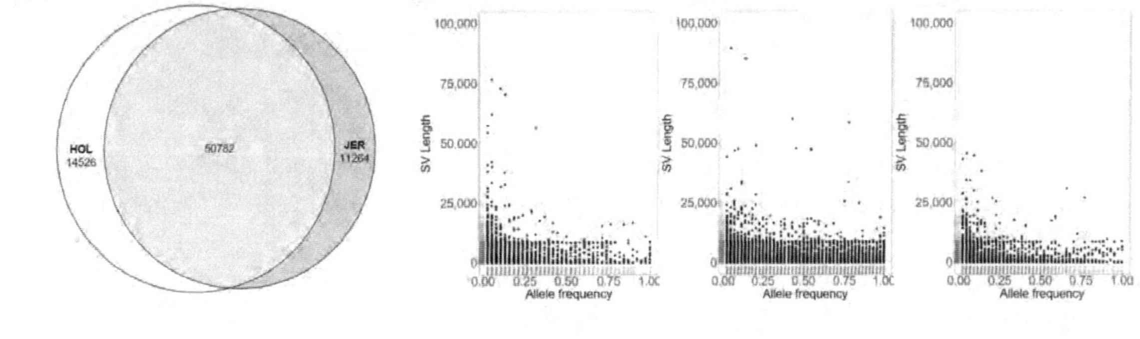

Figure 1. A Venn diagram showing SVs detected across or within Holstein (HOL) or Jersey (JER) breed (A). The relationship between allele frequency and length of SVs for Holstein specific SVs (B), those that occurred in both breeds (C) and Jersey specific SVs (D)

Figures 1B-1D show a trend of longer SVs with lower allele frequencies in the population, for both breed specific as well as across breed SVs. As expected, high allele frequency SVs were more likely across breeds. Other studies have estimated the proportion of breed specific SVs at 66% (Boussaha *et al* 2015) when comparing 3 breeds, 15% (Mielczarek *et al.* 2018) in 13 breeds, 48% (Hu *et al.* 2020) in 10 breeds, 54% (Mei *et al.* 2020) in 8 breeds and 76% (Low *et al.* 2020) in 3 breeds. While others found different allele frequencies of the same SV in different populations of taurine, indicus and zebu cattle (Upadhyay *et al.* 2021). This variation reflects the variable power of the different studies, driven by the numbers of samples, breeds included and breed definitions. Large numbers of samples and large numbers of breeds are likely required before we can be certain of the proportion of SVs that are breed specific.

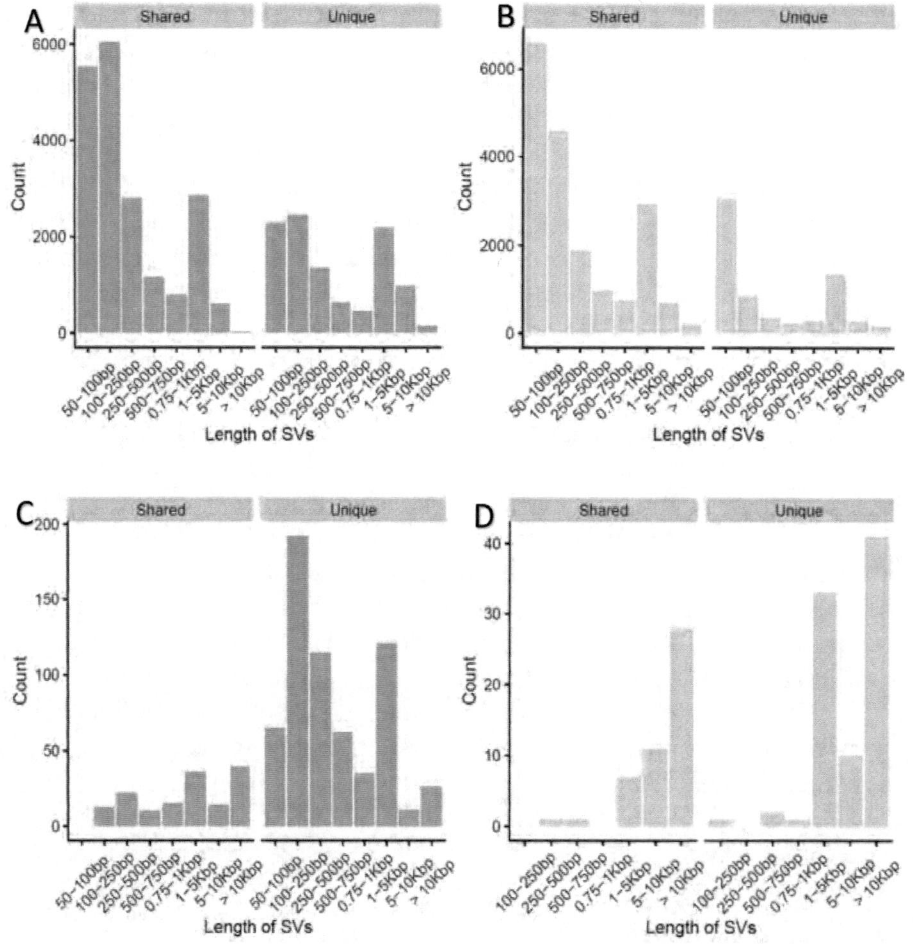

Figure 2. Numbers of deletions (A), insertions (B), inversions (C), and duplications (D) of different length for across breed (shared) and breed specific (unique) SVs. Note x-axis is not to scale

In agreement with other studies (Boussaha *et al.* 2015; Upadhyay *et al.* 2021; Zhou *et al.* 2022) insertions and deletions tended to be smaller and duplications larger (Figure 2). In this study we found insertions and deletions more often occurred across both breeds (Figure 2A and 2B), while inversions were much more often breed specific (Figure 2C). Few duplications were detected but they tended to be slightly more likely to be breed specific. However, reads that span the structural variant are required to call them accurately, therefore this dataset with read N50 of 26-30Kb has limited power to detect very large SVs, likely partially accounting for the low numbers of duplicates found. Other studies also find lower numbers of large duplications compared with insertions and deletions (Mei *et al.* 2020; Zhou *et al.* 2022). It's also likely that many duplications were removed when SVs were merged across animals due to difficulty deciphering breakpoints for SVs. Given the small population size used here, read length N50 and the difficulties associated with accurate annotation of large and complex SVs this study had limited power to detect large and rare SVs.

CONCLUSION

This small pilot study in 2 breeds highlights that it would be beneficial to have a dataset with large numbers of animals and breeds to understand the structural variation that exists in the bovine genome. The BovineLRC has been formed to achieve this. It also highlights that more work is required to accurately annotate and genotype large and complex SVs. Further work is required to understand the impact of the SVs detected in this study on traits important to the dairy industry.

ACKNOWLEDGEMENTS

The authors thank DairyBio, a joint venture project between Agriculture Victoria, Dairy Australia and The Gardiner Foundation, for funding.

REFERENCES

Bellott D.W., Hughes J.F., Skaletsky H., *et al.* (2014) *Nature* 508: 494.
Boussaha M., Esquerré D., Barbieri J., *et al.* (2015) *PLOS ONE* 10: e0135931.
Chaisson M.J.P., Sanders A.D., Zhao X., *et al.* (2019) *Nature Communications* 10: 1784.
Chen L., Pryce J.E., Hayes B.J., *et al.* (2021) *Animals* 11: 541.
Couldrey C., Keehan M., Johnson T., *et al.* (2017) *J. Dairy Sci.* 100: 5472.
Crysnanto D., Leonard A.S., Fang Z.-H., *et al.* (2021) *Proceedings of the National Academy of Sciences* 118:e2101056118.
Daetwyler H.D., Capitan A., Pausch H., *et al.* (2014) *Nature Genet.* 46: 858.
Feuk L., Carson A.R. and Scherer S.W. (2006) *Nature Reviews Genetics* 7: 85.
Hayes B.J. and Daetwyler, H.D. (2019) *Annual Review of Animal Biosciences* 7: 89.
Ho S.S., Urban A.E. and Mills R.E. (2020) *Nature Reviews Genetics* 21: 171.
Hu Y., Xia H., Li M., *et al.* (2020) *BMC Genomics* 21: 682.
Kadri N.K., Sahana G., Charlier C., *et al.* (2014) *PLoS Genet.* 10: e1004049.
Lee Y.-L., Takeda H., Costa Monteiro Moreira G., *et al.* (2021) *PLoS Genet.* 17: e1009331.
Li H. (2018) *Bioinformatics* 34: 3094.
Low W.Y., Tearle R., Liu R., *et al.* (2020) *Nature Communications* 11: 2071.
Mei C., Junjvlieke Z., Raza S.H.A., *et al.* (2020) *Genomics* 112: 831.
Mesbah-Uddin M., Guldbrandtsen B., Iso-Touru T., *et al.* (2017) *DNA Research* 25: 49.
Mielczarek M., Frąszczak M., Nicolazzi E., *et al.* (2018) *BMC Genomics* 19: 410.
Nguyen T.V., Vander Jagt C.J., Wang J., *et al.* (2023) *Genet. Sel. Evol.* 55: 9.
Rosen B.D., Bickhart D.M., Schnabel R.D., *et al.* (2020) *GigaScience* 9: 1.
Rothammer S., Capitan A., Mullaart E., *et al.* (2014) *Genet. Sel. Evol.* 46: 44.
Sedlazeck F.J., Rescheneder P., Smolka M., *et al.* (2018) *Nat Methods* 15: 461.
Upadhyay M., Derks,M.F.L., Andersson G., *et al.* (2021) *Genomics* 113: 3092.
Wick R. Flitlong. https://github.com/rrwick/Filtlong
Zhou Y., Yang,L., Han X., *et al.* (2022) *Genome Res* 32: 1585.

X AND Y CHROMOSOME SNPS AS INDICATORS OF SEX IN QUALITY ASSURANCE CHECKS FOR GENOMIC ANALYSIS IN WAGYU

C. Teseling[1] and D.P. Garrick[2,3,4]

[1]Australian Wagyu Association, Armidale, NSW, 2350 Australia
[2]The Helical Company Limited, Rotorua, BOP, New Zealand
[3]Theta Solutions LLC, Lacey, WA, USA
[4]A.L. Rae Centre for Genetics and Breeding, Massey University, New Zealand

SUMMARY

In 2018 genomic information was incorporated in the Australian Wagyu BREEDPLAN analysis. This development necessitated the implementation of genotype quality assurance (QA) checks to ensure the genotypes which were included in the genomic relationship matrix (GRM) were from samples collected from the same registered animals as the phenotypes in the analysis. One of the quality assurance checks is to compare the recorded sex of the registered animal to that of the predicted sex from the genotype. The Australian Wagyu population developed from a relatively small number of founder animals which limited the genetic diversity of the breed. Lower levels of genetic diversity reduce the usefulness of X chromosome heterozygosity to predict the sex of the animal from which the sample was collected. Modern SNP chips include Y chromosome data, the use of which in sex prediction can greatly increase the accuracy of sex-based quality control checks.

INTRODUCTION

The Wagyu breed in Australia was established from 221 Fullblood foundation animals mostly exported from Japan between 1990 and 1997. In comparison to other breeds, this is a relatively small number of foundation animals which results in the risk of increased subsequent inbreeding and lower levels of genetic diversity (Ferdosi *et al.* 2019). In populations where the effective population size is relatively small, the diversity on the X chromosome could be expected to be half of that of the autosomes (Schaffner 2004), while Mészárosová *et al.* (2022) found that X chromosome heterozygosity could vary significantly in the same population.

In 2018 genomic information was incorporated in the Australian Wagyu BREEDPLAN analysis. An important component of utilising genomic information is to ensure the genotype is associated with the correct registered animal and corresponding phenotypes. The Animal Genetics and Breeding Unit (AGBU) developed a data pipeline which incorporates a range of QA checks to ensure genotype integrity (Connors *et al.* 2017).

An important genotype QA check which should be implemented by genetic analysis service providers is to compare the recorded sex of the registered animal to the sex predicted from the animal's genotype (Connors *et al.* 2017, ICAR Guidelines 2022, McClure *et al.* 2018). Not every commercial chip includes chromosome Y SNPs, however, they typically contain chromosome X markers. Both the X and Y chromosomes contain a pseudo-autosomal region (PAR) and it is important to ensure only SNPs from the non-PAR (nPAR) region are used in sex prediction analysis.

The sex of the genotype can be predicted by evaluating the nPAR X chromosome heterozygosity and/or the presence or absence of calls on nPAR Y chromosome SNPs. Normally females have two copies of the X chromosome while males have one X chromosome and one Y chromosome.

Combining the X and Y chromosome results sometimes lead to conflicting sex prediction outcomes. In the very rare occurrence where an animal has Turner (X0) or Klinefelter's syndrome (XXY), the X and Y chromosome results will conflict. Also, when a semen sample only report X chromosome SNPs, it could indicate a female sex selected semen sample was submitted for genotyping.

When the early bovine chips were manufactured, the sequence information on sex chromosomes were not well assembled, and no Y chromosome SNPs were on the chips. These issues, as well as a low genetic diversity on the X chromosome can result in registered females having low X chromosome heterozygosity and their genotypes incorrectly excluded from the genetic analysis. This paper investigates the use of nPAR X and Y chromosome data in estimating genotype sex and considers results where the X and/or Y predicted sex of the genotype and the recorded sex of the registered animal may conflict. The aim is to determine population specific thresholds for Australian Wagyu to improve accuracy of sex prediction and reduce incorrect exclusion of valid female genotypes from genomic analyses.

MATERIALS AND METHODS

The Australian Wagyu Association (AWA) has more than 325,000 animals genotyped, from more than 3,600 different chips or manifests. These data are stored in their genotype database hosted by the Helical Company (Garrick *et al.* 2023). More than 50% of these genotypes are from research or commercial animals which are not registered in the AWA's registration database hosted by the Australian Business and Research Institute.

The samples can be categorised based on their number of SNPs. Table 1 shows the different categories of chips, the number of animals genotyped for each category, their number of SNPs as well as the numbers of nPAR X and Y chromosome SNPs present.

Table 1. Number of genotypes in each chip category with number of SNPs, number of nPAR X chromosome SNPs and number of nPAR Y chromosome SNP.

Chip Category	#Genotypes	#SNPs	#X Chrom	#Y Chrom
Parentage	59,620	180 to 641	0	0
10K	7,821	6,900 to 10,000	218 to 271	0 to 7
30K	5,444	19,000 to 35,000	634 to 919	0 to 7
50K	86,313	35,000 to 49,000	291 to 1,893	0 to 189
70K	17,730	50,000 to 77,000	288 to 1,561	6 to 239
100K	148,006	93,000 to 96,000	2,091	269
140K	272	137K to 140K	2,015	25
770K	187	777,963	2,821	267

Genotypes were extracted and analysed to compare the accuracy of predicting the sex of the animals from which the sample was collected. Animals genotyped using chips with no nPAR X chromosome SNPs or less than six nPAR Y chromosome SNPs or a call rate lower than 95% were excluded from the analysis. The first analysis included all animals that had nPAR X and at least six nPAR Y chromosome SNPs present on the chip which resulted in a total of 247,057 genotyped animals. The second analysis excluded all genotypes of animals not registered in the AWA Herdbook, which reduced the total to 104,860 genotypes. An animal must be parent verified to both its parents to be registered in the AWA's Herdbook, which ensure a high level of quality assurance.

Traditionally the focus has been on using the nPAR X chromosome to predict the sex of the animal from which the sample was collected. The increasing number of Y chromosome SNPs on more recently developed chips make it possible to consider the number of called nPAR Y chromosome SNPs to improve the accuracy of the prediction. If the genotype has a high proportion of nPAR Y chromosome SNPs reported, it could indicate that the sample was collected from a male while a very low proportion or no called nPAR Y chromosome SNPs would be suggestive the animal was a female.

The sex prediction accuracies of three different methods were assessed by comparing the predicted sex of the genotype with the sex of the associated registered animal. The three methods are presented in Table 2 where Method 1 only used the nPAR X chromosome heterozygosity, Method 2 use both the nPAR X chromosome heterozygosity (Method 1) as well as the number of called nPAR Y chromosome SNPs (McClure et al. 2018, Garrick 2019) and Method 3 use nPAR X and Y chromosome SNPs in a stepwise approach where the proportion of nPAR Y chromosome SNPs are first used and only if that doesn't predict a clear result, the nPAR X chromosome heterozygosity of Method 1 is considered.

Table 2. The criteria used to determine the sex prediction accuracy of three different methods

	Predicted Male	Predicted Female	Ambig.
1	nPAR X $\leq$ 5% Heterozygosity	nPAR X > 5% Heterozygosity	
2	Method 1 + nPAR Y > 5	Method 1 + nPAR Y < 2	X $\neq$ Y
3	nPAR Y < 0.4 or nPAR Y = 0.4 to 0.6 + Method 1	nPAR Y > 0.6 or nPAR Y = 0.4 to 0.6 + Method 1	

RESULTS AND DISCUSSION

Table 3 shows the distributions of the proportion of nPAR Y called SNPsfor all genotypes and registered animals. The results show there is a wide continuum of proportions observed across all the samples tested with no clear cut-off between what may be expected as male vs. female samples for the nPAR Y SNPs used. To determine if earlier chips which tended to have fewer Y SNPs available are disproportionally contributing to this observed variation, genotypes with less than 100 Y SNPs on the chip were removed from the analysis containing all genotypes (96,348 genotypes removed). The results still indicate no clear break in the observed proportions.

Table 3. Distribution of the numbers of animals with various proportions of called Y chromosome SNPs when all genotyped animals (All Genos), only chips with more than 100 Y SNPs (> 100 Y), or only registered animals (Registered) were analysed

	Proportion of called Y chromosome SNPs									
	<0.1	<0.2	<0.3	<0.4	<0.5	<0.6	<0.7	<0.8	<0.9	<1.0
All Genos	141,762	885	315	141	43	92	1,686	101	13,009	89,023
> 100 Y	85,045	601	194	79	24	7	2	10	68	64,679
Registered	78,266	489	189	96	30	13	286	33	4,018	21,440

Figure 1 shows the distribution of the proportion of genotypes after genotypes with nPAR X chromosome heterozygosity of zero (expected males based on X chromosome only) were excluded, resulting in 144,308 and 79,147 from all and registered animal genotypes respectively. The graph on the left in Figure 1 displays the distribution of the number of animals genotyped relative to the heterozygous proportion of nPAR X chromosome. The graph on the right includes only that subset of animals with low X chromosome heterozygosity excluding animals with the proportion of genotypes with nPAR Y > 0.5 (1,725 which are expected to be male). The same reduction was observed when genotypes from registered animals were analysed.

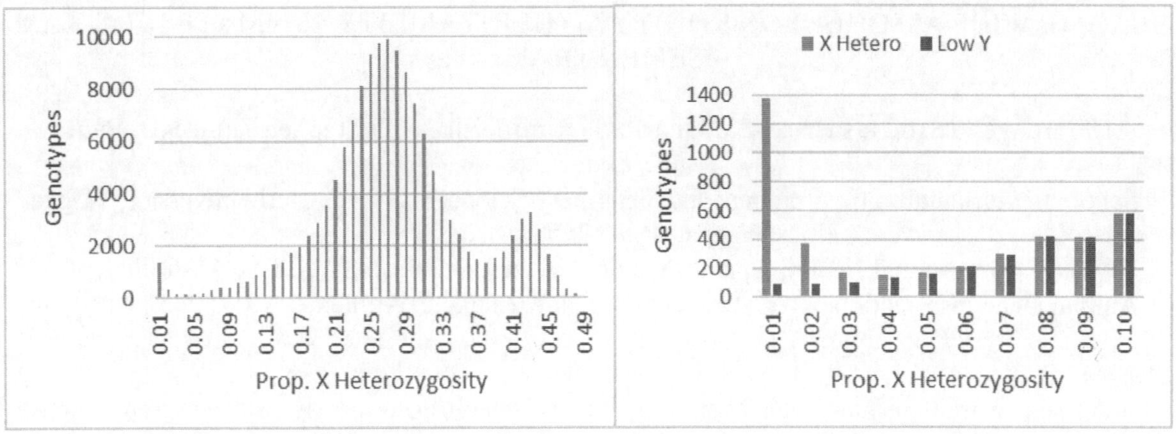

Figure 1. Number of genotyped animals exhibiting X chromosome heterozygosity (left) and animals with <0.1 proportion X chromosome heterozygosity after genotypes with more than 0.5 proportion of called Y chromosomes were excluded (right)

Using the three methods presented in Table 2 to predict the sex of the 104,860 genotypes of registered animals (78,906 females and 25,689 males) found that Method 1 incorrectly predicted 975 (1.23%) females to be males and 165 (0.65%) males to be females. Method 2 predicted 2,955 (3.75%) females to be ambiguous and 513 (0.65%) females to be males while 312 (1.21%) and 147 (0.57%) of males were predicted to be ambiguous and females respectively. Method 3 incorrectly predicted 508 (0.64%) females to be males and 433 (1.6%) males to be female.

Inspection of the "problem" genotypes suggests that some batches of samples exhibited little or no variation in the number of called Y chromosome SNPs, perhaps due to problems with the cluster files used for SNP calling at those loci in those batches.

CONCLUSIONS

Reduced genetic diversity negatively impacts the usefulness of X chromosome heterozygosity as the only criteria to predict animal sex from called SNPs.

Combined use of X and Y chromosome SNPs reduces the number of animals with incorrectly predicted sex. However, consideration of the chip content and careful scrutinization of variation and distribution of results will be required as additional criteria to reduce the percentage of incorrectly predicted sex from genotype calls to less than 0.5% of all animals.

REFERENCES

Connors N., Cook J., Girard C., Tier B., Gore K., Johnston D. and Ferdosi M. (2017) *Proc. Assoc. Advmt. Anim. Breed. Genet.* 22: 317.

Ferdosi M.H., Connors N.K. and Khansefid M. (2019) *Proc. Assoc. Advmt. Anim. Breed. Genet.* 23: 118.

Garrick D.P. (2019) *Proc. Assoc. Advmt. Anim. Breed. Genet.* 23: 23.

Garrick D.P., Howse, J. E. and Garrick, D.J. (2023) Retrieved from https://www.helicalco.com/

ICAR Guidelines (2022) Available online at: https://www.icar.org/Guidelines/04-DNA-Technology.pdf (Accessed February15, 2023).

McClure M.C., McCarthy J., Flynn P., McClure J.C., Dair E., O'Connell D.K. and Kearney, J.F. (2018). *Frontiers in Genetics* 9(84).

Mészárosová M., Moravčíková N., Mészáros G., Pavlík I., Margetín M."and Kasarda R. (2022) https://www.wageningenacademic.com/pb-assets/wagen/WCGALP2022/34_011.pdf

Schaffner (2004) *Nat. Rev. Genet.* 5: 43.

GENOME WIDE ASSOCIATION STUDY AND HERITABILITY ESTIMATES FOR RAM SEMEN TRAITS

M.J. Hodge[1,2], S. de las Heras-Saldana[3], S.J. Rindfleish[2], C.P. Stephen[1], and S.D. Pant[1]

[1] School of Agriculture, Environment and Veterinary Sciences, Charles Sturt University, Wagga Wagga, NSW, 2658 Australia
[2] Apiam Animal Health, Apiam Genetic Services, Dubbo, NSW, 2830, Australia
[3] Animal Genetics Breeding Unit*, University of New England, Armidale, NSW, 2351 Australia

SUMMARY

Ram semen traits influence conception outcomes, which in turn, may influence reproductive efficiency in sheep. As such, this study aimed to estimate genetic parameters and identify Quantitative Trait Loci (QTLs) associated with ram semen traits including volume (VOL), gross motility (GM), concentration (CONC), and percentage post thaw motility (PPTM) in a resource population consisting of five sheep breeds common to Australia. Over 11,000 semen collection records were used to estimate the heritability of semen traits (h^2 = 0.081–0.170). Genome-wide association (GWA) analysis was subsequently performed using a subset of genotyped animals with 5,363 semen collection records. A total of 34 QTLs located on 16 chromosomes were found to be significantly associated with semen traits. Several candidate genes that have previously been linked to male fertility were identified within these QTLs.

INTRODUCTION

Ram semen traits like GM (David *et al.* 2015), CONC (D'Alessandro *et al.* 2001), and PPTM (Morris *et al.* 2001) may influence conception outcomes in sheep following artificial insemination (AI). Moreover, in natural mating, litter size has been reported to be significantly influenced by the ram (Holler *et al.* 2014). Therefore, assessment of semen quality or breeding soundness should be widely practiced by sheep breeders. However, the genetic and physiological drivers contributing to variability in ram semen traits are not fully understood. Identifying genetic determinants that underlie variability in these traits, could aid in better understanding of these traits, and help devise novel strategies to improve conception outcomes.

Past sheep studies using Spanish dairy sheep (Pelayo *et al.* 2019) and Ethiopian rams (Rege *et al.* 2000) have found semen traits which are routinely assessed for use in artificial breeding to be lowly heritable. Comparable studies have not yet been performed in Australian sheep populations. Similarly, only one GWA study has been previously undertaken to identify genomic regions associated with semen traits such as volume, gross motility, and concentration in Assaf rams (Serrano *et al.* 2021). Therefore, the aim of this study was to estimate heritability and identify QTLs associated with ram semen traits in an Australian population comprising of five sheep breeds.

MATERIALS AND METHODS

Phenotypic data. Semen phenotypes for VOL, GM, CONC, and PPTM were provided by an artificial breeding facility for Dohne, Dorper, Merino, Poll Dorset, and White Suffolk rams. Semen collection, assessment, and initial quality control have been previously described (Hodge *et al.* 2022).

Genetic parameter estimation. A total of 11,470 semen collection records from 864 rams were used to estimate genetic parameters, which has been previously described (Hodge *et al.* 2022).

* A joint venture of NSW Department of Primary Industries and University of New England

GWA. Genotype data was available for 330 rams, as such, a subset of 5,363 semen collection records were used to perform a GWA study via R package RepeatABEL (Rönnegård *et al.* 2016). Quality control for genotype data was performed as previously described (Moghaddar *et al.* 2015). Modified Bonferroni was used to identify significant single nucleotide polymorphisms (SNPs), and genes within ±0.5 Mega base (Mb) of these SNPs were identified via the National Centre of Biotechnology Information (NCBI) Genome Data Viewer (Rangwala *et al.* 2021), using the *Ovis aries* genome assembly (Oar_v3.1). SNPs with overlapping ±0.5 Mb regions were considered to represent the same QTL region.

RESULTS AND DISCUSSION

Semen traits were found to be lowly heritable (0.081-0.170) (Table 1), and genetic and phenotypic correlations ranged from -0.630 to 0.321 and -0.074 to 0.347, respectively.

Table 1. Estimated heritability, genetic and phenotypic correlations along with standard errors (heritability in bold on the diagonal, and genetic and phenotypic correlations on upper and lower of the diagonals, respectively)

	VOL	GM	CONC	PPTM
VOL	0.161 (0.041)	-0.071 (0.206)	0.153 (0.227)	-0.262 (0.274)
GM	0.215 (0.020)	0.170 (0.058)	0.321 (0.282)	-0.630 (0.238)
CONC	0.228 (0.022)	0.347 (0.019)	0.089 (0.051)	-0.351 (0.286)
PPTM	0.113 (0.024)	-0.074 (0.024)	0.100 (0.024)	0.081 (0.040)

Overall, the heritability estimates of semen quality traits were found to be low, indicating that environmental variance significantly contributed to phenotypic variance. This is consistent with the fact that the data used in this study was collected over a 20-year period. Furthermore, heritability estimates in the present study largely align with past studies in livestock (Wolf 2009; Berry *et al.* 2019). Ultimately, results of genetic parameter estimation indicate that semen traits have the potential to be improved by selective breeding, as variability in semen traits is partially due to genetics.

A total of 34 QTLs were significantly associated with semen traits, including 8 QTLs for VOL, 9 for GM, 12 for CONC, and 5 for PPTM (Figure 1).

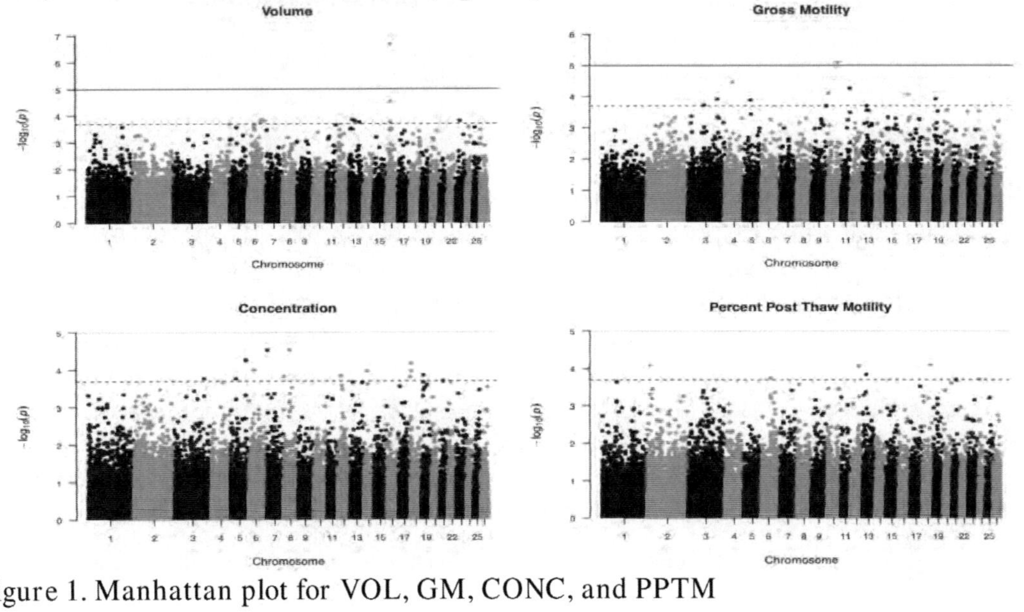

Figure 1. Manhattan plot for VOL, GM, CONC, and PPTM

Several candidate genes were identified within QTLs associated with semen traits (63 for VOL, 105 for GM, 60 for CONC, and 21 for PPTM). Table 2 presents candidate genes identified in the top two most significant QTLs associated with each semen trait. Noteworthy candidate genes found within QTLs associated with VOL, GM, CONC, and PPTM include *ANKRD17*, *LAMB1*, *YTHDC2*, and *CYLC2*, respectively. Proteomic analysis of pig semen found ANKRD17 to be significantly expressed in the post sperm-rich fraction (one of three parts which constitutes pig semen) of higher fertility boars compared to lower fertility (Martine *et al.* 2022). *LAMB1* was identified as a candidate gene associated with progressive motility following a GWA study in pigs (Gao *et al.* 2019). *YTHDC2* is reported as significantly upregulated in murine testes during spermatogenesis (Wojtas *et al.* 2017). Furthermore, *YTHDC2* null mice have significantly smaller testes, which ultimately resulted in infertility (Hsu *et al.* 2017). Finally, *CYLC2* was abundantly expressed in cryopreserved semen from bulls with high conception rates proven by AI (D'Amours *et al.* 2019). Several key positional candidates identified within genomic regions associated with semen traits have also been reported in past studies to have functional roles influencing semen quality, spermatozoal motility, and conception outcomes following AI. Thus, such genes may be important putative candidates for semen quality and potentially influence to conception outcomes in sheep.

Table 2. Top two most significant QTLs and candidate genes associated with different semen traits

Trait	Position (Chromosome:Mb)	Candidate Gene
VOL	6: 87.69 - 88.69	*ANKRD17*^{A}, *ALB*, *AFP*, *AFM*, *RASSF6*, *CXCL8*, *CXCL5*, *PPBP*, *PF4*, *CXCL1*^{A}, *TRNAS-GGA*^{A}, *TRNAC-GCA*^{A}
	16: 23.63 - 25.23	*PPAP2A*^{A}, *SKIV2L2*^{A}, *DHX29*, *CCNO*^{A}, *MCIDAS*^{A}, *CDC20B*^{A}, *GPX8*^{A}, *GZMK*^{A}, *ESM1*, *SNX18*, *HSPB3*, *ARL15*^{A}
GM	4: 48.58 - 49.58	*COG5*, *DUS4L*, *BCAP29*, *SLC26A4*, *CBLL1*, *SLC26A3*^{A}, *DLD*^{A}, *LAMB1*^{A}, *LAMB4*, *NRCAM*
	10: 68.09 - 69.09	*GPC6*, *TRNAE-UUC*
CONC	7: 2.20 - 3.20	*YTHDC2*^{A}, *KCNN2*
	8: 39.75 - 40.75	*KLHL32*^{A}, *NDUFAF4*^{A}, *GPR63*^{A}, *FHL5*, *UFL1*, *FUT9*, *TRNAH-GUG*, *TRNAF-GAA*
PPTM	2: 19.27 - 20.27	*CYLC2*^{A}
	18: 45.66 - 46.66	*PAX9*, *SLC25A21*, *MIPOL1*, *FOXA1*

Note: Genes with superscript ^{A} were previously found to be associated with male fertility

There has only been one past GWA study identifying QTLs associated with semen traits published in sheep. This study involved the analysis of semen traits like VOL, GM, and CONC collected from Assaf rams (Serrano *et al.* 2021). There was no overlap in QTLs significantly associated with any of the semen traits identified in the present study and those genomic regions significantly associated with semen traits in Assaf rams. Unlike the present study, which used Modified Bonferroni, only QTLs associated with GM in Assaf rams passed significance (10% false discovery rate), which is likely due to use of a more stringent significance threshold as well as using semen collected only from Assaf rams. Given Australian sheep breeds exhibit higher rates of linkage disequilibrium (LD) decay (Al-Mamun *et al.* 2015), and single breed resource populations exhibit higher rates of LD (van de Berg *et al.* 2016), the present study used a multi-breed population to identify QTLs significantly associated with ram semen traits in breeds common to Australia.

CONCLUSIONS

Semen quality traits including VOL, GM, CONC, and PPTM are lowly heritable, and as such, may be improved via selective breeding. Furthermore, several of the candidate genes identified in the present study have been previously found to influence spermatogenesis and normal morphological development and may be putative candidates influencing ram semen traits. Thus, validating such genes would be beneficial to determine their impact on semen traits, and in turn potential subsequent influence on conception outcomes and reproductive efficiency in sheep.

ACKNOWLEDGEMENTS

We would like to kindly thank Charles Sturt University, Apiam Animal Health, and Sheep Genetics (Meat and Livestock Australia) for their contributions to this research.

REFERENCES

Al-Mamun H., Clark S., Kwan P. and Gondro C. (2015) *Genet. Sel. Evol.* 47: 1.

Berry D., Eivers B., Dunne G. amd McPharland S. (2019) *Theriogenology* 123: 202.

D'Alessandro A., Martemucci G., Colonna M. and Bellitti A. (2001) *Theriogenology* 55: 1159.

D'Amours O., Calvo É., Bourassa S., Vincent P., Blondin P. and Sullivan R. (2019) *Mol. Reprod. Dev.* 86: 999.

David I., Kohnke P., Lagriffoul F., Praud O., Plouarboué F., Degond P. and Druart X. (2015) *Anim. Reprod. Sci.* 161: 75.

Gao N., Chen Y., Liu X., Zhao Y., Zhu L., Liu A., Jiang W., Peng X., Zhang C. and Tang Z. (2019) *BMC Genomics* 20: 797.

Hodge M., Rindfleish S., Heras-Saldana S.d.l., Stephen C. and Pant S. (2022) *Animals* 12: 2946.

Holler T.L., Dean M., Taylor T., Poole D.H., Thonney M.L., Thomas D.L., Pate J.L., Whitley N. and Inskeep E.K. (2014) *J. Anim. Sci.* 92: 3108.

Hsu P.J., Zhu Y., Ma H., Guo Y., Shi X., Liu Y., Qi M., Lu Z., Shi H., Wang J., Cheng Y., Luo G., Dai Q., Liu M., Guo X., Sha X., Shen B. and He C. (2017) *Cell Res.* 27: 1115.

Martine C., Roca J., Alvarez-Rodriguez M. and Rodrigues-Martinez H. (2022) *Biology* 11: 236.

Moghaddar N., Gore K. and Daetwyler H. (2015) *Genet. Sel. Evol.* 47: 1.

Morris L., Johnson W., Leibo S. and Buckrell B. (2001) *Reprod. Fert. Develop* 13: 193.

O' Meara C., Hanrahan J., Prathalingam N., Owen J., Donovan A., Fair S., Ward F., Wade M., Evans A. and Lonergan P. (2008) *Theriogenology* 69: 513.

Pelayo R., Ramón M., Granado-tajada I., Ugarte E., Serrano M., Gutiérres-gil B. and Arranz J.J. (2019) *Animals* 9: 1147.

Rangwala S., Kuznersov A., Ananiev V., Asztalos A., Borodin E., Evgeniev V., Joukov V., Lotov V., Pannu R. and Rudnev D. (2021) *Genome. Res.* 31: 159.

Rege J., Toe F., Mulasa-Mugerwa E., Tembely S., Anindo D., Baker R. and Lahlou-Kassi A. (2000) *Small Ruminant Res.* 37: 173.

Rönnegård L., McFarlane S., Husby A., Kawakami T., Ellegren H. and Qvarnström A. (2016) *Methods Ecol. Evol.* 7: 792.

Serrano M., Ramón M., Calvo J.H., Jiménez M., Freire F., Vázquez J. and Arranz J. (2021) *Animals* 15: 100065.

van de Berg I., Boichard D. and Lund M. (2016) *J. Dairy Sci.* 99: 8932.

Wojtas M., Pandey R., Mendel M., Homolka D., Sachidanandam R. and Pillai R. (2017) *Mol. Cell* 68: 374.

Wolf J. (2009) *Reprod. Domest. Anim.* 44: 338.

Proc. Assoc. Advmt. Anim. Breed. Genet. 25: 286 - 289

ONLINE MENDELIAN INHERITANCE IN ANIMALS (OMIA) – LOOKING TO THE FUTURE

I. Tammen[1], M. Mather[2], D. Vanichkina[2], J. Nothman[2], Z. Li[2], S. Jufri[2] and F.W. Nicholas[1]

[1]The University of Sydney, Sydney School of Veterinary Science, Sydney, NSW 2006, Australia
[2]The University of Sydney, Sydney Informatics Hub, Darlington, NSW 2008, Australia

SUMMARY

Online Mendelian Inheritance in Animals (OMIA) is a freely available curated knowledgebase that contains information and facilitates research on inherited traits and diseases in animals. For the past 27 years, OMIA has been used by animal geneticists, breeders, and veterinarians worldwide as a definitive source of information. Recent increases in curation capacity and funding for software engineering support have resulted in software upgrades and commencement of several new initiatives, which include the review of variant information and links to human diseases caused by orthologous genes, and the introduction of phenotype and breed ontologies. We provide an overview of current information and recent enhancements to OMIA and discuss how we are expanding the integration of OMIA into other resources and databases via the use of ontologies.

INTRODUCTION

OMIA (https://omia.org) is a freely available, curated, online knowledgebase which provides users with up-to-date summary information on the known harmful and beneficial variants in animals, together with background information on known inherited disorders and beneficial traits. OMIA is modelled on and reciprocally hyperlinked to Online Mendelian Inheritance in Man (OMIM, https://www.omim.org), and provides further links to PubMed and Gene records at the National Center for Biotechnology Information and the European Bioinformatics Institute's Ensembl.).

OMIA focuses on traits and diseases ('phenes') with confirmed or suspected Mendelian modes of inheritance. However, several phenes with unknown or complex modes of inheritance and phenes caused by somatic mutations, genetic modifications or genome editing are also included. Furthermore, OMIA highlights 'landmark' papers (reporting major advances) and lists reviews and papers describing genetic maps and reference genomes. While most OMIA entries are for the major domesticated animal species, more than 370 (mainly vertebrate) animal species have entries in OMIA. Information about humans and model organisms such as mouse, rat, and zebrafish are not included, as they have dedicated species-specific resources.

Since the beginning of 2021, the curation team has increased from one (FWN) to two main curators (FWN and IT) and bequest funding has enabled software engineering support (MM). In this paper we provide an overview of OMIA data and a summary of recent software updates, major enhancements to likely causal variant tables and OMIA-OMIM hyperlinks, and the launch of Pioneers of Mendelian Inheritance in Animals (PMIA). We provide an update on current initiatives that focus on the use of ontologies to expand the interoperability of OMIA with other resources such as the Anstee Hub for Inherited Diseases in Animals (AHIDA, https://ahida.sydney.edu.au/app/home).

MATERIALS AND METHODS

OMIA software upgrade. Since August 2010, the OMIA database and website have been using Django software, a high-level Python Web framework. In July 2021, software packages were upgraded from Python 2 to Python 3 (https://www.python.org/) and from Django 1.9 to 3.2

(https://www.djangoproject.com/), and coding was reviewed and refined with the aim of future-proofing OMIA and improving homepage response times. Furthermore, search options were refined to increase the number of fields that are searched in a 'quick' search to improve user experience.

Likely causal variant tables. 'OMIA variant ID' and 'Source of Genetic Variant' fields were added to variant tables to provide unique numerical variant identification and to facilitate inclusion of variant information for genome-modified or edited variants. In collaboration with many colleagues (see OMIA's acknowledgement page for details: https://omia.org/acknowledgements/) variant information for cats, dogs, cattle, sheep, horse, pigs and goats has been reviewed and updated to Human Genome Variation Society (HGVS) nomenclature (https://varnomen.hgvs.org/). New site administration tools were introduced to facilitate automated 'liftover' of variant information to newer reference genome assemblies, and to enable export of variant information in variant call format (VCF) for submission to the European Variant Archive (EVA, https://www.ebi.ac.uk/eva/).

Review of OMIA-OMIM hyperlinks. Since 1997, OMIA has been reciprocally hyperlinked to OMIM. Links to OMIM are created by OMIA curators when new phenes are entered into OMIA. In the past, this focused on adding OMIM phenotype identifiers (IDs), while OMIM gene IDs were rarely added. OMIM automatically downloads OMIA phene IDs that have an OMIM ID link once a week and updates OMIM accordingly. In OMIA, separate fields for 'OMIM phene' and 'OMIM gene' hyperlinks were recently created, and we reviewed OMIA-OMIM hyperlinks for all OMIA phenes for which a likely causal variant has been identified in at least one species. OMIM links were confirmed, deleted, or added.

Pioneers of Mendelian inheritance in animals (PMIA). In 2022, PMIA was added to OMIA, accessible from the home page. This project comprises a series of commentaries on papers that illustrate the early discoveries of Mendelian inheritance in animals.

Integration of phenotype, disease and breed ontologies. OMIA previously included a home-grown list of 20 'phene categories' that could be used in 'Advanced Search' and in the 'Browse' page to create category-specific phene lists, but many OMIA phenes did not have a phene category specified. To allow for comprehensive categorisation, OMIA's 20 phene categories have been replaced with 28 major biological system headers from the Mammalian Phenotype (MP) Ontology (Smith and Eppig 2009) and two headers from the Mondo Disease Ontology (Mondo, https://mondo.monarchinitiative.org/). The MP ontology headers are included in Mondo, a global disease ontology that aims to harmonise disease definitions across the world. To facilitate this, the 'category' field has been included in phene-species pages and each phene has been linked to at least one category by a curator (IT). In addition, a new field enables inclusion of hyperlinks between OMIA disease entries and the corresponding homologous disease in Mondo.

Recognising the need to replace OMIA's home-grown breed list with a computable comprehensive list of standardised breed names, the OMIA team instigated the creation of the Vertebrate Breed Ontology (VBO, https://github.com/monarch-initiative/vertebrate-breed-ontology) in a project led by the Monarch Initiative (https://monarchinitiative.org/), with key personnel funded by the University of Colorado, in collaboration with colleagues from Iowa State University and with FAO colleagues responsible for the Domestic Animal Diversity Information System (DAD-IS). Curation tools relating to 'breed' in OMIA have been updated to allow inclusion of hyperlinks to VBO.

RESULTS AND DISCUSSION

OMIA is a globally used knowledgebase. Google Analytics user data for 2022 identified 41,803 users (94,579 sessions) from 163 countries. Until recently, curation was predominately conducted single-handedly by one curator, and limited funding restricted access to urgently needed software upgrades and modifications. Increased curation capacity and bequest funding support to upgrade and refine the underlying software is improving curator and user experiences and has resulted in the commencement of several innovations while maintaining ongoing curation activities. In February 2023, OMIA included information on 2,327 phenes across 377 species, contained 4,336 phene-species entries and included a total of 29,453 references. Core statistics for key livestock species are summarised in Table 1.

Table 1. Summary of OMIA information relating to key livestock species (22/3/2023)

	Dog	Cattle	Cat	Pig	Sheep	Horse	Chicken	Goat	All
Total phenes	863	628	404	355	300	259	239	102	4336
Mendelian phenes	397	297	136	133	117	61	135	24	1778
Mendelian phenes with at least one likely causal variant known	336	204	103	65	54	48	56	17	1012
Likely causal variants known	491	269	171	72	86	105	71	30	1486

During 2022 the daily automated PubMed literature search resulted in 17,653 hits, of which 719 papers were identified to be added to OMIA. Additional references were added as part of other curation activities. We are currently trialling other literature search strategies to reduce the number of false-positive 'hits', including use of the machine learning tool LitSuggest (Allot *et al.* 2021) and an AI-based tool developed in house from Microsoft's PubMedBERT (Gu *et al.* 2021).

Likely causal variant tables. We reported the introduction of variant tables in OMIA in 2018 (Tammen and Nicholas 2018) and indicated that the ultimate aim would be to provide an EVA ID for all variants to reduce the need to standardise and update variant information in OMIA. However, EVA does not accommodate all types of variants, very few authors of OMIA-relevant papers submit variant information to EVA, and new EVA IDs are allocated infrequently. With the help of many colleagues (https://omia.org/acknowledgements/) we reviewed and standardised historic variant information in OMIA using HGVS nomenclature, with the aim to report location information based on a recent reference genome assembly where possible. In October 2021, variants listed in OMIA that were lacking an EVA ID but had standardised location information were submitted to EVA using a new OMIA pipeline for export of variant information in VCF for submission to EVA. The need for more standardised nomenclature for variants has been widely discussed to ensure greater transparency in relation to DNA testing. To this end, OMIA numerical variant IDs are now presented in the first column of all OMIA variant tables. An OMIA variant ID provides a unique unchanging ID for each likely causal variant, including those complex variants for which there is no HGVS nomenclature or no EVA ID. Review papers have started to include OMIA variant IDs in their tables.

Review of OMIA-OMIM hyperlinks. For phenes in OMIA that had at least one causal variant identified the review of OMIM links resulted in confirmation of 607 OMIM links, addition of 683 OMIM links and deletion of 46 OMIM links. Most of the added OMIM links were OMIM gene IDs (n=493), as these were in the past not routinely added to OMIA. OMIA currently lists 2424 models of human traits based on links to OMIM. 2119 OMIM entries have a link to OMIA. The revision of OMIA-OMIM hyperlinks will facilitate comparative-medicine-related research approaches. However, a large list of OMIA phenes without known likely causal variants have not yet been reviewed, as it is more speculative to identify homology between human and animal phenes if the underlying genetic cause is unknown.

Pioneers of Mendelian Inheritance in Animals. PMIA was first announced on Mendel Day (8 March) in 2022, and launched as part of OMIA on the 8th of July 2022, two weeks before the bicentenary of Mendel's birth on the 22nd. Currently PMIA includes detailed commentaries by FWN on 15 papers that illustrate the early discoveries of Mendelian inheritance in animals.

Integration of phenotype, disease and breed ontologies. Three major current projects relate to the introduction of phenotype, disease and breed ontologies to OMIA. Ontologies are controlled vocabularies that represent knowledge both by their meaning and their relationship to each other and provide unique numerical identifiers to enable advanced computational analysis. We aim to harmonize breed and disease definitions in OMIA in a computer-accessible format, thus enabling integration with other global online resources and integration with the submission portal of AHIDA, (Tammen *et al.* 2021) which is currently under development.

So far, home-grown OMIA phene 'categories' have been replaced with 28 MP 'major biological system headers' and 2 Mondo categories. These categories are visible on phene-species pages in addition to visibility in advanced searches and in the OMIA browse page (https://omia.org/browse/). At least one category has been added to each OMIA phene, so that it is now possible, e.g., to search OMIA for all entries categorised as 'pigmentation phenotype' (MP:0001186). Further curation work is needed to add multiple categories as required.

Breed information in OMIA phene-species entries and in variant tables has been replaced with links to the VBO. VBO is based on FAO's Domestic Animal Diversity Information System (DAD-IS) breeds list and has been updated (especially for cat and dog breeds) with information from other international organizations, communities, and experts.

Finally, in a second collaboration with the Monarch Initiative, we are working towards integrating OMIA information into Mondo. So far, a new field has been created to enable addition of hyperlinks to Mondo, we have commenced adding Mondo links in OMIA and are currently discussing how to integrate OMIA information into Mondo.

CONCLUSION

Our vision for the future is that in addition to summarising information about inherited conditions in animals, OMIA becomes a global repository for standardised information on likely causal variants for diseases to allow transparent delivery of DNA diagnostics, and in linkage with the currently under-development Anstee Hub for Inherited Diseases in Animals, becomes a tool that enables semiautomated diagnosis for rare or emerging inherited conditions in animals.

A key remaining challenge is how best to harness automation and engage a wider contribution to curation efforts to ensure sustainability for the next 25 years. We always welcome feedback on current information presented in OMIA.

ACKNOWLEDGEMENTS

The Acknowledgements tab on the OMIA home page provides a detailed account of the contributions of the many people who have contributed to the development of OMIA since its inception. Recent software engineering support has been possible due to support of the Sydney Informatics Hub, a core research facility of the University of Sydney, and funding from the Ronald Bruce Anstee Bequest to the Sydney School of Veterinary Science.

REFERENCES

Allot A., Lee K., Chen Q., Luo L. and Lu Z. (2021) *Nucleic Acids Res.* 49: W352.

Gu Y., Tinn R., Cheng H., Lucas M., Usuyama N., Liu X., Naumann T., Gao J. and Poon H. (2021) *ACM Trans. Comput. Healthcare* 3: 1.

Smith C.L. and Eppig J.T. (2009) *Wiley Interdiscip. Rev. Syst. Biol. Med.* 1: 390.

Tammen I. and Nicholas F.W. (2018) *Proc. of the 11th WCGALP*, Auckland, New Zealand.

Tammen I., Wade C., Waud B., Gimeno M., Velie B.D. O'Rourke B. and Nicholas F.W. (2021) *Proc. Assoc. Advmt. Anim. Breed. Genet.* 24: 383.

GENETIC EVALUATION OF COAT TYPE FOR AUSTRALIAN ANGUS

A.M. Samaraweera[1], H. Aliloo[2], A. Byrne[1], C.J. Duff[1] and S.A. Clark[2]

[1]Angus Australia, 86 Glen Innes Road, Armidale, NSW 2350, Australia
[2]School of Environmental and Rural Science, University of New England, Armidale, NSW 2350, Australia

SUMMARY

Animals with sleeker coats are commonly considered to have better heat tolerance, tick resistance, and a lower incidence of dags in feedlot environments. The objective of this study was to estimate genetic parameters for coat type traits and to estimate genetic correlations between coat type and scan and carcass weight traits using single-step methods. Two coat type traits were defined based on the month of scoring where scores recorded in April to October were considered as coat type 1 (CT1) and those recorded in November to March were categorized as coat type 2 (CT2). The coat type traits were moderately heritable, and the heritability of CT1 (0.36 ± 0.04) was higher than CT2 (0.32 ± 0.03). Genetic correlations between coat type traits and steer and heifer ultrasound scan traits (eye muscle area, intramuscular fat) were either low to moderate in strength, but favourable in direction. The outcomes of this study suggest selection for sleeker coat type is possible without any associated detrimental effect on scan and carcase traits.

INTRODUCTION

It is common for domestic livestock to shed their hairy coats for a sleeker coat at the onset of summer in preparation for warmer months. A lower core body temperature and greater perspiration have been previously observed among sleeker coat-typed cattle showing superior heat tolerance ability (Yeates 1955; Dikmen *et al.* 2008). Sleeker and shorter coats are also associated with tick resistance since hairy and thicker coats support tick attachment and prevent the removal of ticks via the animal's self-grooming (Hansen 2004). Therefore, *Bos taurus* cattle with a predominantly sleeker coat are advantaged over hairy coat cattle during summer or in tropical or sub-tropical conditions where the heat and tick infestations are the highest. In addition, beef cattle with sleeker hair types have a lower incidence of dags in a feedlot environment, with less associated challenges of dag removal prior to slaughter, particularly in colder and wetter climates.

Since 2019 Angus Australia has published a research breeding value (RBV/RBVs (plural)) for the coat type through the TransTasman Angus Cattle Evaluation (TACE). This has enabled the selection of a desirable coat type to match with the beef cattle production systems. Preliminary data analyses for coat type in Australian Angus cattle showed that the coat type is moderately heritable, however, the association between coat type and other economically important traits are yet to be explored. Therefore, the objective of this study was to estimate the genetic parameters and genetic correlations between coat type and live-animal ultrasound assessed traits and carcass weight using single-step methods.

MATERIALS AND METHODS

Data and traits. Coat scores were collected assessing the animal's hair length, fibre diameter, and handle based on a one to seven numbered scoring guide according to Turner and Schleger (1960). Scores range from; score one, animals with extremely short sleek hair similar to the hair found in *Bos indicus* to score seven being a very hairy coat.

Coat scores of 6188 animals were extracted from the Angus Australia database for progeny from the Angus Sire Benchmarking Program (Parnell *et al.* 2019). Data was collected year-round, and scores were available from animals born from 2008 to 2021. Since coat type may vary as the animal ages (Durbin *et al.* 2020), coat scores collected after 720 days were excluded from the analyses. Only purebred Angus animals were used in the analyses by selecting animals with an Angus breed percentage of 87.5 or above. Multiple records per animal were excluded by keeping only the earliest record. The contemporary groups were constructed by concatenating the year-month of measurement, herd, and the breeder-defined management group. The contemporary groups with at least ten individuals were selected for the analyses. The data cleaning process resulted in 6177 records for analysis. Individuals scored for coat type were the progeny of 401 sires and included 2115 females and 4062 males. Coat score measures were broadly classified into two traits based on the phenotypic averages and variation in each month of scoring. Accordingly, coat scores recorded from April to October were identified as coat type 1 (CT1) and the scores collected during the rest of the year as coat type 2 (CT2).

Live-animal ultrasound scan traits and carcass weight were also extracted for animals with a coat type record from the Angus Australia database to estimate the genetic correlations with coat type traits. The live-animal ultrasound scan traits extracted were eye muscle area (EMA, measured in cm^2), intramuscular fat (IMF, measured in %), and P8 fat (P8, measured in mm). The data cleaning process was similar to that described for coat type traits. The contemporary groups for the scan and carcass weight traits were formulated as described by Graser *et al.* (2005). The live-animal ultrasound scan traits were separated as steer (S) and heifer (H) traits. Genomic information for animals with a phenotypic record of which were imputed for 45364 markers per genotype (Aliloo and Clark, 2021) was also extracted.

Statistical analyses. The genetic parameters for different traits were estimated using single-step univariate and bivariate animal models. The contemporary group and sex were fitted as the fixed effects, linear and quadratic effects of age were fitted as the covariates, and the animal effect was fitted as the random effect in animal models for coat type traits. Sire by herd interaction was not significant for CT1 and CT2, therefore, was not included in the final models. The model parameters used for scan traits and carcass weight were as described by Graser *et al.* (2005). The variance components were estimated using the single-step method implemented in airemlf90 (Misztal, *et al.* 2018).

RESULTS

Descriptive statistics of different traits used in the analyses are given in Table 1. The mean CT1 was higher than CT2 (2.8 vs. 2.0), and scores for CT1 and CT2 ranged from 1 to 5 and 1 to 4.5, resp-

Table 1. Descriptive summaries of coat type traits, 1 and 2

Trait[a]	No. of records	% Genotyped	Mean	SD	Minimum	Maximum
CT1	2221	98	2.8	0.6	1	5
CT2	3956	98	2.0	0.5	1	4.5
SEMA	3861	99	70.6	11.2	38	100
SIMF	3824	99	6.7	1.2	3	8.3
SP8	3838	99	10.5	4.4	1	22
HEMA	1912	97	61.5	7.3	41	82
HIMF	1901	97	6.2	1.3	2.5	8.3
HP8	1899	97	8.7	3.4	1	19
CWT	4084	99	422.7	68.9	238	607

[a]CT1: coat type scored for 04-10 months; CT2: coat type scored for 11-03 months; SEMA: Scan steer eye muscle area (cm^2); SIMF: Scan steer intramuscular fat (measured in %); SP8: Scan steer P8 fat (mm); HEMA: Scan heifer eye muscle area (cm^2); HIMF: Scan heifer intramuscular fat (ether extract %); HP8: Scan heifer P8 fat (mm); CWT: Carcass weight (kg).

-ectively. There were at least three scores in each coat type trait with more than 18% animals recorded. Therefore, there was an adequate coat score variation and an adequate number of records in each score to fit linear models for each coat type trait. More than 97% of animals recorded for all traits were also genotyped.

The heritability for CT1 (0.36 ± 0.04) was similar to CT2 (0.32 ± 0.03) (Table 2). Across all traits used in this study, the highest heritability was observed for CWT (0.45 ± 0.03) and the lowest was for HEMA (0.25 ± 0.05). The heritabilities of S-scan traits were higher than H-scan traits, and the heritability of scan traits was highest for P8 fat.

The correlation between CT1 and CT2 was 0.76 ± 0.08. The genetic correlation estimates between coat type traits and other traits were negative (*i.e.* favourable) except for CT1in SP8 and CT1 and CT2 in HP8 where a very small positive correlation was obtained (Table 3). The genetic correlation coefficients between coat type and scan and carcass weight traits ranged from -0.26 to 0.03 in CT1 and -0.27 to 0.06 in CT2. The genetic correlations between scan and carcass traits and coat type traits were slightly lower for CT2.

Table 2. Additive genetic (V_a), sire by herd (V_{sxh}), and residual variances (V_e), and heritability ± standard deviations ($h^2 \pm$ SD) from univariate single-step analyses

Trait[a]	V_a	V_{sxh}	V_e	$h^2 \pm$ SD
CT1	0.10	-	0.17	0.36 ± 0.04
CT2	0.06	-	0.13	0.32 ± 0.03
SEMA	6.89	0.81	13.44	0.33 ± 0.03
SIMF	0.13	0.00	0.27	0.32 ± 0.01
SP8	2.20	0.09	2.62	0.45 ± 0.03
HEMA	4.47	0.92	12.62	0.25 ± 0.05
HIMF	0.22	0.00	0.53	0.29 ± 0.01
HP8	1.64	0.11	3.11	0.34 ± 0.05
CWT	449.14	35.48	505.57	0.45 ± 0.03

[a]Traits and units are as given in Table 1.

Table 3. Genetic correlations (± standard deviations) for CT1 and CT2 with steer and heifer live-animal ultrasound scan traits and carcass weight from bivariate single-step analyses

Traits[a]	No. of animals		Genetic correlations	
	CT1	CT2	CT1	CT2
SEMA	4819	5219	-0.22 ± 0.09	-0.26 ± 0.07
SIMF	4793	5208	-0.26 ± 0.01	-0.27 ± 0.01
SP8	4808	5207	0.03 ± 0.08	-0.12 ± 0.07
HEMA	4133	5868	-0.11 ± 0.10	-0.22 ± 0.09
HIMF	4122	5857	-0.06 ± 0.01	-0.13 ± 0.01
HP8	4120	5855	0.06 ± 0.10	0.06 ± 0.08
CWT	4928	5333	-0.25 ± 0.07	-0.25 ± 0.06

[a]Traits and units are as given in Table 1.

DISCUSSION

Coat type traits were moderately heritable in this study, therefore, genetic improvement towards a desired coat type can be achieved in breeding programmes. Angus Australia reports the RBVs for CT2 that are based on the coat scores recorded during the Australian summer. Studies based on a coat scoring system that records the extent of hair shedding at the onset of summer in the United States yielded similar heritability estimates to our study. For example, heritability estimates for

American Angus and Limousin cattle in the United States were 0.34 to 0.40 (Durbin *et al.* 2020) and 0.33 (Williams *et al.* 2006), respectively.

Genetic correlation estimates between coat type traits and scan and carcass traits were favourable in this study. Selecting animals with a sleeker coat using either CT1 and CT2 could result in improvements in CWT, EMA, and IMF in subsequent generations. These results are aligning with the anecdotal feedback from breeders suggesting that sleeker coats are associated with superior performance.The favourable genetic correlations are slightly stronger in CT2 than CT1 for most scan traits other than for heifer P8 fat. Therefore, CT2, which is used to produce an RBV, would be an agreeable trait to select for sleeker coat type while also improving the meat quality and carcass weight. However, this needs further investigations including more animals and estimation of genetic correlations for other production traits including weight and fertility traits.

CONCLUSIONS

Coat type traits were moderately heritable. Selecting animals for a sleeker coat type can lead to simultaneous improvements in both carcass weight and meat quality.

ACKNOWLEDGMENTS

The authors would like to thank the Department of Industry, Science and Resources for their financial support through an Innovation Connections Grant.

REFERENCES

Aliloo H., and Clark S. A. (2021) *Anim. Prod. Sci.* 61: 1958.

Dikmen S.E.R.D.L., Alava E., Pontes E., Fear J.M., Dikmen B.Y., Olson T.A. and Hansen P.J. (2008) *J. Dairy Sci.* 91: 3395.

Durbin H.J., Lu D., Yampara-Iquise H., Miller S.P. and Decker J.E. (2020) *Genet. Sel.* 52: 1.

Graser H.U., Tier B., Johnston D.J. and Barwick S.A. (2005) *Aust. J. Exp. Agric.*45: 913.

Hansen P.J. (2004) *Anim. Reprod. Sci.*82: 349.

Misztal, I., Tsuruta S., Lourenco D.A.L., Masuda Y., Aguilar, I. Legarra, A. and Vitezica. Z. (2018). 'Manual for BLUPF90 family of programs', University of Georgia, Athens, USA.

Parnell P.F., Duff C.J., Byrne A.I. and Butcher N.M. (2019) *Proc. Assoc. Advmt. Anim. Breed. Genet.* 23: 492.

Turner H.G. and Schleger A.V. (1960) *Aust. J. Agric. Res.* 11: 645.

Yeates N.T.M. (1955) *Aust. J. Agric. Res.* 6: 891.

Williams J.L., Garrick D.J., Enns R.M. and Shirley K.L. (2006) *In Proceedings-ASAS Western Section* (Vol. 57, p. 29).

GENETIC PARAMETERS FOR LINEAR TYPE TRAITS IN HUNGARIAN CHAROLAIS BEEF CATTLE

B.J. Crook[1], W.M.S.P. Weerasinghe[1] and M. Torok[2]

[1]Agricultural Business Research Institute, UNE, Armidale, NSW, 2350 Australia
[2]National Association of Hungarian Charolais Cattle Breeders, Miskolc, Hungary

SUMMARY

Genetic parameters are reported for 18 linear type traits recorded by the National Association of Charolais Breeders in Hungary: back width (BW), chest width (CW), rump width (RW), shoulder width (SW), back-loin length (BL), rump length (RL), thigh length (TL), canon girth (CG), heart girth (HG), width of hip bones (HW), width of thigh (TW), roundness of thigh (RT), loin thickness (LT), development/frame (DF), top line straightness (TS), muzzle width (MW), fore legs (FL) and hind legs (HL). All traits were scored by Association staff using a linear scale from 1 to 10 based on the system developed by the Institut de L'Elevage in France. Animals averaged 508 days in age when scored. Bivariate models with weight at weaning (WW) were investigated for each trait using BLUPF90, with inclusion of genomic data. There were 2,524 animals with linear trait scores, 42,442 animals with a weight at weaning and 7,660 animals with a genotype. Of animals with trait scores, 88% had a weaning weight and 61% were genotyped. Six generations of pedigree were used, giving a total of 55,928 animals in the analyses. Trait heritability ranged from 0.11 for FL to 0.41 for CG. Genetic correlations with WW ranged from -0.36 for HL to +0.50 for RL. These traits could be incorporated into future genetic analyses of the breed.

INTRODUCTION

The National Association of Hungarian Charolais Cattle Breeders (MCTE) was officially formed in 1992 and currently has 236 members registering around 6,000 calves per year on average. Of calves registered, approximately 61% are recorded for weight at weaning (WW). The MCTE have utilised the BREEDPLAN genetic evaluation service provided by the Agricultural Business Research Institute (ABRI) since 2002, with their most recent evaluation representing 80,954 animals, a multi-trait analysis of gestation length, birth weight, post-birth growth (including WW), scrotal and ultra-sound scan records and a separate analysis of calving ease using birth difficulty scores, birth weights and gestation length records. In late 2019, the MCTE launched its Genome Program to members, with genotypes incorporated in their August 2022 BREEDPLAN evaluation using a Single-Step model (Johnston *et al.* 2008).

In 2016, the MCTE initiated a program of assessing Charolais cattle for 18 structural traits related mainly to the muscularity and skeletal attributes of the animal. ABRI reviewed the data to determine what, if any, level of genetic variation was expressed in the linear type scores. This paper provides preliminary estimates of heritability for the 18 linear traits and their genetic correlation with WW.

MATERIALS AND METHODS

Records for WW were extracted from the August 2022 BREEDPLAN evaluation, these having been pre-adjusted to a constant age at weighing (200 days) and constant age of dam (5 years) as outlined by Graser *et al.* (2005). The contemporary group for WW consisted of herd of origin, sex, year of birth, birth number (single vs twin), birth type (natural vs ET), breeder-defined management group and weigh date. Extracted records were pruned to remove single-animal contemporary groups and those comprising ET calves. The final data set contained 42,442 records for WW (mean 234.3 ± 45.8 kg), with contemporary group size ranging from 2 to 284 (mean of 43).

Linear scores were available on 2,524 animals for 18 type traits: back width (BW), chest width (CW), rump width (RW), shoulder width (SW), back-loin length (BL), rump length (RL), thigh length (TL), canon girth (CG), heart girth (HG), width of hip bones (HW), width of thigh (TW), roundness of thigh (RT), loin thickness (LT), development/frame (DF), top line straightness (TS), muzzle width (MW), fore legs (FL) and hind legs (HL). Age at scoring ranged from 176 to 1,334 days (mean 508.0 days, SD 147.2) and a majority (69%) of the animals scored were female. All scoring was undertaken by a single MCTE-approved technician using a linear scale from 1 (small/weak/thin/worst) to 10 (big/strong/wide/best) according to the guidelines approved by the MCTE (Institut de L'Elevage 2014). Contemporary group was defined as herd of origin, year of birth, breeder-defined management group for WW and date of scoring. After the removal of records for single animal contemporary groups (n=10), group size ranged from 2 to 115 (mean of 19). Sex, birth number and age at scoring to be fitted explicitly in subsequent models.

Genotypes on 8,934 animals were available, coming from a 50K SNP panel (BovineSNP50 BeadChip, Illumina Inc., San Diego, CA.). QC of genomic data was conducted using PLINK software (Chang *et al.* 2015), with SNPs removed at a minor allele frequency of <0.05, a deviation from Hardy–Weinberg equilibrium of $p<1E^{-6}$ and call rates <90%. Only those SNPs located on autosomal chromosomes were used. Individual genotypes were excluded if the call rate for all loci was <85%. Sporadic missing SNPs were imputed using FImpute v3 (Sargolzaei *et al.* 2014) and pedigree information for the genotyped population was included. Genotypes were excluded when a parentage conflict was detected. The final data set comprised 7,660 genotypes and 42,854 SNPs. Most of the genotypes (82%) were from females.

Bivariate models comprising each linear type trait and WW were conducted using the AIREMLF90 program in the BLUPF90 family of software (Misztal *et al.* 2018). The model for type traits included scoring contemporary group, sex and birth number (single or twin) as fixed effects and age at scoring as a linear covariate, with the variance being partitioned into additive genetic and residual components. The model for WW included WW contemporary group only as a fixed effect, with variance being partitioned into additive genetic, maternal genetic (uncorrelated) and residual components. Preliminary analysis of WW fitting an additional random effect for the dam's permanent environment suggested a small variance component (54.8±10.2) that was dropped from subsequent bivariate models. Six generations of pedigree were included, giving 55,928 animals in each analysis. A genotype file and associated map file were included in the analysis, with 20% of genotyped animals having a linear score record and 82% having a WW record. 88% of scored animals were recorded for WW. Default values were used in creating the H matrix (Aguilar *et al.* 2010).

RESULTS AND DISCUSSION

Average scores for type traits ranged from 4.72 for Thigh Length (TL) to 6.09 for Top Line Straightness (TS), with the standard deviation in scores ranging from 0.93 for Heart Girth (HG) to 1.20 for Roundness of Thigh (RT). No scores of 10 were allocated. Score distributions approximated normality within trait, suggesting a linear analysis of scores was appropriate.

The additive genetic variance and heritability for each linear type trait are summarised in Table 1. Most traits related to the muscularity and skeletal attributes were associated with moderate heritability, while functional traits like FL and HL were low. These are comparable to estimates reported by Doyle *et al.* (2018) for a range of subjectively assessed muscularity, skeletal and functional traits in Irish Charolais cattle. The genetic variances reported by Berry *et al.* (2019) were higher for a range of similar traits in a large population comprising 3 European and 2 British breeds, yet the direct heritability estimates were similar to those presented here.

Variance estimates for WW averaged over the 18 bivariate analyses were 581.7, 183.4 and 819.5 kg^2 for the direct genetic, maternal genetic and residual components, respectively. Both the phenotypic variance (1,584.6 kg^2) and the direct heritability (0.37) seem inflated compared to estimates from considerably larger Charolais datasets (Donoghue and Betrand 2004; Phocas and Laloe 2004). In contrast, estimates ranging from 0.30-0.39 were reported for smaller populations of Charolais (El-Saied *et al.* 2006; Herrera-Ojeda *et al.* 2019; Rezende *et al.* 2022) and may partially reflect the heterogeneity of variances reported for WW in the Charolais breed (Quintanilla *et al.* 2002; Donoghue and Bertrand 2004). The maternal genetic heritability (0.12) obtained in this study agrees with estimates reported by others.

The genetic correlations between WW and each linear trait are given in Table 1. Most traits were positively correlated with WW, in the order of 0.30 to 0.50. The correlation for TS, MW and FL was close to zero, while for HL was negative. It is not surprising that positive correlations with body weight are evident in this population, given that most of the linear traits relate to body size and dimensions. Strongly positive genetic correlations between live weight and a range of muscularity and skeletal traits were reported by Berry *et al.* (2019).

Table 1. Estimates of additive variance (V_A) and direct heritability (h^2) for 18 linear type traits and the genetic correlation (r_G) between each trait and weight at weaning

Trait	V_A	h^2	r_G
Back width (BW)	0.269 ± 0.046	0.305 ± 0.049	0.228 ± 0.080
Chest width (CW)	0.181 ± 0.035	0.227 ± 0.042	0.386 ± 0.089
Rump width (RW)	0.278 ± 0.047	0.296 ± 0.046	0.345 ± 0.082
Shoulder width (SW)	0.249 ± 0.045	0.268 ± 0.046	0.343 ± 0.086
Back loin length (BL)	0.236 ± 0.051	0.206 ± 0.043	0.451 ± 0.087
Rump length (RL)	0.196 ± 0.043	0.207 ± 0.043	0.498 ± 0.089
Thigh length (TL)	0.279 ± 0.047	0.295 ± 0.047	0.274 ± 0.083
Canon girth (CG)	0.298 ± 0.040	0.412 ± 0.049	0.395 ± 0.066
Heart girth (HG)	0.144 ± 0.034	0.199 ± 0.045	0.364 ± 0.097
Width of hip bones (HW)	0.260 ± 0.044	0.291 ± 0.046	0.396 ± 0.081
Width of thigh (TW)	0.300 ± 0.053	0.278 ± 0.046	0.381 ± 0.083
Roundness of thigh (RT)	0.306 ± 0.052	0.291 ± 0.046	0.322 ± 0.083
Loin thickness (LT)	0.236 ± 0.046	0.259 ± 0.047	0.345 ± 0.088
Development/frame (DF)	0.207 ± 0.046	0.213 ± 0.046	0.342 ± 0.091
Top line straightness (TS)	0.135 ± 0.044	0.123 ± 0.040	-0.013 ± 0.034
Muzzle width (MW)	0.230 ± 0.046	0.246 ± 0.047	0.152 ± 0.090
Fore legs (FL)	0.094 ± 0.033	0.109 ± 0.039	-0.170 ± 0.144
Hind legs (HL)	0.161 ± 0.047	0.142 ± 0.041	-0.357 ± 0.130

These results indicate that subjectively assessed muscularity and skeletal traits have potential use in Hungarian Charolais breeding programs where the breeding goal includes improvements in the physical appearance of animals. Linear scores for front and hind leg structure show less utility, a similar outcome reported in other EU populations (Doyle *et al.* 2018). While the moderate genetic correlations with live weight suggest that selection for improved growth rate may bring some improvements in animal appearance, there is sufficient scope for gains in muscularity to be achieved without pursuing growth. Berry *et al.* (2019) reported genetic correlations in the order of 0.44 to 0.66 between muscularity traits in registered live animals and carcase conformation in commercial cattle. Positive correlations were also evident between muscularity traits and carcase primal cut yields. Similar results were reported by Bonfatti *et al.* (2013) for Italian Piemontese cattle, with live animal scores for muscularity type traits having a positive genetic correlation with European carcase conformation grades. The genetic correlations reported by Bouquet *et al.* (2010) for Blonde d'Aquitaine and Limousin cattle were considerably stronger (0.54-0.78), suggesting the use of linear muscularity trait scores as indirect criteria for genetic improvements in carcase conformation grade.

CONCLUSION

There is evidence for genetic variation being expressed in the linear type traits recorded in the Hungarian Charolais population - particularly those relating to the muscularity and skeletal attributes – that could facilitate genetic improvements in animal appearance. Although moderately correlated with live weight, these linear type traits may also provide indirect predictors of genetic merit for carcase conformation. This might allow Hungarian breeders to better target the European carcase grading system beyond a weight-based breeding goal.

ACKNOWLEDGMENTS

The authors acknowledge the National Association of Hungarian Charolais Cattle Breeders and their members in providing access to the data used.

REFERENCES

Aguilar I., Misztal I., Johnson D.L., Legarra A., Tsuruta S. and Lawlor T.J. (2010) *J. Dairy Sci.* 95: 743.

Berry D.P., Pabiou T., Fanning R., Evans R.D. and Judge M.M. (2019) *J. Anim. Sci.* 97: 2329.

Bonfatti V., Albera A. and Carnier P. (2013) *J. Anim. Sci.* 91: 2057.

Bouquet A., Fouilloux M.N., Renand G. and Phocas F. (2010) *Livest. Sci.* 129: 38.

Chang C.C., Chow C.C., Tellier L.C., Vattikuti S., Purcell S.M. and Lee J.J. (2015) *GigaSci.* 4: 13742.

Donoghue K.A. and Bertrand J.K. (2004) *Livest. Prod. Sci.* 85: 129.

Doyle J.L., Berry D.P., Walsh S.W., Veerkamp R.F., Evans R.D. and Carthy T.R. (2018) *J. Anim. Sci.* 96: 1628.

El-Saied U.M., de la Fuente Ruiz L., Tirados F.S.P. and Rodríguez R. (2006) *Span. J. Agric. Res.* 2: 146.

Graser H., Tier B., Johnston D., and Barwick S.A. (2005) *Anim. Prod. Sci.* 45: 913.

Herrera-Ojeda J.B., Parra-Bracamonte G.M., Lopez-Villalobos N., Martínez-González J.C., Magaña-Monforte J.G., Morris S.T. and López-Bustamante L.A. (2019) *J. Appl.Gen.* 60: 385.

Institut de L'Elevage (2014) *Guide pratique du pointage des bovins de race a viande, du servrage a l'age adulte.*

Johnston D.J., Ferdosi M.H., Connors N.K., Boerner V., Cook J., Girard C.J., Swan A.A. and Tier B. (2018) *Proc. World Congr. Genet. Appl. Livest. Prod.* 19: 269.

Misztal I., Tsuruta S., Lourenco D.A.L., Masuda Y., Aguilar I., Legarra A. and Vitezica Z. (2018) Manual for BLUPF90 family of programs, University of Georgia.

Phocas F. and Laloë D. (2004) *Livest. Prod. Sci.* 89: 121.

Quintanilla R., Laloe D. and Renand G. (2002) *Proc. World Congr. Genet. Appl. Livest. Prod.* 11: 367.

Rezende M.P.G., Malhado C.H.M., Biffani S., Carrillo-Tabakman J.A., Fabbri M.C., Crovetti A., Carneiro P.L.S. and Bozzi R. (2022) A*nimal* 16: 100528.

Sargolzaei M., Chesnais J.P. and Schenkel F.S. (2014) *BMC Genom.* 15: 1.

VALIDATION OF BREEDING VALUES FOR ROBUSTNESS IN AUSTRALIAN MERINOS

D.L. Waters[1], J.H.J. van der Werf[1], D.J. Brown[2], S.F. Walkom[2] and S.A. Clark[1]

[1]School of Environmental & Rural Science, University of New England, Armidale, NSW, 2351 Australia
[2]Animal Genetics and Breeding Unit*, University of New England, Armidale, NSW, 2351 Australia

SUMMARY

Livestock production often involves raising animals in environments which can vary substantially between locations and years. It could be beneficial to select animals that have genetic merit which is more robust to environment variation, rather than animals that are more sensitive. This study attempts to validate breeding values for robustness estimated using reaction norm models. Reaction norm models were used to regress breeding values for body weight across different growth environments in the Information Nucleus Flock. The same model was fit to MERINOSELECT data, and the rank-correlation for EBVs of sires with progeny in both datasets was calculated. The pattern of genetic variance and heritability across environments was very similar between datasets. The rank correlation of breeding values for a subset of sires with the best distribution of progeny in both populations was 0.60, 0.22 and 0.17 for the intercept, slope and scale-corrected slope, respectively. The results indicated that the genetic variation in robustness across growth environments was, to some extent, repeatable across the two datasets. Genotypes that re-ranked more in the INF/RF also tended to re-rank more in MERINOSELECT, although the relationship was weak. The analysis could benefit from the inclusion of genomic data to increase linkage across environments and between datasets.

INTRODUCTION

Genotype-by-environment (GxE) interactions occur when the effect of an animal's genotype is dependent on the environment it exists in. This can result in variation between individuals in the robustness of their genetic effect to different environments. In extensive livestock systems where environments can vary substantially between years, genotypes that consistently rank highly for important traits across environments (i.e., robust genotypes) could be more valuable than sensitive genotypes who tend to change in rank.

Reaction norm (RN) models have been used widely to study GxE and rank individuals based on their robustness to environmental variation. Unlike univariate models, RN models allow the estimated breeding value (EBV) of genotypes to change as a function of an environmental covariable (EC), which describes the quality of the environment. When a linear function is used, the change in EBV across the EC is given by the slope. The slope can be directly used as an EBV for how robust the performance of a genotype is across the EC, while the slope can be 'scale-corrected' to yield an EBV for how much a genotype re-ranks across the EC (Waters *et al.* 2022).

Although some research has demonstrated RN models can increase the accuracy of phenotypic predictions (Oliveira *et al.* 2018; Mota *et al.* 2020), it could also be useful to explore whether RN variance components and individual breeding values (e.g., intercept, slope and scale-corrected slope) are repeatable in independent, but genetically linked populations. This would provide some guidance on how reliable EBVs based on linear RN models might be if applied in practice to select for robustness. The Australian Sheep CRC Information Nucleus Flock (INF) and the Meat and

* A joint venture of NSW Department of Primary Industries and the University of New England

Livestock Australia Resource Flock (RF) contain the progeny performance of sires across several locations representative of the Australian sheep environments (Van der Werf *et al.* 2010). Many of these same sires also have progeny recorded across several locations and years in the wider MERINOSELECT population (Brown *et al.* 2007). This data structure presents the opportunity to investigate such a question.The aim of the study was to investigate whether RN breeding values for the robustness of performance in post-weaning weight to different growth environments in the INF/RF can be validated in the MERINOSELECT data.

MATERIALS AND METHODS

The analysis consisted of two parts. The environmental covariable (EC) was first estimated for each animal in the INF/RF and MERINOSELECT data. Reaction norm models were then fit separately to both datasets to estimate breeding values for robustness across growth environments, which were then compared for sires with the most progeny in both data sets. All models were fit using ASReml 4.2 (Gilmour *et al.* 2021).

Estimate the EC. MERINOSELECT and INF/RF animals with a weaning weight (WWT) recorded between 50-120 days of age and a post-weaning weight (PWT) recorded between 120-329 days of age, along with a recorded sire and dam were extracted. Animals were excluded from the analysis if they were born or reared as quadruplets or greater, and if the age of dam was more than 12 years old at the time of recording. Contemporary groups were formed based on a flock × year × management group combination and required at least 15 animals from at least 3 different sires.

The best linear unbiased estimation (BLUE) of the post-weaning growth rate (PWGR) of each contemporary group was used as the EC for each animal. PWGR was calculated as the difference between PWT and WWT measurements, divided by the number of days between the measurements and expressed in grams per day. Animals with less than 40 days between WWT and PWT measurements were removed, along with animals deviating more than 3 standard deviations (SD) from their contemporary group mean for PWGR. This left 12,087 and 277,060 animals in the INF/RF and MERINOSELECT populations respectively. An animal model with PWGR as the response variable was fit jointly to the INF/RF and MERINOSELECT data to obtain a BLUE of PWGR for each contemporary group, forming the EC. The EC was centred to a mean of zero.

Independent reaction norms. Contemporary groups more than 3 SD from the population mean EC, and individuals more than 3 SD from their contemporary group mean PWT were excluded from the analysis. To reduce the number of uninformative animals in the MERINOSELECT data, contemporary groups were only included if they contained 1) at least one direct progeny of a sire with progeny or grand-progeny in the INF/RF, or 2) more than 25% of the animals were related to the INF/RF, with the minimum relationship being a grandsire with grand-progeny in the INF/RF. This left 11,638 and 206,733 animals in the INF/RF and MERINOSELECT data, respectively.

The linear RN models were of the form:

$$\mathbf{y} = \mathbf{Xb} + \mathbf{Z_1 a_0} + \mathbf{Z_2 a_1} + \mathbf{Z_3 c} + \mathbf{Q_g} + \mathbf{e} \qquad (1)$$

Where $\mathbf{y}$ is the vector of PWT records, $\mathbf{X}$ is an incidence matrix for the fixed effects $\mathbf{b}$, $\mathbf{Z_1}$ and $\mathbf{Z_2}$ are matrices relating records to the additive genetic effects for the intercept ($\mathbf{a_0}$) and slope ($\mathbf{a_1}$) respectively, $\mathbf{Z_3}$ is an incidence matrix relating records to the additive maternal effects ($\mathbf{c}$), which were estimated independently of the additive genetic effects, $\mathbf{Q}$ is a matrix of the proportion of each animal's genome originating from 451 genetic groups, $\mathbf{g}$ is the vector of random genetic group effects, and $\mathbf{e}$ is the vector of the residual effects. Fixed effects included age at measurement, birth type and rear type interaction, sex, and contemporary group. The residual variance was estimated independently at four and six intervals along the EC for the INF/RF and MERINOSELECT population respectively. The variance of the intercept and slope was modelled as follows:

$\begin{bmatrix}\mathbf{a_0}\\ \mathbf{a_1}\end{bmatrix} \sim N(0, \mathbf{A} \otimes \mathbf{K})$ where $\mathrm{K}=\begin{bmatrix}\sigma^2_{a0} & \sigma_{a1a0}\\ \sigma_{a0a1} & \sigma^2_{a1}\end{bmatrix}$ and **A** is the pedigree relationship matrix. The genetic variance across the EC was obtained using $G = \boldsymbol{\Lambda}\mathrm{K}\boldsymbol{\Lambda}'$, where $\boldsymbol{\Lambda}$ contained two columns; the first was a vector of 1's, and the second was a vector of EC values. The heritability of PWT at a given EC level was obtained by dividing the genetic variance at the EC by the sum of the genetic, maternal and residual variance. Scale-corrected EBVs for the slope were estimated using a genetic regression (Waters *et al.* 2022), which makes the slope EBVs independent of the intercept EBVs.

RESULTS AND DISCUSSION

Animals were normally distributed across the EC in both datasets (Figure 1), although there was a larger range in MERINOSELECT. The pattern of genetic variance and heritability across the EC was very similar between the two datasets (Figure 1), although the genetic variance and heritability were slightly lower in MERINOSELECT. Overall, the RN models estimated very similar levels of GxE in both populations.

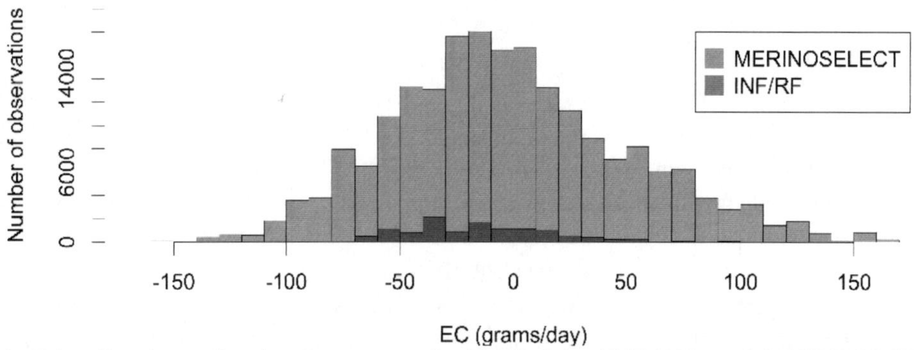

Figure 1. Distribution of animals across the EC in the INF/RF and MERINOSELECT data sets

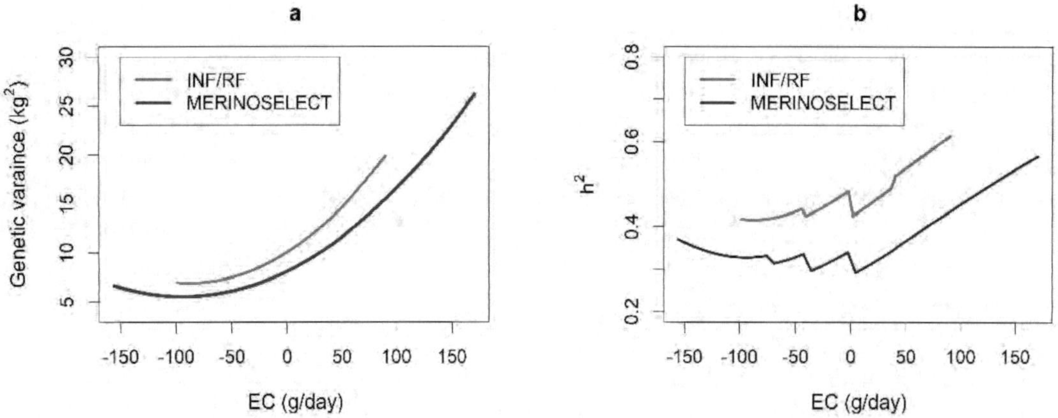

Figure 2. Genetic variance (a) and heritability (b) of PWT across the EC in INF/RF and MERINOSELECT data estimated using the independent reaction norm models

The rank-correlation of EBVs for sires with progeny in both populations was small but positive for the slope and scale-corrected slope (Table 1), and higher for the intercept. The difference in correlation between the intercept and slope is likely a function of the accuracy of the EBVs, as the intercept in generally easier to estimate accurately than the slope. The correlations were considerably higher when considering a subset of 56 sires with the best distribution of progeny in both populations, highlighting the importance of data structure when estimating RN parameters. Unlike the slope EBVs, the scale-corrected slope EBVs were uncorrelated with the intercept, so they represented the slope variation available for selection independent of the overall performance.

Table 1. Rank correlation of RN EBVs for sires with direct progeny in both populations (a), and a subset of 56 sires with progeny in at least 4 contemporary groups ranging by at least 50 g/day (b)

EBV	All sires (a)	Subset Sires (b)
Intercept	0.43	0.60
Slope	0.15	0.22
Scale-corrected slope	0.02	0.17

While these results imply that selection based on RN EBVs could yield a response in the robustness of performance across growth environments while simultaneously increasing the mean (intercept), the relationship between datasets was weak to moderate. This was most likely influenced by the distribution of progeny across the EC. To accurately estimate EBVs for the slope (robustness), sires require progeny across a wide range of EC values (Calus *et al.* 2004). Because only a relatively small number of sires had progeny widely distributed across the EC in both datasets, the power to detect a relationship between robustness in the two datasets was probably limited. Utilising genomic data to increase genetic linkage across the EC could help address this issue.

Overall, it appears that the success of breeding for robustness will be dependent on the structure of data available to estimate it accurately. If robustness is to be considered in genetic evaluations, breeders should be encouraged to ensure even stronger genetic linkage across years and locations. Other traits and environmental descriptors should also be explored to better understand the total variation available for selection of robustness.

CONCLUSION

The analysis demonstrated that the RN models estimated very similar levels of GxE across the two populations. The rank-correlation of EBVs for sires with the best distribution of progeny in both populations was low but positive for the slope EBVs. The results indicated that genetic variation in the RN slope was repeatable the two datasets, so selection based on these EBVs should lead to a response for robustness across growth environments. However, the relationship was not strong. The analysis could be improved by using more accurate EBVs for the slope, which could be achieved by increasing the linkage between environments with genomic data.

REFERENCES

Brown D.J.., Huismanm A.E., Swan A.A., Graser H.U., Woolaston R.R., Ball A.J., Atkins K.D. and Banks, R.G. (2007) *Proc. Assoc. Advmt. Anim. Breed. Genet.* 17: 187.

Calus M.P., Bijma P. and Veerkamp R.F. (2004) *Genet. Sel. Evol.* 36: 5.

Gilmour A.R., Gogel B.J., Cullis B.R., Welham S.J. and Thompson R. (2021) ASReml User Guide Release 4.2 Functional Specification, VSN International Ltd, Hemel Hempstead, HP2 4TP, UK

Mota L.F.M., Fernandes G.A. Jr, Herrera A.C., Scalez D.C.B., Espigolan R., Magalhães A.F.B., Carvalheiro R., Baldi F. and Albuquerque L.G. (2020) *Anim. Genet.* 51: 2.

Oliveira D.P., Lourenco D.A.L., Tsuruta S., Misztal I., Santos D.J.A., de Araújo Neto F.R., Aspilcueta-Borquis R.R., Baldi F., Carvalheiro R., de Camargo G.M.F., Albuquerque L.G., Tonhati H. (2018) *J. Anim. Sci.* 96: 1.

van der Werf J.H.J., Kinghorn B.P. and Banks R.G. (2007) *Anim. Prod. Sci.* 50: 998.

Waters D.L., Clark S.A., Moghaddar N. and van der Werf J.H.J. (2022) *Genet. Sel. Evol.* 54: 4.

MERITS OF USING NEW INTRAMUSCULAR FAT MEASUREMENT TECHNOLOGIES IN GENETIC EVALUATION OF AUSTRALIAN LAMB

P. Alexandri[1,3], S.F. Walkom[1,3], S. Stewart[3,4], P. McGilchrist[2,3], C. Steel[2,3] and D.J. Brown[1,3]

[1] Animal Genetics Breeding Unit*, University of New England, Armidale, NSW, 2351 Australia
[2]School of Environment and Rural Science, University of New England, Armidale, NSW, 2351 Australia
[3]Advanced Livestock Measurement Technologies project, Meat & Livestock Australia, North Sydney, NSW, 2060 Australia
[4]School of Science, Health and Engineering, Murdoch University, WA, 6150, Australia

SUMMARY

This study investigated the genetic association of intramuscular fat predicted with the MEQ probe (MEQIMF) and the SOMA NIR device (SOMAIMF) with Near Infra-Red analysed intramuscular fat (IMF%), tenderness, carcass eye muscle, fat and tissue depth. MEQ and SOMA NIR predicted IMF have only just became available to Australian processors, with data on genetic resources limited to 1,380 and 1,320 records, from research and seedstock flocks, respectively. Genetic analysis showed that MEQIMF has a moderate heritability (0.42 ± 0.1) and a high genetic correlation (0.95 ± 0.07) with chemical intramuscular fat. Similarly, SOMAIMF was estimated to have a moderate heritability (0.42 ± 0.1) and a strong genetic correlation with IMF% (0.94 ± 0.03). The results of the genetic analysis for IMF measured with the new technologies are likely to facilitate identifying the high intramuscular fat carcasses and in turn animals that have genetically superior eating quality.

INTRODUCTION

Eating quality in lamb is positively influenced by intramuscular fat, which has been found to increase tenderness, flavour and juiciness (Stewart *et al.* 2021). It is accepted that animals with higher levels of intramuscular fat produce meat which will be favoured by consumers (Pannier *et al.* 2014). Negative genetic correlations between intramuscular fat and lean meat yield (Gardner *et al.* 2018) also suggest that selection to improve the later needs to be undertaken with consideration for eating quality, because of its genetic correlation with intramuscular fat (Mortimer *et al.* 2018). Unlike beef, there is no visual marble score routinely used in the grading of lamb carcasses, with intramuscular fat percentage records (IMF%) in the national genetic evaluation determined by applying chemical analysis laboratory methods, which are time consuming and expensive. New technologies for measuring intramuscular fat objectively can facilitate adoption of Meat Standards Australia (MSA) grading in lamb (Pannier *et al.* 2014) because they offer fast, cheap, objective, on chain and non-destructive methods to measure the trait. For this study, two new technologies: i) the Meat Eating Quality (MEQ) probe (Carbone 2022), and ii) the SOMA Near Infra-Red (NIR) device were evaluated. The aim was to investigate the genetic relationship between lamb intramuscular fat measurements obtained with the MEQ probe and the SOMA NIR device, with IMF%, and, where possible, with other eating quality metrics (e.g. shear force) and carcass traits.

MATERIALS AND METHODS

Chemical IMF data. Eating quality and carcass traits were collected from 32,735 Merino and Merino-crossed lambs from the MLA Resource Flock (RF) and from seedstock ram breeding flocks. Mean lamb age was 264 (±76) days. Traits included intramuscular fat percentage (IMF%), shear

* A joint venture of NSW Department of Primary Industries and the University of New England

force 5 days after slaughter (SF5), eye muscle (*M. longissimus thoracis et lumborum* (LL)) depth (CEMD), fat at 45 mm from spine midline over the 12th rib (c site, CFAT) and total tissue depth measured at the 12th rib (GRFAT). Carcass traits were measured after slaughter in commercial abattoirs according to the procedure described by Mortimer *et al.* (2018). The percentage of intramuscular fat (IMF%) at the eye muscle was determined using a near infrared procedure (NIR) as described by (Perry *et al.* 2001). Shear force (SF5) at 5 days after slaughter was measured on a section of the LL as described by Hopkins *et al.* (2010).

MEQ probe data. For a subset of 1,380 of the above lambs, intramuscular fat was predicted using the MEQ probe (MEQIMF). MEQIMF was measured on the hot carcass where the MEQ probe was inserted in the area around the 13th rib and scans were completed to estimate intramuscular fat (Carbone 2022). The lambs with MEQIMF measures were born in 2021 and were measured between 2021 and 2022 (mean age at slaughter was 182 ±67 days) and originated from eight different flocks and 95 sires.

SOMA NIR data. SOMA NIR predicted intramuscular fat records (SOMAIMF) were collected from a different subset of the RF animals which included 1,307 lambs born in 2021 and measured between May and July 2022. The lambs were from the MLA resource flock and were progeny of 152 sires. They were slaughtered in commercial abattoirs, carcasses were chilled overnight (3 – 4 ^{o}C) and intramuscular fat was measured with the SOMA NIR device positioned directly over the surface of the loin at a cut between the 12th and 13th rib, based on the procedures described by Stewart *et al.* (2022). The number of animals and mean values for each trait and data set are illustrated in Table 1. Both MEQ probe and SOMA NIR device had previously been validated on independent data, not included in this study.

Table 1. Number of records (N) for each data set and mean trait values (standard deviation). HCWT: hot carcase weight, IMF%: chemical intramuscular fat percentage, MEQIMF: MEQ probe predicted IMF, SOMAIMF: SOMA NIR predicted IMF, SF5: shear force 5 days after slaughter, CEMD: eye muscle depth, CFAT: fat at the c-side, GRFAT: fat at the GR site

Data set	N	HCWT	IMF%	MEQ IMF	SOMA IMF	SF5	CEMD	CFAT	GRFAT
IMF%	32,735	23.63 (4.0)	4.49 (1.2)	-	-	32.40 (11.9)	30.93 (5.0)	4.37 (2.5)	14.02 (6.1)
MEQ probe	1,380	24.95 (4.4)	3.77 (1.0)	3.92 (1.0)	-	37.99 (14.1)	34.14 (4.6)	4.64 (2.2)	14.24 (6.0)
SOMA NIR	1,307	21.41 (3.6)	3.87 (1.1)	-	4.23 (1.1)	-	30.72 (5.0)	3.35 (2.0)	11.91 (5.2)

Statistical analysis. Variance components and genetic parameters for IMF, MEQIMF and SOMAIMF were estimated using a linear mixed model and REML methods with ASReml software (Gilmour *et al.* 2015). Fixed effects included type of birth, contemporary group, age of the animal and the age of dam (in days). The quadratic function of hot carcass weight was included to adjust all traits. The model also included the random effect of animal and genetic group (Swan *et al.* 2016). Maternal effects were not fitted since preliminary analysis showed they were non-significant. For all data sets, contemporary group was defined by breed, flock, management group, sex, date of measurement and kill group (Huisman *et al.* 2008).

To estimate genetic correlation and covariance of MEQIMF and SOMAIMF with other carcass and eating quality traits, a series of bivariate analyses were performed in ASReml. Due to convergence difficulties genetic groups were not fitted in the bivariate analysis and only animal was included in random effects.

RESULTS AND DISCUSSION

Heritability for MEQIMF and SOMAIMF was moderate (Table 2) and thus both traits display genetic variation and can be used effectively in selection. These estimates were similar to the heritability for IMF% data set (Table 1), which was also moderate (0.50 ± 0.03) and similar to estimates previously reported for the trait in Merino and Merino-cross lambs (Mortimer *et al.* 2010; Mortimer *et al.* 2014; Mortimer *et al.* 2018). Variance components of MEQIMF and SOMAIMF were consistent with those for IMF. However, smaller number of records in the MEQ probe and SOMA NIR data sets have limited ability to account for genetic groups. In this case more data is needed to clarify how these effects may impact variance estimates.

Table 2. Estimates of phenotypic ($\hat{\sigma}_p$), additive ($\hat{\sigma}_a$), and residual ($\hat{\sigma}_\varepsilon$) variance and heritability (h^2) for chemical IMF (IMF) and IMF predicted with MEQ probe (MEQIMF) and SOMA NIR device (SOMAINF). Variance components were estimated separately for each data set. Standard error in parentheses

Trait	Data	h^2	$\hat{\sigma}_p$	$\hat{\sigma}_a$	$\hat{\sigma}_\varepsilon$
IMF%	IMF%	0.50 (0.03)	1.12 (0.06)	0.57 (0.02)	0.37 (0.02)
MEQIMF	MEQ probe	0.42 (0.10)	0.61 (0.03)	0.25 (0.06)	0.35 (0.05)
IMF		0.71 (0.10)	0.77 (0.04)	0.55 (0.10)	0.22 (0.10)
SOMAIMF	SOMA NIR	0.42 (0.07)	0.81 (0.03)	0.34 (0.07)	0.47 (0.07)
IMF		0.51 (0.06)	0.93 (0.04)	0.48 (0.07)	0.45 (0.06)

Genetic correlations between MEQIMF and IMF, and between SOMAIMF and IMF were strong and positive (0.95 ± 0.07 and 0.94 ± 0.03, respectively), and suggest that both could be used as objective measurements to select for intramuscular fat in breeding programs.

Table 3. Genetic correlations between MEQIMF, SOMAIMF, IMF and other traits, with standard error in parentheses. MEQIMF: MEQ probe predicted IMF, SOMAIMF: SOMA NIR predicted IMF, IMF: chemical IMF, SF5: shear force 5 days after slaughter, CEMD: eye muscle depth, CFAT: c- side fat, GRFAT: GR site fat

	MEQ probe data	SOMA NIR data	Chemical IMF data
Trait	MEQIMF	SOMAIMF	IMF
IMF	0.95 (0.07)	0.94 (0.03)	-
CEMD	0.06 (0.21)	0.20 (0.11)	0.11 (0.03)
CFAT	0.32 (0.20)	0.40 (0.12)	0.20 (0.03)
GRFAT	0.35 (0.17)	0.22 (0.11)	0.20 (0.03)
SF5	-0.26 (0.17)	-	-0.39 (0.03)

Genetic correlations for MEQIMF and other carcass and eating quality traits in general were aligned to the ones estimated for IMF% (Table 3). Moderate genetic correlations of MEQIMF and

SOMAIMF have been observed for CFAT and GRFAT. These correlations were stronger than the ones estimated on the IMF% data set and higher than the ones previously observed by Mortimer *et al.* (2018) between CFAT, GRFAT and IMF%. The same authors reported slightly negative genetic correlations between IMF% and CEMD. In this study, the genetic correlation between IMFSOMA and CEMD was moderate positive and stronger than the correlation between IMF% and CEMD. On the other hand, the correlation between MEQIMF and CEMD was low but with high standard error, indicating more records are needed to determine the genetic relationship between these two traits. The genetic correlation between intramuscular fat and SF5 was moderate and negative for both the MEQ probe and IMF% data sets (Table3), and similar to estimates between IMF% and SF5 reported in previous studies (Mortimer *et al.* 2014). There was no correlation estimate for SOMAIMF and SF5 due to limited SF5 records for this cohort.

When more data becomes available, the genetic relationship between MEQIMF and SOMAIMF and other traits will be re-estimated, and their suitability to select for intramuscular fat will be re-assessed. More MEQIMF and SOMAIMF data will also help to define the capacity of the different technologies evaluated to predict intramuscular fat.

CONCLUSIONS

New technologies to measure intramuscular fat are becoming available and both MEQ probe and SOMA NIR device provide an opportunity to capture more intramuscular fat phenotypes as they provide a fast, cheaper and non-destructive alternative to laboratory procedures. The genetic variance and heritability of MEQ probe and SOMA NIR predicted intramuscular fat were generally similar to the ones observed for IMF% on the same animals. MEQIMF and SOMAIMF traits were found to be highly genetically correlated with IMF%, which suggests that intramuscular fat measured with the new technologies investigated for this study could be treated as the same trait as IMF% in genetic evaluation. More research is needed to determine the genetic association between MEQIMF and SOMAIMF, and other traits.

ACKNOWLEDGEMENTS

This work was part of the Advanced Measurement Technologies for globally competitive Australian meat project, funded by the Australian Government Department of Agriculture and Water Resources and a large number of industry partners. The authors thank the teams behind the Sheep CRC Information Nucleus Flock and MLA resource flock.

REFERENCES

Carbone R. (2022) Available at: https://www.mla.com.au/research-and-development/reports/2022/real-time-meat-eating-quality-probe-technology-refinement-and-commercialisation.

Gardner G.E., Starling S., Charnley J., Hocking-Edwards J., *et al.*. (2018) *Meat Sci.* 144: 91.

Gilmour A.R., Gogel B.J., Cullis B.R., Welham S.J. and Thompson R. (2015) *VSN International Ltd, Hemel Hempstead, UK*

Huisman A.E., Brown D.J., Ball A.J. and Graser H.-U. (2008) *Aus. J. Exp. Ag.* 48: 1177.

Mortimer S.I., Fogarty N.M., van der Werf J.H.J., Brown D.J., *et al.* (2018) *J. Ani. Sci.* 96: 3582.

Mortimer S.I., van der Werf J.H.J., Jacob R.H., Hopkins D.L., *et al.* (2014) *Meat Sci* 96: 1016.

Mortimer S.I., van der Werf J.H.J., Jacob R.H., Pethick D.W.,*et al.* (2010) *Ani. Prod. Sci.* 50: 1135.

Pannier L., Gardner G.E., Pearce K.L., McDonagh M., *et al.* (2014) *Meat Sci.* 96: 1076.

Stewart S.M., Corlett M., McGilchrist P., Steel C. and Gardner G.E. (2022) *An Advanced measurement technologies for globally competitive Australian meat project. 3.24.3.*

Stewart S.M., Lauridsen T., Toft H., Pethick D.W., *et al.* (2021) *Meat Sci.* 181: 108358.

Swan A.A., Brown D.J. and van der Werf J.H.J. (2016) *Ani. Prod. Sci.* 56: 87.

MERITS OF USING DEXA TO MEASURE LEAN MEAT YIELD FOR THE GENETIC EVALUATION OF AUSTRALIAN LAMB

S. F. Walkom[1,3], P. Alexandri[1,3], S. Connaughton[2,3], G. Gardner[2,3], A.Williams[2,3] and D.J. Brown[1,3]

[1]Animal Genetics Breeding Unit*, University of New England, Armidale, NSW, 2351 Australia
[2]School of Science, Health and Engineering, Murdoch University, WA, 6150, Australia
[3]Advanced Livestock Measurement Technologies project, Meat & Livestock Australia, North Sydney, NSW, 2060 Australia

SUMMARY

Dual-energy X-ray absorptiometry (DEXA) is rapidly gaining acceptance as a reference method for analysing body composition. Since initial developments in 2017, as part of the Advanced Livestock Measurement Technologies project, there has been an influx of DEXA measurements and some additional computed tomography (CT) measurements on genetically informative animals via the MLA funded Resource Flock and companion industry satellite flocks. Although more data is required, the results suggest that the DEXA lean meat yield is likely to be the same genetic trait as the CT measured lean meat yield. These results are promising and plans regarding the utilisation of DEXA data within Sheep Genetics national evaluation should begin. However, improving hook tracking technologies and data transfer pathways concurrently is also required.

INTRODUCTION

The financial value of a carcase is influenced by its saleable meat yield, which differs across supply chains, markets and cutting specifications. Historically, consumer preferences in domestic and international markets has driven the industry to produce meat cuts that are larger and leaner (Banks 2002). Terminal sheep breeders in Australia have been able to sustain genetic gains over a long period (Swan *et al.* 2017), partly due to breeding objectives targeting increased growth and lean meat yield. These traits can be accurately evaluated from a young age using selection indexes based on body weight, along with eye muscle and fat depth scanned on live animals (Swan *et al.* 2015). Due to a limited supply of carcase recording in seedstock flocks, the majority of genetic gain achieved in lamb lean meat yield has been reliant on a correlated response from index selection (Swan *et al.* 2015). This has driven interest and research funding to develop carcase based lean meat yield measuring technology within the supply chain.

Dual-energy X-ray absorptiometry (DEXA) has recently been accredited for commercial use in Australian lamb abattoirs for predicting carcase lean %. This accreditation is based upon its capacity to predict the carcase lean% reference standard measured using computed tomography (CT). This has excellent synergy with the existing Sheep Genetics databases in Australia which offer a lean meat yield breeding value that is also based upon the CT measurement of whole carcase lean% and is more cost effective and easier to implement within the processing environment.

Since initial developments in 2017, part of the Advanced Livestock Measurement Technologies (ALMTech) project (Gardner *et al.* 2021), DEXA technology and the algorithms behind the conversion of the DEXA image to measures of lean, bone and fat have been updated (Connaughton and Gardner 2023). Coinciding with these recent developments, there has been an influx of DEXA measurements and some additional CT measurements on genetically informative animals via the MLA funded Resource Flock (van der Werf *et al.* 2010) and companion industry satellite flocks.

---

* A joint venture of NSW Department of Primary industries and the University of New England

This study determines the genetic variation and the suitability of using DEXA lean meat yield as part of the National Genetic Evaluation, alongside or in conjunction with current CT lean meat yield records.

MATERIALS AND METHODS

Data. The analysis utilised carcase and lean meat yield records collected as part of the MLA funded Resource Flock and its previous iteration, the Information Nucleus Flock (van der Werf *et al.* 2010). As part of the broader Resource flock project, data was also collected on commercial (seedstock, non-research) animals as satellite flocks to the MLA Resource flock (Alexandri *et al.* 2022). This process involved animals from two sources: i) surplus animals – animals not selected for breeding based on phenotypic or genetic performance, and ii) structured progeny test – where dams were joined to sires to generate progeny for phenotyping. Consequently, to date, approximately 44 thousand lambs, of primarily a Merino ewe base but including both pure maternal and terminal breeds and their Merino cross progeny have been slaughtered and phenotyped. Carcases were measured for key carcase characteristics, including but not limited to carcase eye muscle depth (CEMD), carcase fat depth at the c-site (CFAT) and chemical intramuscular fat (CIMF) (Table 1).

As a component of the larger project, a sub-section of lambs was measured for lean meat yield via CT and/or DEXA. The CT records on lambs were primarily observed on a subset of the Resource Flock animals recorded since 2007, for a total of 3,646 carcases. The CT scanned lamb carcases represented 22 different sire breeds and 936 sires, with a mean CT lean of 57.8% (SD = 3.5). DEXA measurements were primarily collected on the accompanying satellite flocks. Consequently, only 1,018 carcases (320 sires represented) from the Resource Flock had both a CT lean and DEXA lean record. There were 4,104 lamb carcases recorded via DEXA representing 750 sires and 22 sire breeds. The mean lean meat yield from DEXA was 55.7 (SD = 5.4).

Table 1: Summary of carcase and lean meat yield records analysed within this study (count of contemporary groups = CGs)

Trait	Records	CGs	Sires	Mean	SD
Carcase Eye Muscle Depth (mm)	37,278	1,341	2,646	31.0	5.1
Carcase Fat Depth (mm)	36,624	1,328	2,623	4.3	2.4
Chemical Intramuscular Fat (%)	33,874	1,298	2,634	4.5	1.2
CT lean meat yield (%)	3,646	212	936	57.8	3.5
DEXA lean meat yield (%)	4,104	86	750	55.7	5.4

Statistical Analysis. The DEXA and CT lean meat yield records were analysed using univariate models in ASReml (Gilmour *et al.* 2015). Genetic correlations between the lean meat yield technologies (DEXA and CT) with a subset of carcase traits, carcase eye muscle depth (CEMD), carcase c-site fat depth (CFAT) and intramuscular fat (IMF), were estimated from a series of bi-variate models in ASReml.

The analyses were carried out with an animal model that incorporated all pedigree available on phenotyped animals within the LAMBPLAN database (Brown *et al.* 2007). Maternal effects were not fitted within this analysis, as is the standard approach for carcase traits in the LAMBPLAN analysis. Fixed effects in the model included birth type, age, age of dam (linear and quadratic covariates) and sire breed. The bi-variate analysis between trait pairs were completed with hot carcase weight fitted as a covariate to all carcase traits. Contemporary group was fitted as a sparse fixed effect and defined by flock, management group, sex, date of measurement and kill group (Huisman *et al.* 2008). The model did not include genetic group effects to avoid issues with analyses converging due to the small number of records. Due to the low number of records and the diversity of breeds and genetic makeup represented in the sires, the inability to correctly account for genetic group effects is likely to lead to some inflation of the heritability estimates.

RESULTS AND DISCUSSION

Heritability estimates for the DEXA and CT lean meat yield measures were high and similar, 0.51 and 0.50, respectively (Table 2). Including hot carcase weight as a covariate resulted in a slight increase in heritability for both traits. Estimates within this study are consistent with previous heritability estimates of CT measured lean meat yield, where moderate to high heritabilities were reported in Charolais (0.47), Suffolk (0.45), Texel (0.46; Jones *et al.* 2004), Norwegian White (0.57; Kvame and Vangen 2007) and Scottish Blackface (0.48; Karamichou *et al.* 2006). Heritability for CT lean meat yield, in a smaller subset of this population, has previously been reported as 0.53 (0.63 if carcase weight fitted as a covariate) (Walkom *et al.* 2021). Unfortunately, whilst the heritability and variances observed are similar (Table 2) between the two technologies, the small number of animals recorded with both is a limitation, and further examination is required to be able to declare that lean meat yield technologies are interchangeable in the genetic evaluation.

Table 1. Estimates of phenotypic (σ_p^2), additive (σ_a^2) and residual (σ_e^2) variance and heritability (h^2) for DEXA and CT recorded lean meat yield (LMY). Standard error in parentheses

Trait	Model	h^2	σ_p^2	σ_a^2	σ_e^2
DEXA LMY		0.51 (0.06)	5.36 (0.13)	2.71 (0.34)	2.65 (0.29)
DEXA LMY	HCWT co-variate	0.58 (0.06)	3.99 (0.10)	2.32 (0.27)	1.67 (0.22)
CT LMY		0.50 (0.06)	5.82 (0.15)	2.88 (0.39)	2.94 (0.34)
CT LMY	HCWT co-variate	0.54 (0.06)	4.90 (0.13)	2.63 (0.33)	2.27 (0.29)

The phenotypic correlation between DEXA and CT lean meat yield was 0.81 ± 0.01, but as highlighted, this is based on only 1,018 carcases. The corresponding genetic correlation between lean meat yield recorded with the two technologies was 0.87 ± 0.03 (Table 3). The correlation is very high but significantly different from each other, suggesting that there may be differences in how lean meat yield is measured across the two technologies despite the fact that DEXA has been trained to predict the CT measurement. However, this discrepancy may also be due to differences in samples measured by each method and the low number of sires with significant numbers of progeny recorded for both traits.

Genetic correlations between the two lean meat yield measures and a subset of key carcase traits are relatively consistent between the two technologies for estimating LMY (Table 3). The similarity of genetic correlations with the other carcase traits suggests that whilst the two technologies have primarily been recorded on separate sub-populations, they seem to capture the genetic (co)variation in lean meat yield consistently.

To make use of commercially available DEXA data it will be crucial to ensure that these records are correctly linked to the corresponding animal. This can be challenging in an abattoir environment where routine processing practices (ie. retain for trimming) can affect carcase sequences and identification. Therefore, until hook tracking is reliably implemented in plants with DEXA, collection of these data should be observed by technical staff to ensure animals' identities are correctly linked to the carcase and DEXA data. To assist with quality control, all consignments should have pre-slaughter weights and condition scores immediately prior to the kill.

Table 3. Genetic correlations between DEXA LMY and CT LMY and other carcass and meat quality traits. Standard error in parentheses. HCWT: hot carcase weight, IMF: chemical intramuscular fat, CEMD: eye muscle depth, CFAT: fat at the c-site

Trait	DEXA LMY	CT LMY
CT LMY	0.87 (0.03)	-
CEMD	0.36 (0.06)	0.46 (0.05)
CFAT	-0.60 (0.05)	-0.63 (0.04)
IMF	-0.34 (0.05)	-0.37 (0.04)

CONCLUSION

Although more data is required, very high genetic correlations suggest that the DEXA lean meat yield is likely to be the same trait as the CT measured lean meat yield. These results are promising and plans regarding the utilisation of DEXA data within Sheep Genetics national evaluation should begin. Rigorous data collection protocols are also required to ensure efficient collection of accurate data.

ACKNOWLEDGEMENTS

This work was part of the Advanced Measurement Technologies for globally competitive Australian meat project, funded by the Australian Government Department of Agriculture and Water Resources and many industry partners. The authors thank the teams behind the Sheep CRC Information Nucleus Flock, MLA resource flock and industry flocks which contributed to this study.

REFERENCES

Alexandri P., Walkom S.F., Swan A.A., van der Werf J.H.J. and Brown D.J. (2022) *Proc. World Cong. Genet. App. Livest. Prod.* 12

Banks R.G. (2002) *Wool Tech. Sheep Breed.* 50: 584.

Brown D.J., Huisman A., Swan A.A., Graser H-U, Woolaston R., Ball A., Atkins K.D. and Banks R.G. (2007) *Proc. Assoc. Advmt. Anim. Breed. Genet* 17: 187.

Connaughton S. and Gardner G.E. (2023) Report on DXA Accreditation algorithm update. *An Advanced measurement technologies for globally competitive Australian meat project. 3.11.5c*

Gardner G.E., Apps R., McColl R. and Craigie C.R. (2021) *Meat Sci.* 181: 108556.

Gilmour A., Gogel B., Cullis B., Welham S. and Thompson R. (2015) *Hemel hempstead: VSN International Ltd*

Huisman A., Brown D.J, Ball A. and Graser H.-U. (2008) *Aus. J. Exp. Ag.* 48: 1177.

Jones H., Lewis R., Young M. and Simm G. (2004) *Livest. Prod. Sci.* 90: 167.

Karamichou E., Richardson R., Nute G., McLean K. and Bishop S. (2006) *Anim. Sci.* 82: 151.

Kvame T. and Vangen O. (2007) *Livest. Sci.* 106: 232.

Pietrobelli A., Wang Z., Formica C. and Heymsfield S. B. (1998) *Am. J. of Phys. Endocrinol. Metab.* 274: E808.

Scholz A.M., Bünger L., Kongsro J., Baulain U. and Mitchell A.D. (2015) *Anim.* 9: 1250.

Swan A.A., Banks R.G., Brown D.J. and Chandler H.R. (2017) *Proc. Assoc. Advmt. Anim. Breed. Genet* 22: 326.

Swan A.A., Brown D.J. and van der Werf J.H.J. (2015) *Anim. Prod. Sci.* 56: 87.

van der Werf J.H.J., Kinghorn B.P. and Banks R.G. (2010) *Anim. Prod. Sci.* 50: 998.

Walkom S.F., Gardner G.E., Anderson F., Williams A. and Brown D.J. (2021) *Meat Sci.* 181: 108524.

PROGRESS OF THE SOUTHERN MULTIBREED RESOURCE POPULATION: HARD-TO-MEASURE PHENOTYPES TO DRIVE GENOMIC SELECTION

B.J. Walmsley[1,2], K.L. Moore[1], S.F. Walkom[1], S.A. Clark[3], T. Granleese[4] and K.A. Donoghue[5]

[1]Animal Genetics Breeding Unit*, University of New England, Armidale, NSW, 2351 Australia
[2]NSW Department of Primary Industries, Livestock Industries Centre, Armidale, NSW, 2351 Australia

[3]School of Environmental and Rural Science, University of New England, Armidale, NSW, 2351 Australia

[4]NSW Department of Primary Industries, Grafton Primary Industries Institute, Grafton, NSW, 2460 Australia

[5]NSW Department of Primary Industries, Agricultural Research Centre, Trangie, NSW, 2823 Australia

SUMMARY

This paper describes the progress in the first half of a large, 5-year breeding project run across New South Wales involving five temperate beef breeds and the Brahman breed. The project's purpose is to generate up to 8,000 progeny that allows the benefits of genomic selection to be captured, particularly for traits that are lowly recorded due to being difficult or costly to record or which are yet to be routinely included in genetic evaluations, e.g., fertility, health, and resilience. The project has generated 4,886 progeny from three cohorts, with another 1,990 females to calve in mid-2023. Cohort one, born in 2020, has now had all steers complete feedlot finishing with carcass traits recorded, with the heifers having completed their first calving and subsequent rebreed. Details concerning the recording of hard-to-measure traits to this point in the project are provided. The high-density SNP genotypes collected, and the recording of these traits will contribute to the genomic reference populations and BREEDPLAN evaluations of the breeds involved.

INTRODUCTION

The potential exists to significantly increase profitability in the Australian beef industry using EBVs and selection indexes by capturing the benefits of genomic selection. In 2018, the BREEDPLAN genetic evaluation system implemented single-step GBLUP (Johnston *et al.* 2018), which was a significant step toward realising these gains. Achieving the full benefits of genomic selection is contingent on a number of other factors. The impacts effective population size, relative size of the reference population, trait heritability (Goddard and Hayes 2009), the relatedness within the reference population and its relatedness to selection candidates (Pszczola *et al.* 2011) have on the success of genomic selection have been well described. To successfully improve profitability, genomic selection must provide predictive accuracy for all traits that impact profitability and form the basis of current and future selection indexes. For this reason, the size of the reference population required is a function of not only the number of animals with genotypes and phenotypes but also the types of phenotypes recorded. This is particularly the case for traits that are difficult or costly to record or are economically important but yet to be routinely included in genetic evaluations, e.g., fertility, health, and resilience traits. The Southern Multi-breed project (SMB: Walmsley *et al.* 2021), initiated in 2020, and the Repronomics™ project (Johnston *et al.* 2017), initiated in 2013, are two industry research initiatives that have been developed to address

* A joint venture of NSW Department of Primary Industries and the University of New England

these needs across multiple breeds. The relatedness of the animals in the SMB project to the wider respective Australian breed populations has been examined by Moore *et al.* (2023). This paper provides a brief update on the progress of the SMB Project with a focus on the hard-to-measure and lowly recorded BREEDPLAN traits that are being evaluated as part of SMB.

BRIEF OVERVIEW

The SMB project is being conducted across the New South Wales Department of Primary Industries research facilities; Trangie Agricultural Research Centre; Grafton Primary Industries Institute; Tocal Agricultural Centre; Glen Innes Agricultural Research and Advisory Station; Elizabeth MacArthur Agricultural Institute (EMAI); Menangle and the University of New England (UNE) research feedlot, "Tullimba" (Kingstown). The project is focused on the five numerically largest temperate breeds (viz. Angus, Charolais, Hereford, Shorthorn, and Wagyu) in southern Australia and the Brahman breed, which is commercially relevant in the sub-tropics of NSW and links SMB to the Repronomics™ project (Johnston *et al.* 2017). A critical design feature is that all breeds are managed in mixed breed groups, which allows valid breed comparisons to be made. Walmsley *et al.* (2021) describe the design and initiation of the SMB project.

PROGENY GENERATED

The aim was to generate up to 8,000 progeny across the project's lifetime. To date, the project has generated progeny in 2020, 2021, and 2022, with a fourth cohort due to begin calving in mid-2023. The number of calves in cohorts 1 to 3 are shown in Table 1. The progeny were generated by a combination of artificial insemination (AI) programs followed by natural mate back-ups over the base cows and natural mate bulls over all female progeny generated by the project. In total, AI programs included 265 sires (Angus=74; Brahman=30; Charolais=30; Hereford=55; Shorthorn=33; Wagyu=43) with 141 natural mate sires used (Angus=46; Brahman=10; Charolais=16; Hereford=34; Shorthorn=15; Wagyu=20).

Table 1. Number of progeny generated by year (cohorts 1 to 3) and sex, in the Southern Multibreed project, and the number of pregnancies for cohort 4

	Progeny generated				Confirmed Pregnancies
Year	2020	2021	2022	Total	2023
Steers	710	784	965	2459	-
Heifers	694	758	975	2427	-
Overall	1404	1542	1940	4886	1990

TRAITS RECORDED

All calves are intensively recorded from birth following BREEDPLAN protocols for the current standard BREEDPLAN traits. These include birth (BW), weaning (WW), yearling (YW) and finished (FW) weights, calving ease (CE), ultrasound scan traits (eye muscle area (EMA), rib fat (RIB), rump fat (RUMP) and intramuscular fat (IMF)), carcass traits (Carcass Weight, EMA, RIB, RUMP and IMF) including retail beef yield (RBY), days-to-calving (DTC) and temperament (TEMP – docility and crush scores). Table 2 presents the number of records for these traits across all breeds for the first three cohorts. Birth weight and calving ease are the first traits recorded in an animal's life and as such have the largest number of records to date (n=4,880 and 4,886, respectively). Both traits relate to the probability of calf survival through the birthing process and as such have important impacts on profitability. Mean birth weight was 38 kg (±7.9 SD) and ranged between 8.5 kg and 69.5 kg, with 88% of records between 25 and 50 kg. Although the majority of calving ease scores were category 1, all five categories have been observed.

In addition to those in Table 2, other traits are important for improving profitability. The increasing importance of animal welfare and pressure to reduce dehorning means that poll status has the potential to be an important economic trait. Horn/poll status has been recorded at marking (n = 4,545), with assessments also conducted at weaning (n = 2,700) to capture late developing horns/scurs. Phenotypes have been observed for all horn/poll classifications described by Connors *et al.* (2021). Animal health traits that impact welfare and productivity have also been recorded. Worm egg counts (WEC) have been measured at weaning (n = 2,681) as well as prior to heifer joining and steers entering the feedlot (n = 2,487) in the first two cohorts. Figure 1 shows the average cube root transformed WEC for sites and years at weaning. Immune competence (Wilkie and Mallard, 1999) has been proposed as a trait that could increase general disease resistance through selection to reduce the incidence of diseases such as bovine respiratory disease. Currently, only cohort two has been recorded for immune competence (n = 1,412) at weaning.

Table 2. Number of progeny per cohort recorded* for current BREEDPLAN traits in the Southern Multibreed project for the first three cohorts

Cohort	BW	CE	WW	YW	FW	Scan	Carc.	RBY	DTC	TEMP.
2020	1402	1403	1291	1278	1248	1278	628	157	505	1282
2021	1541	1543	1412	1394	170	868	-	-	-	1138
2022	1937	1940	-	-	-	-	-	-	-	-

*See text for trait descriptions.

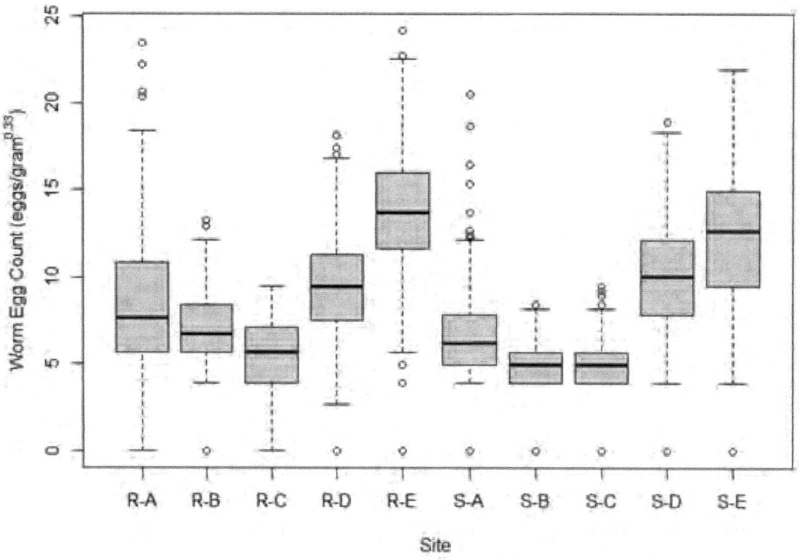

Figure 1. Average weaning worm egg count (cube root transformed) for each research site (designated A to E) and years (R-2020 and S-2021) for the first two cohorts of Southern Multibreed calves (n=2,681)

The importance of fertility as a driver of profitability, and the relatively low levels of DTC recording in the beef industry (Gudex and Millen, 2019) have created the need to examine new traits for inclusion in genetic evaluations. Regular ovarian scans have been conducted using real-time ultrasound to identify puberty in heifers or return to oestrous in first-lactation females by the presence of a *corpus luteum*. Currently, 1,321 heifers have had puberty assessments conducted, with 465 first-lactation females assessed for return to oestrous. Analysis of ovarian activity records is described by Donoghue *et al.* (2023). Prior to joining, heifers and first-lactation females have had body condition score assessments, hip height measurements and ultrasound body composition scans taken, following the protocols described by Wolcott *et al.* (2023), for evaluation as potential indicators of the capacity to maintain body condition during periods of high energy demand. McKiernan (1990) described a muscle score scale that has a significant association with RBY and can be used to assess live animals for increasing profitability. Progeny (n = 2,703) were assessed using this scale when ultrasound scanning has been conducted. Feed is the major direct cost in beef production meaning that profitability is a direct function of changes in feed efficiency. In an effort to address this steer feed intake has been recorded (n = 628) in the feedlot (Torres-Varquez *et al.* 2018).

CONCLUSIONS

The SMB project has produced 4,886 progeny in the first 2.5 years with another 1,990 pregnant females to calve in mid-2023. An extensive recording program has focused not only on traits routinely recorded in BREEDPLAN but also those which are difficult or costly to record or are yet to be routinely included in genetic evaluations, e.g., health and resilience. A significant body of high-quality data is being produced from the investment made by industry and government. This represents a valuable resource to benchmark across-breed performance and capture the benefits of genomic selection, particularly for hard-to-measure traits. As such, the project will enable more effective selection for those traits contributing to value chain profit.

ACKNOWLEDGEMENTS

The Southern Multi-Breed project is an initiative co-funded by the NSW Department of Primary Industries, the University of New England, and the Meat and Livestock Australia Donor Company. The authors give thanks to the efforts of the research station managers, and their staff, our skilled team of technicians, ultrasound scanners, AI, DNA lab, data, project managers, and scientists. Individual breeders, participating breed societies, and the beef industry are thanked.

REFERENCES

Connors N.K., Walmsley B.J. and Donoghue K.A. (2021) *Proc. Assoc. Advmt. Anim. Breed. Genet.* 24: 70.

Donoghue K.A., Rippon R., Wolcott M.L., Moore K.L., Clark S.A. and Walmsley B.J. (2023) *Proc. Assoc. Advmt. Anim. Breed. Genet.* 25: *These Proceedings.*

Goddard M.E. and Hayes B.J. (2009) *Nat. Review. Genet.* 10: 381.

Gudex B.W. and Millen C.A. (2019) *Proc. Assoc. Advmt. Anim. Breed. Genet.* 23: 488.

Johnston D.J. (2017) *Proc. Assoc. Advmt. Anim. Breed. Genet.* 22: 385.

Johnston D.J., Ferdosi M.H., Connors N.K., Boerner V., Cook J., Girard C.J., Swan A.A. and Teir B. (2018) *Proc. 11th World Congr. Genet. Appl. Livest. Prod.* Paper 269.

McKiernan W.A. (1990) *Proc. Assoc. Advmt. Anim. Breed. Genet.* 9: 447.

Moore K.L., Walkom S.F., Siddell J.P. and Walmsley B.J. (2023) *Proc. Assoc. Advmt. Anim. Breed. Genet.* 25: *These Proceedings.*

Pszczola M., Strabel T., Mulder H.A. and Calus M.P.L. (2012) *J. Dairy Sci.* 95: 389.

Torres-Varquez J.A., van der Werf J.H.J. and Clark S.A. (2018) *J. Anim. Sci.* 96: 4521.

Walmsley B.J., Donoghue K.A., Johnston D.J., Clark S.A., Siddell J.P., Walkom S.F., Granleese T. and Arthur P.F. (2021) *Proc. Assoc. Advmt. Anim. Breed. Genet.* 24: 423.

Wilkie B. and Mallard B. (1999) *Vet. Immunol. Immunopathol.* 72: 231.

Wolcott M.L., Johnston D.J., Jeyaruban M.G. and Girard C.J. (2023) *Proc. Assoc. Advmt. Anim. Breed. Genet.* 25: *These Proceedings.*

ESTIMATE THE GENETIC PARAMETERS AND ANALYSIS OF CULLING REASONS IN IRANIAN HOLSTEIN DAIRY CATTLE

H. Keshavarzi[1], E. Dehnavi[2] and A. Small[1]

[1]Agriculture and Food, Commonwealth Scientific and Industrial Research Organization (CSIRO), Armidale, NSW, 2350 Australia
[2]Animal Genetics Breeding Unit*, University of New England, Armidale, NSW, 2351 Australia

SUMMARY

Many animals are culled from the herd on dairy farms annually due to health problems, and this involuntary culling causes significant economic losses to the dairy industry. This study aimed to identify the trend of culling reasons and lifespan and estimate their genetic parameters. The cow history records and pedigree files of 11 large commercial dairy farms with cows culled from 1995 to 2015 were used. It is estimated that 18.6% of cows are culled voluntarily by farmers, while 81.4% of cows leave the herd involuntarily. Three main reasons for involuntarily culling were reproductive problems (25.9%), death and others (16.7%), and infectious diseases (14.3%). Over time, the distributions of culling reasons have altered with a reduction in "death and others", suggesting a better or more precise diagnosis of culling reasons and improvement in dairy farm recording systems. The average lifespan of cattle was 4.42 years with heritability of 0.14. The heritabilities of culling reasons were very low and ranged from 0.03±0.02 (metabolic and digestive disorders) to 0.08±0.03 (mastitis and udder problems). The significance of the maternal effect for some traits like mastitis indicates that it may be possible to improve an individual's health and, therefore, farm profitability genetically.

INTRODUCTION

Dairy cows are expected to remain economically useful in their herds for a much shorter period of time than the natural lifespan of cows, which is approximately 20 years. The length of lifespan from birth until culling (Hu *et al.* 2021) varies from 4.9 years in US (De Vries 2017), to 6.3 years in UK (Pritchard *et al.* 2013) or 6.75 years in Australia (Wondatir Workie *et al.* 2021). An increase in lifespan can increase profitability by reducing the annual costs of replacement of cows, which indicates the economic importance of lifespan for dairy farmers.

Farmers have several reasons for culling cows from their herds which can be generally classified as voluntary or involuntary culling (Weigel *et al.* 2003). Compared to voluntary culling, which is based on optimal economic decisions, the involuntary culling occurs when farmers have to remove their productive, profitable cows due to illness, injury, infertility, or death (Wondatir Workie *et al.* 2021). Due to improvements in genetic trends for fertility and adding health disorders such as mastitis as well as longevity in the selection index for dairy cattle, culling rate is expected to change over time. In addition, it might be possible that some genes related to health disorders that lead to culling in dairy cattle pass over the generations. The objective of this study was to estimate the trend and genetic parameters of lifespan and culling reasons of dairy cattle.

MATERIALS AND METHODS

The cow history records of 11 commercial dairy farms in Iran, which included cows culled between 1995 to 2015 were extracted from an on-farm record-keeping software. The variables extracted included herd, parity number, cow ID, birth date, culling date, culling reason, and the ID of sire and dam (for known parents). Data was edited by SQL Server Management Studio

---

* A joint venture of NSW Department of Primary Industries and the University of New England

(Microsoft, 2012). Cows with missing parity numbers, birth dates, culling dates, unknown dam, milk period >10, date of birth greater than their dams' birth date, or missing culling details were removed from the original dataset. Culling reasons were categorized into seven groups as described in Table 1. Cow lifespan was calculated as the interval between birth date and culling date. The final dataset used for this study was 67,287 records of 13,616 heifers and 53,671 cows. For the analysis of culling reasons and estimation of the genetic parameters, each culling reason was considered as a different phenotype as a binary variable (1 or 0) indicating whether a cow left the herd for that reason or not. The trend of culling reasons over time was plotted in R using ggplot2 package (Wickham 2016).

Using ASReml (Gilmour *et al.* 2015), a binomial model with a logit link function was applied to the dataset to estimate the genetic parameters of each culling reason. For lifetime, however, a continuous model was used in which the data were tested for normality using Shapiro-Wilk test and then log- transformed to approach normality. A range of systematic effects, including herd, year of birth, the season of birth, year of culling, season of culling and their interactions and milk period were tested for significance for each trait (results not shown). Four combinations of random effects for direct genetic, maternal genetic, and maternal permanent environmental effects were compared via univariate analysis for each trait separately. The covariance between direct genetic and maternal genetic effects was ignored. The four models including 1) direct animal genetic effect, 2) direct and maternal effect, 3) direct and maternal permanent environmental effect, and 4) model including all above random effects were tested and then compared using likelihood ratio tests (LRT) between the full and reduced models.

Table 1. Description of the culling reason (and their proportion) of used dairy cattle in this study

Group	Descriptions	Proportion (%)
Voluntary	Low milk production, old age, dairy purpose	18.6
Reproductive problems	Infertility, recurrent abortions, mummy (wax) abortion, stillbirth, ovarian cysts, uterus problems (rupturing, bleeding, infections, and diseases)	25.9
Feet and leg disorders	Lameness, joint infection, dislocation and fracture of the hands, legs and hip, crippling, hoof diseases and spinal cord injuries	7.8
Mastitis and udder problems	Mastitis, protracting and rupturing ligaments of the gland, complete teat-cistern obstruction, udder gangrene and bleeding	8.3
Metabolic and digestive disorders	Bloating, acidosis, ketosis, fatty liver, milk fever, displaced abomasum, obstruction and twisting of gut, omasum accumulation, abomasum and rumen, diarrhea	8.4
Infectious diseases	Leucosis, foot-and-mouth disease, brucellosis, pneumonia, tuberculosis, black leg, Bovine Johne's disease, lung and liver infections/abscess, Bovine Viral Diarrhea Virus (BVDV)	14.3
Death and others	Death, peritonitis, injury, blindness, toxication	16.7

RESULTS AND DISCUSSION

According to data available over 21 years, 18.6% of cows were culled voluntarily by farmers. This was less than that reported (27.1%) by Ghaderi-Zefrehei *et al.* (2017), who studied the culling reason in one farm in Iran. The main reason for involuntarily culling was reproductive problems (RP) which accounted for almost a quarter (25.9%) of culling reasons (Table 1). The other major involuntary reasons for culling the cows from the herd were "death and others" (D&O) (16.7%), and infectious diseases (14.3%) (Table 1). RP and infectious diseases have been reported as the most significant reasons of culling the dairy cows in Iran (e.g., Ghaderi-Zefrehei *et al.* 2017). The remarkable percentage of culling due to RP can be explained by the genetic selection performed on milk yield for many years and negative association exits between these traits (De Vries and Risco 2005). Regarding infectious diseases, Holstein cows are expected to be sensitive to some pathogens in Iran. Furthermore, this study is also included the heifers that have not calved and are mainly culled due to infectious diseases, and reproductive abnormalities not becoming evident until after first calving. The average lifespan was 4.42 years, which is close to US (4.9 years, De Vries 2017) and German dairy cattle (~5 years, Martens and Bange 2013), but lower than Australian cows (6.75, Wondatir Workie *et al.* 2021).

The trend of major involuntary culling reasons for Holstein cows over the period of 21 years is presented in Figure 1. Although there were fluctuations, the level of culling for RP remained high throughout the whole study period. Over the time, culling for D&O showed a downward trend, suggesting better, or more precise, diagnosis of culling reasons and improvement in dairy farm recording systems. There was an increase in involuntary culling of animals due to infectious diseases over time, with a sudden rise in 2002. Factors that may have led to this observation may be increasing the density of animals, which may result in disease spread; improved diagnosis of the culling reason over time (part of this group might come from the D&O group); decreased immunity caused by selection for low somatic cell counts; and the emergence of new diseases.

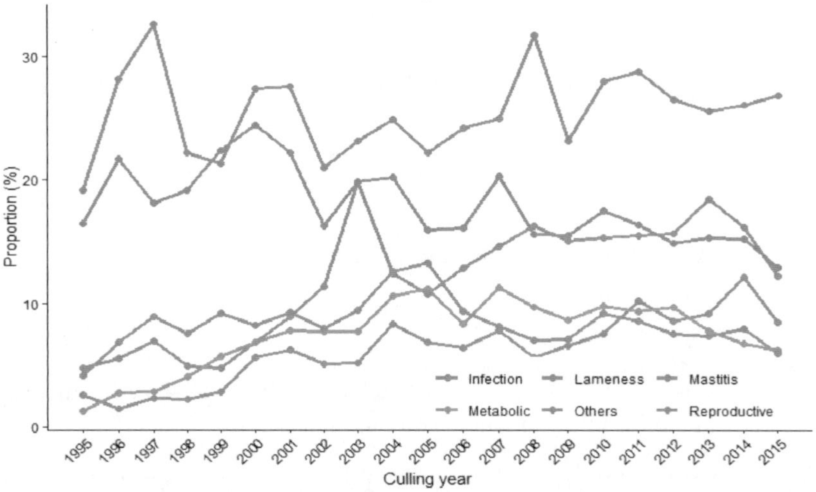

Figure 1. Proportion of involuntary culling reasons by year of culling

For all traits, maternal permanent environment effects were not significant (Table 2). For culling due to "mastitis and udder problems"(M&U), RP and D&O, the maternal genetic effect had small effects with significant likelihood ratio tests (LRTs), and because of the nature of dam effects on reproduction traits, it was retained in the model. Based on the results, heritabilities of culling reasons were low, ranging from 0.03 (metabolic and digestive disorders) to 0.08 (M&U) (Table 2). The heritability for lifespan was higher (0.14) which agrees with Van Pelt *et al.* (2015), however the definition of this trait differed (time from first calving to the last test date for milking production in Van Pelt *et al.* (2015) and time from birth to culling in this study). There is a lack of study on genetic parameters for culling reasons, however the heritability of some of these traits like clinical mastitis (0.01 to 0.42; Nash *et al.* 2000) and lameness (0.15 to 0.22; Weber *et al.* 2013) has been reported.

CONCLUSIONS

This study shows that 81.4% of culling is out of the farmer's control (involuntary culling). Over time, culling reasons have altered with a reduction in "death and others" suggesting the better or more precise diagnosis of culling reasons and improvement in dairy farm recording systems. Despite fluctuations, the rate of culling for Reproductive problems remained high throughout the study indicating the need for improving fertility management and consequently reproductive efficiency. Although the heritabilities of culling reasons were low, our results suggest that some opportunity may exist for genetic improvements in individual's health (e.g., mastitis and reproductive problems) in Iranian Holsteins and therefore improve animal welfare and farm profitability.

Table2. Genetic variance (σ^2_g), maternal variance (σ^2_m), direct heritability (h^2), maternal heritability (m^2) (and their standard error (SE)) and likelihood ratio tests (LRT) and degrees of freedom (df) for the selected model when running a univariate animal model

Trait [1]	σ^2_g (SE)	σ^2_m (SE)	h^2 (SE)	m^2 (SE)	LRT [2]	df [3]
F&L disorders (%)	0.17 (0.04)		0.05 (0.03)		0	0
M&U problems (%)	0.27 (0.04)	0.01 (0.03)	0.08 (0.03)	0.00 (0.02)	213 ***	1
M&D disorders (%)	0.11 (0.03)		0.03 (0.02)		0	0
Reproductive problems (%)	0.15 (0.02)	0.01 (0.01)	0.04 (0.02)	0.01 (0.01)	116 ***	1
Infectious diseases (%)	0.15 (0.02)		0.04 (0.02)		0	0
Voluntary (%)	0.30 (0.03)		0.08 (0.02)		0	0
Death and others (%)	0.16 (0.02)	0.02 (0.02)	0.05 (0.02)	0.00 (0.02)	149***	1
Lifespan (yrs)[4]	0.02 (0.00)		0.14 (0.01)		0	1

[1] All traits except lifespan were fitted in the binomial model on the logit scale (σ2e=3.29). F&L= Feet, and leg; M&U= Mastitis and udder; M&D = Metabolic and digestive.

[2] *** P <0.001, ** P <0.01, * P <0.05, P <0.1, ns or non-significant.

[3] df- the difference in the number of parameters between full and reduced models as 0, 1, 2 –for the base model (direct random effect), when maternal genetic effects or maternal permanent environmental effects or both were added.

[4] Lifestyle is reported as genetic standard deviation (σ_g) instead of genetic variance (σ^2_g).

REFERENCES

De Vries A. (2017) *J. Dairy Sci.* 100: 4184.

De Vries A., and Risco C.A. (2005) *J. Dairy Sci.* 88: 3155.

Ghaderi-Zefrehei M., Rabbanikhah E., Baneh H., Peters S.O. and Imumorin I.G. (2017) *J. Appl. Anim. Res.* 45: 524.

Gilmour A.R., Gogel B.J., Cullis B.R., Welham S.J., and Thompson R. (2015) ASReml user guide release 4.2 structural specifications. VSN International Ltd: Hemel Hempsted, HP1 1ES, UK.

Hu H., Mu T., Ma Y., Wang X., and Ma Y. (2021) *Front Genet.* 12: 695543.

Martens Ha. and Bange C. (2013) *Lohmann inf.* 48: 53.

Microsoft, 2012. http://www8.cs.umu.se/education/examina/Rapporter/EsquiviasFlarup.pdf.

Nash D.L., Rogers G.W., Cooper J.B., Hargrove G.L., Keown J.F. and Hansen L.B. (2000). *J. Dairy Sci.* 83: 2350.

Pritchard T., Coffey M., Mrode R. and Wall E. (2013) *Animal* 7: 34.

Weber A., Stamer E., Junge W. and Thaller G. (2013) *J. Dairy Sci.* 96: 3310.

Weigel K., Palmer R. and Caraviello D. (2003) *J. Dairy Sci.* 86: 1482.

Wickham H. (2016) 'ggplot2: Elegant Graphics for Data Analysis.' (Springer International Publishing: Imprint: Springer: Cham).

Wondatir Workie Z., Gibson J.P. and van der Werf J.H.J. (2021) *Anim. Prod. Sci.* 61: 680.

GENETIC PARAMETERS AND LAMBDA VALUES FOR POST-WEANING PRODUCTION TRAITS IN MERINO SHEEP

S. de las Heras-Saldana, P. Gurman, A.A. Swan and D.J. Brown

Animal Genetics Breeding Unit*, University of New England, Armidale, NSW 2351 Australia

SUMMARY

Appropriate genetic parameters are essential for accurate selection of animals with improved genetic merit for economically important traits. In recent years, Merino breeders have tended to record animals earlier in life making it important to characterise post-weaning measurements. Additionally, genomic information is used in Australian Merino genetic evaluations to obtain more accurate estimations of genetic merit using single-step GBLUP, utilising a weighting factor to partition polygenic and genomic variance, hereby referred to as lambda (λ). This study aimed to estimate genetic parameters and lambda values for production traits measured at the post-weaning stage in Merino sheep. Phenotypic records were obtained at the post-weaning stage for weight (PWT), eye muscle depth (PEMD), fat depth (PFAT), greasy fleece weight (PGFW), clean fleece weight (PCFW), fibre diameter (PFD), fibre diameter coefficient of variation (PDCV), and staple length (PSL). Genetic parameters were estimated with univariate and bivariate analyses, while a genomic REML analysis was performed to calculate the lambda value for each trait. Moderate to high heritability estimates were observed, ranging between 0.25 to 0.56. Genetic correlations were moderately positive between PWT and PCFW, PGFW, PFD, and PSL and negative for PDCV. Lambda values were on average (0.64) slightly higher than the current value used for genomic evaluation ($\lambda = 0.5$) and ranged from 0.51 to 0.90. Genetic parameters reported in this study are generally consistent with previous studies and will be used to update the genetic parameters used by Sheep Genetics for the MERINOSELECT analyses.

INTRODUCTION

The Australian sheep industry has significantly improved sheep production through the establishment of breeding programs. Genetic parameters are essential to accurately estimate breeding values and to predict the genetic and economic gain of the traits in breeding programs. Previous studies have shown that heritability increases as the age of measurement increases (Brown *et al.* 2013; Mortimer 2017), raising the need to estimate genetic parameters for each relevant stage of the developmental period. Previous studies estimated genetic parameters for live weight, ultrasound fat and muscle and wool traits at different stages in Merino sheep. Heritability estimates for wool traits were moderate for the yearling, hogget, and adult stages (Greeff *et al.* 2008, Brown *et al.* 2013, Mortimer *et al.* 2017), high for ultrasound traits (Mortimer *et al.* 2017) and moderate for live weight (Greeff *et al.* 2008; Mortimer 2017). However, Sheep Genetics recently revised the methods used to classify traits to each stage and redefined more accurate intervals for birth, marking (days 1-39), weaning (days 40-149), post-weaning (days 150-299), yearling (days 300-449), hogget (days 500-659), and adult (days 660-6059). These changes influence how the data is used in the analysis and therefore, it is necessary to estimate new genetic parameters to be used especially for the post-weaning stage as more data are now available.

In recent years, more accurate estimations of the genetic merit have been achieved by including genomic information in a single-step genomic BLUP (ssGBLUP). The use of this method requires a lambda (λ) for partitioning pedigree and genomic information. Moreover, Gurman *et al.* (2021)

* A joint venture of NSW Department of Primary Industries and University of New England

reported lambda values higher than 0.5 for carcass traits were required for ssGBLUP pointing to the importance of further studies. This project aims to estimate the genetic parameters, including heritabilities, correlations and lambda values for live weight, ultrasound, and wool traits recorded at the post-weaning stage in Australian Merino sheep.

MATERIALS AND METHODS

Animals and traits recorded. Flocks with the most complete data recorded were selected from the MERINOSELECT database applying the following thresholds: age of dam known and less than 12 years; date of birth known and with multiple dates recorded within each flock and year; sex; birth type and rear type; more than 5 years of records; flocks with at least 75% of animals with full pedigree; and phenotypes recorded between 2000 and 2022. This selection resulted in 307,815 animals (Table 1) from 175 flocks with measurements at the post-weaning stage (P; between 150 to 299 days of age) for body weight (PWT; kg), live ultrasound eye muscle depth (PEMD; mm) and live ultrasound fat at the C site (PFAT; mm), greasy fleece weight (PGFW; kg), clean fleece weight (PCFW; kg), fibre diameter (PFD; μm), fibre diameter coefficient of variation (PDCV; %), and staple length (PSL; mm). Animals were the progeny of 6,748 sires and 148,420 dams with up to 5 generations of pedigree used in the analysis.

Table 1. Descriptive statistics for live weight, wool and ultrasound traits at the post-weaning stage in Merino sheep

	PWT	PEMD	PFAT	PCFW	PGFW	PFD	PDCV	PSL
Records	210,832	60,363	60,148	77,584	73,755	44,424	43,233	23,085
Genotype	9,446	25,697	8,613	7,657	12,974	6,220	6,207	15,447
Mean	35.1	24.6	2.3	2.2	3.1	16.5	18.3	72.4
SD	4.7	4.3	0.5	0.4	0.6	1.0	2.6	14.3
Min	12.4	9.7	0.5	0.4	0.8	11.9	10.6	25
Max	63.6	42.7	5	4.4	6.7	21.9	32	140
CV (%)	13.5	17.3	22.2	19.2	17.9	6.3	14.4	19.8

* For the trait abbreviations, see text.

Genetic parameters. For the univariate analysis, a linear mixed animal model was fitted in ASReml v4.2 (Gilmour *et al.* 2015) with fixed effects as birth type (4 levels), rear type (4 levels), age, sex (female and male), age of dam (12 levels), contemporary groups (between 344 to 2,266 levels), and weight fitted for PFAT and PEMD. The random effects consisted of genetic groups (defined by flock and time period as per MERINOSELECT), animal genetic, maternal genetic and permanent environmental.

Genome-base restricted maximum likelihood (GREML). The variance components were estimated using only the animals with genotype information (imputed 60k SNP chip) in a univariate GREML via MTG2 software (Lee and van der Werf 2016). The model included adjusted phenotypes for fixed effects and contemporary groups, with random effects fitted for pedigree, genetic groups and genomic relationship matrices. Lambda was calculated as the ratio of $\boldsymbol{\lambda} = \frac{\sigma_G}{\sigma_G + \sigma_{A22}}$; where σ_G is the genetic variance and σ_{A22} is the variance explained by the numerator relationship matrix.

RESULTS AND DISCUSSION

Genetic variances and heritabilities. Moderate to high heritabilities were estimated for live weight, ultrasound and wool traits ranging between 0.25 (0.01) and 0.56 (0.01) (Table 2). The heritabilities for PFAT (0.25), PEMD (0.27) and PWT (0.32) were consistent with previously reported estimates for PEMD (0.20 to 0.25), PFAT (0.15 to 0.22) adjusted with weight and PWT (0.31) (Mortimer *et al*, 2014 and 2017; Huisman *et al.* 2008). However, a lower permanent environmental effect was observed for PWT (0.05) compared with the 0.11 reported by Mortimer *et al.* (2017). The heritability estimates for post-weaning wool traits were moderate to high, ranging from 0.29 to 0.56, agreeing with the estimates reported previously at the hogget stage ranging between 0.27 to 0.60 (Greeff *et al.* 2008). Fibre diameter had a higher heritability at post-weening (0.56; Table 2), similar to previous studies at hogget and yearly (0.60 to 0.61; Greeff *et al.* 2008; Brown *et al.* 2013) but lower than the reported by Mortimer *et al.* (2017) at the yearling stage (0.74). There was a low maternal permanent environment effect for PGFW and PCFW (0.04 to 0.05), which was also observed by Mortimer *et al.* (2017) at the yearling stage.

Table 2. Estimates of phenotypic (σ^2_P) variance, heritabilities (h^2) and ratios of maternal genetic (m^2) and maternal permanent environmental effect (Pe^2) variances, and the ratio of genetic group:**additive variance ($\sigma_{GG:G}$)** for live weight, wool, and ultrasound traits in Merino sheep (standard error)

Trait	σ^2_P	h^2	m^2	Pe^2	$\sigma_{GG:G}$
PWT	19.9(0.08)	0.32(0.01)	0.06(0.0)	0.05(0.0)	1.14(0.17)
PEMD	3.69(0.02)	0.27(0.01)			0.18(0.08)
PFAT	0.21(0.0)	0.25(0.01)			0.29(0.12)
PCFW	0.15(0.0)	0.29(0.02)	0.03(0.01)	0.05(0.01)	0.27(0.08)
PGFW	0.25(0.0)	0.32(0.01)	0.03(0.01)	0.04(0.01)	0.06(0.07)
PFD	1.11(0.01)	0.56(0.01)			0.37(0.1)
PDCV	4.3(0.03)	0.29(0.01)			0.04(0.04)
PSL	69.45(0.82)	0.47(0.02)			0.65(0.21)

* For the trait abbreviations, see text.

Table 3. Phenotypic (below diagonal) and genetic (above diagonal) correlations between live weight, wool and scan traits in Merino sheep

	PWT	PEMD	PFAT	PCFW	PGFW	PFD	PDCV	PSL
PWT		-0.03	0.05	0.21	0.23	0.25	-0.17	0.14
PEMD	0.12		0.48	-0.13	-0.13	0.09	-0.16	0.12
PFAT	0.12	0.32		-0.17	-0.16	0.12	-0.29	0.06
PCFW	0.45	-0.02	-0.09		0.89	0.38	0.11	0.55
PGFW	0.42	-0.01	-0.03	0.91		0.34	0.08	0.40
PFD	0.21	0.07	0.11	0.25	0.26		-0.13	0.26
PDCV	-0.13	-0.09	-0.09	-0.01	0.02	-0.10		-0.11
PSL	0.17	0.07	0.07	0.32	0.30	0.25	-0.12	

* For the trait abbreviations, see text. Standard errors ≤ 0.01 and 0.02 to 0.05 for phenotypic and genetic correlations, respectively.

Genetic and phenotypic correlations. Among the wool traits, PGFW and PCFW were highly genetically correlated (0.89), while PGFW had a small genetic correlation with PDCV (0.08). Moderate and positive genetic correlations were observed between PWT with PCFW (0.21),

PGFW (0.23), and PFD (0.25), whereas PWT was negatively correlated with PDCV (-0.17). These genetic correlations suggest that selection for higher live weight will result in an increase in PCFW, PGFW and PFD, but a decrease in PDCV. Mortimer *et al.* (2017) reported higher genetic correlations between PWT with yearling GFW (0.46), CGW (0.46) and SL (0.21). Ultrasound traits (PEMD and PFAT) had moderate phenotypic (0.32) and genetic (0.48) correlations. Low negative genetic correlations were observed between ultrasound traits and PCFW, PGFW and PDCV, consistent with the negative correlations observed by Mortimer *et al.* (2014) and Huisman and Brown (2009) between yearling GFW and PFAT (-0.26 to -0.48) and PEMD (-0.06 to -0.26). The phenotypic correlations were higher for PWT with the other traits but lower for ultrasound and wool traits.

Genomic REML. Heritability and lambda values were also estimated for all traits (Table 4). Lambda values averaged 0.70 but ranged from 0.51 to 0.90. Heritabilities ranged from 0.28 to 0.56 for the traits slightly differing from the heritabilities estimated from the pedigree models. Overall, these results suggest that lambda of $\lambda = 0.5$ used in the routine analyses could be adjusted slightly, but this needs to be investigated further for a greater range of traits.

Table 4. Estimation of heritabilities, phenotypic variances and lambda for live weight, wool and scan traits in Merino sheep

Trait	PWT	PEMD	PFAT	PCFW	PGFW	PFD	PDCV	PSL
Lambda	0.66	0.67	0.76	0.51	0.62	0.86	0.90	0.62
σ^2_p	13.24	4.21	0.14	0.12	0.21	1.06	4.52	66.19
h^2	0.34	0.28	0.33	0.44	0.47	0.56	0.30	0.39

* For the trait abbreviations, see text. σ^2_p: phenotypic variance; h^2: heritability.

CONCLUSIONS

This study provides estimates of genetic parameters and correlations between economically essential traits such as live weight, wool, and ultrasound traits at a post-weaning stage. The genetic parameters described in this study can be incorporated into the routine evaluation. Lambdas differed from 0.5, indicating that further research will be needed to investigate new strategies to incorporate this information in the ssGBLUP analysis, its impact on prediction accuracies and its use for multi-breed evaluations.

ACKNOWLEDGMENTS

This research was funded by Meat and Livestock Australia project L.GEN.2204. The authors acknowledge the contributions of the MERINOSELECT ram breeders and other industry-funded research and sire evaluation flocks who contributed data to this study.

REFERENCES

Brown D.J., Swan A.A. and Gill, J.S. (2013) *Proc Assoc Advmt. Anim. Breed. Genet.* 20: 110.

Gilmour A.R., Gogel B.J., Cullis B.R., Welham S.J. *et al.* (2015) ASReml User Guide, UK

Greeff J.C., Safari E., Fogarty N.M., Hopkins ,D., *et al.* (2008) *J. Anim. Breed. Genet* 125: 205.

Gurman P., Li L., Swan A.A., *et al.* (2021) *Proc. Conf. Assoc. Adv. Anim. Breed. Genet.* 24: 135.

Huisman A.E., Brown D.J., Ball A.J. and Graser H.U. (2008) *Aust. J. Exp.Agric.* 48: 1177.

Huisman A. E. And Brown D.J. (2009). *Anim Prod. Sci.* 49: 283.

Lee S. and van der Werf J.H.J. (2016) *Bioinformatics* 32: 1420.

Mortimer S.I., Swan A.A., Brown D.J. and van der Werf J.H.J. (2014) *Proc. World Congr. Genet. Appl. Livest. Prod*: 17.

Mortimer S.I., Hatcher S., Fogarty N.M, van der Werf, J.H.J., *et al.* (2017) *J. Anim. Sci.* 95: 1879.

LONGEVITY OF REFERENCE POPULATIONS IN A TRANS-TASMAN GENETIC EVALUATION: REVIEW OF THE ANGUS SIRE BENCHMARKING PROGRAM

S.F. Walkom[1], C.J. Duff [2], C. Girard[1] and K. Moore[1]

[1]Animal Genetics Breeding Unit*, University of New England, Armidale, NSW, 2351 Australia
[2]Angus Australia, 86 Glen Innes Road, Armidale, NSW, 2350 Australia

SUMMARY

The Angus Sire Benchmarking Program (ASBP) remains the cornerstone genomic reference behind Angus Australia's TransTasman Angus Cattle Evaluation (TACE). The success of industry funded genomic reference populations depends on the ability to maintain a strong relationship of the seedstock population with the sires selected for the reference population. Results from a review of the ASBP show that, for hard to measure traits (eg. feed intake), the ASBP is influencing the accuracy of breeding value estimation across the registered population. However, the evolution of the genetic make-up of the Trans-Tasman herd means that the continued collection of hard to measure phenotypes via the ASBP or similar programs is essential.

INTRODUCTION

In recent decades, Angus breeders in Australia have achieved genetic improvement in profitability through the application of performance-based selection programs, using a highly effective genetic evaluation pipeline underpinned by BREEDPLAN software (Graser *et al.* 1995). Coinciding with the emergence of genomic technology and the foreseen transition to a genomically enhanced evaluation, Angus Australia commenced the Angus Sire Benchmarking Program (ASBP) in 2010 (Parnell *et al.* 2019). Since then 12 cohorts (11 cohorts have provided data to date) of sires have produced progeny to help build a relevant genomic reference for Australian and New Zealand Angus Cattle. To capture all of the potential value genomic selection presents, genomic reference populations should have a low average relationship between the reference animals, while ensuring that the relationship between the reference population and the animals being evaluated is high (Clark *et al.* 2012; Pszczola *et al.* 2012). A key design feature of the ASBP has been the development of a genomic reference of 4,000 – 6,000 animals recorded for hard to measure traits, with reference sires refreshed annually (Parnell *et al.* 2019) to account for the decay in linkage disequilibrium over time (Porto-Neto *et al.* 2014). The Trans-Tasman Angus population is managed by a multitude of breeders predominantly spread across southern Australia and New Zealand, encapsulating a diversity of environments, production systems and breeding objectives. Consequently, without a nucleus breeding program controlling the dissemination of genetic material, the sires represented in the ASBP needs to align with the past and future selection decisions of Angus breeders. Consequently, this paper endeavours to quantify the importance of an evolving reference population which changes to reflect current (and future) genetics each year.

MATERIALS AND METHODS

Angus Sire Benchmarking Program. The key objective of the ASBP was to establish a contemporary reference population, and the associated genotypes and phenotypes for economically important traits to facilitate the application of genomic selection for the Angus breed. Parnell *et al.* (2019) described the initiation of the ASBP, which commenced in 2010, with 35 Angus bulls

---

* A joint venture of NSW Department of Primary Industries and University of New England

joined by fixed-time AI to 1,640 cows across 5 co-operator herds. Subsequently, between 21 to 47 bulls have been joined to 1,000 to 2,500 cows annually. For each year's matings (subsequently referred to as a cohort, with cohort 1 identifying matings from 2011 and so on), a genetically diverse range of bulls were nominated by breeders from all states of Australia and New Zealand. Sires from the USA and the UK were also included in some cohorts. Sires represented in each cohort were predominately young bulls (2 – 3 years of age), with some older influential sires also included.

Relationship metrics. Numerator relationship matrices (NRM) were constructed with unpublished AGBU nrmblock software as per Moore *et al.* (2022), based on algorithms by Aguilar *et al.* (2011) and Sargolzaei *et al.* (2005). For each sire in the breed's pedigree which produced progeny in each year from 2010 to 2021, the relatedness to animals generated for the eleven ASBP cohorts was calculated based on the off-diagonal elements of the NRM. This component of the study focused on three relationship metrics, 1; the sires' relationship to their closest relative, 2; the sires' average relationship with their 10 closest relatives, and 3; the sires' average relationship with the animals in the reference population cohort. Summary statistics across sire groups were weighted by the number of progeny sired by the individual within the Trans-Tasman pedigree.

Accuracy estimates. Breeding values for Angus Australia's TransTasman Angus Cattle Evaluation (TACE) are estimated using BREEDPLAN software which applies ssGBLUP models as per Johnston *et al.* (2018), with the accuracy estimations for this study based on the BREEDPLAN methodology reported by Li *et al.* (2017). To test the influence of ASBP data on the accuracy of breeding values for sires represented in the TACE pedigree, a series of modified evaluations were conducted where the genetic evaluation was completed with subsets of the ASBP data excluded based on the TACE pedigree, genotypes and data available in August 2022. The analyses were 1; no ASBP data, 2; Cohort 1-3 data only, 3; Cohort 1-6 data only, 4; Cohort 1-9 data only, and 5; All ASBP data.

RESULTS AND DISCUSSION

The relationship of the progeny in Cohort 1 with the sires which had progeny present in the TACE pedigree declined over time. The average relationship remained reasonably consistent between the cohort progeny and the industry sires (blue line, Figure 1), and this is a by-product of the effective population size and that the top 10 genetically influential ancestors explain 42% of the genetic diversity in the population (Clark *et al.* 2019). However, whilst the average relationship remains relatively constant, the relationship metrics focusing on the strength of the relationship with the closest relatives were shown to noticeably decline (Figure 1). This rate of decline, while not uniform, was relatively consistent across all the cohorts. This suggested that the evolution of the Trans-Tasman Angus population is largely constant as a result of the effective population size and limitations on sourcing outside genetics. The merit of the ASBP ultimately depends on its ability to produce accurate breeding values for hard to measure and economically important traits among future selection candidates.

The importance of the ASBP reference population to the accuracy of selection candidate estimated breeding value (EBV) accuracy is largely governed by the baseline accuracy which, in turn, is driven by the size of the reference population and the effective population size (Clark *et al.* 2012). It should be noted that within a ssGBLUP analysis the reference expands beyond the ASBP and includes all animals from the broader industry which have both phenotypes and genotypes. Consequently, for highly recorded traits like 400-day weight, the contribution of the ASBP data is minimal. For the sires used across the Angus breed in 2012, 2016 and 2020, the mean change in accuracy was less than 1% (Figure 2). In contrast, for carcase intramuscular fat the mean impact of the ASBP data for single trait accuracy of the sires from the same three years was an accuracy increase of 5.7%, 7.5% and 8.2% (Figure 2), respectively. These estimates are inclusive of the contribution to EBV accuracy of correlated traits, which is a feature of the BREEDPLAN multi-trait analysis. After accounting for this, the impact of the ASBP data to carcase intramuscular fat EBV accuracy was reduced for the three drops to +1.5%, +2.4% and +3.8%, respectively. The value of the ASBP data was most noticeable for net feed intake, where there is minimal recording outside of the reference, with the ASBP data leading to an average change in single trait accuracy (BREEDPLAN reported multi-trait analysis in brackets) of +8.7% (+2.0%), +10.3% (+3.2%) and +11.3% (+4.8%) for the 2012, 2016 and 2020 sires (Figure 2), respectively.

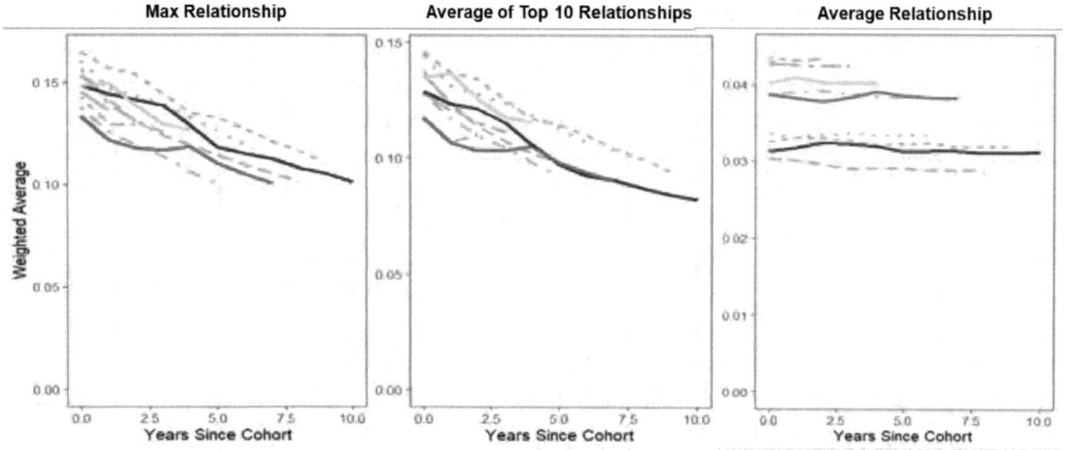

Figure 1. The average relatedness metrics, weighted by the sires progeny count within year, between ASBP cohort progeny and sires of calves born *n* years after the cohort mating: Cohort 1 (2011) = blue, Cohort 4 (2014) = red, Cohort 7 (2017) = green with other cohorts in grey

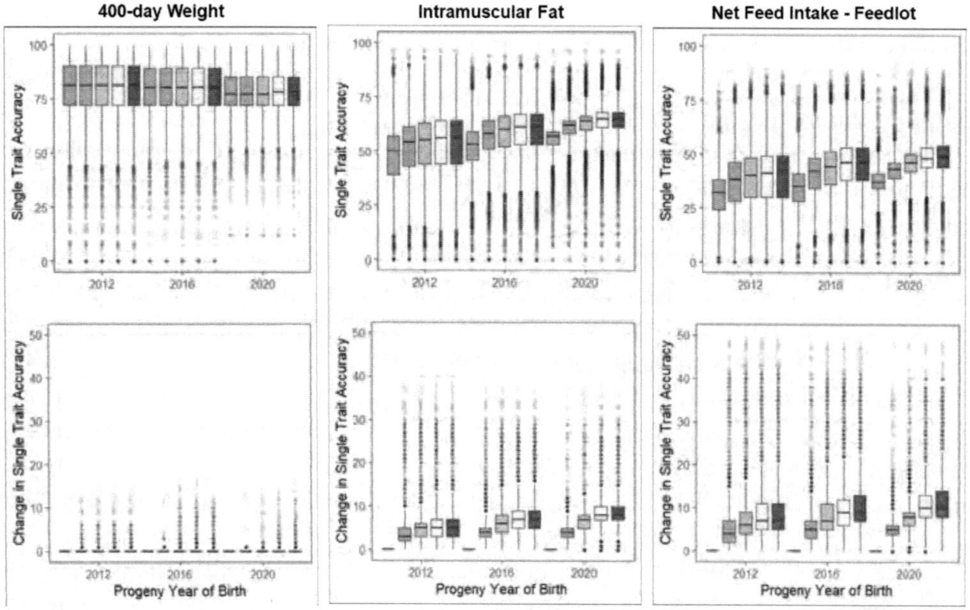

Figure 2. Impact of including ASBP phenotypes from Cohorts 1-3 (purple), Cohorts 1-6 (orange), Cohorts 1-9 (yellow) and All Cohorts (blue), compared to when no ASBP phenotypes (green) are available on the single trait accuracy of breeding values of the sires of the 2012, 2016 and 2020 progeny for 400-day weight, carcass Intramuscular Fat and Net Feed Intake – Feedlot

The impact on EBV accuracy of the decline in relatedness between ASBP cohorts and sires appearing in the TACE pedigree in later years is most clearly observed for net feed intake (Figure 2). For industry sires used in 2012, the inclusion of ASBP data provided an extra 8.7% accuracy, however if the ASBP had concluded after either the 3rd, 6th or 9th cohort this gain would have only been +5.6%, +7.5% and +8.5%, respectively. As expected, the majority of the accuracy gain observed in the 2012 sires comes from the earlier cohorts with cohorts 1-3 accounting for 67% of the overall accuracy improvement. In comparison, for 2020 sires, cohorts 1-3 only provide 46% (+5.2%) of the overall accuracy improvement observed when including the ASBP data, with 94% of the gains in accuracy achieved from cohorts 1-9 data. This suggests that, for traits which Angus breeders aren't able to readily measure on farm, the ASBP recording makes a valuable contribution and shows that investment in the reference needs to continue to reflect the diversity of genetics represented in the current selection candidates.

CONCLUSIONS

To maximise the contribution to EBV accuracy provided by reference population projects, this study demonstrates that relationships between reference animals should be low, but that they need to be sufficiently genetically diverse that their relationship to the broader population is high. As relatedness between ASBP cohorts and subsequently used industry sires declined, there was a corresponding fall in accuracy gains from the ASBP phenotypes. This shows that for traits which are lowly recorded in the broader Angus population, the ASBP remains highly valuable. It also clearly demonstrates that investment in reference populations needs to be ongoing to reflect the diversity of genetics represented within selection candidates.

ACKNOWLEDGMENTS

This research was partially funded by Meat and Livestock Australia project L.GEN.2007. The authors acknowledge Angus Australia for the provision of the required data and their breeder members who participate in the Angus Sire Benchmarking Program and provide data for the TransTasman Angus Cattle Evaluation.

REFERENCES

Aguilar I., Misztal I., Legarra A. and Tsuruta S. (2011) *J. Anim. Breed. Genet.* 128: 422.

Clark S.A., Granleese T. and Parnell P. (2019) *Proc. Assoc. Advmt. Anim. Breed. Genet.* 23: 75.

Clark S.A., Hickey J.M., Daetwyler H.D. and van der Werf J.H.J. (2012) *Genet. Sel. Evol.* 44: 4.

Graser H. U., Goddard M. E., and Allen J. (1995) *Proc. Assoc. Advmt. Anim. Breed. Genet.* 11: 56.

Johnston D.J., Ferdosi M., Connors N., Boerner V., Cook J., Girard C., Swan A. and Tier B. (2018) *Proc. World Congr. Genet. Appl. Livest. Prod.,* 19: 269.

Li L., Swan A. and Tier B. (2017) *Proc. Assoc. Advmt. Anim. Breed. Genet* 22: 89.

Moore K., Ferdosi M., Girard C., Walkom S. and Johnston D. (2022) *Proc. World Cong. Genet. App. Livest. Prod.* 12

Parnell P., Duff C., Byrne A. and Butcher N. (2019) *Proc. Assoc. Advmt. Anim. Breed. Genet.* 23: 492.

Porto-Neto L. R., Kijas J.W. and Reverter A. (2014) *Genet. Sel. Evol.* 46: 1.

Pszczola M., Strabel T., Mulder H.A. and Calus M.P.L. (2012) *J. Dairy Sci.* 95: 389.

Sargolzaei M., Iwaisaki H. and Colleau J.J. (2005) *J. Anim. Breed. Genet.* 122: 325.

ANGUS BREEDCHECK – VALIDATION USING INDUSTRY DATA

C.J. Duff[1], A.M. Samaraweera[1], A.I Byrne[1], A.B. Ingham[2], P.A. Alexandre[2], L.R. Porto-Neto[2] and A. Reverter[2]

[1]Angus Australia, 86 Glen Innes Road, Armidale, NSW 2350, Australia
[2]CSIRO Agriculture and Food, 306 Carmody Rd., St Lucia, QLD 4067, Brisbane, Australia

SUMMARY

Angus BREEDCHECK is a genomic based tool that predicts breed composition for 11 breeds with a focus on Angus content. In this study we compare the Angus BreedCHECK genomic breed composition (GBC) estimates to pedigree-based breed content estimates (PBC) for five animal classes (AC) recorded on the Angus Australia database. The AC populations being Herd Book Register (HBR), Angus Performance Register (APR), Angus Commercial Register (ACR), Angus HeiferSELECT (AHS) and the Multi Breed Register (MBR), including 143,879, 75,369, 6,379, 25,710, and 2,780 animals, respectively. Additionally, comparisons were made within a subset of Angus cross *Bos indicus* (n=1,201) and Angus cross Hereford (n=365) cattle, as determined by PBC, from the MBR. Across the 254,117 animals in this study, there is close alignment in the mean and standard deviation of Angus content as derived by GBC and PBC, with a mean of 99.3% and 99.4% and standard deviation of 3.6 and 4.1, respectively. While 97.7% of the study animals fell within ±10% in Angus content when comparing GBC to PBC. Within the AC populations, and across the sub-set of Angus cross *Bos indicus* and Angus cross Hereford cattle, close alignment was also observed in the comparative statistics. Using a large industry dataset, this study has validated the precision of Angus BreedCHECK to estimate beef cattle breed content, with an emphasis on Angus content.

INTRODUCTION

Understanding breed composition is important in many beef cattle breeding programs, such as those linked to beef supply chains with branded beef schemes or with structured crossbreeding programs. This information is also important for understanding the potential effectiveness of genomic breeding values that are based on breed specific reference populations, like those provided in the Angus HeiferSELECT product (Alexandre *et al.* 2022; Angus Australia 2023).

Breed composition of beef cattle is historically assessed by documenting the breed of foundation sires and dams (usually pure-bred), which facilitates the calculation of breed composition in subsequent generations through simple mode of inheritance i.e. 50% of the breed from the sire and 50% breed from the dam (Sölkner *et al.* 2010). More recently, genomic prediction of breed composition, based on a breed-based genomic reference population, has allowed for breed composition estimation where breed composition is unknown through documentation, particularly where no or limited pedigree is available (i.e., commercial animals or beef products). Importantly, current studies have shown that genomic prediction can offer precision to breed composition estimation in livestock (Sölkner *et al.* 2010; Gurman *et al.* 2017; Reverter *et al.* 2020: Ryan *et al.* 2022).

Angus BREEDCHECK is a genomic based tool that predicts breed composition for 11 breeds with a focus on Angus content. It was developed by Angus Australia in collaboration with the CSIRO, Australia's national science agency. It is currently available via Angus HeiferSELECT which is a genomic selection tool to help inform the selection of Angus replacement females (of 87.5% Angus content or greater) in a commercial beef breeding operation.

The objective of this study is to compare the breed composition values from Angus BreedCHECK on a large industry dataset of genotyped Angus influenced animals, to their known breed background.

MATERIALS AND METHODS

Data for the 254,117 animals in this study was accessed from the Angus Australia database. For each animal the data included genomic estimates of breed composition from Angus BreedCHECK (GBC), pedigree estimated breed composition (PBC) and the animal class (AC).

The method used to estimate the GBC values (%) is described in detail by Reverter *et al.* (2020). In short, a linear regression model was used to estimate the GBC of individuals where the SNP genotypes are regressed on the allele frequencies from a reference population of 11 breeds. More specifically, this is based on a genomic profile for each animal containing 45,364 autosomal SNPs and a breed based genomic reference population including Angus (n=868), Brahman (n=330), Charolais (n=71), Hereford (n=111), Holstein (n=144), Limousin (n=53), Murray Grey (n=62), Santa Gertrudis (n=53), Shorthorn (n=88), Simmental (n=27) and Wagyu (n=43).

The method used within the Angus Australia database to estimate the PBC values (%) involves breeders and Angus Australia staff documenting the breed of foundation sires and dams in the pedigree, followed by calculating breed composition in subsequent generations by summing 50% of the breed content inherited from the sire and dam.

The AC categories extracted for this study are Herd Book Register (HBR), Angus Performance Register (APR), Angus Commercial Register (ACR), Angus HeiferSELECT (AHS) and Multi Breed Register (MBR). The ACs are applied within the Angus Australia database to primarily cater for service delivery flexibility, however they also broadly categorise the levels of expected Angus breed purity. For example, HBR animals, considered the highest purity of Angus, can only bred from HBR sires and dams. APR or ACR animals can be bred from foundation (or base) Angus animals, AHS are wholly commercially bred Angus, while MBR, as the name suggest, includes components of non-Angus breeds.

To validate Angus BreedCHECK, this study compared the GBC values to the PBC values for the different AC populations. Additionally, a similar comparison was made in a subset of Angus cross *Bos indicus* cattle (n=1,201) and Angus cross Hereford cattle (n=365), as determined by PBC, recorded on the MBR.

RESULTS AND DISCUSSION

Across the 254,117 animals in this study, there is close alignment in the mean and standard deviation of Angus content as derived by GBC and PBC, with a mean of 99.3% and 99.4% and standard deviation of 3.6 and 4.1, respectively (Table 1). This also highlights the high Angus content represented in the overall study population. Additional to the summary statistics, the proportion of animals with an Angus content difference equal to or less than 10%, when comparing GBC to PBC estimates, were calculated. Accordingly, 97.7% of the study animals fell within ±10% when comparing Angus contents.

Close alignment is also observed for mean and standard deviation values between GBC and PBC within the AC groups (Table 1). The largest difference being between the MBR GBC and PBC Angus content means of 84.2% and 79.8%, respectively, but with similar standard deviations. This may be explained by limitations in the current breed reference population underpinning the GBC estimates or, more likely, inaccurate PBC values stemming from incorrect foundation breed allocations for some animals in the MBR study group.

When comparing the mean GBC and PBC values by AC (Table 1), the findings follow industry expectations of the HBR being the highest mean Angus content followed closely by the APR. The ACR and AHS have marginally lower means, and are in close alignment with one another, which is expected given the commercial nature of both animal classes (i.e., non-seedstock). Also as expected, the MBR, which is a multi-breed population, has the lowest mean Angus content and largest standard deviation by both GBC and PBC estimation methods. This is further outlined in Table 2 showing 99.9% of HBR animals are at least 87.5% Angus content by GBC estimation. Conversely, and as expected, a significantly lower 46.6% of MBR animals are categorised in the highest Angus content level.

Table 1. Angus content statistics by animal class (AC) and estimation methods of genomic breed content (GBC) and pedigree breed content (PBC)

AC[a]	# Animals	GBC (%) Mean	GBC SD	PBC (%) Mean	PBC SD	Difference (<±10%)
HBR	143,879	99.8	1.2	99.9	1.0	99.8 %
APR	75,369	99.5	2.5	99.1	5.5	96.4 %
ACR	6,379	98.2	6.9	99.2	3.5	91.1 %
AHS[b]	25,710	98.0	6.1	-	-	-
MBR	2,780	84.2	14.3	79.8	14.9	81.2 %
All	254,117	99.3	3.6	99.4	4.1	97.7 %

[a]AC: Animal Class, HBR: Herd Book Register, APR: Angus Performance Register, ACR: Angus Commercial Register, AHS: Angus Heifer Select, MBR: Multi Breed Register.
[b]PBC is not calculated on AHS animals.

Table 2. Proportion of animals by animal class (AC) and Angus content levels from genomic breed content estimation (GBC)

	Angus Content Levels			
AC[a]	**≥87.5%**	**≥75%**	**≥50%**	**<50%**
HBR	99.9%	100.0%	100.0%	0.0%
APR	99.1%	99.9%	100.0%	0.0%
ACR	95.5%	98.5%	99.6%	0.4%
AHS	94.8%	98.4%	99.8%	0.2%
MBR	46.6%	75.4%	98.6%	1.4%
All	98.5%	99.5%	100.0%	0.0%

[a] See Table 1.

There was also close alignment of mean and standard deviation values when comparing GBC to PBC estimates for Angus (Table 3), *Bos indicus* or Hereford (Table 4) content within the subset of MBR animals. For example, in the Angus cross *Bos indicus* group the Angus breed content mean by GBC and PBC was 78.0% and 77.1% respectively (Table 3). For the *Bos indicus* component (Table 4) in the same animals, the means were 22.9% and 21.4% respectively. Similar results were observed in the Angus cross Hereford population.

Additionally, most animals (93%) had Angus breed content estimates that fell within the ±10% difference range (Table 3). A similar result was observed for the *Bos indicus* and Hereford content estimates (Table 4) with 90.2% and 99.7% respectively falling with the ±10% difference range. The correlations presented (Table 3 and 4) between the GBC and PBC estimates also support general alignment with the values being moderate to strong and positive in direction.

Table 3. Angus content statistics by estimation methods of genomic breed content (GBC) and pedigree breed content (PBC) for multi breed register (MBR) animals

Breed[a]	# Animals	GBC (%) Mean	GBC SD	PBC (%) Mean	PBC SD	Difference (<±10%)	Correlation
AA*BI	1201	78.0	9.1	77.1	7.4	89.3%	0.69
AA*HH	365	91.8	3.0	93.9	3.5	99.4%	0.66
All	1566	81.2	10.0	81.0	9.8	93.0%	0.86

[a]AA*BI: Angus Cross *Bos indicus* (Brahman or Santa Gertrudis), AA*HH: Angus Cross Hereford.

Table 4. Bos Indicus or Hereford content statistics by estimation methods of genomic breed content (GBC) and pedigree breed content (PBC) for multi breed register (MBR) animals

Breed[a]	# Animals	GBC (%) Mean	GBC SD	PBC (%) Mean	PBC SD	Difference (<±10%)	Correlation
AA*BI	1201	21.4	7.4	22.9	8.7	90.2%	0.72
AA*HH	365	6.9	3.8	6.1	3.5	99.7%	0.63

[a] See Table 3.

CONCLUSIONS

This study has validated the precision of Angus BreedCHECK to estimate beef cattle breed content, with a close alignment of the comparative statistics when comparing GBC to PBC estimates, as well as an alignment with industry expectations of the Angus content differences across the ACs from the Angus Australia database. Therefore, Angus BreedCHECK provides potential value as a tool for the estimation of breed content in Angus or Angus influenced breeding programs, particularly commercial herds, or within Angus beef supply chain initiatives. Angus BreedCHECK can also be used in the assessment of the effectiveness of the genomic breeding values provided from Angus HeiferSELECT.

REFERENCES

Alexandre P.A, Porto-Neto L.R, Hine B, Ingham A, Samaraweera M, Duff C and Reverter A. (2022) *12th World Congress on Genetics Applied to Livestock Production. 1930–1933*

Angus Australia (2023). Available at: https://www.angusaustralia.com.au/content/uploads/2022/05/HS-TechDoc_A42-1.pdf

Angus Australia (2023a). Available at: https://www.angusaustralia.com.au/member-services/registrations

Gurman P.M, Swan A.A and Boerner V. (2017). *Proc. Assoc. Advmt. Anim. Breed. Genet.* 22: 341.

Reverter A, Hudson N.J, McWilliam S, Alexandre P.A, Li Y, Barlow R, Welti N, Daetwyler H, Porto-Neto L.R and Dominik S. (2020). *J. Anim. Sci.* **98**: 1.

Ryan C.A, Berry D.P, Pabiou T and Purfield D.C. (2022) *12th World Congress on Genetics Applied to Livestock Production* 12: 745.

Sölkner J, Frkonja A, Raadsma H.W, Jonas E, Thaller G, Gootwine E, Seroussi E, Fuerst C, Egger-Danner C and Gredler B (2010) *Interbullbull Bulletin.* 42: 62.

QUANTIFYING THE LINKAGE BETWEEN GENETICS REPRESENTED IN THE SOUTHERN MULTI-BREED PROJECT AND THE WIDER BEEF POPULATIONS

K.L. Moore[1], S.F. Walkom[1], J.P. Siddell[2] and B. Walmsley[1,2]

[1] Animal Genetics Breeding Unit*, University of New England, Armidale, NSW, 2351 Australia
[2] NSW Department of Primary Industries, Livestock Industries Centre, Armidale, NSW, 2351 Australia

SUMMARY

Southern Multi-Breed (SMB) is a landmark five-year project, collecting high-quality phenotypes and genotypes on animals from six breeds, managed in multi-breed groups at five NSW Department of Primary Industries research herds. Data collected will enhance genetic evaluations and facilitate the development of a multi-breed genetic evaluation. The project design focussed on selecting foundation cows and sires to represent the breed's populations. This paper aimed to quantify the linkage between the genetics represented in the SMB project and the breed populations. Within each breed, the average relationship coefficient of each animal to SMB foundation cows and sires and all animals in the breed was calculated and plotted to form a visual metric of the linkage. Regression slopes between 1.11 and 1.37 and correlations between 0.86 and 0.99 were calculated from the plots. The visual and quantitative metrics indicated that the genetics in SMB represent the breed populations. Therefore, the reference data collected as part of SMB will benefit the broader industry.

INTRODUCTION

The potential benefits of genomic selection are directly impacted by the size of the reference population, trait heritability, and effective population size (Goddard and Hayes 2009), relatedness amongst the reference animals, and relatedness to selection candidates (Pszczola *et al.* 2011). Therefore, when designing reference data projects, multiple design principles must be balanced to maximise the value of the collected data.

Southern Multi-Breed (SMB) is a landmark five-year reference data project involving six beef cattle breeds. Calves are born and managed in mixed-breed groups across five NSW Department of Primary Industries (DPI) sites (Walmsley *et al.* 2021). Progeny is intensively performance recorded for BREEDPLAN traits and other traits of economic importance (Donoghue *et al.* 2001, Walmsley *et al.* 2023). The over-arching goal of SMB is to collect high-quality reference data - particularly for hard to measure traits - to enhance genetic evaluations and facilitate the development of a multi-breed genetic evaluation. As such, the design of the SMB project is of critical importance, not just to ensure fair head-to-head across breed comparisons but also to ensure that the generated reference data genetically represents the breed populations. Moore *et al.* (2022) presented a metric to describe and compare the relatedness of reference populations to a whole breed. A key element of the SMB project design was the selection of foundation cows and sires to maximise relationships between SMB and the wider breed. This paper aims to assess if the choice of foundation cows and sires has been effective in ensuring that SMB data is genetically linked to the whole population within the respective breeds.

MATERIALS AND METHODS

Commencing in 2020, three cohorts of SMB progeny have now been born across five DPI

* A joint venture of NSW Department of Primary Industries and University of New England

locations (Trangie Agricultural Research Centre, Trangie; Grafton Primary Industries Institute, Grafton; Tocal Agricultural Centre, Tocal; Glen Innes Agricultural Research and Advisory Station, Glen Innes; Elizabeth MacArthur Agricultural Institute (EMAI); Menangle) from six purebred breeds (Angus, Brahman, Charolais, Hereford, Shorthorn and Wagyu). In addition to the purebred matings, at Grafton, a small amount of cross-breeding involving Brahman reciprocal matings to Angus and Hereford also occurred. Walmsley *et al.* (2021; 2023) provide details about the SMB project design, and Walkom *et al.* (2021) outline the mating strategy within the project. At the commencement of SMB, foundation cows were purchased from industry seedstock herds. These were identified as herds that BREEDPLAN recorded and were influential in the breed (either using a wide range of sires or selling their genetics to other seedstock herds). Groups of cows were sourced from these herds. All cows were BREEDPLAN performance recorded with pedigree information and were selected to be representative of the national population (assessed via 400-day weight and reproduction EBVs), but especially if their sires were current influential sires (i.e. a large number of progeny). Angus foundation cows were also retained from the NSW DPI muscling (McKiernan and Robards 1997) and feed efficiency selection (Arthur 1997) herds. Female progeny are retained in the project, with foundation cows exiting the project as the number of project-born females increases. Project sires were also BREEDPLAN performance recorded with pedigree information. Natural mate sires were purchased from industry herds, and nominations were sought by the industry for artificial insemination sires. In both sire mating types, sires were selected to represent the breed, with an emphasis on using current or immerging influential sires. This involved studying the pedigrees to identify sire lines not already represented in the SMB foundation animals and undertaking MateSel (Kinghorn 2011) analysis to identify new and important genetics to include in the project. Several sires were also used that provide genetic links with other reference data projects (past and present, i.e. Repronomics, Beef CRC and existing within-breed reference data projects). Sires were used across sites and years, with new sires also purchased each mating. This study considered the cows and sires that produced the first two cohorts of calves, with Donoghue *et al.* (2021) providing details on the first two cohorts of calves produced.

Moore *et al.* (2022) described a methodology to assess how related reference populations are to a wider population. This method was used to assess how the SMB foundation cows and sires are related to the breed population for five of the six breeds represented in the project. All known pedigree information was available for breeds A, B, C and E, but pedigree was only available for a subset of breed D. Of the 267 breed D foundation animals, 116 foundation cows and sires were present in the available pedigree subset, and these animals were considered in the current study. No pedigree was available for the breed not included in this study. A whole breed numerator relationship matrix was constructed for each breed in the study based on the breed's recorded pedigree. The average relationship coefficient for each animal in the breed was calculated with 1) SMB foundation cows and sires and 2) all animals within the breed. A visual metric (Figure 1) was produced for each breed, where the average relatedness to SMB animals (y-axis) was plotted against the average relatedness to all animals (x-axis) in the breed. The regression slope and Pearson correlation coefficient were also calculated for each plot to quantify the relationships between SMB and the whole breed population.

RESULTS AND DISCUSSION

Table 1 summarises by breed the 1,149 foundation cows and 277 sires. Cows were from similar age structures and sourced from 5 to 13 herds per breed. The number of foundation cows per breed varied depending on the number of sites the breeds were present at. Breed C had the largest number of cows (n=370) due to being present at four of the sites. Breeds B, D and E were located at two or three project sites, while breed A was located at only one site. The number of sires per breed depended on the number of cows and sites the breeds were present at, and mating was from natural mating or artificial insemination.

Figure 1 describes how animals in the breed were related to SMB foundation cows and sires and the breed population. The average relatedness to all animals in the breed ranged between 0.0 and 0.05. The exception was breed E, where the average relatedness to all animals was much higher (0.0 to 0.2), reflecting breed E's reduced diversity due to being founded from a limited number of animals. These plots demonstrate that the genetics represented in SMB are well linked to the breed population for all five breeds considered in this study. The shapes of the plots indicate that the animals with higher average relationship coefficients to all animals in the breed also demonstrated higher average relationship coefficients to SMB animals. Data points in the plots generally followed the 1:1 line marked, although they tended to be slightly above the line. The 1:1 line is the expected relationship between both axes when the reference population is the same as the whole population (Moore *et al.* 2022).

Table 1. A summary of the SMB project foundation cows and sires

Breed	A	B	C	D	E
Number of site locations	1	2	4	2	3
Number of foundation cows	166	182	370	219	212
Number of herds	7	5	13	10	9
Year of birth range	2008-18	2009-19	2009-18	2010-17	2010-18
Number of project sires	41	36	89	48	63

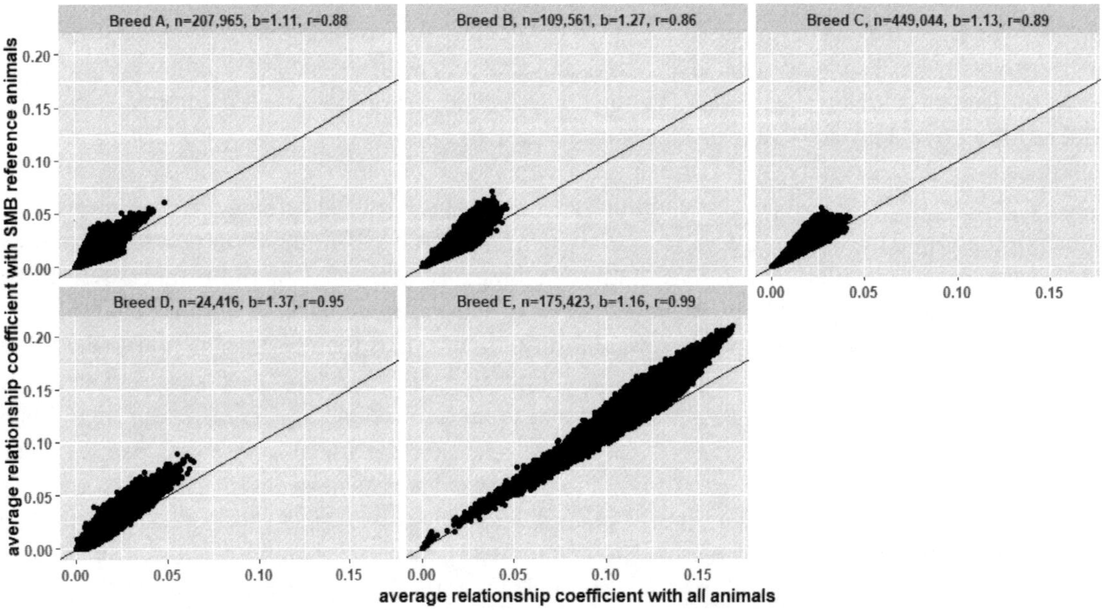

Figure 1. The average relatedness to all animals (x-axis) compared with animals in the SMB reference population (y-axis), regression coefficient (b) and correlation (r) for all animals born after 2010 (n) in five of the breeds represented in SMB

Regression slopes were estimated to be 1.11, 1.27, 1.13, 1.36 and 1.16, respectively, for breeds A, B, C, D and E. Moore et al. (2022) reported that a regression slope close to 1 was considered optimum. For each breed, the regression slope was slightly greater than 1. A regression slope above 1 indicates that the reference population (in this case, SMB founder cows and sires) contains a higher proportion of animals with high relatedness to all animals. The above 1 regression coefficients found align with the strategy to target high-impact and diverse genetics when sourcing foundation cows and sires and suggest that when selecting future project sires, more emphasis can be placed on increasing diversity as high-impact animals are currently represented. These regression slope values were comparable to those found for the live weight reference populations assessed by Moore *et al.* (2022). Live weight is generally one of the better-recorded traits, but only between 30 and 74% of the breed population was recorded for live weight. Selection strategies for SMB cows and sires required that animals were BREEDPLAN recorded. Similar regression slopes for the SMB foundation animals and the breed's live weight references appear sensible. Pearson correlations of 0.88, 0.86, 0.89, 0.95 and 0.99 were estimated for breeds A, B, C, D and E, respectively. These correlations indicate a very strong relationship between the relatedness of SMB and the breed population. This correlation was especially high for breeds D and E, and this could suggest that foundation animals contain a more even representation of different relatedness levels to the whole breed. Lower correlations can be seen in the plots (particularly for breeds A and B) where animals that were the highest related to all animals showed a narrower range in relatedness values to SMB, i.e. the width of the cluster was smaller for higher related animals, and this aligns with the strategy of targeting cows with influential sires, and current or emerging influential sires themselves.

CONCLUSIONS

SMB is a landmark project collecting reference data to enable across-breed comparisons and provide valuable reference data for within-breed genomic evaluations. As part of the project design, foundation cows and sires were identified to be representative of the breed. This study confirmed that these foundation cows and sires used in SMB are related to the breed population. Therefore, the reference data collected will benefit the development of multi-breed genetic evaluations and within-breed genomic selection programs.

ACKNOWLEDGEMENTS

The Southern Multi-Breed project is an initiative co-funded by NSW Department of Primary Industries, University of New England and Meat and Livestock Australia Donor Company. The authors give thanks to the efforts of the research station managers and their staff, our skilled team of technicians, ultrasound scanners, AI, DNA lab, data, project managers and scientists. Individual breeders, participating breed societies and the southern beef industry are thanked for their support. The authors also acknowledge MLA project L.GEN.2007 which funded this study of SMB data.

REFERENCES

Arthur P.F., Archer J.A., Herd R.M., Richardson E.C., Exton S.C., Wright J.H., Dibley K.C.P. and Burton D.A. (1997) *Proc. Assoc. Advmt. Anim. Breed. Genet.* 12: 234.

Donoghue K.A., Walmsley B.J., Siddell J.P., Granleese T., Penrose L. and Arthur P.F. (2021) *Proc. Assoc. Advmt. Anim. Breed. Genet.* 24: 98.

Goddard M.E. and Hayes B.J. (2009) *Nat. Review. Genet.* 10: 381.

Kinghorn B. P. (2011) *Genet. Sel. Evol.* 43: 4.

McKiernan W.A. and Robards G.E. (1997) *Proc. Assoc. Advmt. Anim. Breed. Genet.* 12: 77.

Moore K.L., Ferdosi M.H., Girard C.G., Walkom S.F. and Johnston D.J. (2022) *Proc. of the 12th WCGALP*, Rotterdam, The Netherlands.

Pszczola M., Strabel T., Mulder H.A. and Calus M.P.L. (2012) *J. Dairy Sci.* 95: 389.

Walkom S.F., Donoghue K.A., Arthur P.F., Clark S.A. and Walmsley B.J. (2021) *Proc. Assoc. Advmt. Anim. Breed. Genet.* 24: 419.

Walmsley B.J., Donoghue K.A., Johnston D.J., Clark S.A., Siddell J.P., Walkom S.F., Granleese T. and Arthur P.F. (2021) *Proc. Assoc. Advmt. Anim. Breed. Genet.* 24: 423.

Walmsley B.J., Moore K.L., Walkom S.F., Clark S.A., Granleese T. and Donoghue K.A. (2023) *Proc. Assoc. Advmt. Anim. Breed. Genet.* 25: *These proceedings.*

Proc. Assoc. Advmt. Anim. Breed. Genet. 25: 334 - 337

UTILITY OF POOLED DNA SAMPLES FOR ESTIMATING A FLOCK PROFILE

P.M. Gurman, K. Gore and D.J. Brown

Animal Genetics Breeding Unit*, University of New England, 2351, Armidale, Australia

SUMMARY

The flock profile product by Sheep Genetics allows commercial Merino breeders to benchmark their flock's genetic merit based on the genotypes of 20 animals. Sheep breeders collect DNA samples from their sheep using Tissue Sampling Units, which are then sent to the DNA laboratory and converted into genotypes for the 20 animals, which are used to calculate individual animal breeding values. The final reported value provided to the breeder is the mean of the estimated ASBVs for the 20 animals. This study documents an in-silico investigation to determine if the individual animal genotypes can be combined into an allele frequency, which is used instead to estimate the flock profile breeding value. The mean correlation across traits was 0.99999, while the mean regression slope was 0.9999 These results show that it is possible to calculate the flock profile breeding values based on the allele frequencies. Further research is now required to research and develop procedures on a commercial scale and examine the correlation between a genotype from a pooled sample and the allele frequencies calculated from individual genotypes at this scale.

INTRODUCTION

The flock profile test is a genomic test offered to Australian Merino sheep breeders, which provides a benchmark of their flock's genetic merit compared to the MERINOSELECT analysis (Swan *et al.* 2018). This product requires that DNA samples are collected using Tissue Sampling Units (TSU) on 20 randomly selected sheep from the most recent drop, which are then sent to a genotyping laboratory and analysed as 20 individual animals. The resulting genotypes are then used to calculate Australian Sheep Breeding Values (ASBVs) for each animal based on the reference population of genotyped and phenotyped animals from the MERINOSELECT single-step analysis (Swan *et al.* 2018), assuming unknown pedigree. The ASBVs for the individual animals are then averaged to estimate the flock profile. This process results in ASBVs that are directly comparable to ASBVs reported in the full MERINOSELECT single-step analysis and validated by leaving the data of one flock out of the analysis at a time and estimating breeding values from the remaining data (Swan *et al.* 2018). This service has been used since its inception in 2016 for over 600 commercial flocks.

Currently, the cost of a flock profile includes the cost of genotyping 20 animals. One option for reducing the cost of this product and increasing its adoption is to pool the DNA from the 20 animals. The pooled sample can then be processed by the genotyping laboratory to obtain the dosage/allele frequency based on these 20 animals. For this to be a viable option, the ASBV estimated from a pooled sample needs to be equivalent to the ASBV calculated from the mean of the 20 animals calculated separately. This study examines if the mean of the 20 animal's ASBVs as is currently done to calculate a flock profile is sufficiently like the ASBV calculated from the mean of the individual genotypes (allele frequency) from the 20 animals, which would be available from a single genotype from a pooled DNA sample in practice.

MATERIALS AND METHODS

In this study, previous flock profile tests (n flocks = 673, n animals = 13,017) were used, extracting the genotypes for each individual animal from the MERINOSELECT analysis. These

* A joint venture between NSW Department of Primary Industries and University of New England

genotypes have previously been cleaned (individual call rate 90+%, heterozygosity <=50%) and imputed to fill in sporadic missing SNP calls using all available genotypes for the chip on which the animal was genotyped. The genotypes were then imputed to the 4 other separate SNP chips that have significant reference populations (n>10000 Australian and New Zealand genotypes for each reference set). The separate imputation results were then combined into a set of 60,410 SNP genotypes, starting from the original genotype and adding the SNPs from the other chips that were not already present. All imputation was performed using Beagle (Browning *et al.* 2018; 2021). Genotypes for the animals included in the MERINOSELECT analysis were then used to calculate SNP effects based on their ASBVs. The reference population for the traits analysed ranged from 11,192 to 143,356 genotyped and phenotyped animals with a mean of 74,338 animals. Genotypes for each flock profile were then used to calculate the mean of the genotypes for each flock profile, i.e. twice the allele frequency for each flock profile, and the resulting genotype values as double precision floating point values between 0 and 2 were used to calculate an ASBV based on Swan *et al.* (2018). These new pooled results were then compared to the traditional method as part of the current MERINOSELECT analysis. Analyses were performed for all traits which are reported for flock profile tests and traits used in current selection indexes, (body weight at weaning, post-weaning, yearling and adult age stages; greasy fleece weight at adult and yearling, clean fleece weight at yearling; fibre diameter, its coefficient of variation, staple length and curvature at yearling and adult; carcase fat and eye muscle depth at post-weaning and yearling; early breed cover and breech wrinkle and late body wrinkle. Metrics examined between the two sets of ASBVs included Pearson correlations, dispersions calculated as $cov(\boldsymbol{u}_{current}, \boldsymbol{u}_{pooled})/var(\boldsymbol{u}_{pooled})$ and the scaled bias as $\frac{\overline{u_{pooled}-u_{current}}}{\sigma_g}$. Data preparation, calculation of EBVs and statistical analysis of the results was performed using Python 3.10 and the Pandas library 1.5.0.

RESULTS AND DISCUSSION

The mean correlation across traits between the current flock profile ASBVs and those obtained from allele frequencies was $0.999985 \pm 6.17 \times 10^{-5}$, with these correlations presented in Figure 1. The outlier trait was post-weaning faecal egg count (PWEC), which had a correlation of 0.9997. The mean dispersion was 0.9999 ± 0.003. For most traits, there was a slight increase in the dispersion of the ASBVs estimated, with the dispersions presented in Figure 2. PWEC was again the outlier with a lower variation in the ASBVs estimated from the allele frequency. Finally, the mean scaled bias was -0.0180 ± 0.108, though this deviation from zero was largely driven by the PWEC bias value (-0.67). The scaled biases are presented in Figure 3. These results show little difference between the ASBVs estimated from the mean of the ASBVs from individual animals and those estimated from the allele frequencies. This is not surprising as the calculation of breeding values is a linear function of the SNP effects. The reason for the slightly reduced precision in PWEC is likely, in part, due to the non-normality of the phenotypic distribution of PWEC data. While the transformation of the data used, largely addresses this problem, the slightly lower precision is unlikely to have a realised effect on selection decisions on farm.

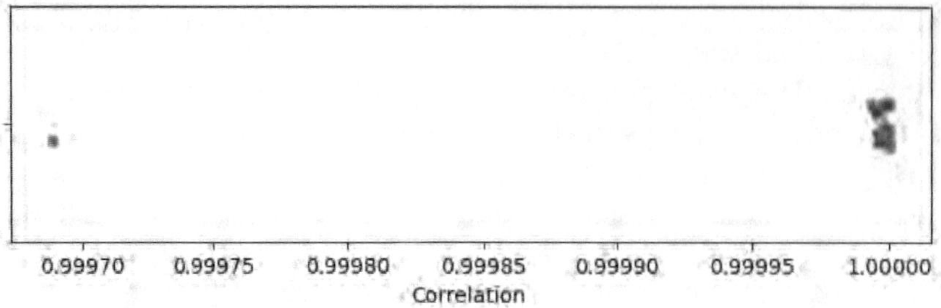

Figure 1. Correlations between ASBVs obtained from allele frequencies and from the mean of the individual animal genotypes for all flock profile traits

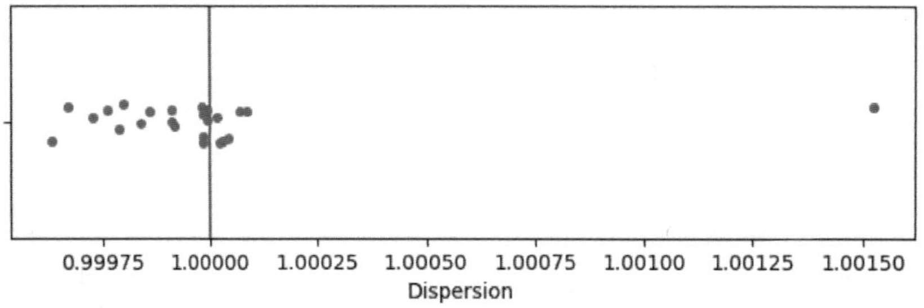

Figure 2. Dispersions between ASBVs obtained from allele frequencies and from the mean of the individual animal genotypes for all flock profile traits

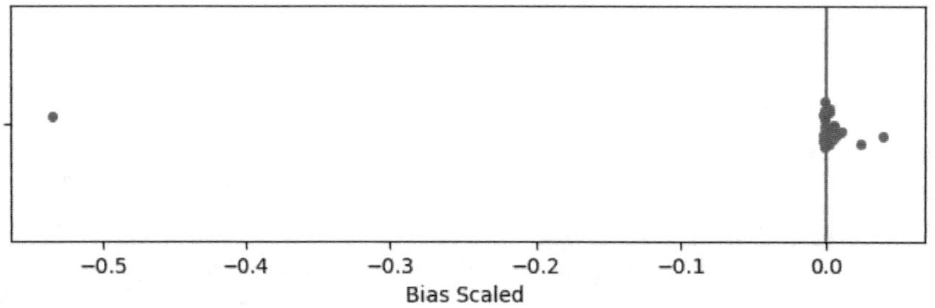

Figure 3. Biases scaled by the genetic standard deviation between ASBVs obtained from allele frequencies and from the mean of the individual animal genotypes for all flock profile traits

Various potential issues arise in processing the DNA for a pooled sample. One issue is having all individuals being represented equally within the pooled sample. High volume genotyping labs don't normalize the concentration of DNA when processing individuals (Neogen Australia, pers comms). The additional cost of normalization of DNA concentrations before pooling would mean that a direct 95% reduction in price of the flock profile would not be feasible. We expect that the cost reduction, would still be at a point where it would be beneficial, as other uses for DNA pooling have demonstrated (Bell *et al.* 2017; Aldridge *et al.* 2022). This could also allow for a larger proportion of the flock to be included, rather than the current 20 individuals, which would potentially be a better representation of that flock.

While in this paper we have used the mean of the genotype, extracting the frequency from the data generated by the genotyping platform may not be as straight forward. Janicki *et al.* (2008) present multiple methods for extracting or calculating the SNP allele frequencies. One method indicates that the Illumina Genome Studio Genotyping Module (Illumina Inc) automatically produces the B allele frequency in its reporting which was demonstrated to be acceptable as the frequency. Alternative platforms (e.g. Affymetrix or genotyping by sequence) would need to explore alternative methods.

Imputation of the pooled genotype is another issue. The current version of the Beagle software requires genotypes as input, coded 0,1,2. Version 4 of Beagle is capable of accepting a genomic likelihood, which may be usable for imputing the pooled genotype, and providing a genomic probability (Browning *et al.* 2016). Wen *et al.* (2010) have also presented algorithms specifically for pooled genotype data.

This paper demonstrates that flock profile ASBVs may be able to be calculated from a pooled genotype, however validation of the pooled sample methods would require individual and pooled genotyping results. The most cost-effective way of achieving this would be to resample existing flock profile animals using the pooling process, or to attempt this new method on breeder submitted flock profile tests alongside the current individual animal genotyping process.

CONCLUSION

This research suggests that collapsing genotypes down to the mean of the genotypes has little impact on the ASBV calculated for a flock profile. Further research is needed to determine if the pooling of DNA samples before genotype estimation can be used to reduce the costs of calculating a flock profile, including challenges of application in a commercial environment.

ACKNOWLEDGEMENTS

Funding for this research was provided by Meat and Livestock Australia under L.GEN 2204 and L.GEN 2203.

REFERENCES

Aldridge M.N., Marjanovic J., Henshall J.M., de Klerk B., Peeters K. and de Haas Y. (2022) *Proc. 12th World Congr. Genet. Appl. Livest. Prod.* 12: 781.

Bell A.M., Henshall J.M., Portro-Netro L.R., Dominik S., McCulloch R., Kijas J. and Lehnert S.A. (2017) *Genet. Sel. Evol.* 49: 1.

Browning B.L. and Browning S.R. (2016) *Am. J. Hum. Genet.* 98: 116.

Browning B.L. Tian X., Zhou Y. and Browning S.R. (2021) *Am. J. Hum. Genet.* 108: 1880.

Browning B.L., Zhou Y. and Browning S.R. (2018) *Am. J. Hum. Genet.* 103: 338.

Janicki P.K. and Liu J. (2008) *Internet J. of Genom. and Proteomics* 5: 1.

Swan A.A., Gurman P.M., Boerner V., Brown D.J., Clarke S., Gore K., Granleese T. and van der Werf J.H.J. (2018) *Proc. 11th World Congr. Genet. Appl. Livest. Prod.* 11: 346.

Wen X. and Stephens M. (2010) *Ann. Appl. Stat.* 4: 1158.

Proc. Assoc. Advmt. Anim. Breed. Genet. 25: 338 - 341

ASSESSING THE INFLUENCE OF BAYESR AND GBLUP ON SNP EFFECTS USED IN THE CORRELATION SCAN METHODOLOGY

B.S. Olasege[1,2], I. van den Berg[3], E.J. Breen[3], M. Haile-Mariam[3,6], P.N. Ho[3], L.R. Porto-Neto[2], B.J. Hayes[4], M. Goddard[3,5], J.E. Pryce[3,6] and M.R.S. Fortes[1,4]

[1]The University of Queensland, School of Chemistry and Molecular Biosciences, QLD 4072, Australia
[2]CSIRO Agriculture and Food, St Lucia, QLD 4067, Australia
[3]Agriculture Victoria, AgriBio, Centre for AgriBioscience, VIC. 3083, Australia
[4]The University of Queensland, Queensland Alliance for Agriculture and Food Innovation, St Lucia, QLD 4072, Australia
[5]Faculty of Veterinary and Agricultural Sciences, The University of Melbourne, Parkville, VIC. 3052, Australia
[6]School of Applied Systems Biology, La Trobe University, VIC. 3083, Australia

SUMMARY

The Correlation Scan (CS) identifies local genomic regions that disproportionately contribute to the genetic correlation between traits using SNP effects generated from GBLUP. BayesR, has been shown to precisely localise SNP effects, and the BayesR SNP effects size are often less shrunk than GBLUP. Therefore, we aimed to compare the SNP effects generated from GBLUP and BayesR models on the resulting localised genomic regions using the CS method. Single-trait and bivariate models were used to analyse fertility data from Brahman cows (age at detection of first corpus luteum; 996 animals) and bulls (insulin-like growth factor measured from blood; 1022 animals) genotyped with the Illumina BovineHD (770K) SNP chip. We observed that the local correlation (r) estimates were larger with GBLUP than BayesR. There were considerable differences in the r estimates on chromosome 5, 14, and X. Further analysis into the distribution of the SNP effects of a QTL region on chromosome 14 highlights the effect that each method had on CS results. GBLUP spread the effect across neighbouring SNPs, while BayesR localised the effect to a small number of SNPs, reducing the r estimates. The differences between GBLUP and BayesR were reduced with BayesR bivariate model. As BayesR bivariate model tended to select common SNPs as having non-zero effects on both traits compared to BayesR single-trait, the patterns of the r estimates were larger in the bivariate model. Other metrics from the BayesR bivariate model identified similar regions as the GBLUP in CS results. Our results showed that BayesR SNP effects can be used in our CS, but the bivariate model is recommended.

INTRODUCTION

Estimated genetic correlations between traits are useful parameters for developing and optimising animal breeding programs (Petrini *et al.* 2016). However, little is known about the local genomic regions that disproportionately contribute to these overall genetic correlations. With the widespread use of genomic data, the knowledge of local regions affecting trait correlations could allow breeders to make a more targeted genomic selection. The Correlation Scan (CS) identifies local genomic regions that contribute to estimates of the genetic correlations between traits (Olasege *et al.* 2022). The CS framework was developed using SNP effects generated from GBLUP, but it is possible to extend it for Bayesian approaches. BayesR has been shown to precisely localise SNP effects and the effect sizes are less shrunk than GBLUP (Kemper *et al.* 2015). Therefore, we used BayesR (single and bivariate models) to generate the SNP effects for the CS and compared the observed results with those obtained from GBLUP.

MATERIALS AND METHODS

The two traits used for this study were age at detection of first corpus luteum in cow (AGECL, n=980) and blood concentration of insulin-like growth hormone measured in bulls (IGF1b, n=964) from a Brahman population. A detailed description of the traits is provided by Olasege et al. (2021). The estimated genome-wide genetic correlation between these traits was -0.65 (Olasege *et al.* 2021).

SNP effects for the CS were calculated using single-trait and bivariate GBLUP (Olasege *et al.* 2022) and BayesR (Breen *et al.* 2022) models, with BovineHD 770K SNP chip. The posterior inclusion probability (PIP) and Q2 probability (the probability that the SNPs are associated with either of the traits) were also obtained from bivariate BayesR. Then local correlations (r) were estimated using the SNP effects using each model. The method to estimate r (correlation of 500 SNP effects in sliding windows between the two traits) has been previously detailed by Olasege *et al.* (2022).

RESULTS AND DISCUSSION

Single- and bivariate r estimates for the BayesR model are presented in Figure 1. The GBLUP single-trait result has been published (Olasege *et al.* 2022). The bivariate result for the GBLUP model looks identical to the single trait (result not shown). GBLUP yielded larger r estimates than BayesR. While both models identified similar windows, there were considerable differences in the r estimates on chromosome 5, 14, and X. For example, a QTL region including *PLAG1* (Fortes *et al.* 2012; Hawken *et al.* 2012) on chromosome 14 was not identified by the BayesR single-trait model. However, with BayesR bivariate model, this region was signalled.

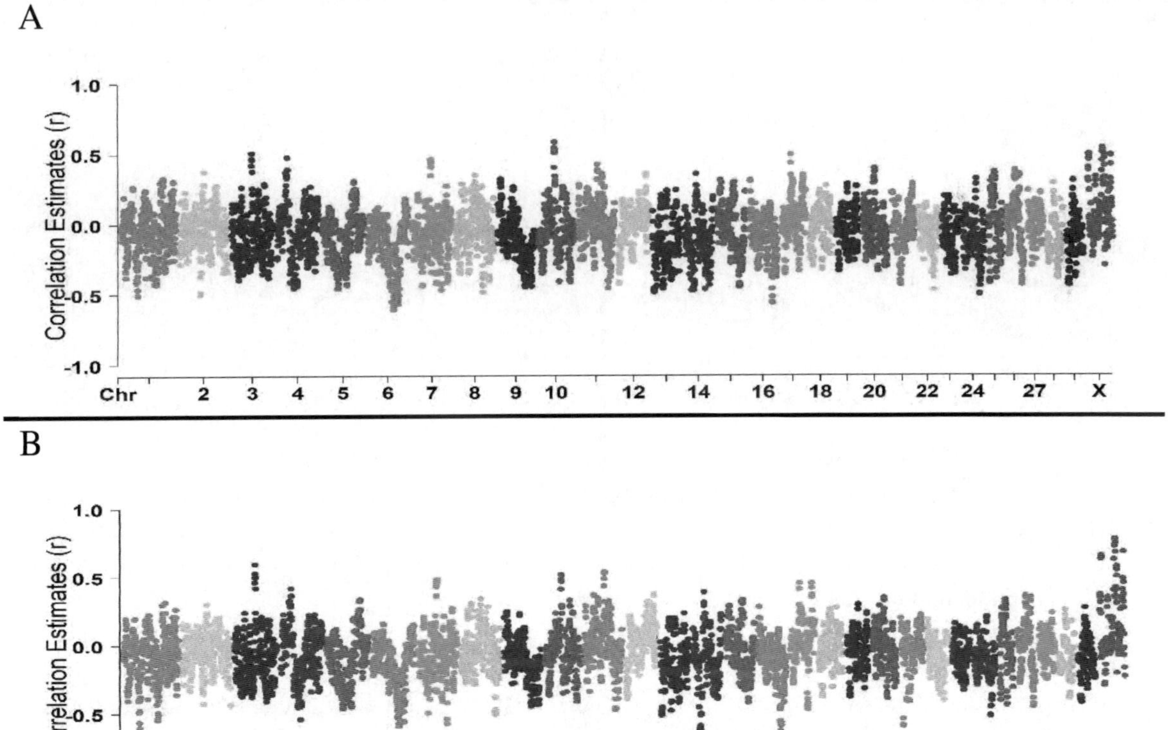

Figure 1. Genome-wide plots of the local correlation (r) estimates for age at first corpus luteum and blood concentration of insulin growth hormone for BayesR model using SNP effects from single-trait (A) and bivariate model (B)

By investigating the 100 SNP effects surrounding the *PLAG1* region between GBLUP (single-trait) and BayesR (single- and bivariate model), we found that GBLUP (Figure 2A; r = 0.96) spread the effect across neighbouring SNPs, while BayesR SNP effects were localised to a small number of SNPs. BayesR bivariate (Figure 2C; r = 0.76) identified similar SNPs for each trait as having non-zero effects whereas BayesR single trait (Figure 2B; r =0.23) often picked different sets of SNPs. Leveraging on the PIP and Q2 probability from BayesR bivariate model, the regions identified as the most significant from GBLUP CS were also signalled using Q2*PIP, showing that these two metrics could complement the CS method (Figure 3).

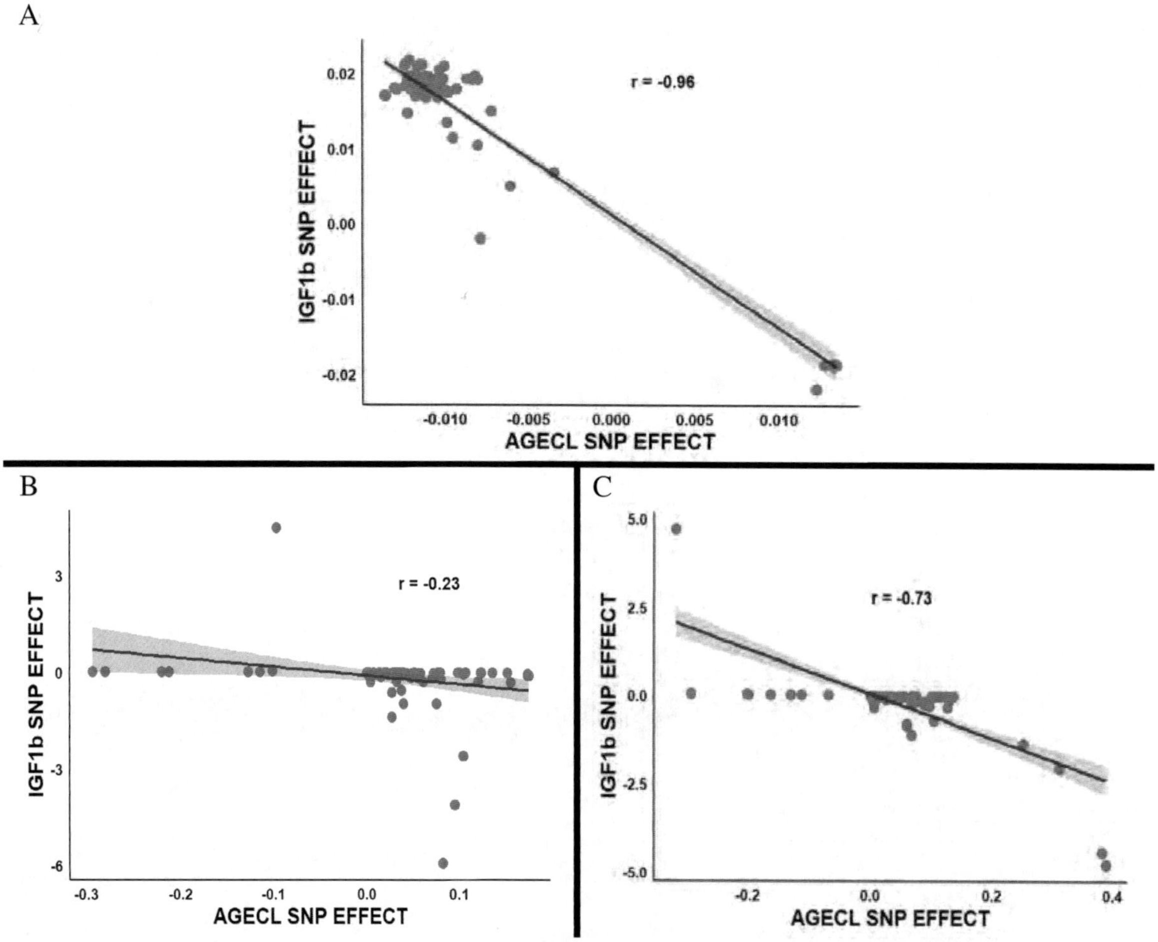

Figure 2. The regression of the distribution of the 100 SNP effects within the boundary of the *PLAG1* gene between age at first corpus luteum and (AGECL) and blood concentration of insulin growth hormone (IGF1b) using GBLUP single-trait (A), BayesR single-trait (B), and BayesR bivariate Model (C)

CONCLUSIONS

The differences in model assumptions led to differences in local correlations estimated using GBLUP and BayesR. GBLUP spreads the effect across neighbouring SNPs, whereas BayesR localised the effect to a small number of SNPs. With bivariate BayesR, SNP effects tend to be allocated to common SNPs across the traits, while BayesR single trait may select different SNPs for each trait, resulting in reduced r estimates. Our results showed that BayesR SNP effects can be used for the CS, but the bivariate model is recommended. Q2 and PIP from BayesR bivariate model could complement the CS method for insights into important QTLs.

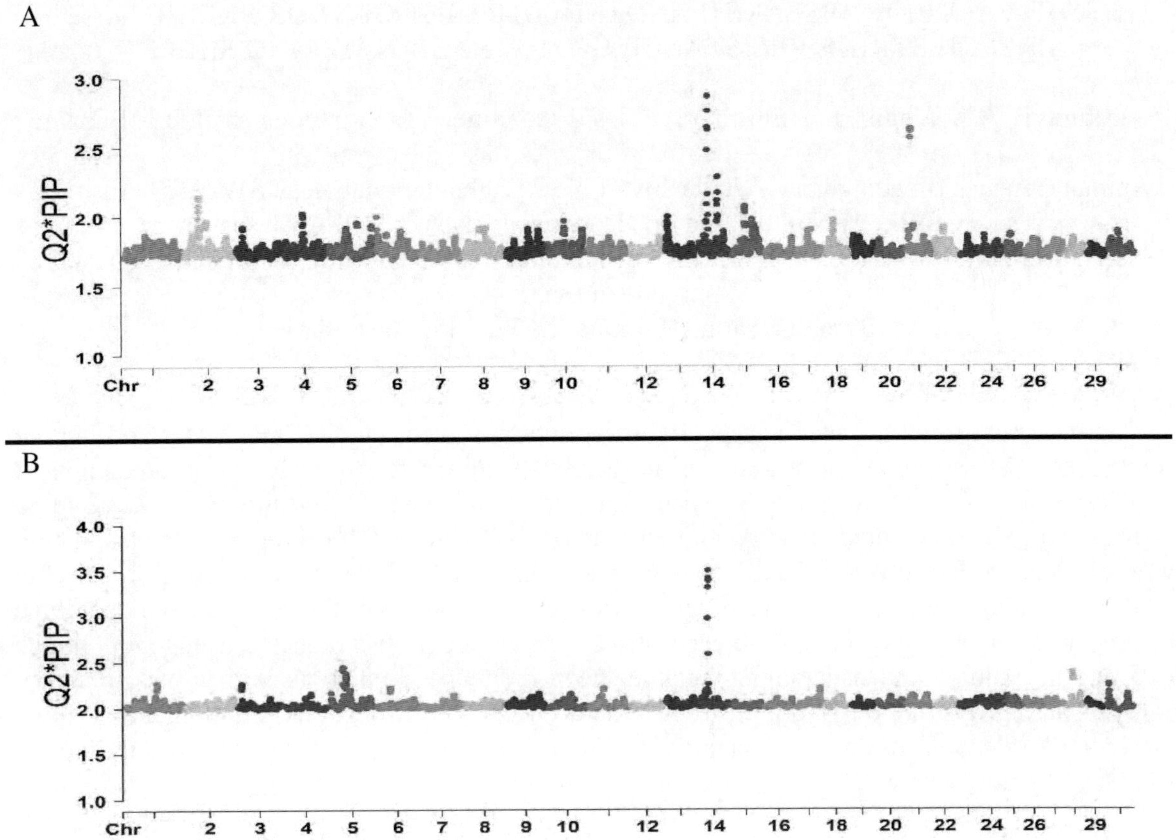

Figure 3. The posterior inclusion probability (PIP) weighted by the Q2 probability for age at detection of first corpus luteum (A) and blood concentration of insulin growth hormone (B) from BayesR bivariate model

ACKNOWLEDGEMENTS

This research was conducted using the legacy database of the Cooperative Centre for Beef Genetics Technologies and their core partners, including Meat and Livestock Australia. The authors acknowledge all the Cooperative Centre for Beef Genetics Technologies participants for their efforts in conducting the field trials and recording the phenotypes.

REFERENCES

Breen E.J., MacLeod I.M., Ho P.N., Haile-Mariam M., Pryce J.E., Thomas C.D., Daetwyler H.D., and Goddard, M.E. (2022) *Comm. Bio.* 5: 661.

Fortes M.R., Reverter A., Hawken R.J., Bolormaa S. and Lehnert S.A. (2012) *Bio. Reprod.* 87: 58.

Hawken R.J., Zhang Y.D., Fortes M.R.S., Collis E., Barris W.C., Corbet N.J., Williams P.J., Fordyce G., Holroyd R.G., Walkley J.R.W. and Barendse W. (2012) *J. Anim. Sci.* 90: 1398.

Kemper K.E., Reich C.M., Bowman P.J., Vander Jagt C.J., Chamberlain A.J., Mason B.A., Hayes B.J. and Goddard M.E. (2015) *Genet. Sel. Evol.* 47: 1.

Olasege B.S., Porto-Neto L.R., Tahir M.S., Gouveia G.C., Cánovas A., Hayes B.J. and Fortes M.R. (2022) *BMC Genom.* 23: 1.

Olasege B.S., Tahir M.S., Gouveia G.C., Kour J., Porto-Neto L.R., Hayes B.J. and Fortes M.R. (2021) *Anim. Prod. Sci.* 16: 1863.

Petrini J., Iung L., Rodriguez M.A., Salvian M., Pértille F., Rovadoscki G.A., Cassoli L.D., Coutinho L.L., Machado P.F., Wiggans G.R. and Mourão, G.B. (2016) *J. Anim. Breed. Genet.* 133: 384.

THE VALUE OF RESEARCH AND INDUSTRY FLOCKS FOR PREDICTING BREECH STRIKE RESISTANCE IN AUSTRALIAN MERINO SHEEP

E. Dehnavi[1], A.A. Swan[1], J.L. Smith[2], T.L. Bird-Gardiner[3], G. Burbidge[4] and D.J. Brown[1]

[1] Animal Genetics Breeding Unit*, University of New England, Armidale, NSW, 2351 Australia
[2] CSIRO Agriculture and Food, Armidale, NSW, 2350 Australia
[3] NSW Department of Primary Industries, Agricultural Research Centre, Trangie, NSW, 2823 Australia
[4] Burbidge Farms, Tarcutta, NSW, 2652 Australia

SUMMARY

This study used data from the Sheep CRC Information Nucleus Flock (INF), a Merino Lifetime Productivity (MLP) flock, and a ram breeding flock (Connemara) to evaluate the prediction of breeding values for breech strike between and within datasets. Cross-validation was used to evaluate the accuracy, predictability and dispersion of estimated breeding values. Validation between datasets had low predictability due to low linkage (pedigree-based) across flocks, but validation within datasets was encouraging. Considering the poor linkage between the three datasets and the low incidence of breech strike across flocks, the industry needs to continue investing in building and maintaining suitable sheep reference populations with a wide range of traits, including flystrike observations, to develop accurate predictions required to underpin direct and indirect selection. In addition, quantifying the value of genomic information to improve the accuracy of predictions will be the subject of ongoing research.

INTRODUCTION

Flystrike is estimated to be the fifth highest cost to the Australian sheep industry ($170 million per year, Lane *et al.* 2015), with breech strike identified as the most common type. Resistance to flystrike is a priority research area for Australian Wool Innovation (AWI). To make genetic progress in flystrike resistance, accurate and standardised data collection of phenotypes for flystrike, probably combined with genotyping is the first step. Establishing a well-designed sheep reference population is a crucial step (van der Werf *et al.* 2010) for developing Australian sheep breeding values (ASBVs; Brown *et al.* 2010), especially considering the different incidence rates of flystrike in various environments (Bird-Gardiner *et al.* 2013; Greeff *et al.* 2014; Smith *et al.* 2009). Therefore, this study used data from the Sheep CRC Information Nucleus Flock (INF), one of the Merino Lifetime Productivity (MLP) flocks, and a ram breeding flock (Connemara) to estimate the accuracy, predictability and dispersion of pedigree-based breeding values within and across datasets.

MATERIALS AND METHODS

Data. A phenotype for breech strike was defined as a binary trait with 0/1 indicating "struck"/"not struck" within a defined shearing period (described in detail by Dehnavi *et al.* 2023). Three datasets including animals phenotyped for breech strike (struck or non-struck) were used for this study. The first dataset was from the Sheep CRC Information Nucleus Flock (INF), including 1,335 Merino lambs born between 2008 and 2011, recorded across six research stations (Trangie, NSW; Cowra, NSW; Rutherglen, VIC; Hamilton, VIC; Struan, SA and Turretfield, SA). A second dataset with 2,115 animals from 28 sires from the New England sire evaluation site hosting a Merino Lifetime Productivity (MLP) flock at the CSIRO "Chiswick" research station at Uralla,

* a joint venture of NSW Department of Primary Industries and University of New England

NSW. The MLP animals were born in 2017 and 2018. The last dataset was 1,941 lambs born between 2017 and 2021 in the "Connemara" Merino ram breeder flock, which were the progeny of 31 sires. Routine screening for flystrike is done primarily from birth to crutching time (6-7 months) for the Connemara flock but continued for a longer period in the other flocks. For this study, all animals were assessed and considered for flystrike up to yearling age. Pedigree data for animals with phenotypes were extracted from the Sheep Genetics MERINOSELECT database.

Cross-validation analyses. Breeding values (EBVs) were estimated using a binomial model with a probit link function in ASReml (Gilmour *et al.* 2015). Models included contemporary groups (CGs) and the interaction of birth type and rearing type as significant fixed effects, and the direct animal genetic effect was considered a random effect (Dehnavi *et al.* 2023). To estimate the differences in the accuracy of predictions, an internal cross-validation procedure within each dataset and external cross-validation between datasets were tested as described by Legarra and Reverter (2018). The MLP and Connemara (CON) datasets were separated into four cross-validation groups, and INF data were grouped into three groups of approximately the same size. All animals were randomly assigned to subgroups based on their CGs (Table 1).

Table 1. The number of animals (N), sires (Sire), sires per contemporary group (Sire/CG), average incidence (Mean) and standard deviation (SD) for breech strike (0/1) for subgroups used in the cross-validation analysis

Group	N	Sire	Sire/CG	Mean	SD
INF dataset					
INF1	473	66	15.67	0.19	0.40
INF2	383	80	13.89	0.07	0.26
INF3	479	109	16.10	0.13	0.34
MLP dataset					
MLP1	579	28	11.91	0.05	0.21
MLP2	459	28	12.20	0.05	0.23
MLP3	692	28	11.09	0.05	0.22
MLP4	385	28	10.55	0.09	0.28
Connemara dataset					
CON1	368	23	6.25	0.24	0.43
CON2	562	20	6.75	0.21	0.41
CON3	511	28	7.50	0.25	0.43
CON4	500	21	7.67	0.16	0.37

Prior to generating EBVs, variance components were computed separately within each dataset, and in the combined dataset. These components were then used for the best linear unbiased prediction (BLUP) analysis and calculation of accuracy. Breeding values were estimated in the full dataset using pedigree and phenotype information for all animals. Following the analysis of the full dataset, six validation scenarios were investigated (Table 2). First, EBVs were calculated for each internal validation group (Table 1) after their phenotypes were removed, using data from the other groups of that dataset as a training population (INF – INF analysis, replicated three times; MLP – MLP and CON – CON analyses, replicated four times). Second, the prediction of each dataset was carried out using two other grouped datasets as a training population (INF+MLP – CON analysis and INF+CON – MLP analysis, each replicated four times; MLP+CON – INF analysis, replicated three times).

For each scenario, validation metrics were calculated and averaged across replicates. Accuracy and dispersion metrics were computed using the LR method (Legarra and Reverter 2018). The accuracy (LR_{acc}) was calculated as the covariance between EBVs from the full and part analysis corrected for kinship matrix and genetic variance. The dispersion (LR_{slop}) was calculated as regression slopes between the EBVs from each of the analyses (partial) with EBVs from the full analysis (whole) in the validation set. The LR_{slop} parameter is expected to have a value close to one if there is no over or under-dispersion. Pearson correlation between partial and whole EBVs was also considered as an indicator of the predictability of EBVs.

RESULTS AND DISCUSSION

Validation results followed similar patterns for two of the three different internal validation scenarios (Table 2). For the CON – CON and MLP – MLP scenarios, correlations between EBVs of each analysis and EBVs of full analysis in the validation set were 0.89 and 0.83, respectively and for the INF – INF scenario the correlation was 0.53. The LR_{acc} was low ranging from 0.14 for INF – INF to 0.37 for MLP – MLP internal scenarios. The LR_{slop} for all internal validation scenarios was more than one. However, validation within the INF and MLP datasets was closer to one (1.06 and 1.05, respectively). CON and MLP could not predict breech strike in the INF dataset accurately. This scenario had a low correlation (0.35), very low accuracy (0.08) and a high LR_{slop} (1.21). INF alone could not predict animals externally (results not shown in Table 2). INF with CON predicted MLP (INF+CON – MLP scenario) with a correlation of 0.51, accuracy of 0.13 and dispersion of 0.83 compared to the prediction of CON using INF and MLP (INF+MLP – CON scenario) with a correlation, accuracy and dispersion of 0.16, 0.06 and 0.59, respectively (Table 2).

Table 2. The number of records (N_{Train} and N_{Valid}), the percentage of progeny in the training group having common sires with the validation group (F_{Prog}), genetic variance (σ^2_g), Pearson correlation, linear regression coefficient (LR_{slop}) and accuracy (LR_{acc}) for each validation scenario (training – validation) averaged across replicates

Scenario	N_{Train}	N_{Valid}	F_{Prog}	σ^2_g	Correlation[1]	LR_{acc}[1]	LR_{slop}[1]
INF – INF	890	445	65.76	0.12	0.53 (0.09)	0.14 (0.01)	1.06 (0.14)
CON – CON	1456	485	71.40	0.09	0.89 (0.04)	0.37 (0.03)	1.12 (0.06)
MLP – MLP	1586	529	100	0.09	0.83 (0.02)	0.24 (0.01)	1.05 (0.02)
MLP+CON – INF	4056	445	1.25	0.10	0.35 (0.07)	0.08 (0.03)	1.21 (0.35)
INF+MLP – CON	3450	485	4.45	0.10	0.16 (0.12)	0.06 (0.03)	0.59 (0.05)
INF+CON – MLP	3276	529	4.49	0.10	0.51 (0.10)	0.13 (0.02)	0.83 (0.11)

[1] Standard deviation for evaluation metrics is presented within parenthesis.

The internal-validation scenarios for MLP and CON resulted in higher prediction accuracy compared to the INF dataset. This may be because the INF dataset consists of different flocks subjected to different fly control regimes across a range of environments with a large degree of between-strain genetic variances (Swan *et al.* 2016), whereas the other two scenarios were performed within one flock (Connemara and New England sites), and in the case of CON, without pre-emptive fly control. However, the genetic variance of breech strike was low for all datasets with slightly more variation for INF (Table 2). Additionally, INF had a lower percentage of link progeny from common sires between training and validation data (66% for INF compared to 71% and 100% for CON and MLP, respectively; Table 2).

The accuracy of genomic predictions (Habier *et al.* 2010) and parameter estimation (van der Werf *et al.* 2010) can benefit from larger reference populations. Accurate and consistent data recording in seed stock flocks can contribute to establishing a reference population for the industry (Alexandri *et al.* 2022). In this study, there were low levels of linkage which contributed to low correlations and accuracy when predicting breeding values between datasets.

Overall, the predictability of breeding values for validation animals was lower between datasets than within datasets. This shows the necessity of strategic data collection, especially from flocks that are well-linked externally to be able to predict animals across flocks with different incidence rates accurately. It is important to note that the effectiveness of data also depends on the quality of the trait measured, its incidence rate and diversity within and between flocks as well as the influence of environmental effects recorded on the flock. Genomic information can fill the gaps in the pedigree-based relationship matrix and this is likely to lead to better genetic connections between data sets. Therefore, investigating the impact of genomic versus pedigree information on predictions between datasets will be a focus of ongoing research.

CONCLUSIONS

This study demonstrated that flystrike was predictable within each of the three datasets used for this study, but predictions between datasets were not feasible due to the low genetic linkage established through pedigree alone. In order to build a reference population for predicting flystrike it is critical to establish well-linked flocks across environments. The ideal flock has accurate and standardised data collection including phenotypes for different flystrike types (breech and body strike), along with phenotypes for production and indicator traits and genotypes for a large number of animals.

ACKNOWLEDGEMENTS

This work was funded by Australian Wool Innovation Ltd (project ON-00775) and used data provided by the Sheep CRC, Merino Lifetime Productivity project and Burbidge Farms. The Australian Government supports the research, development and marketing of Australian wool through Australian Wool Innovation Ltd.

REFERENCES

Alexandri P., Walkom S. F., Swan A. A., van der Werf J.H.J. and Brown D. J. (2022) *Proc.* 12th WCGALP, Rotterdam, The Netherlands.

Brown D.J., Swan A.A. and Gill J.S. (2010) *Anim. Prod. Sci.* 50: 1060.

Bird-Gardiner T.L., Brown D.J., Smith J. and Mortimer S.I. (2013) *Proc. Assoc. Advmnt. Anim. Breed. Genet.* 20: 183.

Dehnavi E., Swan A.A., Ramsay A.M.M., Bird-Gardiner T.L. and Brown D.J. (2023) *Proc. Assoc. Advmnt. Anim. Breed. Genet.* 25: *These Proceedings.*

Gilmour A. R., Gogel B. J., Cullis B. R., Welham S.J. and Thompson R. (2015) ASReml User Guide Release 4.2 structural specifications. In: *VSN International Ltd: Hemel Hempsted*, UK.

Greeff J.C., Karlsson L.J.E. and Schlink A.C. (2014) *Anim. Prod. Sci.* 54: 125.

Habier D., Tetens J., Seefried F.R., Lichtner P. and Thaller G. (2010) *Genet. Sel. Evol.* 42: 5.

Lane J., Jubb, T., Shephard R., Webb-Ware J. and Fordyce G. (2015) Final report of project B.AHE.0010, Meat and Livestock Australia Ltd.

Legarra A and Reverter (2018) *Genet. Sel. Evol.* 50: 53.

Smith J.L., Brewer H.G. and Dyall T. (2009) *Proc. Assoc. Advmnt. Anim. Breed. Genet.* 18: 334.

Swan A.A., Brown D.J. and van der Werf J.H.J. (2016) *Anim. Prod. Sci.* 56: 87.

van der Werf J.H.J., Kinghorn B.P. and Banks R.G. (2010) *Anim. Prod. Sci.* 50: 998.

DEVELOPMENT OF A NEW BREEDPLAN OBJECTIVE BODY COMPOSITION EBV TO ALLOW SELECTION TO IMPROVE COW SURVIVAL

M.L. Wolcott, D.J. Johnston, M.G. Jeyaruban and C.J. Girard

Animal Genetics Breeding Unit*, University of New England, Armidale, NSW, 2351 Australia.

ABSTRACT

This study presents a new method to describe body composition in lactating cows which is under examination as the basis for a new BREEDPLAN EBV. A phenotypic prediction model was applied to generate predicted cow body composition (CBC), fitting body condition score in 1252 lactating Brahman females at their second mating as the dependent variable, with cow liveweight weight (WT), hip height (HH), scanned P8 fat depth (P8) and eye muscle area (EMA), as predictors, along with significant fixed effects. All main effects were significant in the final model as were the effects of P8*P8 and LWT*EMA. The final model included these terms, along with significant fixed effects, and had an r^2 of 0.82. CBC was calculated applying coefficients generated from the final model, when fitted with animal as random to account for genetic effects. Heritabilities for objective cow body composition traits ranged from 0.43 to 0.75 and CBC had a heritability of 0.52. This was substantially higher than the heritability estimated for cow body condition score submitted to BREEDPLAN for lactating Brahman cows at weaning of their calves (0.16). CBC presents a new opportunity to include a trait in the BREEDPLAN evaluation to describe the genetic difference in body composition for breeding females, and an indirect means for selection to improve cow survival.

INTRODUCTION

Australia's beef producing environments are characterised by seasonal feed quality and quantity, which can see females enter the mating season in sufficiently low body condition to impact reproductive performance and, in extreme situations, survival. Fordyce *et al.* (1990), in a rare study of actual cow survival under extreme drought conditions in northern Australia, showed a strong phenotypic relationship of lower cow body condition score at the start of supplementary feeding with lower chance of survival ($P < 0.05$) to the end of the study. Results from the Beef CRC (Wolcott *et al.* 2014) showed that cow body condition score assessed by experienced technicians was heritable in Brahman females at first calving, and at the start of their second annual mating as first lactation cows, 3.4 months later ($h^2 = 0.27$ and 0.48 respectively). That study also showed that body condition across the lifetime of a cow was at its lowest at the start of mating 2, with first-lactation cows losing an average of 52kg liveweight, 14cm^2 scanned eye muscle area, 5mm scanned P8 fat depth while gaining 0.6cm hip height from pre-calving measurements to the start of their second annual mating. The inclusion of cow body condition in the BREEDPLAN evaluation has been a topic of research for some time, and early results (Johnston *et al.* 1996) showed the trait was moderately heritable ($h^2 = 0.14$ to 0.21) when assessed by breeders scoring lactating Angus and Hereford females recorded at the weaning of their calves. As a consequence, breeders submitting mature cow weight records, at the weaning of their calves, for BREEDPLAN analysis have been encouraged to collect and submit body condition score (evaluated on a five point, 1 – 5 scale) at the same time. The study also concluded that including objective cow fat depth may be a better means of describing cow body composition than condition score for genetic evaluation. More recently, Granleese and Clarke (2019) evaluated body condition scores submitted by Angus breeders at the weaning of their calves, and reported a very similar heritability ($h^2 = 0.16$), and concluded that adequate genetic variation existed for the trait to be improved by selection in that breed.

---

* A joint venture of NSW Department of Primary Industries and University of New England

This study aimed to develop a new means of describing cow body composition based on objective measurements collected in the reference population for Australian beef breeders, and to contrast the genetic parameters for this new trait with those for body condition score assessed by breeders, in lactating females, at the weaning of their calves.

MATERIAL AND METHODS

Reference population cow management and body composition traits. The animals evaluated for this component of the study were from the Beef CRC's Northern Breeding Project, and the subsequent Repronomics™ project (n = 535 and 717 respectively) and comprised lactating Brahman females, as they entered their second annual mating. Breeding and management of Beef CRC females up to their first mating was described by Barwick *et al.* (2009), and Johnston *et al.* (2014) described cow management and traits recorded from their second annual mating, while Johnston *et al.* (2017) described management and recording protocols for Repronomics™ cows. In both projects, females were first mated as two year olds, at an average age of 25 months.

At the start of the second annual mating period (at an average of 37 month of age) objective body composition measurements of liveweight (LWT) hip height (HH), scanned P8 fat depth (P8) and eye muscle area (EMA); along with a subjective body condition score (BCS on a 15 point, 1- to 5+ scale) were recorded for all females (Wolcott *et al.* 2014). Models for cow body composition traits included fixed effects which described cohort (year and location of birth), property of origin, month of birth with the age (in months) and sex of the calf at foot at the time of recording, and all first order interactions. Final models for each trait were determined by sequentially removing non-significant terms (P> 0.05) terms. Variance components were estimated using ASReml (Gilmour *et al.* 2009), fitting animal as random with relationships described using a three generation pedigree.

Predicted cow body composition (CBC). A phenotypic prediction equation was developed in SAS (SAS Institute Inc., Cary, NC, USA). Cow condition score assessed in lactation females at the start of their second mating season (BCS) was fitted as the dependant variable with the initial models including LWT, HH, P8 and EMA as covariates, and their first order interactions. The initial models also included fixed effects, which described the cows' year of birth and the location in which they were managed (cohort), their month of birth and property of origin, along with the month of birth of their calf at foot and its sex. The final model was arrived at by sequentially removing non-significant terms (P> 0.05) terms, and contained the main effects of LWT, EMA, P8 and EMA, and interactions of P8*P8 and LWT*EMA (r^2 = 0.82). Significant fixed effects described the cows' cohort, their property of origin, the month in which they were born, along with the month of birth and sex of their calf at foot, and first order interactions of cohort*month of birth, property of origin*month of birth and cohort*calf month of birth.

This model was fitted in ASReml, with animal as random to account for genetic effects, and with the specification that solutions for fixed effects and covariates be estimated setting the mean to zero. CBC was calculated applying the resulting solutions for LWT, EMA, P8 and EMA, P8*P8 and LWT*EMA to produce a prediction of lactating cow into mating 2 body composition in the units of body condition score. A particular advantage of this method is that it allowed the application of nonlinear relationships of objective traits with body condition score, which would not be accommodated in a multi-trait genetic model where objective traits were included in the evaluation and genetic co-variances allowed to describe their relationship with body condition score.

Industry cow body condition score. Breeders have been encouraged to submit condition score recorded on a six point 1 (poor) to 6 (fat) scale for lactating females at the weaning of their calves (WBCS), to allow genetic parameters for the trait to be estimated (BREEDPLAN 2022). Records for the trait collected from 2010 were extracted from the BREEDPLAN database for these analyses. The records analysed for this component of the study were limited to those assessed and submitted by breeders (N = 1,693), and excluded WBCS recorded in reference and research populations, by more experienced technicians, to specifically describe the genetic parameters for records coming from the industry. An important difference between the mating 2 records analysed for the reference population and those from industry was the range in ages at which the latter were collected, with industry females ranging from 2.5 to 10.5 years of age. A very small proportion of females had multiple records, but those beyond their first record were removed from the analysis as there was insufficient data to run an effective repeatability model. Estimation of genetic parameters for WBCS applied the modelling methods described by Graser *et al.* (2005) for mature cow weight, fitting contemporary group, age at measurement, the age of the cow's dam at her birth, and the age and sex of the calf weaned when the record was collected. Variances were estimated in ASReml (Gilmour *et al.* 2009), fitting animal as random with relationships described using a three generation pedigree.

RESULTS AND DISCUSSION

By focusing on body composition in lactating cows as they enter their second annual mating, the intention was to describe females at the time of greatest challenge to their ability to maintain energy reserves. Table 1 presents descriptive statistics, variance components and the resulting heritability (and its standard error) for the traits examined in this study. Lactating first calf Brahman cows averaged 402.5kg liveweight, had an average of 3.5mm of P8 fat, 43.7cm^2 EMA, and an average BCS of 2.5 at this critical stage in their development.

Table 1. Number of records (N), mean and standard deviation (sd), additive (σ^2_a) and phenotypic (σ^2_p) variances, heritabilities (h^2) and their standard error (se) for predicted mating 2 body composition in lactating first calf Brahman females, and it's component traits, and for industry submitted cow body condition score at weaning (of their calves)

Trait[1]	Units	N	Mean	sd	σ^2_a	σ^2_p	h^2	se
LWT	kg	1,252	402.5	56.5	734.7	1301.9	0.56	0.10
HH	cm	1,252	138.5	7.8	13.2	18.3	0.72	0.09
P8	mm	1,252	3.5	2.6	1.7	3.9	0.45	0.09
EMA	cm^2	1,252	43.7	9.9	16.8	38.9	0.43	0.10
BCS	0-15 score	1,252	2.5	0.6	0.06	0.15	0.43	0.07
CBC	0-15 score	1,252	2.4	0.4	0.05	0.09	0.52	0.10
WBCS	1 to 6	1,693	3.1	0.6	0.05	0.33	0.16	0.06

[1] LWT, HH, P8, EMA and BCS describe measures of liveweight, hip height, ultrasound scanned P8 fat depth and eye muscle area, and body condition score recorded in lactating females as they enter their second annual mating respectively. CBC is predicted cow body composition at mating 2, and WBCS is body condition score submitted by Brahman breeders for lactating females at the weaning of their calves.

Predicted lactating cow into mating 2 body composition. Phenotypic prediction presents opportunities to describe relationships of objective cow body composition traits with BCS which are not available when all traits are included in the genetic evaluation and associations exploited via their co-variances. The most important was the capacity to model the significant non-linear relationships identified for P8 fat depth and the interaction of liveweight with scanned eye muscle area. The coefficients generated to estimate lactating cow body compositon showed that higher WT, EMA and P8 were associated with higher CBC, while the regression cofficients for HH, P8*P8 and LWT*EMA were negative. The magnitude of coefficients meant that the negative solutions for P8*P8 and LWT*EMA had a moderating effect on positive linear effects for the main traits, resulting in a unit increase in LWT, P8 and EMA being associated with greater increases in CBC in animals at the lower end of the distribution, than was the case for heavier, fatter and better muscled cows. A negative coefficient for HH reflects industry perceptions that taller females require greater energy input to maintain condition, and highlight the importance of having some description of frame size in the breeding objective, and the gentic evaluation, for Australia's beef breeders.

Genetic parameter estimates. Genetic parameters for LWT, HH, EMA, P8 and BCS (Table 1) were consistent with those reported by Wolcott *et al.* (2014) (h^2 = 0.65, 0.62, 0.42 0.67 and 0.48 respectively). Johnston *et al.* (1996) reported a heritability of 0.14 to 0.21 for breeder recorded BCS in Angus and Hereford cows at the weaning of their calves, and this was consistent with the result presented by Granleese and Clarke (2019) (h^2 = 0.16), and that reported here for WBCS submitted by Brahman breeders (h^2 = 0.16). CBC was more heritable (h^2 = 0.52) than BCS (h^2 = 0.43), which reflected the higher heritabilities estimated for component HH and LWT traits (h^2 = 0.72 and 0.56). The capacity of lactating females to have adequate body condition at mating is prominent in the breeding objective for almost all beef breeds and production systems in Australia. A description of cow condition that incorporates objective body composition information is very closely aligned to this objective and presents new opportunities for breeders to select and improve genetic gains.

CONCLUSIONS

CBC describes cow body composition in the units of condition score, which is familiar to Australia's beef breeders, while providing a more objective and heritable description of the trait at a critical time for beef females. By making it a trait of lactating cows only, it is independent of the effects of reproduction and, as such, can be a basis for selection to reduce the risk of wet cows falling to critically low body condition. It also presents the opportunity to monitor genetic cow body composition as selection pressure is applied to improve other aspects of female productivity.

ACKNOWLEDGEMENTS

The analyses for this study were undertaken as part of MLA project L.GEN.2204, and phenotypes were generated as part of the Beef CRC and Repronomics™ projects (BN.NBP.0759). The funders of these projects are gratefully acknowledged, as are the technical staff whose efforts made data available for these analyses.

REFERENCES

Barwick S.A., Johnston D.J., Burrow H.M., Holroyd R.G., Fordyce G. Wolcott M.L., Sim W.D., and Sullivan M.T. (2009). *Anim. Prod. Sci.* 49: 367.

BREEDPLAN (2022) Recording cow condition score. Available at: https://breedplan.une.edu.au/recording-performance/recording-cow-condition-score/ (Verified 6 February 2023).

Fordyce G., Tyler R. and Anderson V.J. (1990) *Aust. J. Exp. Agric.* 30: 315.

Gilmour A.R., Gogel B.J., Cullis B.R. and Thompson R. (2009). 'ASReml user guide. Release 3.0.' VSN International: Hemel Hempstead, UK.

Granleese T. and Clarke S.A. (2019). *Assoc. Advmt. Anim. Breed. Genet.* 22: 393.

Graser H.U., Tier B., Johnston D.J. and Barwick S.A. (2005) *Aust. J. Exp. Agric.* 45: 913.

Johnston D.J., Chandler H. and Graser H.U. (1996) *Aust. J. Exp. Agric.* 47: 1251.

Johnston D.J., Barwick S.A., Fordyce G., Holroyd R.G., Williams P.J., Corbet N.J. and Grant T. (2014) *Anim. Prod. Sci.* 54: 1.

Johnston D.J., Grant, T.P., Schatz, T.J., Burns, B.M., Fordyce, G. and Lyons R.E. (2017) *Assoc. Advmt. Anim. Breed. Genet.* 22: 393.

Wolcott M.L., Johnston D.J., Barwick S.A., Corbet N.J. and Williams P.J. (2014) *Anim. Prod. Sci.* 54: 37.

SERUM HEALTH BIOMARKERS SIGNIFICANTLY CORRELATED WITH GENE EXPRESSION IN TRANSITION DAIRY COWS

A. Kudriashova[1,2], A.J. Chamberlain[1,3], J.E. Newton[1], C.R. Bath[1], C.J. Vander Jagt[1], C.M. Reich[1], B.A. Mason[1], J.E. Hemsworth[1] and M.E. Goddard[1,2]

[1] Agriculture Victoria, AgriBio, Centre for AgriBioscience, Bundoora, VIC 3083 Australia
[2] Faculty of Veterinary and Agricultural Sciences, University of Melbourne, VIC 3647 Australia
[3] School of Applied Systems Biology, La Trobe University, Bundoora, VIC, 3083 Australia

SUMMARY

The transition to lactation often results in health issues that impact on longevity of a dairy cow in the herd. Physiological processes involved in energy metabolism and immune response during this period can be measured by blood health biomarkers. These processes are partly genetically driven. In this study, we aim to determine gene expression patterns in circulating leukocytes and investigate associations with serum health biomarkers during the transition period. A single blood sample was collected within 21 days of calving from 110 commercial dairy cows, located on 5 farms in south-eastern Australia. Samples were used for RNA sequencing and serum analysis for glucose, β-hydroxybutyrate (BHB), non-esterified fatty acids (NEFA), urea, albumin, globulin, albumin to globulin ratio (A:G), triglycerides, bilirubin, cholesterol and total protein content. Of the 12470 expressed genes, 2641 were significantly correlated with serum biomarkers. Immune response pathways associated with serum health biomarkers included the chemokine signalling pathway being significantly correlated with total protein and albumin and the NOD-like receptor signalling pathway being significantly correlated with triglycerides. Urea was enriched with the most pathways. We also identified genes previously associated with negative energy balance (NEB) and several genes correlated with multiple biomarkers. This study adds to our understanding of the pathways that are contributing to transition cow health. Future work will identify which of these gene expression changes are under genetic control and the associated variants that can be used in a genomic prediction for transition health.

INTRODUCTION

The transition period from late pregnancy to early lactation for dairy cows coincides with considerable metabolic stress and impacts longevity of a dairy cow in the herd. During this period, the energy requirements of lactating dairy cows cannot be met from feed. This leads to nutrient shortages and NEB when body reserves are mobilised. A prolonged period of NEB is associated with dairy production diseases like mastitis, metritis, retained fetal membranes, abomasum displacement, milk fever and ketosis (LeBlanc *et al.* 2006). The transition period is also often accompanied by overt inflammation response that occurs shortly after parturition and normally is resolved within 3-4 days. However, the immune and inflammatory response is often dysregulated during early lactation. The failure to adapt to metabolic changes and resolve inflammatory reactions may reduce future lactation and reproduction performance (Bradford *et al.* 2015).

Improved animal health and resilience during early lactation could be achieved through genetic selection. Blood is widely used readily accessible multi-organ biofluid. Gold standard blood serum metabolic profile tests include biomarkers associated with energy balance (glucose, BHB and NEFA), immune status (A:G, globulins), protein nutritional status (urea, albumin) (Overton *et al.* 2017), lipid metabolism (triglycerides, cholesterol) and liver function (bilirubin). These biomarker levels are heritable traits and could be used for genomic prediction to improve animal health during the transition period (Luke *et al.* 2019). They also can provide accurate data for both clinical and subclinical health disorders. The mainly positive genetic correlation between the traits suggests that selective breeding can improve the overall health of dairy cows during the transition period (Pryce *et al.* 2016).

Gene expression in blood leucocytes can help to identify biological processes underlying metabolic changes during the transition period. The gene expression pathways can help to identify candidate genes of biological significance for further genome-wide association studies (Pryce *et al.* 2020). The present study was performed to increase the understanding of metabolic adaptation of the dairy cow during the transition period. The aim was to determine gene expression patterns in circulating leukocytes and investigate their associations with serum health biomarkers during the transition period.

MATERIALS AND METHODS

Blood samples were collected within 21 days after calving from 110 multiparous cows in 5 dairy herds in south-eastern Australia. All farms had pasture-based feeding systems with supplementary forages and concentrates fed during milking time. A single blood sample (approx. 8 mL) was collected from the coccygeal vein. Whole blood (0.5 mL) was immediately subsampled into RNAprotect Animal Blood Tubes (QIAGEN) containing RNA protectant. The remaining sample was incubated for 30-60 min at 22^0 C in the dark to optimise clotting, then centrifuged at 1,500 g for 10 minutes and serum retained. Quantification of serum biomarkers was performed in either Regional Laboratory Services (Benalla, Victoria, Australia) or AgriBio (Melbourne, Victoria, Australia). RNA was isolated using the RNeasy Protect Animal Blood Kit (Qiagen), libraries prepared using Nextflex Rapid Directional RNA-Seq Kit 2.0 (Perkin Elmer) and sequenced in a 150 cycle paired end run on the NovaSeq6000 (Illumina Inc).

All paired reads that passed trimming and quality filtering were aligned to the bovine genome ARS-UCD1.2 merged with Btau5 Y and its associated annotations using STAR v2.5.3 (Dobin *et al.* 2013) 2-pass mapping and default settings. Alignment files (.bam) with greater than 15 million read pairs and greater than 83% mapping rate were used for gene count matrix generation. A gene count matrix was generated using Subread v1.5.1 (http://subread.sourceforge.net/). Gene expression data quality was assessed by generating a multidimensional scaling plot. The gene count matrix was normalised with the Bioconductor software package edgeR in R Studio (Robinson *et al.* 2010).

Statistical analysis was performed using R version 4.2.1 (R Core Team 2022). A fixed effect model was fitted to assess the effect of parity, farm, breed, and days in milk on the gene counts and blood biomarkers.

$$y_{ijklm} = \mu + P + F + B + DIM + e_{ijklm},$$

where y is the biomarker concentration (BHB, NEFA, glucose, albumin-globulin ratio, albumin, globulin, total protein, bilirubin, cholesterol, triglycerides, urea), μ is the mean, P is parity (1 to 4 and 5+), F is the effect of farm, B is the effect of breed, DIM is days in milk (from 1 to 21) and e is the random error term. Residuals adjusted for fixed effects of parity, farm, breed and DIM indicated between cow variation in biomarkers and gene expression.

The relationship between serum biomarkers and normalised gene count was investigated by calculating the Pearson correlation between the residuals. As correlation between residuals was not normally distributed, the correlation between the raw gene counts and raw biomarkers were also investigated. Genes with significant correlation with both raw data and the residuals were identified for pathway analysis. Enrichment analyses of biological pathways (KEGG) and gene ontology terms (GO) were conducted using DAVID Bioinformatics resources (https://david.ncifcrf.gov/).

RESULTS AND DISCUSSION

In our study, 2,641 out of 12,470 genes were significantly correlated (P<0.05) with serum biomarkers. Ninety-eight of these genes had unknown function. The highest number of genes was significantly correlated with total protein (830), followed by albumin (458), urea (433), BHB (419),

A:G (394), globulin (359), triglycerides (252) and glucose (207). The lowest number of genes were correlated with cholesterol (109), NEFA (79) and bilirubin (74). The number of biomarkers used for all samples was unequal due to the differences in time of blood collection which might affect the power of different tests (Table 1).

Table 1. Descriptive statistics of the datasets used in this study, including number of samples, mean and standard deviation of serum health biomarkers, number of genes significantly correlated with biomarkers and significant KEGG pathways

Serum biomarkers	Number of samples	Mean (SD)	Number of genes	KEGG pathways (FDR)
BHB (mmol/L)	110	0.78 (0.36)	419	
NEFA (mmol/L)	110	0.57 (0.25)	79	
Albumin-Globulin ratio	82	1.12 (0.22)	394	
Globulin (g/L)	82	32.97 (6.11)	359	
Glucose (mmol/L)	34	2.74 (0.46)	207	
Bilirubin (mmol/L)	34	6.52 (3.12)	74	
Cholesterol (mmol/L)	34	2.14	119	
Triglycerides (mmol/L)	34	0.15 (0.08)	252	NOD-like receptor signalling pathway (<0.01)
Albumin (g/L)	82	35.76 (3.07)	458	Chemokine signalling pathway (<0.05)
Total Protein (g/L)	108	67.98 (6.52)	830	Chemokine signalling pathway (<0.01)
Urea (mmol/L)	110	5.12 (1.60)	433	Cell cycle (< 0.01) p53 signalling pathway (< 0.01) Oocyte meiosis (< 0.05) Progesterone-mediated oocyte maturation (<0.5) Cellular senescence (< 0.01) Human T-cell leukemia virus 1 infection (<0.05) Homologous recombination (<0.05)
All genes			2641	Cell cycle (< 0.01) Chemokine signalling pathway (< 0.01) Oocyte meiosis (< 0.01) Progesterone-mediated oocyte maturation (< 0.01) Osteoclast differentiation (< 0.01)

The 2,570 genes that were correlated with biomarkers were included in an enrichment analysis of KEGG pathways and GO terms. Genes correlated with urea were enriched for cell cycle, p53 signalling pathway and cellular senescence (FDR <0.01), and for oocyte meiosis, progesterone-mediated oocyte maturation (FDR <0.05). Pro-inflammatory chemokine signalling pathway was associated with albumin and total protein and inner immune system NOD-like receptor signalling pathway was associated with triglycerides. This association may indicate the interconnection between lipid mobilisation and immune response during the transition period. In addition, 17 genes correlated with NEFA were enriched in metabolic pathways (FDR < 0.05). Moreover, we identified genes that are known to participate in several metabolic pathways and have been previously identified as important candidate genes for NEB (Soares *et al.* 2021). These genes *(AKT2, CPT1A, CPT1B, PPARA, PPARG, PPP1R3B, PPP2R3C)* are involved in insulin resistance pathway, fatty acids metabolism, PPAR signalling pathway, AMPK signalling pathway, adipocytokine signalling pathway, and glucagon signalling pathway. Several genes in our study were associated with multiple biomarkers. For instance, 52 genes were correlated with both urea and BHB. The correlation between leukocyte gene expression and the levels of serum health biomarkers is not clearly understood and requires further investigation. Presumably, the metabolic changes in transition cow alter the gene expression in leukocytes. In our study, 178 genes negatively correlated with BHB were associated with cell cycle KEGG pathway and RNA binding molecular function which is important in the regulation of gene expression. This is in line with Minuti *et al. 2020* who identified pathways involved in genetic information processes inhibited by BHB.

The results of this study may be limited by the small sample size and unequal number of biomarkers used for all samples.

CONCLUSION

In this study, we examined the correlation between serum health biomarkers and genes expressed in leukocytes during the transition period of dairy cows. The results of this study provide evidence for the hypothesis that serum health biomarkers are significantly correlated with genes expressed in leukocytes during the transition period. This investigation identified 2641 genes significantly correlated with 11 serum health biomarkers. Some genes were correlated with several biomarkers. Significant correlations between genes that have been previously associated with the negative energy balance were found. The findings in this investigation suggest that gene expression analysis can provide a better understanding of physiological processes during NEB. The genes that are correlated with changes in metabolic health were used to identify pathways that may be associated with transition health. Further studies are needed to validate the findings and understand causation and effect of the revealed correlations.

ACKNOWLEDGEMENTS

This study was undertaken as part of the DairyBio program, which is jointly funded by Dairy Australia, the Gardiner Foundation and Agriculture Victoria and was a MAgSc research project at the University of Melbourne.

REFERENCES

Bradford B.J., Yuan K., Farney J.K., Mamedova L.K. and Carpenter AJ. (2015) *J Dairy Sci* 98: 6631.

Dobin A., Davis C.A., Schlesinger F., Drenkow J., Zaleski C., Jha S., Batut P., Chaisson M. and Gingeras T.R. (2013) *Bioinformatics* 29: 15.

Robinson M.D., McCarthy D.J. and Smyth G.K. (2010) *Bioinformatics* 26: 139.

Leblanc S.J., Lissemore K.D., Kelton D.F., Duffield T.F., Leslie K.E. (2006) *J Dairy Sci* 89: 1267.

Luke T.D.W., Nguyen T.T.T., Rochfort S., Wales W.J., Richardson C.M., Abdelsayed M. and Pryce J.E. (2019) *J Dairy Sci* 102: 11142.

Minuti A., Jahan N., Lopreiato V., Piccioli-Cappelli F., Bomba L., Capomaccio S., Loor J.J., Ajmone-Marsan P. and Trevisi E. (2020) *Funct integr genomic* 20: 293.

Overton T.R., Mcart J.A.A. and Nydam D.V. (2017) *J Dairy Sci* 100: 10398.

Pryce J.E., Nguyen T., Ho P., Luke T.D.W., Rochfort S., Wales B., Moate P., Marett L., Nieuwhof G., Abdelsayed M., Axford M., Shaffer M. and Haile-Mariam M. (2020) *Proc. Assoc. Advmt. Anim. Breed. Genet* 23: 390.

Pryce J.E., Parker Gaddis K.L., Koeck A., Bastin C., Abdelsayed M., Gengler N., Miglior F., Heringstad B., Egger-Danner C., Stock K.F., Bradley A.J. and Cole J.B. (2016) *J Dairy Sci* 99: 6855.

Soares RAN, Vargas G, Muniz MMM, Soares MAM, Cánovas A, Schenkel F and Squires EJ. (2021) *J Dairy Sci* 104: 602.

PRELIMINARY GENETIC PARAMETERS FOR FLYSTRIKE AND ITS ASSOCIATION WITH PRODUCTION TRAITS IN AUSTRALIAN MERINO SHEEP

E. Dehnavi[1], A.A. Swan[1], A.M.M. Ramsay[2], G. Burbidge[3] and D.J. Brown[1]

[1] Animal Genetics Breeding Unit*, University of New England, Armidale, NSW, 2351 Australia
[2] Stenhouse Consulting, Brunswick East, VIC, 3057 Australia
[3] Burbidge Farms, Tarcutta, NSW, 2652 Australia

SUMMARY

This study explored the phenotypic and genetic variation in breech flystrike and its relationship with production traits in the Connemara Merino ram breeding flock. Yearling breech strike (struck/non-struck) had a low heritability of 0.09±0.04 when using a binomial model. Heritability estimates for production traits ranged from 0.12±0.04 (for faecal worm egg count) to 0.70±0.07 (for staple length). Genetic correlations between breech strike and production traits varied from negative (-0.25±0.24 for fibre diameter) to high positive (0.70±0.21 for dag score), illustrating both favourable and unfavourable relationships that will have implications for future selection programs incorporating breech strike resistance with production traits.

INTRODUCTION

Flystrike, one of the most significant costs facing the Australian sheep industry, is a parasitic infection caused by the larvae of the sheep blowfly (*Lucilia cuprina*) and can cause production losses, chronic disease, and mortality (Lane *et al.* 2015). Breech flystrike is the most common type of flystrike and is a priority research area for Australian Wool Innovation Ltd (AWI). Australian Sheep Breeding Values (ASBVs) for indicator traits of breech flystrike resistance, such as wrinkle, breech cover, dag and other visual wool traits are available through Sheep Genetics (Brown *et al.* 2010) and provide the industry with tools to improve flystrike resistance through indirect selection. There is also interest in using direct measures of flystrike to further improve the accuracy of selection and also animal welfare. Furthermore, understanding the relationship between flystrike and production traits is also essential for predicting response to selection. Therefore, this study was conducted using an existing Merino sheep industry dataset to explore breech flystrike trait definition, genetic variation and its association with production and visual traits.

MATERIALS AND METHODS

Flystrike dataset. The initial dataset used for this study was 2,692 animals from the Connemara Merino ram breeding flock, located at Tarcutta, NSW, born between 2017 and 2021. There were 3,232 records from 1,364 ewes and 1,328 rams available to check the status of flystrike (including struck and non-struck animals). Routine screening for flystrike is done primarily from birth to crutching time (6-7 months) and mulesing of lambs ceased in this flock in 2018 since then regular monitoring and treatment of flystrike have been undertaken. For this study, all animals were assessed and considered for flystrike from birth to yearling age. Animals who were not affected by flystrike were identified by matching contemporary groups (CG) of affected animals based on wool traits submitted to MERINOSELECT. The contemporary groups (defined as flock, year, sex, date of measurement and management group conducted by Sheep Genetics; Brown *et al.* 2010) of these traits were used as a fixed effect for breech strike. The traits for finding CGs included records for curvature, fibre diameter, fibre diameter CV, and clean and greasy fleece weight traits. The combination of site/flock and year of birth was used to set

*Aa joint venture of NSW Department of Primary Industries and University of New England

minimum thresholds of 0.05 and 20 for frequency of flystrike and number of animals within each class combination, respectively. Further restrictions were applied to remove contemporary groups within the site/flock-years with less than 5 animals and with no variation. After filtering, 1,941 animals with breech strike phenotypes (Table 1) were retained for analysis where most struck animals were observed in summer and early autumn at an average of 184 days.

Flystrike definition. Breech strike, which is an infection of fly larvae in the breech area of sheep, was defined as "struck" or "not struck", coded as 1 or 0 respectively, within a defined shearing period, in this case, birth to yearling (Table 1). Periods were bounded by shearing events and consequently, recorded animals that fall within shearing intervals were assigned a phenotype.

Production traits. Overall, the number of observed records ranged from 1,536 to 5,732 for production and visual traits recorded at the weaning to yearling stages. These traits included breech wrinkle (BWR), dag score (DAG), clean fleece weight (CFW), fibre curvature (CUV), fibre diameter CV (DCV), fibre diameter (FD), faecal worm egg count (FEC), greasy fleece weight (GFW), staple length (SL), staple strength (SS), weaning weight (WWT) and yearling weight (YWT) that were available and sufficient to analyse. Records for production traits were only included for the birth years where breech strike was observed (2017 to 2021). After adding CGs to the fixed part of the model, a range of systematic effects, including birth and rear type (single and multiple), lamb age and age of dam (linear and or quadratic) were tested for significance for each trait and included in the models for variance component estimation accordingly.

Statistical analyses. Results presented in this study were analysed in ASReml (Gilmour *et al.* 2015) using a binomial model with a probit link function for the trait breech "struck" (BRS), and a continuous model for all production traits. Models included random effects for direct genetic and maternal genetic effects, and maternal permanent environmental effects were fitted where they were significant, as shown in Table 2. The univariate models were tested and compared using likelihood ratio tests (LRT) between the full and reduced models. The best models were used for bivariate models between breech strike and other traits.

RESULTS AND DISCUSSION

Breech strike incidence across years of birth indicates that 2020 and 2021 were the lowest (8.0%) and highest (33.7%) flystrike challenge years in this flock, respectively (Table 1). New South Wales had its 6th wettest year on record (720.6mm overall) in 2021, with rainfall 30% above average since 2010 based on the annual climate summary for New South Wales 2023.

Table 1. Descriptive statistics of breech strike across years of birth. %Struck is the percentage of animals with the presence of breech struck (1 vs non-struck, 0) from birth to the yearling shearing

Year of birth	N	SD	Min	Max	%Struck
2017	402	0.46	0	1	30.1
2018	389	0.35	0	1	14.6
2019	399	0.43	0	1	24.6
2020	451	0.27	0	1	8.0
2021	300	0.47	0	1	33.7

The heritability of breech strike for this dataset was estimated as 0.09±0.04 on the probit scale (Table 2) and 0.13 from a threshold model using a Gibbs sampling method (Tsuruta and Misztal 2006; results not shown in Table 2). These estimates were lower than values reported in other studies. Early breech strike heritability (up to 8 months of age) was reported to be 0.32±0.11 (Smith *et al.* 2009) from a linear model on the logarithm-transformed breech strike, and 0.57±0.13

from a threshold model using the Gibbs sampling method (Greeff *et al.* 2014). However later, Greeff *et al.* (2021) reported a heritability of 0.07±0.01 using the linear model on the logarithm-transformed breech strike counts and a heritability of 0.51±0.10 on the logistic underlying scale in adult Merino ewes from the same flock. They also found the heritability of 0.44±0.07 and 0.69±0.18 for Merino ewes at hogget age on the normal and logistic underlying scale respectively, both using a binomial model (Greeff *et al.* 2021). Bird-Gardiner *et al.* (2013) observed heritability of 0.33 ±0.15 for breech strike using a sire linear model in Merinos and 0.30±0.10 across breeds. These differences may be related to the models applied (the binomial model in this study) and scales (the probit underlying scale in this study), and differences between the environmental conditions and management practices across sites contributing to varying levels of expression.

Another aspect to consider is the collection and accuracy of the data, particularly inferences about unaffected animals. In this study, unaffected animals were determined from their contemporary groups (described in the material and method section) recorded for wool traits. Using wool traits is currently the only method for identifying unaffected animals in each group since there is no 'roll-call' as such to identify all animals that were present and the infection by flystrike usually occurs when the animal is in wool. However, this inference may miss animals that disappear and are not recorded for wool traits.

Table 2. Descriptive statistics of traits and direct heritability (h^2), maternal heritability (m^2), and the proportion of maternal permanent environmental to the total variance (Pe^2) using a univariate animal model. Standard errors are in parentheses

Trait[1]	Unit	N	Mean	SD	h^2	m^2	Pe^2
BRS[2]	1/0	1941	0.21	0.41	0.09 (0.04)		
BWR	1-5 score	3896	2.46	0.78	0.27 (0.06)	0.05 (0.03)	
CFW	Kg	4399	1.96	0.51	0.36 (0.06)	0.04 (0.03)	
CUV	Degree/mm	5399	86.73	20.03	0.42 (0.05)		
DAG	1-5 score	1536	1.75	0.84	0.37 (0.08)		
DCV	%	5727	18.38	2.37	0.27 (0.06)	0.06 (0.03)	
FD	µm	5732	15.45	1.40	0.52 (0.07)	0.07 (0.03)	
FEC	Eggs/gm	2851	7.13	4.13	0.12 (0.04)		
GFW	Kg	5711	2.92	0.77	0.39 (0.06)	0.02 (0.02)	
SL	mm	1914	84.97	11.51	0.70 (0.07)		
SS	N/Ktex	2100	30.02	10.05	0.18 (0.05)		
WWT	Kg	3125	19.04	4.09	0.16 (0.06)	0.02 (0.03)	0.12 (0.04)
YWT	Kg	2024	36.71	9.21	0.40 (0.12)	0.01 (0.05)	

[1]Traits are ordered alphabetically and their abbreviations are explained in the text. [2] Breech strike was fitted in the binomial model on the probit scale.

Phenotypic and genetic correlations of breech strike (BRS) with most traits were consistent with Bird-Gardiner *et al.* (2013), except that of Greeff *et al.* (2014) who estimated low and non-significant correlations with wool traits. However, the result of this study was more in line with their later research (Greeff *et al.* 2021). DAG, FEC, and BWR had the highest genetic correlations (>0.60) with BRS supporting the value of these traits as indirect selection criteria in this flock, which is also supported by the other two studies. Among wool traits, clean and greasy fleece weights and fibre diameter CV had moderate positive correlations ($r_g \approx 0.50$) with BRS, and SL had a low positive correlation (r_g=0.27) with BRS, all of which were stronger than other studies (Table 3). The antagonism between fleece weight and breech strike incidence is problematic given the importance of fleece weight in industry breeding objectives and profit. Low, negative genetic correlations for FD, SS, CUR, and YWT (-0.09 to -0.25; P> 0.05) with breech strike indicate that selection for resistance to breech strike may have lower impacts on wool quality and growth traits.

Table 3. Phenotypic (r_p) and genetic (r_g) correlations between breech strike and production traits estimated in this study and in the literature

Trait	This study		Bird-Gardiner *et al.* (2013)		Greeff *et al* (2014) [1]	
	r_p	r_g	r_p	r_g	r_p	r_g
BWR	-0.01 (0.03)	0.64 (0.21)	0.17 (0.03)	0.65 (0.22)	0.20	0.18 (0.17)
CFW	0.04 (0.03)	0.50 (0.18)	0.03 (0.04)	0.28 (0.26)	0.01	0.05 (0.12)
CUV	0.02 (0.03)	-0.09 (0.22)	-0.02 (0.04)	-0.07 (0.28)	-0.08	-0.04 (0.12)
DAG	0.09 (0.04)	0.70 (0.21)	0.08 (0.03)	0.84 (0.49)	0.45	0.81 (0.15)
DCV	0.06 (0.03)	0.59 (0.20)	0.08 (0.03)	0.45 (0.27)	0.05	-0.27 (0.13)
FD	-0.04 (0.03)	-0.25 (0.24)	-0.04 (0.03)	-0.05 (0.19)	0.04	0.14 (0.12)
FEC	0.05 (0.03)	0.69 (0.29)	0.06 (0.03)	0.60 (0.30)	0.01	0.27 (0.12)
GFW	0.05 (0.03)	0.49 (0.18)	0.07 (0.03)	0.32 (0.22)	0.02	0.06 (0.11)
SL	0.02 (0.03)	0.27 (0.22)	-0.09 (0.04)	-0.10 (0.20)	-0.05	0.02 (0.14)
SS	-0.07 (0.03)	-0.21 (0.29)	-0.03 (0.03)	-0.05 (0.30)	-0.01	0.15 (0.16)
WWT	0.04 (0.03)	0.07 (0.28)				
YWT	-0.03 (0.14)	-0.15 (0.28)				

[1] Standard errors are in parentheses except those that are not published. Figures are presented only in stages similar to the present study.

CONCLUSIONS

Breech strike had a low heritability while its correlation with production and visual traits varied depending on the trait. Although indirect selection on indicator traits is of value to improve flystrike resistance, direct selection on the trait itself may help to increase response, particularly in combination with genomic information. The results of this study can be used to predict changes in flystrike resistance and production traits in response to selection for a range of breeding objectives in industry ram breeding programs. Further analysis has commenced combining data from this flock, the Sheep CRC Information Nucleus Flock (INF), the AWI Breech Flystrike Genetics flocks, sire evaluation flocks and other available research and industry flocks.

ACKNOWLEDGEMENTS

This analysis was funded by Australian Wool Innovation Ltd (project ON-00775) and data collection by Burbidge Farms. The Australian Government supports the research, development and marketing of Australian wool through Australian Wool Innovation Ltd.

REFERENCES

Bird-Gardiner T.L., Brown D., Smith J. and Mortimer S.I. (2013) *Proc. Assoc. Advmnt. Anim. Breed. Genet*. 20: 183.

Brown D.J., Swan A.A. and Gill J.S. (2010) *Anim. Prod. Sci.* 50: 1060.

Annual climate summary for New South Wales (2023) Product code IDCKGC55R0.

Gilmour A.R., Gogel B.J., Cullis B.R., Welham S.J. and Thompson R. (2015) ASReml user guide release 4.2 structural specifications. VSN International Ltd: Hemel Hempstead, HP1 1ES, UK.

Greeff J.C., Karlsson L.J.E. and Schlink A.C. (2014) *Anim. Prod. Sci.* 54: 125.

Greeff J.C., Schlink A.C. and Karlsson L.J.E. (2021) *Anim. Prod. Sci.* 61:1893.

Lane J., Jubb T., Shephard R., Webb-Ware J and Fordyce G. (2015) Final report of project B.AHE.0010, Meat and Livestock Australia Ltd.

Smith J.L., Brewer H.G. and Dyall T. (2009) *Proc. Assoc. Advmnt. Anim. Breed. Genet.* 18: 334.

Tsuruta S and Misztal I (2006) *Proc. 8th WCGALP*, Belo Horizonte, Minas Gerais, Brazil. 27.

MILK PRODUCTION AND FERTILITY OF SPRING-CALVED HOLSTEIN-FRIESIAN, JERSEY AND CROSSBRED COWS MILKED ONCE DAILY OR TWICE DAILY IN NEW ZEALAND

J.M.D.R. Jayawardana[1], N. Lopez-Villalobos[1], R.E. Hickson[2] and L.R. McNaughton[3]

[1]School of Agriculture and Environment, Massey University, Palmerston North 4410, New Zealand
[2]Focus Genetics, 17C Mahia St, Ahuriri, Napier 4144, New Zealand
[3]Livestock Improvement Corporation, Private Bag 3016, Hamilton 3240, New Zealand

SUMMARY

In New Zealand about 55% of dairy herds are milked twice daily (TAD) and about 9% of herds are milked once daily (OAD) for their entire lactation, with the balance of herds using variable milking frequencies across lactation. The objective of this study was to investigate fertility of spring-calved Holstein-Friesian (F), Jersey (J) and crossbred of Holstein-Friesian × Jersey cows (F×J) milked either OAD or TAD from 2015-2016 to 2017-2018 in New Zealand using data provided by Livestock Improvement Corporation. The dataset comprised 113 OAD and 531 TAD herds. Eight fertility traits were evaluated: submission in the first 3 weeks (SR21) and 6 weeks (SR42) of mating, in-calf in the first 3 weeks (PR21) and 6 weeks (PR42) of mating, conception to first service (PRFS), not in-calf at end of the breeding season (NIC), 3-week calving (CR21) and 6-week calving (CR42) rates. Cows milked TAD produced greater milk, fat, protein and lactose yields than cows milked OAD, but fat (FP) and protein percentages (PP) were lower in cows milked TAD. Cows milked OAD had better fertility with a higher SR21, PR21, PR42, PRFS, CR21, CR42 and a lower NIC than cows that were milked TAD. Breeds differed in fertility traits in both milking regimens. Jersey and F×J cows had higher SR21, SR42 and PRFS than F cows in OAD milking herds, whereas J cows were mated earlier in the mating season than F and F×J cows in TAD. Fertility of F×J cows was better than purebred cows in both milking populations, evidenced by these cows having the highest PR21, PR42, PRFS, CR21, CR42 and lowest NIC. Once daily milking herds benefited from higher FP and PP and better fertility than TAD herds.

INTRODUCTION

Once-daily milking is becoming popular among New Zealand dairy farmers because it benefits the farmers lifestyle, animal welfare, management of feed shortfalls and reduces the cost of labour (Bewsell *et al.*, 2008). In the production year 2015-2016, about 9% of herd-tested herds milked OAD for the whole lactation in New Zealand (Edwards 2018). Improved reproductive performance was reported with spring-calved cows milked OAD for the entire lactation compared to cows milked TAD for the entire lactation in New Zealand (Jayawardana *et al.* 2022).

The New Zealand dairy herd is comprised of crossbreed of F×J cows (49.6%), F (32.5%), J (8.2%), other breeds and crosses (9.3%) and a small proportion of Ayrshire (0.4%), (LIC and DairyNZ, 2021). Grosshans *et al.* (1997) reported breed differences in reproductive performance of New Zealand dairy cows, including shorter intervals from calving and mating to first service and conception, and higher 6-wk in-calf rates in J cows compared to their F counterparts. Lembeye *et al.* (2016) reported J cows milked OAD for their entire lactation were more efficient per kg of live weight than F and F×J cows milked OAD, but suggested that F×J cows are more suitable for OAD milking due to greater total milk solids production, intermediate feed conversion efficiency and biological efficiency. The information of breed differences in reproductive performance in OAD milking system is scare at cow level. The objective of present study was to evaluate the reproductive performance of F, J and crossbred F×J cows milked either OAD or TAD in New Zealand using data from a national herd-testing database.

MATERIALS AND METHODS

Data. Herd test milk yields, calving, mating, and pregnancy diagnosis information of spring-calved dairy cows from 2015-2016 to 2017-2018 production seasons were obtained from the animal database of Livestock Improvement Corporation. Selected herds had at least 50 cows per herd, herd tested four or more times per lactation, pregnancy test results recorded for at least 80% of cows that calved in the 12-month period, and "early aged pregnancy testing" (tested on or between 35 and 122 days of pregnancy) and fetal age estimated for at least 80% of cows in the herd. Herd test-day milking frequency was used to classify herds into OAD or TAD. If more than 90% of the tested cows on a herd-test date were milked either OAD or TAD in a herd, then it was classified as OAD or TAD milking herd on that herd-test date. Likewise, all herd tests were classified. If all herd tests were OAD throughout the season, then the herd was identified as an OAD milking herd. Likewise, if all herd tests were classified as TAD, then the herd was identified as a TAD herd. Finally, 113 OAD and 531 TAD herds were identified. Herds that were OAD at some herd tests and TAD at other herd tests were excluded.

Breeds. Information of breed composition (expressed in sixteenths) for each cow was used to classify the cows into 3 breed categories; F, J, and crossbred of F×J. Herds without F, J or crossbred of F×J cows were excluded. Cows with either less than 100% known breed proportions or more than 12.5% of any breed other than F or J were excluded. Cows were classified as F or J if they had breed compositions of F ≥14/16 or J ≥14/16, respectively and remaining cows were classified as crossbred of F×J.

Production traits. Milk production data included yields of (MY), fat (FY), protein (PY), lactose (LY) and percentages of fat (FP), protein (PP), and lactose (LP). Lactation records with days in milk ranging 150-305 days were analysed.

Fertility traits. Eight fertility traits were defined: submission for artificial insemination in the first 3 weeks (SR21) and 6 weeks (SR42) of the breeding season, in-calf in the first 3 weeks (PR21) and 6 weeks (PR42) of the breeding season, conception to the first service (PRFS), not in-calf at end of the breeding season (NIC), calving by first 3 weeks (CR21) and 6 weeks (CR42) from the planned start of the calving. Conception dates were calculated as the date of pregnancy diagnosis minus the estimated foetal ages with a pregnancy status of 'pregnant'. If the estimated foetal ages were not available but cows calved in the following season, their conception date was calculated as calving date in the following season minus 282 days. Submission by 3 weeks or 6 weeks of the breeding season was coded as 1 if the first mating date was in the first 21 days or 42 days from the start of mating date, respectively, otherwise coded as 0. Likewise, in-calf by 3 weeks or 6 weeks of the breeding season was coded as 1 if the cow was pregnant in the first 21 days or 42 days from the start of breeding season, respectively, otherwise coded as 0. The variable PRFS was only calculated for cows whose first service was to artificial breeding, and was coded as 1 for cows where date of first service equalled date of conception, and 0 otherwise. Pregnancy status at the last pregnancy testing after the end of the breeding period was used to classify the NIC, cows with pregnancy status 'empty' were coded as 1 whereas cows with pregnancy status 'pregnant' were coded as 0. Cows with last pregnancy status as doubtful but calved in the subsequent season were coded as 0, otherwise 1. Planned start of calving date was obtained for a herd by adding 282 d to the herd's mating start date in each calving season. If a cow calved in the first 3 weeks or 6 weeks from the planned start of calving date then it was coded as 1, otherwise 0. The detailed description of editing the fertility traits and calculation of conception in the present study was described in Jayawardana *et al.* (2023). Cows in their first four parities were considered separately and cows of parity five and above were combined into one category.

Statistical analysis. The statistical analyses were undertaken using SAS version 9.4 software. The production traits were analysed using HPMIXED procedure and all fertility traits with binomial distribution were analysed using the GLIMMIX procedure after a logit transformation. Contemporary groups were defined as the group of cows calving in the same herd and year. The model included the fixed effects of milking regimen, breed, parity, interaction of milking regimen and breed, linear and quadratic effects of deviation of calving date from the herd median calving date (within-herd in each calving year) as covariates, and the random effects of herd-year and residual. Least-squares means with logit scale were back-transformed into the nominal scale for interpretation of the results.

RESULTS AND DISCUSSION

Cows milked TAD produced greater yields of milk, fat, protein and lactose and higher LP than cows milked OAD, but had lower FP and PP and poor fertility outcomes. Results indicate that a higher proportion of cows milked OAD were mated (by 4.6%) in the first 3 weeks of the breeding season, conceived in the first 3 weeks (by 10%) and 6 weeks (by 8.6%) of the breeding season, pregnant to their first service (by 6.8%), calved by 3 weeks (by 6.2%) and 6 weeks (by 4.6%) of the following calving season with a lower percentage not in-calf (by 3.7%) at end of the breeding season compared with TAD milking cows. The better reproductive performance of OAD milking cows is hypothesised to be due to OAD milking reducing the extent of negative energy balance in the early lactation cows (Kay *et al.* 2013).

Table 1: Least-squares means of milk production and fertility traits of Holstein-Friesian (F), Jersey (J) and their crossbred cows (F×J) milked in once daily (OAD) or twice daily (TAD)

Traits[1]	OAD	TAD	OAD			TAD			P-value		
			F	J	F×J	F	J	F×J	MF	Breed	MF × Breed
MY(kg)	3291	4708	3595[a]	2936[c]	3368[b]	5115[d]	4102[f]	4828[e]	<.001	<.001	<.001
FY(kg)	170.4	228.9	171.1[b]	165.2[c]	175.9[a]	230.1[e]	220.4[f]	235.2[d]	<.001	<.001	<.001
PY(kg)	134.6	182.1	141.2[a]	125.3[c]	138.8[b]	190.6[d]	166.7[f]	187.1[e]	<.001	<.001	<.001
LY(kg)	162.1	236.4	176.4[a]	145.5[c]	166.4[b]	255.6[d]	207.1[f]	242.6[e]	<.001	<.001	<.001
FP(%)	5.29	4.98	4.82[c]	5.68[a]	5.28[b]	4.55[f]	5.51[d]	4.93[e]	<.001	<.001	<.001
PP(%)	4.13	3.91	3.95[c]	4.29[a]	4.14[b]	3.75[f]	4.10[d]	3.89[e]	<.001	<.001	<.001
LP(%)	4.94	5.03	4.90[c]	4.96[a]	4.94[b]	5.00[f]	5.06[d]	5.04[e]	<.001	<.001	<.001
SR21(%)	85.3	80.7	83.9[b]	86.6[a]	86.8[a]	78.5[e]	82.5[c]	81.2[d]	<.001	<.001	0.30
SR42(%)	93.9	93.7	92.4[b]	94.8[a]	94.7[a]	92.4[e]	94.9[c]	94.0[d]	0.65	<.001	0.11
PR21(%)	55.4	45.4	53.7[b]	55.7[b]	57.8[a]	43.8[e]	45.1[d]	47.1[c]	<.001	<.001	0.94
PR42(%)	76.5	67.9	74.1[c]	77.1[b]	78.9[a]	65.9[f]	67.8[e]	69.7[d]	<.001	<.001	0.28
PRFS(%)	62.1	55.3	60.0[b]	62.9[a]	63.5[a]	54.4[d]	54.5[d]	56.6[c]	<.001	<.001	0.06
NIC(%)	9.8	13.5	10.5[a]	9.5[a]	8.6[b]	14.8[c]	13.4[d]	12.4[e]	<.001	<.001	0.92
CR21(%)	64.2	58.0	64.0[a]	64.1[a]	65.8[a]	57.0[c]	57.0[c]	59.6[b]	<.001	<.001	0.69
CR42(%)	86.6	82.0	85.7[b]	86.5[b]	87.9[a]	81.0[d]	82.2[c]	82.9[c]	<.001	<.001	0.35

[a-f] Means with different superscripts in the same row are significantly different across milking regimen and breeds ($P < 0.05$).

[1]MY = milk yield; FY = fat yield; PY = protein yield; LY = lactose yield; FP = fat percentage; PP = protein percentage; LP = lactose percentage; SR21 = cow inseminated in the first 3 weeks from the start of mating; SR42 = cow inseminated in the first 6 weeks from the start of mating; PR21 = cow conceived in the first 3 weeks from the start of mating; PR42 = cow conceived in the first 6 weeks from the start of mating; PRFS = cow conceived to first service; NIC = cow not in-calf at end of the breeding season; CR21 = cow calved in the first 3 weeks from the planned start of the calving; CR42 = cow calved in the first 6 weeks from the planned start of the calving; MF= milking frequency.

In both milking systems, F cows had greater MY, PY and LY compared with J and F×J cows, but FY was higher for crossbred F×J cows than purebred F and J cows. Jersey cows were less affected than F and F×J cows by OAD milking with a reduction in MY, PY and FY ranged 25-28%, whereas in F and F×J cows the reduction ranged between 26% to 30%. Fat, protein and lactose percentages were higher in J cows and lowest in F cows in both milking populations. Sneddon *et al.* (2015) reported that J milk was most valuable per litre in New Zealand under the milk product portfolios of whole-milk powder, skim-milk powder, cheese and butter. Milk from J milked OAD has the highest value per litre, due to the increase in fat and protein percentage. Significant interactions were found between milking frequency and breed for milk production traits in this study. However, no milking frequency × breed interactions were detected for any fertility traits. Similarly, in the experimental study by Clark *et al.* (2006), J and crossbred F×J cows were submitted for mating at similar rates, but F×J cows had superior PR42 and NIC rates than J cows. This suggests that conception rates were higher in F×J cows than J cows. Jayawardana *et al.* (2023) reported that heterosis effects of F×J cows for SR21 was lower than PR21 and PR42 in OAD (SR21=2.8% vs PR21=5.5% and PR42=4.1%) and TAD (SR21=3.1% vs PR21 and PR42=5.8%) milking systems. Across both milking regimens crossbred F×J cows had the best overall reproductive performance, and F cows had the worst reproductive performance.

CONCLUSIONS

Cows milked OAD for the entire lactation had higher FP, PP and better fertility outcomes than cows milked TAD during the entire lactation. Fertility differed among breeds in both milking systems. Jersey cows were presented earlier for mating than F cows. Crossbred F×J cows had better fertility than purebred F and J cows, they became pregnant sooner in the mating season, and calved earlier in the following season than F and J cows in both milking populations.

ACKNOWLEDGEMENTS

The first author was funded AHEAD project by the Sri Lankan Government under the funds of the World Bank. The authors acknowledge Katie Carnie for extracting the data provided by the Livestock Improvement Corporation.

REFERENCES

Bewsell D., Clark D. A. and Dalley D.E. (2008) *J Agric Educ Ext* 14: 69.

Clark D.A., Phyn C.V., Tong, M.J., Collis, S.J. and Dalley D.E. (2006) *J Dairy Sci.* 89: 1854.

Edwards J.P. (2018) *Anim. Prod. Sci.* 59: 570.

Grosshans T., Xu Z.Z., Burton L.J., Johnson D.L. and Macmillan K.L. (1997) *Livest Prod Sci.* 51: 41.

Jayawardana, J.M.D.R., Lopez-Villalobos N., Hickson R.E. and McNaughton L.R. (2023) *J. Dairy Sci.* 106: 364.

Jayawardana, J.M.D.R., Lopez-Villalobos N., McNaughton L.R. and Hickson R.E. (2022) *J. Dairy Sci.* 105: 8911.

Kay J.K., Phyn C.V., Rius A.G., Morgan S.R., Grala T.M. and Roche J.R. (2013) *J. Dairy Sci.* 96: 6274.

Lembeye F., López-Villalobos N., Burke J.L., Davis S.R., Richardson J., Sneddon N.W., and Donaghy D.J. (2016) *NZ J. Agric. Res.* 59: 351.

LIC and DairyNZ. (2021) New Zealand Dairy Statistics 2020-21, Hamilton, New Zealand.

Sneddon N.W., López-Villalobos N., Hickson R.E., Shalloo L. and Garrick D.J. (2015) *Proc. NZ Soc. Anim. Prod.* 75: 48.

Proc. Assoc. Advmt. Anim. Breed. Genet. 25: 362 - 365

CONCEPTION-BASED FERTILITY TRAIT FOR GENETIC EVALUATION OF NEW ZEALAND DAIRY CATTLE

K. Stachowicz, E.C. Ooi, B. Santos and P.R. Amer

AbacusBio Limited, PO Box 5585, Dunedin 9058, New Zealand

SUMMARY

Genetic evaluation of dairy cow fertility in New Zealand is currently based on calving season day, defined as the number of days from planned start of calving for the herd to cow calving date. This phenotype has gestation length embedded in it. Recently, a concern has been raised that shortened gestation lengths are the driving force behind good reproductive performance, as opposed to the cow's ability to conceive in a timely manner. Therefore, the goal of this research was to explore a range of possible alternative fertility phenotypes to find a replacement for calving season day that would be, at least on a phenotypic level, independent from gestation length. Using data from herds with good data quality, alternative conception-based fertility trait definitions were evaluated and compared. Variance components were estimated using ASReml software. Binary six-week in-calf rate was suggested as the best trait definition due to the relatively high genetic variance, desirable genetic and residual correlations with other fertility traits evaluated, and practicality of data recording. Further testing and validation are planned before a new conception-based fertility trait is finalised for inclusion in routine genetic evaluation.

INTRODUCTION

Over the past decade, research into calving date-based approaches to the NZ fertility evaluation has demonstrated a substantial improvement in validation accuracy for a wide range of key fertility metrics of commercial relevance to NZ dairy farmers, relative to previous prediction methods. However, the implementation of the new fertility breeding value has not been fully endorsed by all industry partners. Of particular concern was an increased role of short gestation length (GL) in driving the superior fertility predictions for new animals. When information is scarce or inaccurate on submission and conception rates, fertility breeding value estimates are likely to be dominated by GL. It can be argued that GL is not a true fertility trait, and some concerns exist about deployment of strong selection pressure for short GL. Therefore, it would be advantageous to separate fertility breeding values into three separate components: (1) resumption of cyclicity and oestrous expression (submission rate); (2) probability of getting pregnant (conception rate); and (3) GL. Submission rate (PM21) and GL are currently being evaluated in NZ. The goal of this research was to examine alternative definitions of conception-based fertility traits and recommend one that would be most suited for New Zealand dairy farming systems.

MATERIALS AND METHODS

Data. Fertility phenotypic data were extracted from New Zealand's national dairy database. This data included records from 2005 to 2014 calving seasons. Mating and calving records from the first five parities were considered. Extensive data filters were applied to obtain data from herds with good recording practices and sufficient animals. A random sample of around 30,000 cows with phenotypic records was drawn from herds meeting these criteria. Data edits and current fertility trait (CSD0 - heifer, CSD - cow and PM21) definitions were described in detail by Stachowicz *et al.* (2014). Ten conception-based fertility phenotypes were derived for testing (Table 1). They incorporate a variety of attributes, including the timing of conception (i.e., continuous - CR1 and CR2; binary versions – CR7 to CR10), the number of inseminations required to achieve conception (i.e., CR3), and conception outcomes associated with various categories of insemination (i.e., CR4 to CR6). Two versions of each phenotype were tested; one as defined as per Table 1, and one which included penalties for 'poor' fertility cows – i.e., carryover cows and cows that had been culled for infertility. For continuous traits, penalties were defined as the maximum value for the contemporary group, plus an additional 21-day oestrus cycle (for interval traits) or insemination (for number of inseminations), whereas cows with a binary trait penalty were set to 0. Conception confirmation is currently defined by non-return and the presence of a subsequent calving rather than the use of pregnancy diagnosis data; this is likely to change in the future because of industry data coordination initiatives.

Table 1. Conception-based fertility traits definitions

Trait	Trait name	Definition	Unit	Min.	Max.
CR1	Time of conception day	Days from PSM to conception	days	-21	100
CR2	Interval from first to last insemination	Days from first insemination to conception	days	0	100
CR3	Number of inseminations	Number of inseminations within season	score	1	10+
CR4	Pregnant to first service	Confirmed pregnant to first service	binary	0	1
CR5	Pregnant to any service	Confirmed pregnant to any service	binary	0	1
CR6	Pregnant to AI	Confirmed pregnant to AI	binary	0	1
CR7	Three-week in-calf rate	Confirmed pregnant within three weeks of PSM	binary	0	1
CR8	Six-week in-calf rate	Confirmed pregnant within six weeks of PSM	binary	0	1
CR9	Three-week in-calf rate	Confirmed pregnant within three weeks of first mating	binary	0	1
CR10	Six-week in-calf rate	Confirmed pregnant within six weeks of first mating	binary	0	1

Genetic analysis. Variance components estimation was carried out using ASReml software (Gilmour *et al.*, 2009). Traits that are currently evaluated in the New Zealand genetic evaluation of fertility traits were analysed using models described by Amer *et al.* (2016) and Stachowicz *et al.* (2015, 2021). Conception-based fertility traits were analysed with a repeated records animal model, which in a simplified linear version can be represented as:

$$CR1\text{-}10 = CG + Age*Breed + Age^2*Breed + TR + FR + HO + Inbr + Het + Rec + a + pe + e$$

where:

- *CG* is the fixed contemporary group effect of herd-year-parity,
- *Age*Breed & Age²*Breed* are the fixed linear and quadratic regressions of age at calving nested within breed,
- *TR* is the fixed effect of pregnancy termination reason (normal, abortion, induction, premature),
- *FR & HO* are fixed linear regressions of New Zealand Friesian and foreign Holstein breed composition,
- *Inbr* is fixed linear regression of inbreeding,
- *Het* is fixed linear regression of heterosis,
- *Rec* is fixed linear regression of recombination loss,
- *a* is a random animal effect,
- *pe* is a random permanent environmental effect,
- *e* is a random error term.

Each of the conception-based traits was first analysed with a univariate model. Next, five traits were chosen for further work and were analysed using pairwise bivariate models with traits from the current evaluation system (CSD0, CSD, PM21). Finally, three traits of interest were analysed in three-trait models with CSD0 and PM21, with these new conception traits considered as alternative potential replacements for CSD.

RESULTS AND DISCUSSION

The results of initial univariate analysis of conception-based fertility traits defined with and without penalties are presented in Table 2. Heritabilities were consistently higher for phenotypes with penalties applied compared to phenotypes without penalties. This is the opposite trend to what was found for calving season day (CSD) in the past (Stachowicz *et al.*, 2014). We hypothesise that using penalties to account for carryovers and cows that were culled due to fertility issues leads to higher estimates of genetic variance. Traits derived using planned start of mating as opposed to using a cow's first mating as a base had higher heritability. This is consistent with observations from seasonal calving herds in Ireland (Stachowicz *et al.*, 2022). Based on the univariate results, five traits were chosen for bivariate runs. Genetic correlations were estimated between those traits and traits in the current genetic evaluation of fertility (CSD0 - heifer, CSD - cow, PM21; Table 2). Pregnant to first service (CR4) had the lowest genetic correlations with CSD and PM21, whereas the remaining traits had values ranging from 0.90-0.96. Genetic correlations with CSD0 ranged from 0.45-0.63. Phenotypic correlations (data not shown) between PM21 and conception-based traits were much lower than genetic correlations. This suggests that the extra records from conception phenotypes should add value, over and above submission data, when bulls have lower numbers of daughters.

Table 2. Heritabilities (h^2) and repeatabilities (rep) of conception-based fertility traits with (*) and without penalties, and their genetic correlations (r_G) for a subset of 5 selected traits with calving season rate heifer (CSD0), calving season day cow (CSD) and submission rate (PM21)

Trait	h^2	rep	h^{2*}	rep*	r_GCSD0	r_GCSD	r_GPM21
CR1	0.018	0.089	0.030	0.139	0.56	0.96	-0.90
CR2	0.008	0.054	0.017	0.093			
CR3	0.008	0.060	0.014	0.081			
CR4	0.012	0.047	0.014	0.046	-0.50	-0.86	0.63
CR5	0.007	0.007	0.012	0.054			
CR6	0.013	0.058	0.014	0.068	-0.47	-0.96	0.91
CR7	0.020	0.060	0.028	0.066	-0.63	-0.95	0.95
CR8	0.011	0.064	0.020	0.081	-0.45	-0.96	0.91
CR9	0.008	0.045	0.014	0.053			
CR10	0.006	0.025	0.013	0.042			

Three traits (CR1; timing of conception and CR7/CR8; three- and six-week in-calf rates) were chosen as potential replacements for CSD and included in three-trait variance components estimation with CSD0 and PM21. This decision was based on estimates of genetic and residual (data not shown) correlations as well as on within-season data availability and naming conventions already used by farmers. Results are presented in Table 3. With multiple trait models, estimates of heritabilities tend to increase compared to univariate runs. Genetic correlations between conception-based traits and CSD0 were comparable to current estimates for CSD (Amer *et al.*, 2016). Three-week in-calf rate had the highest genetic correlation with PM21 (0.94) compared to timing of conception and six-week in-calf rate (0.91). this indicates that three-week in-calf rate would be least preferred conception-based phenotype as it would provide less additional information on top of three-week submission rate compared to the other definitions. The binary six-week in-calf rate would likely be preferable to the continuous timing of conception trait because as soon as the six-week period from planned start of mating is complete the entire contemporary group's phenotypes are available and can be used immediately for evaluation. However, there may be a trade-off between timely data availability and potential biases introduced to evaluation if phenotypes of poor fertility contemporaries were not included in evaluation of the binary trait.

Table 3. Heritabilities (repeatabilities; on diagonal) and genetic correlations (off diagonal) for conception-based fertility traits (with penalties (*) with calving season rate heifer (CSD0) and submission rate (PM21)

	CSD0	CR1*	PM21		CSD0	CR7*	PM21		CSD0	CR8*	PM21
CSD0	0.023			CSD0	0.022			CSD0	0.021		
CR1*	0.64	0.048 (0.13)		CR7*	-0.66	0.044 (0.07)		CR8*	-0.59	0.033 (0.09)	
PM21	-0.58	-0.91	0.067 (0.16)	PM21	-0.56	0.94	0.065 (0.16)	PM21	-0.55	0.91	0.063 (0.16)

CONCLUSION

The goal of this research is to construct a more accurate conception-based fertility trait, as well as to determine whether greater overall economic advantage could be achieved with inclusion of this new trait in an economic index. This requires a more comprehensive definition of how the different components of fertility genetics contribute to farm profitability than is available in the current genetic evaluation system so they can be weighted accordingly. Based on our results, the continuous time of conception trait and binary three- and six-week in-calf rates are recommended for further testing in full GE univariate and multivariate runs. Next steps will include validation work, where phenotypes of the validation cow cohort are set to missing in prototype genetic evaluations, with the predictive ability of test models then evaluated across a range of fertility phenotypes, including the impact on GL. High genetic correlations between conception traits and CSD indicate that there might still be GL effects embedded in the new conception-based fertility trait. Correlations between conception-based fertility traits and GL EBVs will be assessed during validation and testing work. After the final conception-based fertility phenotype is chosen, the next step will be to incorporate the new trait in the economic index alongside GL which will have a non-linear economic weight to help ensure that any further shortening trend in GL will not pass the point after which short GL might have negative impacts on calf health and survival.

ACKNOWLEDGEMENTS

Funding and data for this research were provided by NZAEL, the subsidiary of DairyNZ Ltd.

REFERENCES

Amer P.R., Stachowicz K., Jenkins G.M. and Meier S. (2016) *J. Dairy Sci.* 99: 8227.

Gilmour A.R., Gogel B.J., Cullis B.R. and Thompson R. (2009) VSN International Ltd, Hemel Hempstead, HP1 1ES, UK. www.vsni.co.uk.

Stachowicz K., Jenkins G.M., Amer P.R. and Meier S. (2014) *Interbull Bull.* 48: 89.

Stachowicz K., Jenkins G.M., Amer P.R., Bryant J.R. and Meier S. (2015) *Proc. Assoc. Advmt. Anim. Breed. Genet.* 21: 370.

Stachowicz K., Amer P.R., Falconer B. and Garrick D.P. (2021) *Proc. Assoc. Advmt. Anim. Breed. Genet.* 24: 370.

Stachowicz K., Amer P.R., Kelleher M.M., Ring S.C. and Evans R.D. (2022). *Proc. World Congr. Genet. Appl. Livestock Prod.* 12.

ATTRIBUTES OF LACTATING COWS RANKED DIVERGENTLY FOR FERTILITY USING A MILK MID-INFRARED SPECTROSCOPY BASED PREDICTION MODEL

A.R. Bird[1], J.E. Newton[2] and P.N. Ho[2]

[1] University of Melbourne, Parkville, Victoria 3052, Australia
[2] Agriculture Victoria, AgriBio, Centre for AgriBioscience, Bundoora, Victoria 3083, Australia

SUMMARY

Predicting the fertility of dairy cows is a powerful tool for the dairy sector, giving farmers more confidence in making breeding and culling decisions. This study aimed to determine how animals ranked as high probability of conception to first insemination by a previously published model (i.e., MIR fertility model) differed from that of low-ranking individuals for various phenotypic and genotypic traits.

Data including MIR spectral data, herd test data, and genotypic breeding values were obtained for 18,831 cows located in Australia. Cows were ranked based on the probability of conception estimated by the MIR fertility model, and subsets of animals were created from the top and bottom 10% of animals. The mean values for phenotypic and genotypic traits of each group were compared using two-way t-tests. High MIR-predicted fertility (MFERT) cows were found to have significantly better reproductive performance and reduced somatic cell counts. High MFERT animals were found to have lower 305-day milk, fat, and protein yields compared to average, but this difference was reduced by correcting for parity. Conversely, high MFERT cows had higher milk production ABVs compared to average. Finally, high MFERT cows showed improved balanced performance index and health weighted index scores compared to the data average, and the low MFERT cows. Future studies should investigate how high and low MFERT cows differ in terms of other health traits, and how the model performs in conjunction with other breeding and management tools.

INTRODUCTION

Historical selection for milk production has resulted in poor fertility among the Australian dairy herd, with factors such as nutritional subfertility, where a cow must dedicate her energy resources either to her current calf through milk production or her future calf by maintaining body condition, as one of the possible contributors (Friggens *et al.* 2010). Whilst genetic selection has allowed for improvement in fertility, predicting the likelihood of pregnancy early on in the joining period would be helpful for farmers in making informed breeding decisions. Ho *et al.* (2019) developed an MIR model as a means of predicting an animal's likelihood of conception to first insemination. To offer more insight into the predicted outcomes of using the MIR model, this study aimed to determine the phenotypic and genotypic differences between animals with a high likelihood of conception as predicted by the model, and animals with a low likelihood of conception.

MATERIALS AND METHODS

Data collection. Data from 18,831 cows, collected between 2016 and 2020, from 50 farms in Victoria, Tasmania, and New South Wales were obtained from DataGene (Bundoora, Victoria, Australia). Herd size ranged from 69 to 1,217 head. Breed proportions were 59.80% Holstein, 4.88% Jersey, and 35.32% other breeds. Data included milk production parameters from 1st herd test after calving and 305-day cumulative lactation estimates, calving age, calving to first AI interval, calving date, pregnancy outcomes and date of birth, BPI, HWI, and Australian breeding values (ABVs) for milk production traits, overall type, condition score, survival, calving ease, somatic cell count, daughter fertility, feed saved, heat tolerance, and gestation length. MIR spectral data were obtained directly from herd test centres. For animals with multiple previous lactations prior to the insemination event, data from the first record were used.

Data analysis. The MIR fertility model uses MIR spectral data and other herd testing data to derive a fertility prediction (pMIR) for each cow. Further details of this model have been described by Ho *et al.* (2019), but briefly, a training population of cows were categorised as having good or poor fertility based on whether they conceived to their first insemination. This training population was then used to train the model using partial least squares discriminant analysis. The model was used to assign the pMIR values. Data were split into 96 herd-year groupings to minimise the potential effects of environmental and management factors. Within each herd-year, cows were ranked on pMIR, and the most fertile top 5, 10, or 20% of animals and least fertile, bottom 5, 10, and 20% of animals were placed in subgroups. The mean and standard deviations of the ABVs, indices and phenotypic traits for each of the subgroups, as well as the overall dataset were calculated. To compare potential biological differences between cows predicted to have poor and good fertility, two-way t-tests were conducted in R (R Core Team 2021) between the top subgroups and the average, the bottom subgroups and the average, and the top and bottom subgroups. Results with a p-value of less than 0.05 were deemed statistically significant. Similar differences between the high fertility and low fertility animals were seen regardless of what proportion of the population was selected. Therefore, data for the top and bottom 10% of animals are presented here, as this proportion has the highest prediction accuracy (Ho and Pryce 2020).

RESULTS AND DISCUSSION

This study found that groups of dairy cows predicted as having a high likelihood of conception by the MIR fertility model (high MFERT) differed from animals predicted as having a low likelihood of conception (low MFERT) on several traits.

Relationship between MIR fertility ranking and phenotypic performance. High MFERT animals had significantly higher mean pregnancy rates compared to the population average whereas low MFERT cows had significantly lower mean pregnancy rates (Table 1), indicating that high MFERT cows had a higher chance of reproductive success. This is consistent with the findings of Ho *et al.* (2019) when the model was first developed and tested. High MFERT cows also had decreased calving to first and second AI intervals compared to the average (Table 1). This is highly desirable in a seasonal production system as these traits allow cows to become pregnant early in the joining season and subsequently calve early in the season (Berry *et al.* 2013).

Table 1. Mean and standard deviation of phenotypic traits related to fertility for high MFERT subgroup, average (all animals) and low MFERT subgroup

	High	Average	Low
Pregnant after joining period rate	0.84 ± 0.37[a]	0.75 ± 0.43[b]	0.53 ± 0.5[c]
Pregnant after 1st AI rate	0.48 ± 0.5[a]	0.42 ± 0.49[b]	0.30 ± 0.46[c]
Pregnant after 2nd AI rate	0.71 ± 0.46[a]	0.63 ± 0.48[b]	0.43 ± 0.5[c]
Calving to first AI interval (days)	81.46 ± 26.05[a]	93.68 ± 49.93[b]	114.72 ± 82.8[c]
Calving to last AI interval (days)	125.1 ± 90.04[a]	143.46 ± 105.53[b]	163.29 ± 122.4[c]

a,b,c denotes where means are significantly different (P<0.05)

Low MFERT cows had a significantly elevated SCC (774,640±1,560,500 cells/mL, P<0.001) compared to the average (176,840±594,300 cells/mL), whereas high MFERT cows had a significantly lower SCC (56,890±83,700, P<0.001). Lomander *et al.* (2013) reported an increased calving to first AI interval of cows with high SCC compared to their contemporaries, corresponding with the findings above.

There was no significant difference between the 305-day milk, and milk fat and protein yields for low MFERT cows and the average, whereas high MFERT animals had a significantly lower 305-day yield compared to the data average. However, these differences were no longer significant after correcting the data for parity.

Table 2. Mean and standard deviation of phenotypic traits related to milk production for high MFERT subgroup, average and low MFERT subgroups

	High	Average	Low
305-day milk yield (litres)	6304.6 ± 2007.8[a]	6794.3 ± 2138.0[ab]	7094.7 ± 2122.9[b]
305-day fat yield (kg)	256.1 ± 73.8[a]	270.06 ± 80.05[ab]	282.84 ± 81.36[b]
305-day protein yield (kg)	220.1 ± 67.4[a]	231.6 ± 70.6[ab]	238.5 ± 69.0[b]

a,b,c denotes where means are significantly different ($p<0.05$)

Relationship between MIR fertility ranking and genetic merit. Apart from comparing the phenotypic performance of high versus low MFERT cows, we also analysed how genetic merit differed. There was no significant difference in daughter fertility ABV in any of the subgroups, which may be due to the low heritability of fertility (Tenghe *et al.* 2015). However, high MFERT cows were found to have a significantly lower gestation length ABV (-0.84±2.24, $P<0.05$) than average (-0.55±2.28), and low MFERT animals had a higher gestation length ABV (-0.16±2.19, $P<0.001$) compared to the average. The gestation length BV has been shown to be correlated with rate of conception to first AI (Vieira-Neto *et al.* 2017), which was the value being predicted by the MIR fertility model.

High MFERT cows were shown to have a significantly increased SCC ABV (119.9±16.85, $P<0.001$) compared to average (115.3±17.99), whereas low MFERT cows had a significantly lower SCC ABV (107.73±17.11, $P<0.001$) than average, consistent with the phenotypic SCC values presented above. Favourable genetic relationships have been previously shown between SCC and various fertility traits, including calving interval and days to first service (Wall *et al.* 2003), which are closely related to rate of conception to first AI.

High MFERT cows were found to have significantly higher production ABVs compared to average, whereas low MFERT cows had significantly lower production values compared to average (Table 3). This demonstrates genetic potential for the high MFERT cows to produce a high quality and quantity of milk, which further demonstrates that the phenotypic differences in production presented above are more likely the result of differences in parity. The ASI, which combines protein, fat, and milk yield ABVs based on their economic value, was also included in the analysis. High MFERT cows in were found to have a significantly higher ASI than the average, whereas low MFERT cows had a significantly lower ASI.

Table 3. Mean and standard deviation of breeding values related to milk production for high MFERT subgroup, average (all animals) and low MFERT subgroups

	High	Average	Low
Milk ABV	107.3 ± 460.8[a]	89 ± 453.12[a]	36.32 ± 453.17[b]
Fat ABV	12.88 ± 15.01[a]	7.56 ± 15.38[b]	4.08 ± 15.39[c]
Protein ABV	11.39 ± 10.42[a]	7.51 ± 10.89[b]	4.55 ± 11.29[c]
ASI	91.11 ± 71.31[a]	56.67 ± 73.89[b]	35.09 ± 74.24[c]

a,b,c denotes where means are significantly different ($p<0.05$)

Australia's two main selection indices were used to look at the combined effects of health, fertility, and production ABVs on the genetic merit of high and low MFERT cows. This study found that high MFERT cows had a significantly higher mean BPI (123.6±79.16, P<0.001) compared to the average (82.9±82.58), whereas the BPI for low MFERT animals was significantly lower than average (45.6 ± 78.59, P<0.001). Animals with increased BPIs have been shown to live longer productive lives and have greater chance of conception which lowers AI costs (Newton *et al.* 2017). As a result, cows in the high MFERT group could generate an average additional profit of $41 per cow per year compared to the average. High MFERT animals also had a significantly higher mean HWI value of 91.35±61.41 (P<0.001) compared to the average (64.28±63.04). The low MFERT animals had a mean HWI of 35.81±58.96, which was significantly lower than the data average (P<0.001). This was expected due to the heavy weighting of health and fertility in the HWI, but was included in the study to quantify the difference in profitability of low and high MFERT cows.

Applications. These findings should give farmers confidence to use the MIR fertility model along with other existing tools (e.g., daughter fertility breeding values or BPI) to make decisions on farm, particularly to improve fertility without affecting milk production. For example, the model can be used as an additional tool to support optimised semen allocation, where high predicted fertility cows can be assigned sexed semen, and low predicted fertility cows can be assigned beef or conventional semen (Newton *et al.* 2021). Additionally, farmers could use the model to identify high BPI cows with low predicted fertility, and implement management strategies such as nutritional adjustments to improve the profitability of these cows before the mating period starts.

CONCLUSIONS

This study shows that high MFERT cows as ranked by the MIR fertility model had improved reproductive performance, lower SCCs, and higher BPI, ASI, and HWI values, showing potential for improved lifetime profitability. There were no significant differences in milk production between groups, but high MFERT cows had above average milk solid percentages. Further studies should be undertaken into how the model can be used in conjunction with other commonly used breeding and management tools, as well as how high and low MFERT animals differ with regard to other health traits.

ACKNOWLEDGEMENTS

This study was undertaken as part of the DairyBio research program, which is funded by Dairy Australia (Melbourne, Australia), the Gardiner Foundation (Melbourne, Australia), and Agriculture Victoria (Melbourne, Australia). It formed part of A. Bird's MAgSc at University of Melbourne.

REFERENCES

Axford M., Santos B., Stachowicz K., Quinton C., Pryce J.E. and Amer P. (2021) *Anim. Prod. Sci.* 61: 1940.

Berry D.P., Kearney J.F., Twomey K. and Evans R.D. (2013) *Irish. Agr. Food Res.* 52: 1

Friggens N., Disenhaus C. and Petit, H. (2010) *Animal* 4: 1197.

Ho P., Bonfatti V., Luke T. and Pryce J. (2019) *J. Dairy Sci.* 102: 10460.

Ho P. and Pryce J. (2020) *J. Dairy Sci.* 103: 11535.

Lomander H., Svensson C., Hallén-Sandgren C., Gustafsson H. and Frössling J. (2013) *J. Dairy Sci.* 96: 6315.

Newton J.E., Ho P.N. and Pryce J.E. (2021) *Proc. Assoc. Advmt. Anim. Breed. Genet.* 24: 263.

Tenghe A., Bouwman A., Berglund B., Strandberg E., Blom J. and Veerkamp R. (2015) *J. Dairy Sci.* 98: 5763.

Vieira-Neto A., Galvão K., Thatcher W. and Santos J. (2017) *J. Dairy Sci.* 100: 316.

Wall E., Brotherstone S., Woolliams J.A,, Banos G. and Coffey M.P. (2003) *J. Dairy Sci.* 86: 4093.

APPROPRIATENESS OF COMBINING CARCASS DATA FROM ANGUS SIRE BENCHMARKING PROGRAM AND BREEDER HERDS IN A SINGLE GENETIC EVALUATION

A.M. Samaraweera, A. Byrne and C.J. Duff

Angus Australia, 86 Glen Innes Road, Armidale, NSW, 2350 Australia

SUMMARY

The objective of this study was to investigate whether the two different sources of abattoir carcass phenotypes that are currently submitted for inclusion in the TransTasman Angus Cattle Evaluation are genetically the same trait, being abattoir carcass phenotypes measured on cattle in the Angus Sire Benchmarking Program (ASBP), and abattoir carcass phenotypes measured on Angus animals in breeder herds. The abattoir carcass traits used were carcass MSA marble score (CMMS), carcass fat depth at p8 rump site (CP8, measured in mm), and dressed carcass weight (CWT, measured in kg). Additive genetic correlations between the same traits across the two sources were estimated with bivariate animal models. The additive genetic correlations for CP8, CMMS, and CWT were 0.99 ± 0.17, 0.84 ± 0.24, and 0.73 ± 0.23, respectively. Therefore, the two different sources of abattoir carcass phenotypes can be considered genetically to be the same trait and can be included in a unified genetic evaluation as the same trait.

INTRODUCTION

Variations in breeding goals and data collection processes can yield different heritability estimates for the same trait even though the same phenotype is collected. Further, the genetic correlations between the same trait from two different data sources can be low raising concerns regarding the genetic similarity of the traits and the validity of combining data from various sources in a single genetic evaluation. Currently, phenotypes from two different sources are submitted for inclusion in the TransTasman Angus Cattle Evaluation (TACE), i.e., phenotypes from the Angus Sire Benchmarking Program (ASBP) (Parnell *et al.* 2019), and phenotypes measured on Angus animals in breeder herds. This study was formulated to investigate whether the carcass phenotypes collected on the two different data sources are genetically the same trait. Therefore, the objective of this study was to estimate the heritability and additive genetic correlations among the same traits between ASBP and breeder herds to determine the suitability of combining abattoir data from both sources in the TACE.

MATERIALS AND METHODS

Data. Phenotypic records of carcass traits and pedigree were extracted from the Angus Australia database. Among the different carcass traits, carcass MSA marble score (CMMS), carcass fat depth at p8 rump site (CP8, measured in mm), and dressed carcass weight (CWT, measured in kg) were selected based on the availability of an adequate number of records in both ASBP and breeder herds (Table 1).

Extracted carcass records were used in the analyses if, they were pure Angus i.e., Angus percentage is higher than or equal to 87.5 %; the animals were born after 2010; the age at slaughter is available for carcass weight, and a carcass weight record is available for other carcass records; the sire is known, and if the observations are within three standard deviations from each trait mean which is calculated within each data source. Furthermore, contemporary groups with less than five animals and single-sire contemporary groups were discarded from the analyses. Contemporary groups were formed as described by Graser *et al.* (2005). The number of contemporary groups formed in CWT for ASBP and breeder data were 75 and 271, respectively. The average number of individuals within a contemporary group was greater in ASBP data than breeder data (39 vs. 16 for CWT). The data cleaning process excluded ASBP data by 16% and breeder herd data by 29%, respectively. After data cleaning, a total of 3,041 animals from ASBP herds originating from 329 sires and 2,502 dams, and 4621 animals from breeder herds originating from 352 sires and 3,796 dams were used in the study.

Estimation of genetic parameters. Genetic parameters for each trait were estimated with univariate animal model, and the model fitted was as follows.

$$\boldsymbol{y} = \boldsymbol{Xb} + \boldsymbol{Za} + \boldsymbol{e}$$

Where $\boldsymbol{y}$ is the vector of observations for CMMS, CP8, and CWT, $\boldsymbol{b}$ is the vector of fixed effects of contemporary group and linear and quadratic effects of the slaughter age (for CWT) or linear and quadratic effects of CWT (for CMMS and CP8) as covariates, $\boldsymbol{a}$ is the vector of random animal effects, $\boldsymbol{e}$ is the vector of random residual effects, and $\boldsymbol{X}$ and $\boldsymbol{Z}$ are design matrices which relate records to fixed effects and random animal effects, respectively. The covariates fitted in the model, slaughter age and CWT were adjusted for 750 days of slaughter age or 400 kg of CWT as specified in TACE (Angus Australia 2023). The variance components for the random effects were denoted as $Var(\boldsymbol{a}) = \boldsymbol{A}\sigma_a^2$ and $Var(\boldsymbol{e}) = \boldsymbol{I}\sigma_e^2$, where $\boldsymbol{A}$ is the numerator relationship matrix, σ_a^2 is the additive genetic effects variance of the animal, $\boldsymbol{I}$ is the identity matrix, and σ_e^2 is the residual variance. The pedigree consisted of only the animals with records for analysis plus the previous four generations. Additive genetic correlations between the same trait across the two data sources were estimated with bivariate animal models. Variance components for univariate and bivariate models were estimated using the WOMBAT software (Meyer 2007).

RESULTS

The descriptive statistics of the carcass traits are shown in Table 1. On average, animals in breeder herds were slaughtered 26 days later, and they were 7 kg heavier than ASBP animals. Marbling trait mean (CMMS) was also higher for animals in breeder herds by comparison to ASBP animals but not for CP8.

Table 1. Descriptive statistics of four carcass traits in Angus Sire Benchmarking Program (ASBP) and breeder herds

Traits[1]	No of animals	Mean	SD	CV	Min	Max	Mean age	Mean weight
ASBP								
CMMS	3026	523.98	120.82	0.23	160	890	757	453
CP8	3017	22.94	6.42	0.28	4	42	758	453
CWT	2940	457.73	36.91	0.08	315	571.5	762	458
Breeder								
CMMS	2011	568.57	101.31	0.18	230	900	774	449
CP8	4454	21.43	6.07	0.28	6	40	785	464
CWT	4520	464.96	45.21	0.10	279.4	580.5	788	465

[1]CMMS, carcass MSA marble score; CP8, carcass p8 fat (measured in mm); CWT, dressed carcass weight (measured in kg).

The additive genetic variances and heritability estimates for all carcass traits were higher in ASBP herds than breeder herds (Table 2). The additive genetic correlations between the same trait across ASBP and breeder herds were highest for CP8 (0.99 ± 0.17), followed by CMMS (0.84 ± 0.24) while the additive genetic correlations for CWT was the lowest (0.73 ± 0.23, Table 3). The additive genetic correlations between the two data sources increased with an increasing number of common sires between the two herds.

Table 2. Additive genetic (σ_a^2) and residual (σ_e^2) variances, and heritability ± standard errors ($h^2 \pm SE$), from univariate analyses of carcass traits for Angus Sire Benchmarking Program (ASBP) and breeder herds

Traits[1]	ASBP			Breeder		
	σ_a^2	σ_e^2	$h^2 \pm SE$	σ_a^2	σ_e^2	$h^2 \pm SE$
CMMS	5218	6388	0.45 ± 0.06	2234	4730	0.32 ± 0.08
CP8	14.85	15.24	0.49 ± 0.06	12.28	14.68	0.46 ± 0.06
CWT	676	374	0.64 ± 0.07	297	624	0.32 ± 0.05

[1]Traits and units are as given in Table 1.

Table 3. Additive genetic correlations ± standard errors (SE), number of common sires, and number of offspring per sire between ASBP and breeder herds

Traits[1]	Genetic correlations ± SE	No. of common sires	No. of offspring per sire	
			ASBP	Breeder
CMMS	0.84 ± 0.24	8	98	59
CP8	0.99 ± 0.17	11	133	152
CWT	0.73 ± 0.23	10	124	156

[1]Traits and units are as given in Table 1.

DISCUSSION

This study aimed to evaluate whether the phenotypes collected for carcass traits in ASBP and breeder herds are genetically the same trait by estimating additive genetic correlations among the same traits between ASBP and breeder herds. If the additive genetic correlations are closer to one, then the traits from two herds can be declared as being sufficiently similar for analysis as the same trait in the TACE. In this study, the proportion of additive genetic variance shared between the same trait across two data sources were high and significantly different from zero indicating that the genetic influences on the same trait across two herds are almost identical. Therefore, CWT, CP8, and CMMS are genetically the same trait across ASBP and breeder herds.

The variance components and heritability estimates are higher than 0.32 for all carcass traits across the two herds and consistent with past studies on carcass traits (Duff *et al.* 2021; Samaraweera *et al.* 2021; Torres-Vázquez *et al.* 2018; Börner *et al.* 2013; Reverter *et al.* 2000). Accordingly, the heritability for CWT varies from 0.41 ± 0.04 (Börner *et al.* 2013) to 0.75 ± 0.06 (Duff *et al.* 2021); for CP8 ranges from 0.36 ± 0.04 (Börner *et al.* 2013) to 0.56 ± 0.06 (Duff *et al.* 2021) for Australian Angus.

In this study, the additive genetic variances and the heritability estimates were higher in ASBP herds than breeder herds particularly for CWT and CMMS. Similar to this study, higher values have been reported for ASBP collected data, for example, an additive genetic variance of 709 and a heritability of 0.75 ± 0.06 were reported for CWT by Duff *et al.* (2021). This may be a result of the project design of minimal harvesting prior to slaughter (i.e. whole contemporary group) and the general focus on phenotype quality. Conversely, the lower additive genetic variance in the carcass data from breeder herds may be explained by general pre-slaughter harvesting or the commercial nature of the phenotype collection.

The TACE utilizes the variances estimated across all herds, and the herd origin is accounted for in the contemporary group formation, hence in the genetic model as described by Graser *et al.*

(2005). The genetic similarity of the carcass traits between ASBP and breeder herds used in this study further confirms the ability to use them as the same trait in TACE.

CONCLUSION

Phenotypically collected abattoir carcass traits are genetically consistent across ASBP and breeder herds based on the genetic correlations between the same trait across the two herds. Therefore, the abattoir carcass traits used for this study can be treated as identical traits in the TACE.

ACKNOWLEDGEMENTS

The authors would like to thank Angus Australia members and co-operator herds in the ASBP for providing data.

REFERENCES

Angus Australia (2023) www.angusaustralia.com.au/education/breeding-and-genetics/understanding-ebvs. (Accessed May 2023).

Börner V., Johnston D.J. and Graser, H.U. (2013) *Anim. Prod. Sci.*, 53: 1075.

Duff C.J., van ver Werf J.H.J., Parnell P.F. and Clark S.A. (2021) *Anim. Prod. Sci.* 61: 1837.

Graser H.U., Tier B., Johnston D.J. and Barwick S.A. (2005) *Aust. J. Exp. Agric.* 45: 913.

Malau-Aduli A.E.O., Edriss M.A., Siebert B.D., Bottema C.D.K. and Pitchford W.S. (2000). *J. Anim. Physiol. Anim. Nutr.* 83: 95.

Meyer K. (2007) *J. Zhejiang Univ. Sci.*8: 815.

Parnell P.F., Duff C.J., Byrne A.I and Butcher N.M. (2019) *Proc. Assoc. Advmt. Anim. Breed. Genet.* 23: 492.

Reverter A., Johnston D.J., Graser H.U., Wolcott M.L. and Upton W.H. (2000) *J. Anim. Sci.* 78: 1786.

Samaraweera A.M., Torres-Vazquez J.A., Jeyaruban M.G., Johnston D.J. and Boerner V. (2021) *Proc. Assoc. Advmt. Anim. Breed. Genet.* 24: 345.

Torres-Vázquez J.A., van der Werf J. H.J. and Clark S.A. (2018). *J. Anim. Sci.* 96: 4521.

Proc. Assoc. Advmt. Anim. Breed. Genet. 25: 374 - 377

INFOHERDS, GENOMIC SELECTION FOR DAIRY CATTLE IN NEW ZEALAND

B. Santos[1], N. Howes[1], S. Harburg[1], E. Ooi[1], S. Meier[2], A. Fear[3] and P. Amer[1]

[1]AbacusBio Limited, PO Box 5585, Dunedin 9058, New Zealand
[2] DairyNZ Ltd., PO Box 3221, Hamilton, New Zealand
[3] NZAEL, PO Box 3221, Hamilton, New Zealand

SUMMARY

The New Zealand (NZ) dairy industry is serviced by two key sources of genetic evaluations; firstly, the national evaluation delivered by DairyNZ via its subsidiary NZAEL, and secondly, the private evaluations delivered via the major breeding companies (LIC and CRV). The genetic evaluation ecosystem is changing, with genomic selection (GS) enhancing the accuracy of predictions across a wider range of traits. Improving rates of genetic progress and pursuing outcomes which assist with the industry's social license to operate, such as animal welfare and environmental impacts, via genetic pathways are some of the potential benefits of genomics for the dairy industry.

This study presents some of the benefits of an industrywide GS strategy underpinned by a modern and independent pipeline that integrates a source of genomic and phenotypic data. This source – known as the Infoherds programme – would act as a genomic reference population, with dairy herds managed under commercial conditions supplying data to improve prediction accuracy and to facilitate the development of novel traits within existing genetic evaluations.

The results demonstrate that GS could unlock between NZ$185 and NZ$245M in additional value of genetic gain per year to the NZ dairy industry – a 60-80% increase in the value of historic rates of improvement. Realising this potential requires a strategy for successfully implementing genomics via an independent information infrastructure co-ordinated by NZAEL.

INTRODUCTION

DairyNZ via its subsidiary NZAEL is investigating a programme to engage industry information herds (Infoherds) as a source of data contributing to the future genomic evaluations of strategically important traits, including those to be incorporated in the national economic selection index, Breeding Worth (BW). To develop a road map for rolling out the programme, we must value the implementation of genomic selection (GS) to the NZ dairy industry, and then identify potential data sources and integration needs.

The successful implementation of GS depends on the existence of a reference population. Information herds can be large and complex projects requiring an overarching strategic vision to ensure they achieve their key objectives. Disparate priorities, such as building farmer confidence in GS by improving accuracy of prediction, the introduction of novel traits, and support of on-farm management decisions, must all fit into an overarching goal of building a platform for industry genetic evaluations and shared data.

In many countries, the integration of genomic information into genetic evaluation systems underpinned by the collection of phenotypic and genotypic data has improved and accelerated genetic trends (García-Ruiz *et al.* 2016). However, this has not been observed in New Zealand, which might reflect the substantial breed admixture in the national herd as well as the pasture-based production, not common in most countries. Until the industry can integrate genomic information into a standard industrywide platform, the potential value of genomics to the broader industry is unlikely to be fully exploited. This potential value comprises a combination of:

1) Increased genetic progress via the adoption of superior young genomic sires instead of older daughter-proven (DP) sires,

2) Increased prediction accuracy through the estimation of genomic variant effects and more reliable parentage information, and
3) Increased industry engagement, introduction of new traits, cost-effective use of novel phenotypes in selection and decision making on-farm, and improved R&D collaboration.

In New Zealand, Infoherds would contribute to the national Dairy Industry Good Animal Database genetic evaluations as well as to the private evaluations delivered via major breeding companies (LIC and CRV). These companies will also benefit from Infoherds, as an additional source of phenotypic and genotypic data to their traditional herd testing programs and sire proving schemes. The information herd project therefore aligns with these breeding programs and creates a platform for better industry integration to support farmers.

This study presents some of the benefits of an industrywide GS strategy underpinned by a modern and independent pipeline that integrates a source of genomic and phenotypic data. This source – known as the Infoherd programme – would act as a genomic reference population, with dairy herds managed under commercial conditions supplying data to improve prediction accuracy and to facilitate the development of novel traits within existing genetic evaluations.

MATERIALS AND METHODS

This paper uses a geneflow modelling framework to analyse the combined effects of increased adoption of young genomic sires (by 5, 10 or 30% per year) in conjunction with an increased selection differential between GS and DP sires (20 and 40 BW units until 90% adoption). These parameters were used because commercial GS sires had an average initial superiority of 30 BW units over DP sires available in the same year (2021). The difference at 20 BW was deliberately chosen to be conservative, given the inflation issues around GS sires that have been experienced in many dairy industry evaluations around the world. The proportional use of GS sires in 2021 was set at 30%, with increments of 5% from 2016 to 2021. The increased adoption of GS sires and increase in the SD of the BW from new trait additions enabled by Infoherds, were assumed to be intertwined.

GS increases prediction accuracy for young sires, enabling farmers to select younger sire teams and avoid the need for progeny testing of prospective sires within sire proving herds. This shortens the generation interval and subsequently accelerates the rate of genetic progress (Lush 1937).

To estimate the economic impact of a higher adoption rate of young GS sires in the dairy industry, we created a model following the principles of recursive geneflow model methodology outlined by Matthews *et al.* (2019) and Fetherstone *et al.* (2021). It assumes that the average merit of calves born in a given year is half the merit of the sires available for use that year and half the merit of the cow herd, where the sire merit is a weighted combination of DP and GS sires, and the cow merit is set as a weighted combination of calf merit from 2 to 9 years prior.

The model was parameterised with the current BW genetic trend in dairy cows and a base status quo scenario where there is a 30% adoption rate of young GS sires. The results are calculated as the net present value (NPV) of the cumulative increase in merit of the cow herd over the base year value. The NPV was calculated using a discount rate r of 5%, and 4.9M breeding cows. The annualised NPV of genetic improvement was calculated as the cumulative NPV divided by the sum of the discount factors from year 0 to 20.

RESULTS AND DISCUSSION

The results of the geneflow modelling showed that the current uptake in adoption of young GS sires should lead to an 18% increase in genetic trend and a 39% increase in the annualised benefits after 20 years. However, a more aggressive increase in adoption and higher selection differential between DP and GS sires through improved evaluation systems could result in a 50% increase in genetic trend over historic and annualised benefits of genetic improvement of up to $495M.

Genetic trend in cows increased from 12.5 BW/year in the *status quo* to 13.3 and 15.8 BW/year after 20 years for the two levels of bull selection differential, 20 and 40 BW units/year, respectively.

The higher level of trend was attained earlier with faster adoption leading to increased benefits (Figure 1), resulting in total cumulative benefits for genetic improvement over the 20-year timeframe modelled of over $5.8B (differential of 20 BW/year) and $6.6B (40 BW/year). Relative to the *status quo* base scenario, the increase in total cumulative benefits ranged between $89M and $243M with different rates of adoption of GS bulls at the same 20 BW differential. This resulted in annualised benefits between $423M and $434M. The higher initial selection differential between GS and DP sires (40 BW) led to an increase in cumulative benefits between $658M and $1,065M, and annualised benefits between $465 and $495M (Table 1), depending on the adoption rate.

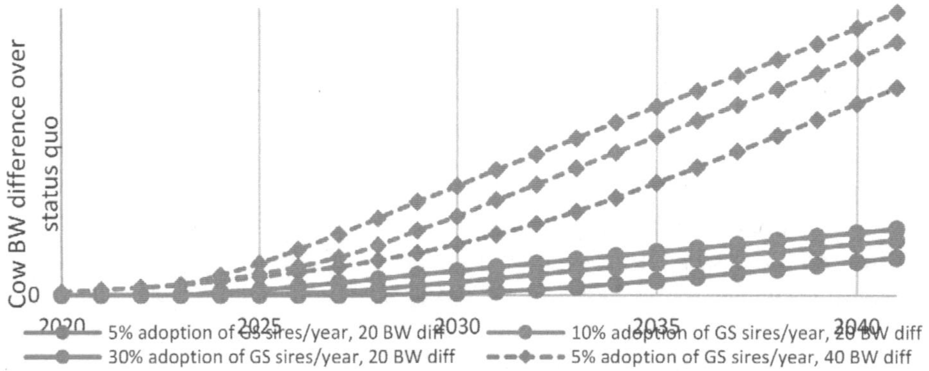

Figure 1. Difference in mean cow BW over the status quo base scenario for each of the increased adoption scenarios, under 20 and 40 BW differential scenarios

Table 1. Value of increased adoption of young sires and higher BW to the NZ dairy industry

Scenario	Genetic trend after 20 years	NPV Benefits after 20 years (NZ$M)	NPV of cumulative benefits (NZ$M)	Annualised benefits (NZ$M)
Historic	10.6	$4,038.66	-$1,569.64	$300.00
Base: max adoption of 50% of GS sires, 20 BW difference	12.5	$5,608.31		$416.60
Adoption of GS sires by 5% per year up to 90%, 20 BW diff	13.3	$5,697.30	$89.00	$423.21
Adoption of GS sires by 10% per year up to 90%, 20 BW diff	13.3	$5,778.74	$170.44	$429.26
Adoption of GS sires by 30% per year up to 90%, 20 BW diff	13.3	$5,851.73	$243.43	$434.68
Adoption of GS sires by 5% per year up to 90%, 40 BW diff	15.8	$6,266.39	$658.09	$465.48
Adoption of GS sires by 10% per year up to 90%, 40 BW diff	15.8	$6,481.26	$872.95	$481.44
Adoption of GS sires by 30% per year up to 90%, 40 BW diff	15.8	$6,673.98	$1,065.67	$495.76

The impact of improving fertility and survival evaluations, as well as adding clinical mastitis and lameness in BW, and recording high-quality phenotypes with the Infoherds structure, were incorporated into the model. The benefits of increased adoption of young GS sires, the addition of new traits, and enhanced accuracy of existing traits to BW, could unlock \$185-\$245M in additional value of genetic gain per year, a 60 to 80% increase on the value of genetic improvement.

Berry (2019) identified areas of benefit with the integration of genomics to livestock breeding programs. For the NZ dairy industry, potential areas of benefit from adoption of GS, supported by Infoherds, involve improved rates of genetic gain and industry engagement. These represent key areas of interest for NZAEL alongside prompt development of systems to monitor genetic diversity, inbreeding and recessive gene frequencies across the entire population, and developing strategies to support the sustainable management of the national genetic resource.

Implementation of genomics within NZAEL also provides several intangible benefits. This includes an independent benchmark of sire offerings from each breeding company. It also enables farmers to review breeding value data on their commercial milking cows and heifers. However, in the current absence of a national genomic evaluation platform, the independent benchmark function is severely compromised until young bulls marketed based on genomic information have daughter records included in the conventional pedigree based NZAEL evaluation. Thus, these intangible benefits comprise a) helping NZAEL to fulfil a crucial role as an independent source of genetic improvement information for the dairy industry, which at the moment does not own, control or have access to a genomic data pipeline to support its R&D programmes; b) supporting industry compliance to underpin more robust assessments of individual farm emissions, shifting from generic parameter estimates of emissions to estimates based on the actual genetic profile of individual herds; and c) industry-wide added value from GS, as the broader industry fully benefits from the genomic resource only when the data and information is available for use by all herds and service operators.

CONCLUSIONS

Whilst genomics has been implemented within the NZ dairy industry for approximately 10 years, the potential value of the technology is yet to be fully realised. This might be reflecting the substantial breed admixture in the national herd as well as the pasture-based production which requires a specific reference population to explore the full effectiveness of genomic selection. A concerted collaborative effort at national level to overcome these limitations is now essential to ensure that the full industry benefits are realised. Additional benefits have also been identified from supporting sustainable management of genetic diversity within the national herd.

It is currently uncertain how funding and data flows will be managed, but stakeholders and potential partners must be engaged for the project to succeed. The value proposition for Infoherds has been previously identified. Communicating this clearly to potential stakeholders, as well as the opportunity cost associated with inaction, will be essential for building productive partnerships.

ACKNOWLEGEMENTS

The authors would like to acknowledge contribution and funding from the 'Resilient Dairy Programme - Innovative breeding for a sustainable dairy future' to develop this work.

REFERENCES

Berry D.P. (2019). *Livestock* 24 (1).

Fetherstone N, Hely F.S., McHugh N., McGovern F.M. and Amer P.R. (2021) *Genet Sel Evol* 53: 5.

Garcia-Ruiz A., Cole J.B., VanRaden P.M., Wiggans G.R., Ruiz-López F.J. and Van Tassell C.P. (2016) *Proc. Nat. Acad. Sci. USA* 113: E3995.

Lush J.L. (1937) *Animal Breeding Plans*. Ames, Iowa: Iowa State Press.

Matthews D., Kearney J.F., Cromie A.R., Hely F.S. and Amer P.R. (2019) *Genet. Sel Evol* 51: 40.

VALIDATION OF CALVING EASE EBVS EXAMINING THE IMPACT OF GENETIC GROUPS AND SINGLE-STEP ON PREDICTIVE ABILITY

P.M. Gurman, L. Li, M.G. Jeyaruban, D.J. Johnston, C.J. Girard, and A.A. Swan

Animal Genetics Breeding Unit*, University of New England, Armidale, NSW, 2351 Australia

SUMMARY

Calving difficulty scores recorded in beef cattle are challenging to analyse due to low frequency of difficult births and the scored nature of the trait, requiring analysis in a threshold model, typically in combination with two linear traits, birth weight and gestation length. Previous software to calculate estimated breeding values (EBVs) for calving ease was not able to include single-step methods or fit genetic groups in models of analysis. In this study, we examined the value of including genetic groups and genomic information via single-step genomic BLUP (ssGBLUP) in the TransTasman Angus Cattle Evaluation (TACE) BREEDPLAN and Hereford BREEDPLAN analyses, by forward-validation in genotyped animals. The greatest improvements in accuracy were observed when including genomic information, with increases of 0.169 and 0.106 in the Angus and Hereford analyses respectively. Adding genetic groups to models had no impact on accuracy, but increased the bias of CE EBVs in ssGBLUP analyses for both breeds.

INTRODUCTION

Traits that are measured as scores are often difficult to analyse, especially if the distribution of the scores is skewed. A linear model can be used in some cases if the scores approximate normality, but a threshold model is typically used to address the imbalance in measurement between categories (Hoeschele *et al.* 1995; Gilmour *et al.* 1998). Mixed-model threshold analyses add extra complexity to solving for fixed and random effects due to the requirement of estimating both the threshold values and the weights to apply to each categorical phenotype.

Calving difficulty scores in BREEDPLAN analyses are characterised by low frequencies of difficult births. Analyses of this trait are performed using a categorical threshold model with birth weight and gestation length included as correlated linear traits to improve prediction for overall calving ease. Since 2017, BREEDPLAN analyses for most traits have been transitioning to ssGBLUP. In November 2019, a ssGBLUP implementation for calving ease was developed in new software for the BREEDPLAN component of the TransTasman Angus Cattle Evaluation (TACE, herein Angus), including genetic groups.

As part of the process of developing these enhancements, the utility of genetic groups came into question. The addition of genetic groups was observed to substantially increase convergence times of the model in the Angus evaluation, and when applied to Hereford BREEDPLAN, resulted in changes in EBVs that were difficult to interpret.

This paper examines the predictive ability of threshold model calving ease EBVs in Angus and Hereford BREEDPLAN with the inclusion of genetic groups and single-step using forward validation procedures.

MATERIALS AND METHODS

Calving ease (CE) data from the March 2022 Angus and May 2021 Hereford BREEDPLAN calving ease analyses after cleaning were used in this study. CE is scored as 1: no assistance required, 2: easy pull, 3: hard pull. Genetic parameters used for these models were adapted from Jeyaruban *et al.* (2015), with the genetic group variance assumed to be equal to the genetic variance. Genetic

* A joint venture between NSW Department of Primary Industries and University of New England

groups were fitted as routinely constructed in BREEDPLAN based on year window, breed, and country for the main analysis traits, with 20 groups for Angus and 16 for Hereford. There groups were included to improve prediction for animals with missing pedigree. Each data set was split into two groups, "training" and "validation", based on year of birth. The training set included animals born before 2019, while the validation set included animals with phenotypes born from 2019 onwards. BLUP analyses were performed in a factorial design, with and without genetic groups, and with and without genotypes. These four analyses were performed, first using all phenotypes, with the resulting EBVs for validation animals denoted as $\hat{\boldsymbol{u}}_w$. Phenotypes for the validation animals were then removed and the analyses repeated, with the resulting EBVs denoted as $\hat{\boldsymbol{u}}_p$. The subscripts "w" and "p" refer to "whole" and "partial" analyses respectively, with the partial EBVs of validation animals ($\hat{\boldsymbol{u}}_p$) informed through their pedigree and genomic relationships with the training animals. Maternal effects were fitted as routinely calculated in BREEDPLAN, but were not examined in the cross-validation, because the validation animals were not chosen to remove all phenotypes connected to the dam. EBVs were analysed on the underlying scale.

Correlations were used to examine the change in EBVs between each analysis. Cross-validation metrics were calculated using the method of Legarra *et al.* (2018). Traditional phenotype-based cross-validation metrics were not considered for this analysis due to the categorical nature of the calving ease trait. Accuracies were calculated by the formula

$$acc = \sqrt{\frac{cov(\hat{\boldsymbol{u}}_w, \hat{\boldsymbol{u}}_p)}{(\overline{diag(\boldsymbol{K})} - \overline{\boldsymbol{K}})\sigma^2_{u,\infty}}}$$

where $\boldsymbol{K}$ is the appropriate relationship matrix for the validation animals with phenotypes for each trait and $\sigma^2_{u,\infty}$ is genetic variance in the validation animals, assumed to be the genetic variance. The dispersion was estimated by $disp = cov(\hat{\boldsymbol{u}}_w, \hat{\boldsymbol{u}}_p)/var(\hat{\boldsymbol{u}}_p)$ and the bias was estimated as $bias = (\overline{\hat{\boldsymbol{u}}_p} - \overline{\hat{\boldsymbol{u}}_w})/\sqrt{\sigma^2_u}$, which was modified by Legarra *et al.* (2018) to allow for comparison between traits. While the validation animals included both genotyped and pedigree-only animals, metrics calculated only included genotyped animals due to computational difficulties. Metrics were also only calculated on direct effects, without consideration of maternal effects. Analyses were performed with the AGBU commercial solver on a computer with 2 x Intel(R) Xeon(R) E5-2697 v3 CPUs.

Table 1. Summary of the data used in the cross-validation studies

	Angus	Hereford
# animals in pedigree	3,006,655	2,247,767
# animals genotyped	200,259	34,585
# phenotypes		
Birth weight (BWT)	1,707,804	781,505
Calving difficulty score (CDS)	482,565	325,978
Gestation length (GL)	519,274	119,468
# validation animals with phenotypes		
Birth weight (BWT)	125,780	48,064
Calving difficulty score (CDS)	37,383	23,818
Gestation length (GL)	47,865	8,345
Proportion of CDS scores: 1,2,3	96.1, 2.7, 1.2	93.2, 4.7, 2.1

RESULTS AND DISCUSSION

A summary of the data used in the forward cross-validation is presented in Table 1. The correlation between EBVs from pedigree models with and without genetic groups for all animals was 0.912 for both the Angus and Hereford analyses. When considering animals born from 2019 onwards, this correlation increased to 0.995 and 0.990 for the Angus and Hereford analyses, respectively. For the models without genetic groups, the correlations between pedigree and ssGBLUP models were 0.994 and 0.990 for the Angus and Hereford analyses, respectively. This decreased for the 2019-born animals to 0.886 and 0.961 for the Angus and Hereford, respectively. For recent animals most likely to be used for selection, inclusion of genomic information had a larger impact on changes in EBVs than inclusion of genetic groups.

Table 2. Cross-validation metrics for the Angus and Hereford analyses calculated based on genotyped animals born in 2019 or later

	EBV	n	Pedigree	Pedigree GG	Single-Step	Single-Step GG
			Angus			
Accuracy	BWT	45,613	0.475	0.475	0.840	0.839
	CE	14,606	0.340	0.340	0.533	0.534
	GL	19,351	0.441	0.442	0.672	0.676
Dispersion	BWT	45,613	0.983	0.982	1.030	1.029
	CE	14,606	0.997	0.999	1.025	1.026
	GL	19,351	0.941	0.950	0.992	0.995
Bias	BWT	45,613	0.002	-0.024	-0.002	-0.033
	CE	14,606	-0.013	-0.051	-0.010	-0.056
	GL	19,351	0.021	-0.029	0.001	-0.057
			Hereford			
Accuracy	BWT	10,285	0.677	0.672	0.869	0.863
	CE	5,715	0.401	0.413	0.516	0.526
	GL	2,670	0.555	0.646	0.655	0.718
Dispersion	BWT	10,285	0.968	0.965	1.010	1.012
	CE	5,715	0.942	0.917	1.008	0.992
	GL	2,670	1.149	1.019	1.127	1.052
Bias	BWT	10,285	-0.014	-0.003	-0.015	0.006
	CE	5,715	0.015	0.017	0.001	0.014
	GL	2,670	0.138	0.038	0.133	0.029

The forward cross-validation results for the Angus and Hereford analyses are presented in Table 2. For the Angus analyses, adding genetic groups to either pedigree or ssGBLUP models had virtually no impact on accuracy. Adding genomic information on the other hand improved accuracy substantially over pedigree-only analyses, by 0.365, 0.194, and 0.231 for BWT, CE and GL EBVs respectively in the ssGBLUP model without genetic groups. Little change was also observed in the dispersion, with all analyses close to the expected value of 1, indicating little evidence of over- or under-prediction. An increase in bias was observed for genetic group models for all traits, especially CE, with the bias increasing from -0.013 to -0.051 in the pedigree model, and from -0.01 to -0.056 in the ssGBLUP model.

For the Hereford analysis, the addition of genetic groups to the pedigree model increased accuracy for CE and GL EBVs, respectively, but as with the Angus analysis, adding genomic information had the largest impact on accuracy. Dispersion was improved for ssGBLUP models, with evidence for over-prediction in pedigree models (regressions < 1). The pattern of changes in bias was not consistent across traits and analyses, but for the CE trait itself, the ssGBLUP model without genetic groups had the least bias.

Based on these validation results, the inclusion of genomic information in ssGBLUP had a large benefit to prediction by increasing accuracies, and in some cases correcting dispersion and minimising bias. Similar benefits were not apparent from the addition of genetic groups, which had no or minor benefit for accuracy and increased bias in CE EBVs for both analyses. Dispersion was largely unaffected by the model, but there was evidence for over-prediction for pedigree models in Herefords. While the pedigree accuracy for Angus is lower than Hereford, this is likely due to differences in the data structure for the two validation groups and warrants further investigation. While large increases in accuracies were observed for the genotyped validation animals, a smaller increase in accuracy is expected for the non-genotyped animals. Animals directly related to a genotyped animal will experience the greatest benefit from single-step, while animals less related will derive a lower benefit. It should also be noted that these validation metrics reflect the expected change for animals without a phenotype, and that individual animal results will vary. These results need to be verified for maternal effects but will require modifications to the validation set design.

Computation times for the models, including genomic information or genetic groups, had a large impact on the commercial viability of these analyses. For the Angus analyses, the model without genomics or genetic groups took 10,377 iterations to converge and 2.03 hours. The addition of genetic groups to this model required 19,986 iterations and 5.9 hours. The genomic model without genetic groups required 11,375 iterations and 20 hours to converge, while the addition of genetic groups increased this to 20,038 iterations and 37.34 hours. While the increase in computation times from the addition of genomic information is large, there is a corresponding increase in accuracies. The addition of genetic groups had no benefit to accuracies and almost doubled the number of iterations required. Therefore, inclusion of genetic groups constructed with the current strategy in this analysis is not recommended.

Calculating the mean of the $\boldsymbol{K}$ matrix for each trait makes using the Legarra *et al.* (2018) method challenging for pedigree-only animals when validating a single-step analysis. While an algorithm exists for calculating the diagonal of the ssGBLUP relationship matrix $\boldsymbol{H}$ (Legarra *et al.* 2020), summary statistics for blocks of $\boldsymbol{H}$ are a challenge. For genotyped animals, the block of the $\boldsymbol{H}$ matrix required is a sub-matrix of the genomic relationship matrix $\boldsymbol{G}$, which can be calculated easily, but the other subblocks of $\boldsymbol{H}$ are more complex. One approach could be to solve the equation $\boldsymbol{v}'\boldsymbol{H}^{-1}\boldsymbol{v}$ by conjugate gradient, where $\boldsymbol{v}$ is a vector of zeros, except in the positions of the validation animals, which are set to $1/n$, where n is the number of validation animals.

CONCLUSION

Clear improvements in predictive ability were obtained for genotyped animals with the addition of genomic information in ssGBLUP models. However, the addition of genetic groups did not provide any improvements in calving ease direct predictions. Given the significant increase in computation time required to add genetic groups to the model, this term can be left out of the model without impact on recently born animals who are candidates for selection.

ACKNOWLEDGEMENTS

This research was partially funded by Meat and Livestock Australia project L.GEN.2204. The authors acknowledge the Agricultural Business Research Institute, Angus Australia and Herefords Australia for the provision of and access to the data for this study.

REFERENCES

Gilmour A.R. and Thompson R. (1998) *Proc. 6th World Congr. Genet. Appl. Livest. Prod.* 20: 613.

Hoeschele I., Tier B. and Graser H.-U. (1995) *J. Anim. Sci.* 73:1609.

Jeyaruban M.G., Johnston D.J., Tier B., Graser H.-U., Jeyaruban M.G., Johnston D.J., Tier B. and Graser H.-U. (2015) *Anim. Prod. Sci.* 56: 927.

Legarra A., Aguilar I. and Colleau J.J. (2020) *J. Dairy Sci.* 103: 3363.

Legarra A. and Reverter A. (2018) *Genet. Sel. Evol.* **50:** 53.

Proc. Assoc. Advmt. Anim. Breed. Genet. 25: 382 - 385

REMODELLING THE GENETIC EVALUATION OF NFI IN BEEF CATTLE – PART 1: DEVELOPING AN EQUIVALENT GENETIC MODEL

L.Vargovic[1], K.L. Moore[1], D.J. Johnston[1], G.M. Jeyaruban[1], C.J. Girard[1], J. Cook[1], J.A. Torres-Vázquez[2] and S.P. Miller[1]

[1] Animal Genetics Breeding Unit*, University of New England, Armidale, NSW, 2351 Australia
[2] Department of Animal and Avian Sciences, University of Maryland, College Park 20742 USA

SUMMARY

Net feed intake (NFI) is the residual portion of daily feed intake (DFI) not explained by growth or maintenance requirements. The NFI phenotype (NFIp) is based on a 70-day test period where DFI and fortnightly weights (to calculate average daily gain (ADG) and maintenance as metabolic mid-weight (MMWT)) are measured. Recording NFIp is costly, and shortening the test length would be advantageous. However, research has shown that ADG cannot be accurately measured from a shortened test. Genetic NFI EBVs (NFIg) were calculated using DFI EBV adjusted for ADG and MMWT EBV and were shown to have a Pearson correlation of 0.99 with the NFIp EBV from 3,088 Angus steers. The regression slope between NFIg and NFIp EBVs was 1.14. Alternative NFIg models where growth and maintenance requirements were obtained from BREEDPLAN live weight traits instead of live weights recorded in the test period, demonstrated high Pearson correlations (r=0.87 to 0.93) and regression slopes between 0.63 and 0.97 with NFIp EBVs. Results suggest that genetic NFI EBVs can be obtained, with growth and maintenance requirements being determined from BREEDPLAN live weight traits. This provides the opportunity to determine if the length of the test to measure DFI can be shortened, reducing the cost of recording NFI per animal.

INTRODUCTION

Net feed intake (NFI) measures feed efficiency and is the residual portion of daily feed intake (DFI) adjusted for growth and maintenance (Koch *et al.* 1963). Over a 70-day feed efficiency test, individual DFI and fortnightly live weight have been recorded in beef cattle in Australia (Arthur *et al.* 2001). Recording NFI is costly due to the test length. Previous studies (Culbertson *et al.* 2015; Clark and van der Werf 2017) demonstrated that DFI could be measured from a shorter test, but to measure average daily gain (ADG), a minimum of 56 test days was required (Archer *et al.* 1997; Culbertson *et al.* 2015), and this represents a limiting factor to reducing the test length. Estimating genetic NFI EBVs (NFIg) has been proposed as an alternative to EBVs based on NFI phenotype (NFIp) (Kennedy *et al.* 1993; MacNeil *et al.* 2011). The method utilises genetic (co)variances and EBVs from tri-variate (DFI, ADG and metabolic mid-weight (MMWT)) analysis to construct a genetic NFI EBV. This study aimed to develop NFIg EBVs to compare against NFIp EBVs and assess if NFIg models may be a suitable alternative for genetic evaluation of feed efficiency when NFIg EBVs were developed using growth and maintenance recorded from a 70-day test or derived from BREEDPLAN live weight traits.

MATERIALS AND METHODS

Data preparation. Feed intake data and the NFI contemporary group (NFI CG; defined as the birth herd, birth year, sex, trial cohort, and previously recorded BREEDPLAN trait CG (i.e. 200-day live weight)) were extracted from the Angus Australia BREEDPLAN database for 3,215 steers. The data was recorded at Tullimba Research Feedlot (Torryburn, NSW) between 2012 and 2021. Individual DFI was measured using the VYTELLE-SENSE system, formerly known as GrowSafe

* A joint venture of NSW Department of Primary Industries and University of New England

Feeders, and the Ruddweight system (www.vytelle.com/vytelle-sense). Animals were fed a standard feedlot diet (energy level of 12 MJ/kg). DFI was recorded over 70 – 77 days, after an initial 21-day acclimatisation period. Weight was recorded fortnightly, up to six times, during the test period. The average DFI, ADG and MMWT were computed across the full test period. ADG during the feed intake test was computed as the linear regression across all trial weights, and MMWT $=((\frac{ADG*test\ length}{2}) + start\ weight)^{0.73}$. Animals with fewer than 56 days of feed intake data over the 70-day test duration, or with fewer than four live weights recorded during the test were removed (n=31). Animals in an NFI CG with fewer than three animals (n=96) were also removed. The final dataset included 3,088 steers from 35 trial groups with DFI, ADG and MMWT records. Steer age at the start of the test (SAGE) was on average 513 ± 73 days (range: 373 – 767 days). These were the progeny of 327 sires and 2,523 dams, and an up to five-generation pedigree containing 9,497 animals built.

To develop NFIg EBVs when ADG and MMWT are not available from the test period (i.e. from a proposed shortened test), BREEDPLAN live weight traits (W200, W400 and W600) and their respective CG at 200, 400 and 600 days of age were extracted from the Angus Australia BREEDPLAN database for all animals in the final NFI dataset. Where there were multiple weights per trait, the weight closest to the target age (i.e. 200 days) was used. These live weight traits were pre-adjusted for heifer factor, animal age and dam age using standard BREEDPLAN procedures (Graser *et al.* 2005). Using BREEDPLAN live weight phenotypes, three ADG terms were constructed to represent gain between 200 and 600 days of age, 200 and 400 days of age and 400 and 600 days of age, i.e. $ADG_{200-600}$ = (W600 – W200) /400. At 200, 400 and 600 days of age, metabolic weights (MWT) were constructed based on BREEDPLAN live weight phenotypes, i.e. $MWT_{200} = WT200^{0.73}$. Descriptive statistics for the traits used in this study are shown in Table 1.

Statistical analyses. The NFI phenotype (NFIp) was calculated by adjusting DFI for growth and maintenance, as per Koch *et al.* (1963.). NFIp = DFI - μ – (β_{adg} x ADG) – (β_{mmwt} x MMWT), where μ, β_{adg} and β_{mmwt} were regression coefficients obtained from the model DFI = μ + NFI CG + (β_{adg} x ADG) + (β_{mmwt} x MMWT) + SAGE. Linear mixed animal models were fit in ASReml (Gilmour *et al.* 2015) to estimate variance components and animal solutions (EBVs). The model to predict NFIp EBVs fitted SAGE and NFI CG as fixed effects in the model.

Genetic NFI (NFIg) EBVs were derived following the procedure of Kennedy *et al.* (1993) using the EBVs of DFI, ADG and MMWT from a tri-variate model. NFIg EBV = DFI EBV – (β_{adg} x ADG EBV) – (β_{mmwt} x MMWT EBV). The genetic regression coefficients $\begin{bmatrix} \beta adg \\ \beta mmwt \end{bmatrix} = G^{-1}c$, where G was the genetic covariance matrix of ADG and MMWT from the tri-variate model and c was the vector of the genetic covariance of DFI with ADG and MMWT.

Six alternative NFIg EBVs (denoted A through to F) were derived from replacing feed test ADG and MMWT with alternative measures derived from BREEDPLAN weight records. The same procedure described for NFIg EBVs was used for these alternative genetic NFI EBVs, and Table 2 describes these models. To avoid autocorrelation issues for the alternative NFIg EBVs, BREEDPLAN-derived MWT recorded at the end of the specified ADG period was not considered. The alternative NFIg models were evaluated using Pearson correlation coefficients (r), regression slope of NFIg EBV on NFIp EBV (b) and difference of means ($\bar{u} - \hat{\bar{u}}$) between NFIp EBV and NFIg EBVs, and these were reported in Table 2.

RESULTS AND DISCUSSION

A Pearson correlation of 0.99 was calculated between NFIp and NFIg EBVs, and EBV means was similar with a difference of means of 0.01 (Table 2). This indicates that NFIg EBVs were unbiased, but the standard deviation was smaller than NFIp EBVs. The regression slope was 1.14, suggesting that the spread of NFIg EBV is narrower and animals at the edge of the distribution may be overestimated. These results demonstrate that NFIg was an equivalent model to NFIp when the same ADG and MMWT terms were modelled. These results agreed with Hoque and Oikawa (2004), who estimated a correlation of 0.97 for a similar comparison in Wagyu cattle.

Table 1. Descriptive statistics for average daily feed intake (DFI), average daily gain (ADG) and metabolic mid-weight (MMWT) measured from a 70-day feed intake test and ADG representing different periods between 200 and 600 days of age (ADG200-600, ADG200-400 and ADG400-600) and metabolic weight at 200 (MWT200), 400 (MWT400) and 600 (MWT600) days of age from BREEDPLAN records for Angus steers

Trait	Unit	N	Mean	SD	Range
1) Feed test records					
DFI	kg/d	3,088	14.8	1.99	6.89 – 22.6
ADG	kg/d	3,088	1.61	0.34	0.52 – 2.90
MMWT	$kg^{0.73}$	3,088	104	6.83	81.8 - 135
2) BREEDPLAN data					
$ADG_{200-600}$	kg/d	2,915	1.02	0.16	0.58 – 1.65
$ADG_{200-400}$	kg/d	2,360	0.82	0.29	0.07 – 1.90
$ADG_{400-600}$	kg/d	2,198	1.30	0.21	0.72 – 2.25
MWT_{200}	$kg^{0.73}$	3,078	55.8	6.01	31.9 – 74.0
MWT_{400}	$kg^{0.73}$	2,367	80.5	8.59	58.0 – 109.5
MWT_{600}	$kg^{0.73}$	2,919	114.0	8.16	86.4 – 141.5

Table 2. Models for genetic NFI (NFIg) EBVs when growth and maintenance were derived from feedlot weights or BREEDPLAN live weights, EBV summary statistics (mean and SD), the Pearson correlation (r), regression slope (b) and difference of means ($\bar{u} - \bar{\hat{u}}$) between NFIp EBV and genetic NFI EBVs

	EBV				
NFI EBV*	Mean	SD	r	b	$\bar{u} - \bar{\hat{u}}$
NFIp	0.02	0.41			
NFIg = EBV(DFI) – β_1 x EBV(ADG) – β_2 x EBV(MMWT)	0.02	0.37	0.99	1.14	0.01
NFIgA = EBV(DFI) – β_1 x EBV($ADG_{200-600}$) – β_2 x EBV(MWT_{200})	0.01	0.44	0.93	0.89	0.02
NFIgB = EBV(DFI) – β_1 x EBV($ADG_{200-600}$) – β_2 x EBV(MWT_{400})	0.01	0.45	0.91	0.86	0.01
NFIgC = EBV(DFI) – β_1 x EBV($ADG_{400-600}$) – β_2 x EBV(MWT_{400})	0.02	0.46	0.90	0.84	0.00
NFIgD = EBV(DFI) – β_1 x EBV($ADG_{400-600}$) – β_2 x EBV(MWT_{200})	0.01	0.44	0.87	0.84	0.01
NFIgE = EBV(DFI) – β_1 x EBV($ADG_{200-400}$) – β_2 x EBV(MWT_{200})	0.04	0.59	0.88	0.63	-0.01
NFIgF = EBV(DFI) – β_1 x EBV($ADG_{200-400}$) – β_2 x EBV(MWT_{600})	0.01	0.40	0.93	0.97	0.01

* see Table 1 for abbreviations; β_1 and β_2 were estimated for each model and varied across the models

Six alternative NFIg models were considered using ADG and MWT derived from BREEDPLAN live weight traits (Table 2). For all alternative models, the Pearson correlation coefficient with NFIp EBVs ranged from 0.87 (model D) to 0.93 (models A & F). The highest correlations were observed when $ADG_{200-600}$ and MWT_{200} or $ADG_{200-400}$ and MWT_{600} were modelled. Models A, B, C and D all showed similar regression slopes (b=0.84 to 0.89) and a slightly higher standard deviation than the NFIp EBV. This suggests that EBVs for animals at the edge of the distribution may be over-estimated. For all NFIg EBVs the bias was small (-0.01 to 0.02). The mean and standard deviation of the alternative NFIg EBVs were generally similar, although the NFIgE EBV showed a higher mean and standard deviation. NFIgE fitted $ADG_{200-400}$ and MWT_{200}. The regression slope was also much lower (0.63) than in other models. The reasons for these differences are not clear. $ADG_{200-400}$ had the smallest gain (0.82 kg/day) compared with $ADG_{200-600}$ (1.02 kg/day) and $ADG_{400-600}$ (1.30

kg/d) (Table 1). NFIgF also modelled $ADG_{200-400}$, with MWT_{600} fitted instead of MWT_{200}. The standard deviation of NFIgF EBVs was similar to NFIp EBVs with a regression slope of 0.97; this suggests that it is important to include WT600 in either the ADG or MWT term in the alternative NFIg model. Model F fitting $ADG_{200-400}$ and MWT_{600} yielded the alternative NFIg EBV with the highest Pearson correlation with NFIp EBV, a regression slope close to 1 and EBVs with similar means and standard deviations compared with NFIp EBVs.

Genetic NFI EBVs where the test period ADG and MMWT was replaced with ADG and MWT derived from BREEDPLAN live weight traits have shown potential as an alternative approach to computing feed efficiency. The next step of this research is to explore if the length of the test period to record DFI can be shortened and genetic NFI EBVs computed using the proposed alternative genetic NFI models. If the test length is reduced, this could lead to more animals being recorded for DFI, reduced cost of recording per animal, and an overall increase in selection response due to a larger number of recorded animals. Further research will be needed to investigate the method in a larger dataset where BREEDPLAN live weights from the whole breed will influence the component EBVs used in this study. In this study, the majority of animals had BREEDPLAN live weights recorded. If genetic NFI EBVs use BREEDPLAN live weights to model growth and maintenance, and especially if the feed intake test length is reduced, animals will potentially no longer be weighed at the feedlot, and BREEDPLAN live weights may be unavailable. More testing is required to ensure that the proposed approach for selecting for feed efficiency is robust for potential scenarios that could occur in practice. The current data structure could not consider maternal effects; with a larger dataset, the potential maternal effects of MWT_{200} and how best to model MWT_{200} can be tested.

CONCLUSIONS

When growth and maintenance terms in the NFIg model were the same as NFIp, phenotypic and genetically derived NFI EBVs were shown to be equivalent models. This study proves, in principle, that feed intake test ADG and MMWT could be replaced with growth and maintenance derived from BREEDPLAN live weights, and the next step is to test if the DFI test length can also be shortened. Before this research can be implemented into national genetic evaluations, further research will be needed using an expanded dataset and testing how robust the approach is in scenarios likely to occur in the industry if the feed intake test were shortened, i.e. missing weights.

ACKNOWLEDGEMENTS

Funding support was provided by MLA project L.GEN.2204. The authors gratefully acknowledge Angus Australia and ABRI for providing the data required to perform the analyses.

REFERENCES

Archer J.A., Arthur P.F., Herd R.M., Parnell P.F. and Pitchford W.S. (1997) *J.Anim.Sci.* 75: 2024.

Arthur P.F., Archer A.A., Johnston D.J., Herd R.M., Richardson E.C. and Parnell P.F. (2001) *J. Anim. Sci.* 79: 2805.

Clark S. and van der Werf J.H.J. (2017) *Proc. Assoc. Advmt. Anim. Breed. Genet.* 22: 175.

Culbertson M.M., Speidel S.E., Peel R.K., Cockrum R.R., Thomas M.G. and Enns R.M. (2015) *J. Anim. Sci.* 93: 2482.

Gilmour A.R., Gogel B.J., Cullis B.R., Welham S.J. and Thompson R. (2015) VSN International Ltd, Hemel Hempstead, HP1 1ES, UK.

Graser H.U., Tier B., Johnston D. and Barwick S. (2005) *Aust. J. Exp. Agric.* 45: 913.

Hoque A. and Oikawa T. (2004) *Anim.Sci. J.* 75: 201.

Kennedy B., Van der Werf J.H.J. and Meuwissen T. (1993) *J. Anim. Sci.* 71: 3239.

Koch R.M., Swiger L.A., Chambers D. and Gregory K.E. (1963) *J. Anim. Sci.* 22: 486.

MacNeil, M.D., Lopez-Villalobos, N. and Northcutt, S.L. (2011) *J. Anim. Sci.* 89: 3917.

Proc. Assoc. Advmt. Anim. Breed. Genet. 25: 386 - 389

REMODELLING THE GENETIC EVALUATION OF NFI IN BEEF CATTLE – PART 2: SHORTENING THE LENGTH OF THE FEED INTAKE TEST

L.Vargovic[1], K.L. Moore[1], D.J. Johnston[1], G.M. Jeyaruban[1], C.J. Girard[1], J. Cook[1], J.A. Torres-Vázquez[2] and S.P. Miller[1]

[1] Animal Genetics Breeding Unit*, Armidale, NSW, 2351 Australia
[2] Department of Animal and Avian Sciences, University of Maryland, College Park 20742 USA

SUMMARY

BREEDPLAN net feed intake (NFI) EBV is derived from a phenotypic regression based on a 70-day feed intake test. Genetic NFI (NFIg) EBVs have been proposed as an alternative EBV and this recent development may also allow for a shortened feed intake test period. This study used feed intake records of 3,088 Angus steers from the full 70-day test and compared them to daily feed intake (DFI) from shortened test periods. Results showed DFI from shortened test periods had similar means but increased phenotypic variation. Phenotypic correlation with DFI from the full test period decreased as the test period decreased in weekly intervals and ranged between 0.75 and 0.99. NFIg EBVs were predicted using DFI from different length tests. The mean of all NFIg EBVs was close to zero, but the EBV standard deviation increased as the test period decreased. Pearson correlations between NFIg EBVs from a full test period and reduced test periods ranged between 0.73 and 0.99, the regression slope of NFIg from reduced test periods on NFIg from the full test period ranged between 0.73 and 0.95, and the bias ranged between 0.00 and 0.02. These results indicate that as the test period decreases, the spread of EBVs increases, resulting in extreme animals having overestimated NFIg EBVs. A shortened DFI test period could be used to estimate NFIg EBVs.

INTRODUCTION

BREEDPLAN net feed intake (NFI) EBV is derived from a phenotypic regression based on a 70-day test with growth and maintenance measured from fortnightly body weight records during the test period (Koch *et al.* 1963, Arthur *et al.* 2001), which is a costly protocol. Reducing the test period would reduce recording costs per animal and allow more animals to be tested. The current NFI EBV requires average daily gain (ADG) to be recorded at regular intervals for a minimum of 56 days (Clark and van der Werf 2017; Archer *et al.* 1997; Culbertson *et al.* 2015). Kennedy *et al.* (1993) proposed calculating NFI EBVs using genetic regression (NFIg), and Vargovic *et al.* (2023) showed NFI EBVs could be predicted using genetic regression with a Pearson correlation of 0.99 between the current BREEDPLAN NFI EBV and NFIg. Vargovic *et al.* (2023) also explored alternative NFIg models where growth and maintenance traits were derived from BREEDPLAN live weight traits. The most promising alternative NFIg model considered DFI from the full test, ADG between 200 and 400-day live weight and maintenance requirements based on 600-day live weight. For this model, the Pearson correlation was 0.93, and the regression slope was 0.97 between the alternative NFIg EBV and the current NFI EBV. To reduce the test length an alternative model is required. This paper investigated the possibility of shortening the DFI test length using the alternative NFIg model proposed by Vargovic *et al.* (2023).

MATERIALS AND METHODS

Data preparation. Feed intake and BREEDPLAN live weight data were available for 3,088 Angus steers from 35 trial groups, and data preparation details are provided in Vargovic *et al.* (2023).

* A joint venture of NSW Department of Primary Industries and University of New England

A series of average DFI traits were constructed (DFIn), where n was the cumulative number of feed test weeks (i.e. DFI4 was the average DFI recorded over the first four weeks of the test period). DFI10 was the average DFI recorded over the full test period, and was used to compare results when the test period was shortened. Growth and maintenance traits were defined using BREEDPLAN live weight traits where ADG was calculated between 200 and 400 days of age ($ADG_{200\text{-}400}$ = (W400 – W200) /200) and metabolic weight was at 600 days (MWT_{600}) ($MWT_{600} = WT600^{0.73}$).

Statistical analyses. Variance components were estimated for DFIn from univariate linear mixed animal models in ASReml (Gilmour *et al.* 2015), and bivariate models were used to estimate genetic and phenotypic correlations between DFI10 and DFIn traits. For all DFIn traits, the NFI CG defined by Vargovic *et al.* (2023) and test start age centred on the mean were fixed effects, and the animal was fitted as a random effect. Vargovic *et al.* (2023) describe the method for genetically derived NFI EBVs (NFIg) using BREEDPLAN live weight traits to model growth and metabolic weight. From this paper, the NFIgF model from Vargovic *et al.* (2023) was considered; NFIg EBV = EBV(DFIn) – β_1 x EBV($ADG_{200\text{-}400}$) – β_2 x EBV (MWT_{600}). NFIg EBV was calculated for each DFIn phenotype to test the impact of reducing the test period, and the regression coefficients (β_1 and β_2) were calculated for each NFIg model. For each set of NFIg EBVs (DFIn, n=1 to 9), the Pearson correlation ® regression slope (b) and bias of means with NFIg EBVs (DFIn, n=10) were calculated.

RESULTS AND DISCUSSION

Decreasing the length of the feed test period resulted in an increase in DFI phenotypic variance, but the raw mean remained similar (Table 1). Heritability estimates were generally similar for DFI measured between 4 and 10 weeks and were slightly lower for DFI measured between 1 and 3 weeks. The phenotypic and additive variances increased as the test length reduced. Genetic correlations between DFIn (n=1 to 9) and DFI10 were not significantly different from 1indicating that all DFIn traits were genetically the same. Phenotypic correlation with DFI10 decreased as the test period decreased and ranged between 0.75 (DFI1) and 0.99 (DFI9).

Table 1. Phenotypic Mean and standard deviation (SD), additive genetic (V_A) and phenotypic (V_P) variance components, heritability and genetic (r_A) and phenotypic (r_p) correlations with daily feed intake at 10 weeks (DFI10) for DFIn (n=1 to 10 weeks)

Trait	Phenotype Mean (SD)	Va (SE)	Vp (SE)	h^2 (SE)	r_A	r_P
DFI10	14.8 (1.99)	1.10 (0.15)	2.39 (0.07)	0.46 (0.06)		
DFI9	14.9 (1.99)	1.10 (0.16)	2.42 (0.07)	0.46 (0.06)	0.99	0.99
DFI8	15.0 (2.00)	1.14 (0.16)	2.47 (0.07)	0.46 (0.06)	0.99	0.99
DFI7	15.1 (2.02)	1.18 (0.17)	2.52 (0.07)	0.47 (0.06)	0.99	0.98
DFI6	15.1 (2.06)	1.20 (0.17)	2.59 (0.07)	0.46 (0.06)	0.99	0.97
DFI5	15.2 (2.12)	1.24 (0.18)	2.64 (0.07)	0.47 (0.06)	0.99	0.95
DFI4	15.2 (2.12)	1.21 (0.18)	2.71 (0.08)	0.45 (0.06)	0.99	0.93
DFI3	15.1 (2.15)	1.21 (0.18)	2.78 (0.08)	0.43 (0.06)	0.99	0.90
DFI2	15.1 (2.23)	1.22 (0.19)	2.89 (0.08)	0.42 (0.06)	0.99	0.85
DFI1	14.9 (2.29)	1.14 (0.20)	3.22 (0.09)	0.35 (0.06)	1.00*	0.75

*estimate at bounds

The mean of all NFIg EBVs was close to zero, with standard deviations ranging from 0.38 to 0.50 (Table 2). Generally, as the DFI test period reduced, NFIg EBVs showed more variation with standard deviations increasing. Pearson correlations between NFIg EBV (DFIn, n=10) and NFIg EBV (DFIn, n=1 to 9) ranged between 0.73 and 0.99, the regression slope ranged between 0.73 and 0.95, and the difference of means ranged between 0.00 and 0.02. As the test period reduced, the Pearson correlation and regression slopes decreased, and the NFIg EBV standard deviation and the difference of means between NFIg EBV (DFIn, n=10) and NFIg EBV (DFIn, n=1 to 9) increased. These results indicate that the bias increased as the test period decreased, and NFIg EBVs for animals at either end of the distribution were over-estimated.

Table 2. Summary statistics for NFIg EBVs using daily feed intake (DFI) from reduced test lengths and BREEDPLAN live weight records to model growth and maintenance, the Pearson correlation (r), regression slope (b) and bias of means (bias) between NFIg EBVs when DFI was recorded over the full test period, and DFI recorded from a reduced test period

DFIn modelled	NFIg EBV						
	Mean	SD	Min	Max	r	b	bias
DFI10	0.01	0.41	-1.85	1.58			
DFI9	0.01	0.41	-1.85	1.56	0.99	0.95	0.00
DFI8	0.01	0.44	-1.90	1.66	0.99	0.89	0.00
DFI7	0.01	0.46	-1.92	1.74	0.98	0.85	0.00
DFI6	0.00	0.48	-1.99	1.79	0.97	0.80	0.01
DFI5	0.00	0.50	-2.13	1.79	0.95	0.76	0.01
DFI4	0.00	0.49	-2.69	1.67	0.92	0.75	0.01
DFI3	-0.01	0.48	-2.64	1.82	0.89	0.74	0.02
DFI2	-0.03	0.46	-2.33	1.93	0.84	0.73	0.03
DFI1	-0.01	0.38	-1.74	1.99	0.73	0.77	0.02

The results in Table 2 demonstrate that reducing the feed intake test period may be possible using genetic NFI EBVs obtained where growth and maintenance are modelled using BREEDPLAN live weight traits. Results from Table 1 agree with earlier studies by Clark and van der Werf (2017), who suggested that the DFI test length could be reduced to 4 weeks. However, Table 2 showed that NFIg EBVs had a regression slope of 0.75 when DFI was measured from a 4-week test and a larger spread of EBVs, especially for the more feed-efficient animals. Given the strong Pearson correlation, a degree of overestimation of extreme animals may be acceptable. Examination of the animal with the most negative NFIg EBV when DFI was recorded over 4 weeks showed that the animal had lower DFI in the earlier weeks of the feed intake test. DFI in weeks 1 to 4 ranged between 5.32 to 6.11 kg/d and increased to between 10.26 and 10.98 kg/d in weeks 8 to 10. This may be because the animal was ill or took longer than usual to become acclimatised to the feeders. The difference in DFI at 4 and 10 weeks was extreme, across the dataset the difference between DFI at 4 and 10-weeks averaged 0.32 kg/d with a standard deviation of 0.77. Other animals with extreme negative NFIg EBVs consistently had negative NFIg EBVs overall test length periods. Figure 1 plots the NFIg EBVs for sires when the feed intake test period was 4 or 10 weeks and illustrates that although the correlation between EBVs is strong, there are changes in EBV at the distribution edges.

Further research is needed to investigate the impact of reducing the test period in a larger dataset where BREEDPLAN live weights from the whole breed may contribute to the component EBVs used to estimate NFIg EBVs. In this study, the majority of animals had BREEDPLAN live weights recorded. If genetic NFI EBVs use BREEDPLAN live weights to model growth and maintenance, and especially if the feed intake test length is reduced, animals will potentially no longer be weighed at the feedlot, and BREEDPLAN live weights may be unavailable for ages occurring during the feed intake test. More testing is required to ensure that the proposal to shorten the DFI test period and estimate NFIg EBVs is robust for potential scenarios that could occur in practice. A challenge with NFIg EBVs is that variance components were unavailable, so the present study could not calculate correlated selection responses. Correlated selection response is expected to increase with increased selection intensity but will decrease as the genetic correlation between NFIg EBVs decreases. Reducing the test period potentially allows more animals to be tested. Allowing for a 21-day acclimatisation period, reducing the DFI test to between 1 and 9 weeks potentially increases the number of animals recorded for DFI by 1.1 to 3.3 times, which may increase the selection intensity. However, as the test period is decreased, the genetic correlation between NFIg from full or reduced DFI tests also decreases, which may reduce the correlated selection response. Therefore, the increase in selection intensity and the decrease in genetic correlation will need to be balanced when determining the ideal test period, which will allow the financial savings of a reduced test but still allow effective NFIg EBVs to be estimated and genetic improvement for feed efficiency.

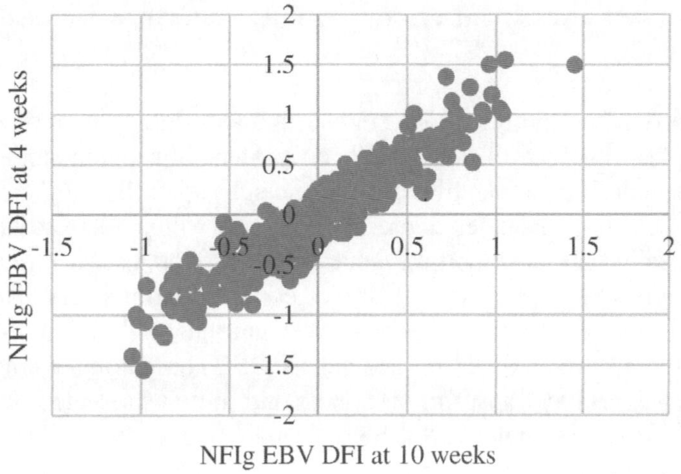

Figure 1. The relationship between NFIg EBVs for sires when daily feed intake (DFI) was recorded from 4 or 10 week feed intake test period

CONCLUSIONS

This study showed that reducing the DFI test period and estimating genetic NFI EBVs with growth and maintenance derived from BREEDPLAN live weights may be possible. Preliminary results are promising, but before implementation, more testing needs to be done and implementation strategies explored.

ACKNOWLEDGEMENTS

Funding support was provided by MLA project L.GEN.2204. The authors gratefully acknowledge Angus Australia and ABRI for providing the data required to perform the analyses and Tullimba feedlot for data collection.

REFERENCES

Archer J.A., Arthur P.F., Herd R.M., Parnell P.F. and Pitchford W.S. (1997) *J.Anim.Sci.* 75: 2024.

Arthur P.F., Archer A.A., Johnston D. J., Herd R.M., Richardson E.C. and Parnell P.F. (2001) *J. Anim. Sci.* 79: 2805.

Clark S. and van der Werf J.H.J. (2017) *Proc. Assoc. Advmt. Anim. Breed. Genet.* 22: 175.

Culbertson M.M., Speidel S.E., Peel R.K., Cockrum R.R., Thomas M.G. and Enns R.M. (2015) *J. Anim. Sci.* 93: 2482.

Gilmour A.R., Gogel B.J., Cullis B.R., Welham S.J. and Thompson R. (2015) VSN International Ltd, Hemel Hempstead, HP1 1ES, UK.

Kennedy B., Van der Werf J.H.J. and Meuwissen T. (1993) *J. Anim. Sci.* 71: 3239.

Koch R.M., Swiger L.A., Chambers D. and Gregory K.E. (1963) *J. Anim. Sci.* 22: 486.

Vargovic L., Moore K., Johnston D., Jeyaruban G., Girard C., Cook J., Torres-Vázquez A. and Miller S. (2023) *Proc. Assoc. Advmt. Anim. Breed. Genet.* 25 *These proceedings.*

ACROSS-BREEDS SYSTEMS BIOLOGY ANALYSIS REVEALS KEY GENES CONTRIBUTING TO FEED EFFICIENCY IN BEEF CATTLE

K. Keogh[1,2], D.A. Kenny[2], P.A. Alexandre[1], M. McGee[3] and A. Reverter[1]

[1]CSIRO Agriculture & Food, Queensland Bioscience Precinct, 306 Carmody Rd., St. Lucia, Brisbane, QLD 4067, Australia
[2]Animal and Bioscience Research Department, Teagasc, Grange, Dunsany, Co. Meath, Ireland
[3]Livestock Systems Research Department, Teagasc, Grange, Dunsany, Co. Meath, Ireland

SUMMARY

Provision of feed in beef production systems is a major determinant of overall profitability as it typically accounts for over 75% of the variable cost. Thus, improving cattle feed efficiency by way of determining the underlying molecular control and subsequently selecting for feed efficient cattle through genomic selection provides a method through which feed costs may be reduced. The objective of this study was to undertake gene co-expression network analysis on RNAseq data generated from *Longissimus dorsi* tissue samples collected from steers divergent for residual feed intake (RFI) within two contrasting breed types (Charolais and Holstein-Friesian). Several gene categories, including differentially expressed genes (DEG) based on the contrasts of both breed and RFI phenotype as well as key transcription factors and proteins secreted in plasma were utilised as nodes of the gene co-expression networks. Significant network connections were identified using an algorithm that exploits the dual concepts of partial correlation and information theory (PCIT). Results revealed 530 and 531 DEG for the RFI and breed contrasts, respectively. PCIT network analysis resulted in the formation of one RFI specific cluster which included genes related to metabolic processes and cell cycle. A second cluster which included genes related to both RFI and breed was enriched for immune-related pathways such as coagulation system and the complement cascade. This latter network was of particular interest due to the potential identification of genes contributing to RFI that are sufficiently robust across breed type. Moreover, genes included within this network also encode proteins secreted in plasma, highlighting the potential use of these genes as blood-based biomarkers for RFI in beef cattle.

INTRODUCTION

Within beef production systems, provision of feed is a major determinant of overall profitability, as it accounts for up to 75% of the total variable costs of production (Kenny *et al.* 2018). Consequently, research related to the identification and subsequent breeding of beef cattle with improved feed efficiency has received attention to alleviate the high input costs and environmental footprint associated with beef production. In particular, residual feed intake (RFI), defined as the difference between an animal's actual and predicted feed intake, has become the feed efficiency measure of choice due to its independence of both growth and body size (Fitzsimons *et al.* 2017). However, despite research efforts aimed at uncovering the molecular control of RFI in cattle, genes which are robust across varying breed type contributing to RFI are yet to be identified (Kenny *et al.* 2018). This is undoubtedly due to the multifaceted nature of the RFI trait as well as the varying experimental parameters employed across different studies, such as breed types, dietary sources and stage of development evaluated, ultimately confounding the subsequent outcome.

Thus, the objective of this study was to undertake gene co-expression network analysis on *Longissimus dorsi* (LD) transcriptomic data collected from steers divergent for RFI within two contrasting breed types (Charolais and Holstein-Friesian). Specifically, differentially expressed genes (DEGs) for both RFI and breed contrasts were used as nodes for the co-expression network analysis. The LD muscle was chosen as a target tissue due to its' economic importance, in addition to its responsiveness to variation in RFI in cattle (Fitzsimons *et al.* 2017).

MATERIALS AND METHODS

All procedures involving animals in this study were reviewed and approved by the Teagasc Animal Ethics Committee and were conducted under an experimental licence issued by the Irish Health Products Regulatory Authority (AE19132/P029).

This experiment was conducted in Ireland under moderate non-extreme climatic conditions as part of a larger research programme designed to examine the within-animal repeatability of intake, growth, and feed efficiency between varying stages of development in Charolais and Holstein-Friesian beef steers (Coyle *et al.* 2016). In total, 167 steers (90 Charolais and 77 Holstein-Friesian) were sourced from commercial farms in Ireland, parentage was included within the animal selection process so as to avoid selecting genetically related animals. At the start of the trial Charolais and Holstein-Friesian steers were on average 283 and 307 days of age, respectively. Following a dietary adaptation period, dry matter intake (DMI) and growth rate were measured over a 70-day feeding trial, during which all steers were offered the same high-energy diet consisting of *ad libitum* concentrates plus a restricted allowance of grass silage daily. Throughout the trial all steers were accommodated indoors, utilising a Calan gate feeding system. The residuals of the regression of DMI on average daily gain (ADG), and mid-test metabolic body weight within each breed were used to compute individual RFI coefficients for each steer (GLM procedure of SAS9.3). Residual feed intake was calculated for each animal as the difference between actual and predicted DMI. Within each breed, steers were ranked for RFI, with high-RFI (feed-inefficient; n=12) and low-RFI (feed-efficient; n=12) steers selected for each breed separately. Samples of LD tissue were collected through punch biopsy from all high-RFI and low-RFI steers following completion of the 70-day dietary trial. Tissue samples were washed with sterile DPBS and immediately snap frozen in liquid nitrogen before subsequent storage at -80°C.

Total RNA was purified from all tissue samples using the Qiagen RNeasy Universal kit (QIAGEN, UK), according to the manufactures instructions as previously described (Higgins *et al.* 2019). The quality of the resultant RNA was assessed using the RNA 6000 RNA Nano Lab Chip Kit and the Agilent Bioanalyser 2100 (Agilent Technologies Ireland Ltd., Dublin, Ireland). All samples passed quality control with RNA integrity numbers greater than 8. The Illumina TruSeq sample preparation kit (Illumina, San Diego, CA) was utilized to construct cDNA libraries for each sample, following which cDNA libraries were sequenced using the Illumina HiSeq 2500 sequencing platform (Illumina, San Diego, CA). Bioinformatic analysis was undertaken as previously described in Higgins *et al.* (2019) including the removal of sequencing adapters and low quality reads using cutadapt (v. 1.13) and quality control of sequencing reads undertaken using FastQC (v. 0.11.5). Trimmed sequencing reads were mapped to the bovine reference genome (ARS-UCD1.2) and also quantified using STAR (v.2.5.1). Differentially expressed genes were detected between each of the two main contrasts: (i) High-RFI versus Low-RFI; and (ii) Charolais versus Holstein-Friesian) using the Bioconductor package, EdgeR (v3.20.9). Gene expression was estimated as Counts Per Million (CPM) and genes were retained for subsequent analysis only when presented in at least 1 CPM in at least half of the samples for each contrast. The top 5% most significant genes (based on Benjamini-Hochberg corrected P-value of differential expression) in each contrast were considered DEG and were selected for subsequent inclusion in the co-expression network analysis. Additionally key transcription factors (TF) and proteins secreted in plasma were also utilised as nodes within the gene co-expression networks. For gene co-expression network analysis, genes selected based on differential expression, as key TF and secreted in plasma were used as nodes and significant edges between nodes identified using the Partial Correlation and Information Theory (PCIT) algorithm (Reverter and Chan 2008). The output of PCIT was then visualised using Cytoscape (V3.9.1)

(Shannon *et al.* 2003) including only significant correlations above 0.9 and their respective genes. Functional enrichment of gene networks was performed using Ingenuity Pathway Analysis (IPA).

RESULTS AND DISCUSSION

A significant difference (P<0.0001) in RFI value was evident for each breed (Charolais: Low-RFI=-0.53, High-RFI=0.55; Holstein-Friesian: Low-RFI=-0.64, High-RFI=0.7). For the RFI and breed contrasts, 530 and 531 DEGs were identified, respectively. Of these 114 genes (12.4%) were common between both contrasts. A total of 1,061 DEG, 292 TF and 405 genes encoding proteins secreted in plasma were identified as associated with variation in both RFI and breed type. Gene co-expression network visualisation of significant correlations between genes above 0.9 equated to 298 genes with 5,625 connections, the main clusters of interest are presented in Figure 1.

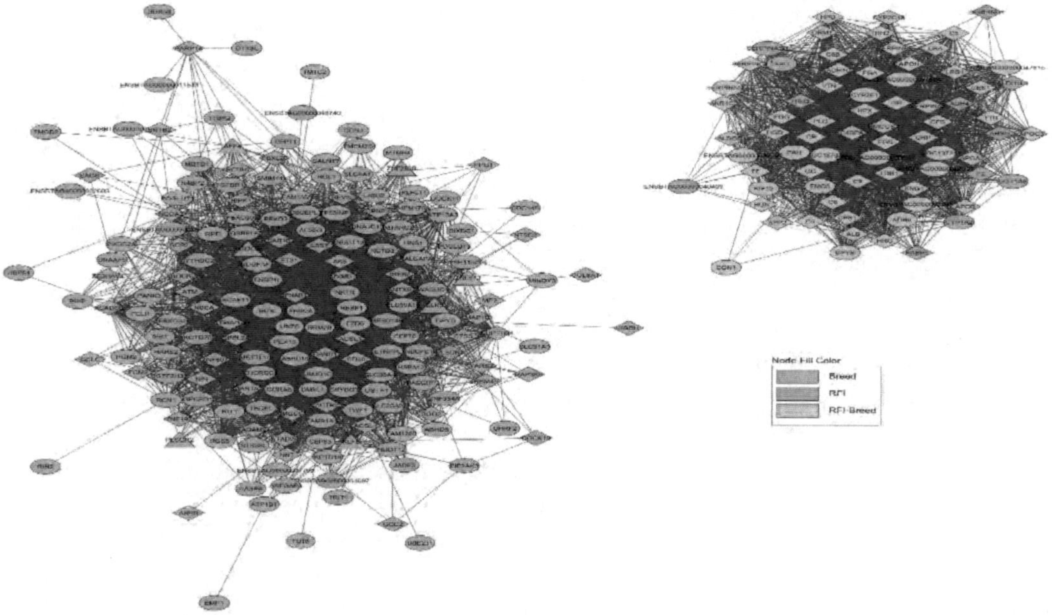

Figure 1. Gene co-expression network of genes related to RFI in Charolais and Holstein-Friesian steers. Node colour are relative to the differential expression contrast: purple represents RFI only; green represents breed only and orange represents genes differentially expressed in both RFI and breed contrasts. Genes encoding transcription factors and proteins secreted in plasma are represented as triangle and diamond shapes, respectively

Network visualisation highlighted a clear cluster of genes specifically related to RFI (purple), whilst a second cluster depicted genes related to both RFI and breed contrasts (RFI-Breed, orange). Functional analysis of the RFI specific cluster of co-expressed genes highlighted pathways related to mitochondrial fatty acid oxidation including fatty acid β-oxidation (adj.P<0.005) and mitochondrial L-carnitine shuttle pathway (adj.P<0.01), suggesting a role for mitochondrial fatty acid oxidation towards variation in RFI in beef cattle. Processes related to fatty acid oxidation have previously been implicated towards divergence in RFI in varying tissue types (subcutaneous adipose: McKenna *et al.* 2018: liver: Taiwo *et al.* 2022), with up-regulation of fatty acid oxidative processes within the feed efficient (low-RFI) cattle apparent in each study. Indeed, McKenna *et al.* (2018) postulated that the increased expression of fatty acid oxidative genes in the low-RFI animals may be due to the efficient cattle directing metabolic processes towards alternative substrate partitioning and fatty acid breakdown in order to facilitate their lower dietary intake.

A potential role for immune processes towards variation in RFI has been established across varying experimental designs (Fitzsimons *et al.* 2017; Kenny *et al.* 2018); however, specific immune related processes are conflicting across experimental designs. Pathway analysis of the network of co-expressed genes related to both RFI and breed revealed an enrichment of immune-related processes including coagulation system and complement cascade ($P<0.001$). Moreover, genes included within this network and pertaining to coagulation (*FGA, FGB* and *FGG*) and complement system (*C3, C5, C9, CFH, CFI* and *CRP*) pathways were previously reported as differentially expressed between cattle divergent for RFI across various breed types including Angus, Nellore, Holstein-Friesian and Charolais (Chen *et al.* 2011; Tizioto *et al.* 2016; Weber *et al.* 2016; Higgins *et al.* 2019). Moreover, the aforementioned genes also encode proteins secreted into plasma, suggesting a potential role for these genes as blood-based biomarkers for RFI in beef cattle.

CONCLUSION

Results from this study provide potential candidate genes, pathways and networks related to feed efficiency in beef cattle. The RFI-breed network is of particular interest for the potential identification of robust genes contributing to the RFI trait irrespective of breed type. Moreover, genes included within this network were also genes coding proteins secreted in plasma, highlighting the genes potential to be explored as blood-based biomarkers for the RFI trait in beef cattle. However, extensive functional experimental validation for the candidate genes and pathways identified in this study is warranted.

ACKNOWLEDGEMENTS

Kate Keogh received funding from the Research Leaders 2025 programme co-funded by Teagasc and the European Union's Horizon 2020 research and innovation programme under Marie Skłodowska-Curie grant agreement number 754380. This research was funded by the Irish Department of Agriculture, Food and the Marine (RSF13/S/519).

REFERENCES

Chen Y., Gondro C., Quinn K., Herd R.M., Parnell P.F. and Vanselow B. (2011) *Anim. Genet.* 42: 475.

Coyle S., Fitzsimons C., Kenny D.A., Kelly A.K. and McGee M. (2016) *J. Anim. Sci.* 94: 719.

Fitzsimons C.F., McGee M., Keogh, K., Waters S.M. and Kenny D.A. (2017) In 'Biology of Domestic Animals', pp.

Higgins M.G., Kenny D.A., Fitzsimons C., Blackshields G., Coyle S., McKenna C., McGee C., Morris D.W. and Waters S.M. (2019) *BMC Genomics* 20: 525.

Kenny D.A., Fitzsimons C., Waters S.M. and McGee M. (2018) *Animal* 12: 1815.

McKenna, C.E., Porter R.K., Keogh, K., Waters S.M., McGee M. and Kenny D.A. (2018) *J. Anim. Sci. Biotech.* 9: 68.

Reverter A. and Chan E.K. (2008) *Bioinformatics*. 24: 2491.

Shannon P., Markiel A., Ozier O., Baliga N.S., Wang J.T., Ramage D., Amin N., Schwikowski B. and Ideker T. (2003) *Genome Res.* 13: 2498.

Taiwo G., Idowu W.D., Wilson M., Pech-Cervantes A., Estrada-Reyes Z.M. and Ogunade I.M. (2022) *Front. Anim. Sci.* 3: 828591.

Tizioto P.C., Coutinho L.L., Oliveira P.S., Cesar A.S., Diniz W.J., Lima A.O., Rocha M.I., Decker J.E., Schnabel R.D., Mourão G.B., Tullio R.R., Zerlotini A., Taylor J.F. and Regitano L.C. (2016) *Sci. Rep.* 6: 39493.

Weber K.L., Welly B.T., Van Eenennaam A.L., Young A.E., Porto-Neto L.R., Reverter A. and Rincon G. (2016) *PLoS One*. 11: e0152274.

Proc. Assoc. Advmt. Anim. Breed. Genet. 25: 394 - 397

IDENTIFICATION OF CLIMATE-RESILIENT MERINO SHEEP USING SATELLITE IMAGES

S. de las Heras-Saldana[1], L. A. Suarez[2], P. K. Wahinya[1], K. L. Bunter[1] and D.J. Brown[1]

[1]Animal Genetics Breeding Unit[1], University of New England, Armidale, NSW, 2351 Australia
[2]Applied Agricultural Remote Sensing Centre, University of New England, Armidale, NSW 2351 Australia

SUMMARY

This study aimed to evaluate the potential use of data from Landsat 5 TM, 7 ETM+, and 8 OLI and meteorology SILO databases to characterise variation in environmental conditions across farms and identify resilient sheep with a low response in performance to changes in the temperature-humidity index (THI) and normalized difference vegetation index (NDVI). A total of 44,848 Merino sheep from 27 farms across Australia were used in this study. The dataset included sheep with complete pedigree and measurements for weaning weight (WWT) and post-weaning weight (PWT). The average NDVI and THI values during the 9 months prior to the phenotypic measurement were used in a linear reaction norm (RN) model with heterogeneous residual variances. The results revealed genotype by environment (GxE) interaction for WWT and PWT between extreme environments with reranking of sires' estimated breeding values along the NDVI gradient. Higher heritability and genetic variances were estimated in favourable environments. Accounting for GxE interactions could lead to a more accurate selection of resilient sheep to changes in climatic and vegetation variables in Australia, and existing environmental data is enabling for this purpose.

INTRODUCTION

The global rise in more variable and extreme climate conditions has demanded the development of strategies to identify and select resilient animals capable of thriving in challenging circumstances. Selection of resilient sheep will help maximise performance across multiple locations with variable conditions. Reaction norm (RN) models relate the genetic merit of animals to the environment, providing estimated breeding values (EBV) for each environmental condition and identifying genotype by environment (GxE) interactions (Schaeffer 2004). The intercept EBVs represent the overall production potential of the animals, while the EBVs for slope indicate their resilience to different environmental conditions. To characterise environments experienced by animals, the temperature-humidity index (THI) has been investigated as a measurement of thermal stress experienced by dairy and beef cattle (Ravagnolo and Misztal 2000; Bradford *et al.* 2016). Similarly, the availability of forage is expected to have a cumulative effect on animal growth, but direct measurements in extensive production systems are challenging (Johnson *et al.* 2018). As such, the normalized difference vegetation index (NDVI) has been used as a proxy of forage coverage (Johnson *et al.* 2018). The impact of thermal stress and forage coverage, as indicated by THI and NDVI, on the growth performance of sheep has not yet been investigated in extensive production systems. This study aimed to identify Merino sheep resilient to changes in THI and NDVI as indicators of environmental conditions.

MATERIALS AND METHODS

* A joint venture of NSW Department of Primary Industries and University of New England

Data. A dataset of 44,848 Australian Merino sheep with weight and pedigree records was used in this study. The dataset includes a full pedigree from 1,420 sires and 20,919 dams spanning four generations. Raw phenotypic measurements, taken between the years 2000 and 2020, included 44,335 measurements for weaning weight (WWT), and 40,265 records for post-weaning weight (PWT). For both traits, records outside four standard deviations from the contemporary group mean were considered outliers and removed from further analysis.

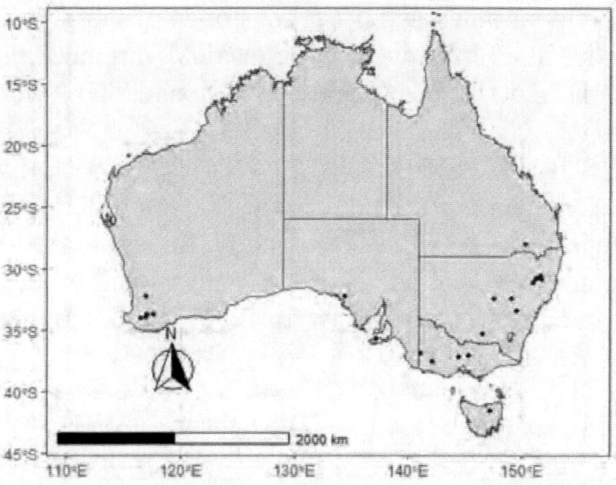

Figure 1. Location of farms across Australia.

Study area. Flocks (N = 27) were located in New South Wales, Queensland, South Australia, Tasmania, Victoria, and Western Australia (Figure 1). Polygons with the coordinates (latitudes and longitudes) were traced on the boundaries of each farm to obtain precise information on climate and forage.

Climatic data. Temperature (°C) and humidity (%) records were obtained for each location from the SILO database (www.longpaddock.qld.gov.au/silo/) to calculate the temperature-humidity index (THI) according to Lallo *et al.* (2018): $THI = (1.8T + 32) - ((0.55\text{-}0.0055Rh)(1.8T\text{-}26))$; where T is the temperature (°C) and Rh the relative humidity (%).

Satellite data. Landsat 5 TM, 7 ETM+, and 8 OLI surface reflectance data (Collection 2, Tier 1) were processed in Google Earth Engine. The cloud mask was applied to all imagery. The red (R) and near-infrared (NIR) bands were used to compute the normalised difference vegetation index as $NDVI = (NIR\text{-}R)/(NIR+R)$. NDVI values were multiplied by 100 to be used in the following analyses.

Since a reliable association of environmental conditions with forage components can be expected only within a season (Johnson *et al.* 2018), we evaluated the daily THI and NDVI records and averaged the values across 9 months prior to the trait measurement to fit as continuous values in the RN model.

Statistical models. Univariate sire reaction norm models were computed in ASReml v4.2 (Gilmour *et al.* 2021) as a linear function of the environment (NDVI or THI averaged across 9 months) and the traits (WWT or PWT) described in a general form by:

$$\boldsymbol{y} = \boldsymbol{\mu} + \boldsymbol{Xb} + \boldsymbol{Z_{Int}} + \boldsymbol{Z_{Slp}} + \boldsymbol{Z_{pm}} + \boldsymbol{e},$$

where y is a vector with weight records, μ is the overall mean, X incidence matrices associating records with the fixed effects, b is a vector of fixed effects solutions for sex (2 levels), birth type (2 levels), rear type (2 levels), age of dam (12 levels), contemporary groups (2,474 levels), covariate (s) for age at measurement, and regression for continuous values of NDVI and THI determined using the average in separate analysis across 9 months prior to the date of phenotypic measurement; Z_{Int} is a matrix linking records to the breeding values to the intercept and Z_{Slp} is an incidence matrix relating records of the breeding values to the slope, Z_{pm} is an incidence matrix associating records with the maternal permanent environmental effect and e is the residual effect. Genetic variances were calculated across either NDVI or THI gradients as $\sigma^2_a = \Phi G \Phi'$, while the breeding values were obtained with $EBV = \Phi E'$; where G is the additive genetic co-variance matrix, E is the matrix with intercept and slope regression coefficients and Φ are the row vectors of a matrix with order one Legendre polynomials (order one) corresponding to the NDVI and THI levels. Genetic correlations across NDVI and THI gradients were defined as $r_{ij} = Cov_{ij}/\sqrt{\sigma^2_{ai} + \sigma^2_{aj}}$. The maternal permanent environmental effect was also modelled using order one Legendre polynomial. Heritability was calculated as $h^2 = \sigma^2_a/\sigma^2_p$ where σ^2_a (sire variance x 4) is the additive genetic variance and σ^2_p corresponds to the phenotypic variance calculated as $\sigma^2_p = \sigma^2_a + \sigma^2_e$. The model fitted heterogeneity in residual variance, allocated based on NDVI (5 levels: 10 to 30, 31 to 40, 41 to 50, 51 to 60, and 61 to 90) and THI (4 levels: 50 to 63, 64 to 67, 68 to 71, and 72 to 82).

RESULTS AND DISCUSSION

Environmental conditions across flocks. Figure 2 depicts changes in NDVI, THI, and temperature between 2000 and 2020 across the studied flocks. Years with favourable conditions are described by relatively high NDVI and moderate temperatures (i.e. 2016), and years with less favourable conditions have relatively high temperatures and low NDVI (i.e. 2019).

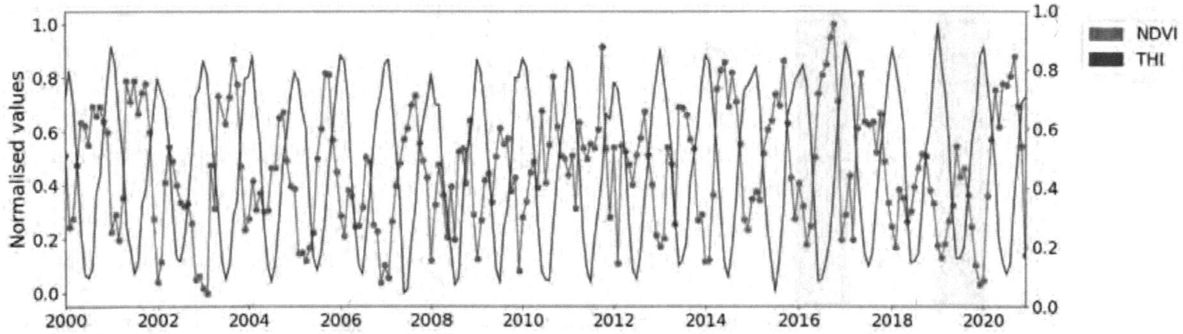

Figure 2. Average monthly temperatures, NDVI, and THI between 2000 and 2020 in 27 flocks located across Australia were normalised for visualisation. *Examples of favourable and unfavourable years are highlighted

Heritability and variances estimated across environmental conditions. Genetic variances (σ^2_a) for WWT and PWT increased with the NDVI (Figure 3A) and decreased for both traits across the THI (Figure 3B). The presence of a scale GxE interaction is evidenced by variation in heritability (h^2) estimates across the environments (Figures 3 C and D). In both traits, h^2 from the RN were higher in favourable conditions (i.e. high NDVI and low THI) and lower in less favourable conditions (i.e. low NDVI and high THI). Similar results were described by Bradford *et al.* (2016) in American Angus cattle with higher heritabilities (WWT) under no heat load (low THI) conditions.

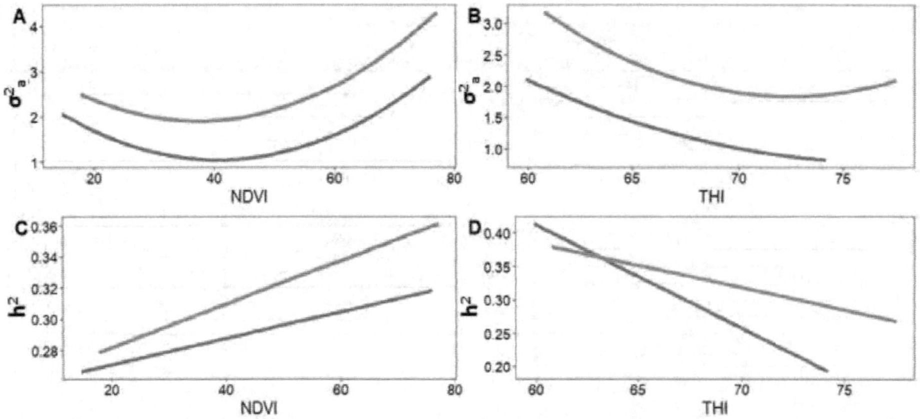

Figure 3. Additive genetic variances (σ^2_a) (A & B) and heritabilities (h^2; linear trend) (C & D) along the NDVI (A & C) and THI (B & D) gradient for WWT (green) and PWT (blue)

The genetic correlations across NDVI and THI gradients were reduced as differences in the environment increased for WWT (Figure 4 A & D) and PWT (plot not shown). The weak genetic correlations between extreme NDVI values contribute to the GxE interactions observed, leading to a higher reranking of sires across NDVI compared to the THI.

Genetic correlations between the intercept and slope of the RN model were positive for WWT (0.17) and PWT (0.27) when NDVI was evaluated. The RN for THI resulted in negative genetic correlations between the intercept and slope for WWT (-0.60) and PWT (-0.24). These results suggested that sheep with higher intercept (EBV) had a more responsive phenotype to the forage coverage as suggested by NDVI. In contrast, such animals (with higher EBV values) exhibited a relatively small response (slope) to THI.

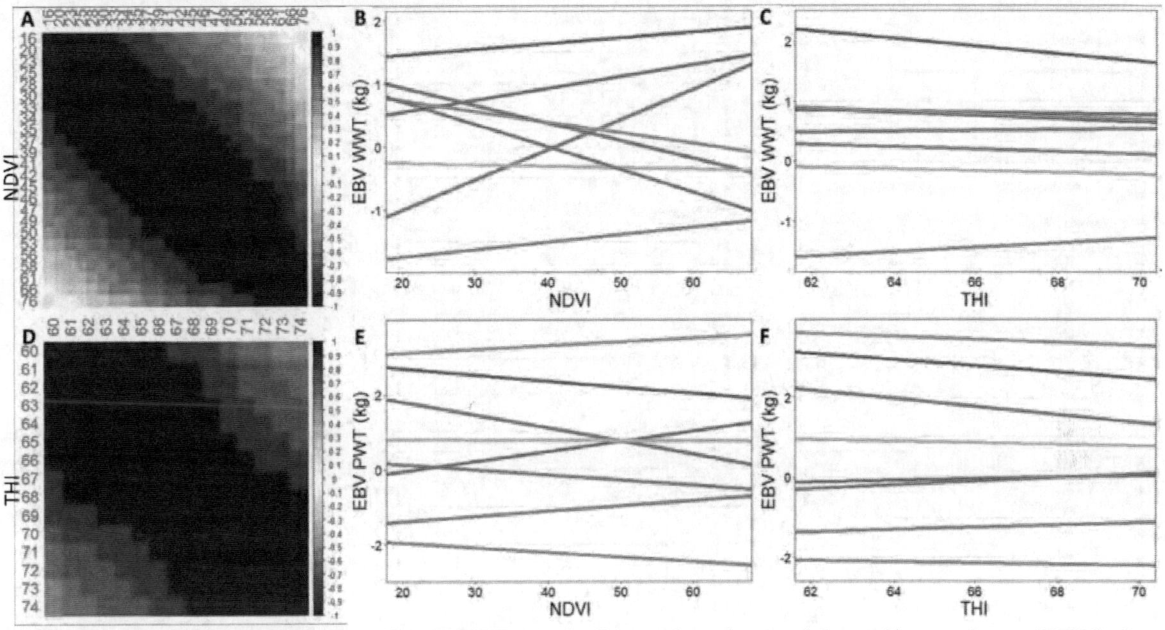

Figure 4. Genetic correlations for WWT (A & D) and estimated breeding values (EBV) for WWT and PWT for eight influential sires (> 100 progeny) along NDVI (A, B & E) and THI (C, D & F) gradients

CONCLUSIONS

This study describes GxE interactions between extreme environments for WWT and PWT in Australian sheep along environmental gradients representing forage coverage (NDVI) and the temperature-humidity index (THI). There was more GxE interactions when NDVI was extremely different, resulting in higher EBV reranking in sires than when THI was considered. Higher genetic variances and heritabilities were estimated in favourable environments. Furthermore, these findings emphasised the opportunity to use climatic and satellite data to describe the environment and identify resilient sheep for THI and NDVI in a national or international context.

REFERENCES

Bradford H.L., Fragomeni B.O., Bertrand J.K., Lourenco D.A.L. and Misztal I. (2016) *J. Anim. Sci.* 94: 4143.

Chadeau-Hyam M., Hoggart C., O'Reilly P., Whittaker J., De Iorio M. and Balding D. (2008) *BMC Bioinform.* 9: 364.

Gilmour A.R., Gogel B.J., Cullis B.R., Welham S.J. and Thompson R. (2021). VSN International Ltd, Hemel Hempstead, HP2 4TP, UK, www.vsni.co.uk

Johnson H.E., Gustine D.D., Golden T.S., Adams L.G., Parrett L.S., Lenart E.A., and Barboza P.S. (2018) *Ecosphere* 9: e02461.

Lallo C.H., Cohen J., Rankine D., Taylor M., Cambell J. & Stephenson T. (2018) *Reg. Environ. Change*, 18: 2329.

Ravagnolo O. and Misztal I. (2000) *J. Dairy Sc* 83: 2126

Schaeffer L.R. (2004) *Livest. Prod. Sci.* 86: 35.